Water Soluble Vitamins

SUBCELLULAR BIOCHEMISTRY

Recent Volumes in this Series

Volume 41	**Chromatic and Disease** Edited by Tapas K. Kundu and Dipak Dasgupta
Volume 42	**Inflammation in the Pathogenesis of Chronic Disease** Edited by Randall E. Harris
Volume 43	**Subcellular Proteomics** Edited by Eric Bertrand and Michel Faupel
Volume 44	**Peroxiredoxin Systems** Edited by Leopold Folhé and J. Robin Harris
Volume 45	**Calcium Signalling and Disease** Edited by Ernesto Carafoli and Marisa Brini
Volume 46	**Creatine and Creatine Kinase in Health and Disease** Edited by Gajja S. Salomons and Markus Wyss
Volume 47	**Molecular Mechanisms of Parasite Invasion** Edited by Barbara A. Burleigh and Dominique Soldati-Favre
Volume 48	**The Cronin Family of Proteins** Edited by Christoph S. Clemen and Ludwig Eichinger
Volume 49	**Lipids in Health and Disease** Edited by Peter J. Quinn and Xiaoyuan Wang
Volume 50	**Genome Stability and Human Diseases** Edited by Heinz-Peter Nasheuer
Volume 51	**Cholesterol Binding and Cholesterol Transport Proteins** Edited by Robin J. Harris
Volume 52	**A Handbook of Transcription Factors** Edited by Tim Hughes
Volume 53	**Endotoxins: Stricture, Function and Recognition** Edited by Xiaoyuan Wang and Peter J. Quinn
Volume 54	**Conjugation and Deconjugation of Ubiquitin Family Modifiers** Edited by Marcus Groettrup
Volume 55	**Purinergic Regulation of Respiratory Diseases** Edited by Maryse Picher and Richard C. Boucher

Olaf Stanger
Editor

Water Soluble Vitamins

Clinical Research and Future Application

Springer

Editor
Olaf Stanger, MD, PhD, MBA
Royal Brompton and Harefield NHS Foundation
Imperial College of Science, Technology
and Medicine
Heart Division
Cardiothoracic Surgery
Sydney Street
London SW3 6NP
United Kingdom
o.stanger@rbht.nhs.uk

ISSN 0306-0225
ISBN 978-94-007-2198-2 e-ISBN 978-94-007-2199-9
DOI 10.1007/978-94-007-2199-9
Springer Dordrecht Heidelberg London New York

Library of Congress Control Number: 2011936736

Printed on acid-free paper

Springer is part of Springer Science+Business Media (www.springer.com)

Preface

For many years, there were vague ideas that some diseases were due to dietary deficiencies. The discovery of vitamins in the early 1900s, their chemical characterization and the clarification of their metabolic functions are sequential aspects of a brilliant chapter in the history of modern nutritional sciences and medicine. The research in vitamins and deficiency-related pathologies has indeed seen a remarkable past, and the path to still better understanding is likely to continue far into the future.

Numerous observations led to the discovery of vitamins and their function, including some truly pioneering investigations that made history.

For example, perhaps one of the first randomized trials was done by James Lind (a British naval surgeon, 1716–1794) in 1747 on board the HMS Salisbury: he randomized 6 groups (2 sailors each) to either "a quart of cyder", or "two oranges and one lemon", or "half a pint sea water", or "twenty five gutts of elixir vitriol", or "two spoonfuls of vinegar", or "the bigness of a nutmeg of an electuray made of garlic, mustard seed, raddish root, Peruvian balsam and gum myrrh", -in addition to the regular diet. The idea was to find out which of the accepted antiscorbutic therapies really worked and the outcome was clear pointing towards the oranges and lemon-treatment (and cyder to some extent).

Another thrilling story refers to the clinical work of William Castle leading to the conclusion that there must be a reaction that takes place between an unknown intrinsic factor (IF) in the gastric juice and an unknown extrinsic factor in beef muscle. That extrinsic antianaemic factor turned out to be cobalamin (vitamin B12) and IF is needed for its absorption. Presumably, the gastric juice in pernicous anemia lacks IF.

Most of the mentioned work was done in the early 1900s. The period from 1912 to 1948 saw the isolation and identification of the variety of compounds and structures collectively termed vitamins. The name, derived from "vital-amines", indicates their elementary metabolic key functions in human metabolism. Indeed, severe deficiency is often associated with malfunction, symptoms and potentially fatal disease. The population-wide impact of overt deficiency-associated disease was so large that key discoveries in the field allowing prevention and leading to therapies were rewarded with several Nobel prizes.

The next logical step was synthetic mass production. Today, the general availability of vitamins to practically everyone and numerous national health programs have saved many lives and prevented complications. True (congenital) metabolic disorders and serious deficiency-associated diseases are rare and in general limited to particular geographic areas and high risk groups. Their identification is of utmost importance to make the best use of the appropriate vitamin's therapeutic potential. When used properly to treat or cure of diseases, vitamins are, indeed, "magical" substances. Due to their efficacy, they should therefore be regarded as drugs with effects and side effects to be weighted against each other. Sometimes a toxic dose is difficult to define and a longstanding misapprehension has led many people to believe that vitamins are harmless in any dose and, particularly in generous dose, can cure almost all of mankind's ills.

Today, it is not the previously fatal deficiency-associated diseases such as pernicious anemia or beri-beri that are in the focus of interest, but rather the relation of suboptimal vitamin bioavailability to chronic disease, which is much more difficult to observe and document. This is complicated by genetic susceptibility, lifestyle, and the presence or absence of health-compromising habits, such as smoking.

Vitamins are truly families of compounds, which include precursors and various bound forms, all with individual roles in metabolism and function. A more recent approach therefore searches for the components and optimal intake of a physiologically complete diet aiming at preventing disease and assuring optimal health. Any state of nutrition can be defined in theory, as deficient, marginal, satisfactory, excessive, or toxic. But in practice, this may prove to be extremely difficult requiring the development of appropriate tests and various status indicators.

In turn, the development and application of new and more sensitive and specific assays further enable us to look more closely into the many functions of vitamins.

At a national level, recommended daily allowances for vitamins become policy statements. Nutrition policy has far-reaching implications in the food industry, in agriculture, and in food provision programs. In fact, the broad availability of vitamins means that a great deal of work concentrates on aspects of overdosage and toxicology, interactions and also on regulations of direct and indirect intake. Water soluble vitamins are complex molecular structures and even today, many areas in vitamin biochemistry are not yet fully understood. Novel effects and functions of vitamins remain and continue to be discovered.

This book in your hands adds fantastic new insights into the biochemistry, metabolism, function, and therapeutic and diagnostic use of water soluble vitamins.

Serious clinical abnormalities including growth retardation, neurological disorders, and dermatological abnormalities occur in conditions of biotin (vitamin H) deficiency. Hamid M. Said (Irvine, California) presents the latest information on the biochemical, physiological, and clinical aspects of biotin nutrition.

Niacin is covered in two very interesting chapters. James B. Kirkland (Guelph, Ontario) explains the association between niacin status and genomic instability in bone marrow cells. The results are important for cancer patients, who tend to be niacin deficient, are exposed to large doses of genotoxic drugs, and suffer short-term bone marrow suppression and long-term development of secondary leukemias.

The recent identification of the nicotinic acid receptor has allowed distinction of the drug-like roles of nicotinic acid from its vitamin functions. The group of Elaine L. Jacobson (Tucson, Arizona) reviews niacin as an antidyslipidemic and cardioprotective drug.

The chapter on thiamin by Derrick Lonsdale (Westlake, Ohio) emphasizes beri-beri as the model for high calorie malnutrition and reviews the biochemistry of the three phosphorylated esters of thiamin and of their transporters. The pathophysiology of thiamin deficiency and the role of synthetic thiamin derivatives as therapeutic agents are discussed.

Riboflavin (vitamin B2) is essential for energy generation in the aerobic cell, through oxidative phosphorylation. Hilary J. Powers and co-workers (Sheffield, U.K.) provide insight into specific functions associated with cell fate determination, and review mechanisms and consequences of riboflavin depletion through effects on the expression of regulatory genes, exerted at both the transcriptional and proteomic level.

Sang Woon Choi (Boston, Mass.) and Simonetta Friso (Verona, Italy) and their co-workers review the evidence indicating a relationship between pyridoxin (vitamin B6) status and cancer as well as cardiovascular disease. Their discussion of the potential mechanisms of action is complemented by the chapter by Georg T. Wondrak (Tucson, Arizona), also from the Jacobson group, presenting insight to the structural basis of pyridoxin activity as a potent antioxidant, metal chelator, carbonyl scavenger and photosensitizer.

Novel aspects of the very complex biochemistry and metabolism of cobalamin (vitamin B12) are elaborated in three chapters. Bernhard Kräutler (Innsbruck, Austria) provides an in-depth survey of the physiological chemistry of cobalamin in the context of the metabolic transformation of cobalamin-derivatives and their use in cobalamin-dependent enzymes, whereas Sergey N. Fedosov (Aarhus, Denmark) reviews the molecular mechanisms of cobalamin transport with emphasis on interaction of corrinoids with the specific proteins and protein-receptor recognition. Practical novel aspects concerning early detection of cobalamin-related disorders are described including medical application of Cbl-conjugates, and purification of corrinoids from biological samples. To round out the subject, Wolfgang Herrmann (Homburg, Germany) suggests a revised definition of cobalamin deficiency.

Ascorbic acid (vitamin C) is covered in three further outstanding chapters. Mario C. De Tullio (Bari, Italy) focuses on largely unknown facets of the role of ascorbic acid in cell metabolism and its involvement in cell signalling and gene expression that are, at least in part, unrelated to antioxidant functions. John X. Wilson and F. Wu (Buffalo, New York) discuss in depth the role of ascorbic acid in correcting microvascular dysfunction and protecting capillary blood flow and arteriolar responsiveness in septic syndromes. Finally, neurons in the central nervous system (CNS) contain some of the highest ascorbic acid concentrations of mammalian tissues. James M. May (Nashville, Tenn.) reviews the role of the specific ascorbate transporter SVCT2 (Slc23a2) in regulating neuronal ascorbate homeostasis and the extent to which ascorbate affects brain function, including antioxidant protection, peptide amidation, myelin formation, synaptic potentiation, and protection against glutamate toxicity in the CNS.

There is increasing research into the importance of folate-derived one-carbon units for DNA and histone methylation reactions, which exert crucial epigenetic control over cellular protein synthesis. Numerous polymorphisms have now been identified in folate related genes and it is thus becoming clear that genetic aspects of folate metabolism are wide-ranging and may touch on events as disparate as prenatal imprinting and cancer susceptibility. This topic is covered by Anne M. Molloy (Dublin, Ireland), who provides a detailed account of genetic aspects of folate metabolism. Willi Wonisch (Graz, Austria) and myself have elaborated on novel enzymatic and non-enzymatic antioxidative effects of folic acid and its reduced derivates that help explain the beneficial effects of folic acid supplementation in various clinical conditions. As an important clinical application J. D. Williams (Tucson, Arizona) reports on recent advances in research on folates in skin cancer and their potential in its prevention. And finally, J. Yang and Philipp Low (W. Lafayette, Indiana), expand upon themes introduced earlier and give a brilliant overview of the biology of the folate receptor (FR) and FR expression in immune cells with relevance for cancer and inflammatory and autoimmune disease. Their description of folate conjugates acting as drug carriers and diagnostic imaging tools holds strong promise for future successful efforts to cure certain types of cancer and chronic inflammatory disease.

This work in sum total so adds a new leaf to the exciting book on vitamin research that today's and tomorrows's scientists will continue to write.

I am deeply thankful to each of the contributing authors. All of them are highly esteemed and leading scientists very much dedicated to the demanding biochemistry of vitamins. They all have willingly agreed to share the latest research results from their respective working groups, laboratories and institutes. I truly thank them for all of their efforts and patience in the preparation of this book.

London, UK Olaf Stanger

Contents

Contributors

Sang Woon Choi Tufts University, Medford/Somerville, MA, USA; Department of Clinical and Experimental Medicine, School of Medicine, University of Verona, Verona, Italy; Vitamins and Carcinogenesis Laboratory, Jean Mayer USDA HNRCA, Tufts University, Boston, MA 02111-1524, USA, sang.choi@tufts.edu

B.M. Corfe Human Nutrition Unit, The School of Medicine, The University of Sheffield, Sheffield S10 2RX, UK, b.m.corfe@sheff.ac.uk

R. Corrocher Tufts University, Medford/Somerville, MA, USA; Department of Clinical and Experimental Medicine, University of Verona, Verona, Italy, roberto.corrocher@univr.it

Mario C. De Tullio Dipartimento di Biologia e Patologia Vegetale, Universita' di Bari, 70125 Bari, Italia, detullio@botanica.uniba.it

Sergey N. Fedosov Protein Chemistry Laboratory, Department of Molecular Biology, University of Aarhus, Aarhus, Denmark, snf@mb.au.dk

Simonetta Friso Tufts University, Medford/Somerville, MA, USA; University of Verona School of Medicine, Department of Medicine, 37134 Verona, Italy, simonetta.friso@univr.it

Wolfgang Herrmann Department of Clinical Chemistry and Laboratory Medicine, University of Saarland, 66421 Homburg, Germany, kchwher@uniklinik-saarland.de

Elaine L. Jacobson Division of Medicinal Chemistry, Department of Pharmacology and Toxicology, College of Pharmacy and Arizona Cancer Center, University of Arizona, Tucson, AZ, USA; Niadyne Development, Inc., Tucson, AZ, USA, jacobse@pharmacy.arizona.edu; elaine.Jacobson@pharmacy.arizona.edu

M.K. Jacobson Division of Medicinal Chemistry, Department of Pharmacology and Toxicology, College of Pharmacy and Arizona Cancer Center, University of Arizona, Tucson, AZ, USA; Niadyne Development, Inc., Tucson, AZ, USA, mjacobson@pharmacy.arizona.edu

H. Kim Division of Medicinal Chemistry, Department of Pharmacology and Toxicology, College of Pharmacy and Arizona Cancer Center, University of Arizona, Tucson, AZ, USA; Niadyne Development, Inc., Tucson, AZ, USA, hkim@pharmacy.arizona.edu

M. Kim Division of Medicinal Chemistry, Department of Pharmacology and Toxicology, College of Pharmacy and Arizona Cancer Center, University of Arizona, Tucson, AZ, USA; Niadyne Development, Inc., Tucson, AZ, USA, mkim@pharmacy.arizona.edu

James B. Kirkland Department of Human Health and Nutritional Sciences, University of Guelph, Guelph, ON, Canada N1G 2W1, jkirklan@uoguelph.ca

Bernhard Kräutler Institute of Organic Chemistry and Centre of Molecular Biosciences (CMBI), University of Innsbruck, Innsbruck, Austria, Bernhard.kraeutler@uibk.ac.at

Derrick Lonsdale Preventive Medicine Group, Westlake, OH 44145, USA, dlonsdale@pol.net

V. Lotto Tufts University, Medford/Somerville, MA, USA; Department of Clinical and Experimental Medicine, University of Verona, Verona, Italy, valentine.lotto@univr.it

Philip Low Purdue University, West Lafayette, IN, USA, plow@purdue.edu

James M. May Departments of Medicine and Molecular Physiology and Biophysics, Vanderbilt University School of Medicine, Nashville, TN 37232, USA, james.may@vanderbilt.edu

Anne M. Molloy School of Medicine, Trinity College Dublin, Dublin 2, Ireland, amolloy@tcd.ie

E. Nakano Human Nutrition Unit, The School of Medicine, The University of Sheffield, Sheffield S10 2RX, UK, e.nakano@sheff.ac.uk

Rima Obeid Department of Clinical Chemistry and Laboratory Medicine, University of Saarland, 66421 Homburg, Germany, rima.obeid@uniklinikum-saarland.de

Hilary J. Powers Human Nutrition Unit, The School of Medicine, The University of Sheffield, Sheffield S10 2RX, UK, h.j.powers@sheff.ac.uk

Hamid M. Said University of California-School of Medicine, Irvine, CA, USA; VA Medical Center, Long Beach, CA 90822, USA, hmsaid@uci.edu

Olaf Stanger Royal Brompton and Harefield NHS Foundation, Imperial College of Science, Technology and Medicine, Heart Division, Cardiothoracic Surgery, Sydney Street, London SW3 6NP, United Kingdom, o.stanger@rbht.nhs.uk

Erina Vlashi University of California, Los Angeles, CA, USA; Purdue University, West Lafayette, IN, USA, evlashi@purdue.edu

J.D. Williams Division of Medicinal Chemistry, Department of Pharmacology and Toxicology, College of Pharmacy and Arizona Cancer Center, University of Arizona, Tucson, AZ, USA; Niadyne Development, Inc., Tucson, AZ, USA, williams@pharmacy.arizona.edu

John X. Wilson Department of Exercise and Nutrition Sciences, University at Buffalo, Buffalo, NY 14214-8028, USA, jxwilson@buffalo.edu

Georg T. Wondrak Department of Pharmacology and Toxicology, College of Pharmacy and Arizona Cancer Center, University of Arizona, Tucson, AZ, USA, wondrak@pharmacy.arizona.edu

Willibald Wonisch Clinical Institute of Medical and Chemical Laboratory Diagnostics, Institute of Physiological Chemistry, Center for Physiological Medicine, Medical University of Graz, A-8010 Graz, Austria, willi.wonisch@aon.at

F. Wu Department of Exercise and Nutrition Sciences, University at Buffalo, Buffalo, NY 14214-8028, USA, fengwu@buffalo.edu

Jun Yang St. Jude Children's Research Hospital, Memphis, TN, USA; Purdue University, West Lafayette, IN, USA, jun.yang@stjude.org

Chapter 1
Biotin: Biochemical, Physiological and Clinical Aspects

Hamid M. Said

Abstract Significant progress has been made in our understanding of the biochemical, physiological and nutritional aspects of the water-soluble vitamin biotin (vitamin H). It is well know now that biotin plays important roles in a variety of critical metabolic reactions in the cell, and thus, is essential for normal human health, growth and development. This is underscored by the serious clinical abnormalities that occur in conditions of biotin deficiency, which include, among other things, growth retardation, neurological disorders, and dermatological abnormalities (reviewed in 1). Studies in animals have also shown that biotin deficiency during pregnancy leads to embryonic growth retardation, congenital malformation and death (Watanabe 1983; Cooper and Brown 1958; Mock et al. 2003; Zempleni and Mock 2000). The aim of this chapter is to provide coverage of current knowledge of the biochemical, physiological, and clinical aspects of biotin nutrition. Many sections of this chapter have been the subject of excellent recent reviews by others (Wolf 2001; McMahon 2002; Mock 2004; Rodriguez-Melendez and Zempleni 2003; Said 2004; Said et al. 2000; Said and Seetheram 2006), and thus, for more information the reader is advised to consider these additional sources.

Keywords Biotin · Vitamin H · Carboxylases · Gluconeogenesis · Fatty acids

1.1 Chemical Structure, Sources and Availability of Biotin

The chemical structure of the biotin molecule (see Fig. 1.1 for the structure of biotin and related/relevant compounds) is composed of two rings to which a valeric acid moiety is attached as a side chain. One of the rings contains a ureido group (N-CO-N) which is involved in the binding of the vitamin to avidin, a glycoprotein found in egg-white that has an extremely high binding affinity toward biotin (Green 1990). The other ring contains a tetrahydrothiophene group to which a valeric acid moiety is attached. While the biotin molecule can exist in eight stereoisomers, the

H.M. Said (✉)
University of California-School of Medicine, Irvine, CA, USA

VA Medical Center, Long Beach, CA 90822, USA
e-mail: hmsaid@uci.edu

O. Stanger (ed.), *Water Soluble Vitamins*, Subcellular Biochemistry 56,
DOI 10.1007/978-94-007-2199-9_1, © Springer Science+Business Media B.V. 2012

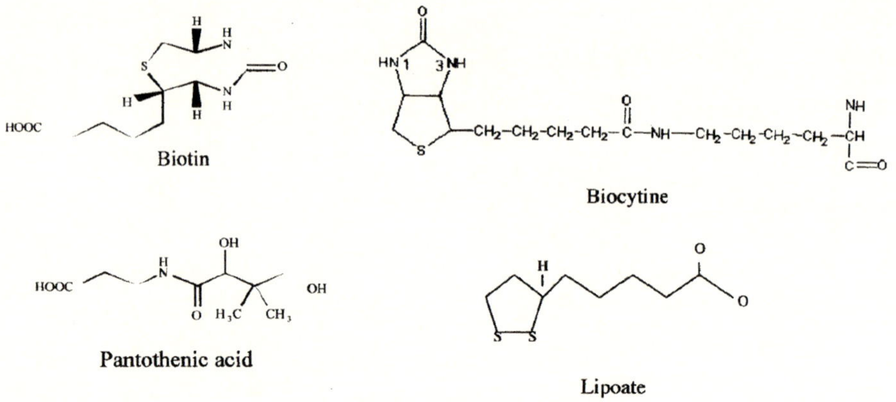

Fig. 1.1 Chemical structure of biotin and related compounds

D-biotin isomer is the only stereoisomer that is biologically active. Biotin has a pKa of 4.5, thus, it exists at physiological pHs mainly in the anionic de-protonated form.

Biotin is essential for all organisms. Animal cells cannot synthesize this micronutrient, but microorganisms (like certain bacteria and yeast) and plant cells can synthesize biotin endogenously (McMahon 2002). Biotin is widely distributed in foodstuff but at a level that is lower than that of other water-soluble vitamins (Combs 1992). Good sources for biotin include organ meat (like liver and kidney), egg yolk, some vegetables and cow's milk; poor sources include lean meat, cereal and fruits (Combs 1992; Harding and Crooks 1961). Bioavailability of dietary biotin varies from one food source to another; it is close to 100% in the case of corn, and as low as 5% in case of wheat (Combs 1992). In addition to the dietary source of biotin, humans have an additional source for this essential micronutrient, which is the bacterial source in the large intestine where the vitamin is produced and can be absorbed (see below). The relative contribution of this latter source of biotin toward overall body requirement of the vitamin is not well defined (Wolf 2001); however, it is becoming increasingly evident that this source of biotin does contribute to the host nutrition, and especially to the cellular nutrition of the localized colonocytes (see below).

1.2 Biotin Requirements and the Incidence of Biotin Deficiency

The recommended daily allowance of biotin has not been established due to the uncertainty about the contribution of the bacterially synthesized biotin in the human large intestine. However, a safe daily intake of the vitamin for both infants and adults has been estimated to be at 35 ug and 150–300 ug, respectively (National Research Council 1980). Biotin appears to be relatively nontoxic even at doses of greater than 60 mg/day for several months (Wolf 2001; Paul 1978; Watanabe 1996).

Biotin-deficiency and sub-optimal levels have been reported with increased frequency in recent years. Deficiency of biotin occurs in patients with inborn errors of biotin metabolism (Wolf 2001; Sweetman and Nyhan 1986), in patients on long-term therapy with anticonvulsant agents (Krause et al. 1982a), and in patients on long-term parenteral nutrition (Krause et al. 1985; Forbes and Forbes 1997). Sub-optimal levels of biotin have been reported during pregnancy (Mock et al. 1997), in substantial numbers of alcoholics (Fennelly et al. 1969; Bonjour 1980), in patients with inflammatory bowel disease (Urabe et al. 1986; Banares et al. 1989), and in patients with seboric dermatitis and Leiner's disease (Nisenson 1957; Messaritakis et al. 1975).

1.3 Metabolic Role of Biotin

In mammals, biotin acts a cofactor for four carboxylases catalyzing the transfer of a carboxyl group to targeted substrates (reviewed in McMahon 2002; Mock 2004; Sweetman and Nyhan 1986; Dakshinamurti and Chauhan 1988). These carboxylases play a critical role in the intermediate metabolism of gluconeogenesis, fatty acid synthesis, and amino acid catabolism (McMahon 2002; Mock 2004; Sweetman and Nyhan 1986; Dakshinamurti and Chauhan 1988). The four biotin-dependent carboxylases are: pyruvate carboxylase (PC; EC 6.4.1.1), propionly-CoA carboxylase (PCC; EC 6.4.1.3), β-methylcrotonyl-CoA caboxylase (MCC; EC 6.4.1.4), and acetyl-CoA carboxylase (ACC; EC 6.4.1.2), with the latter enzyme existing in two genetically distinct forms one of which is in the cytosol (ACC1) and the other is in the mitochondria (ACC2). These carboxylases exist in the inactive apo-forms, which are converted to the active hollo-forms by the action of the enzyme holocarboxylase synthetase (HCLS; EC 6.3.4.10). The conversion occurs via a two-step, ATP-dependent reaction that involves the covalent attachment of a biotin molecule to a lysine residue in the carboxylases. This lysine moiety is located in a highly conserved domain common to all the biotin-dependent carboxylases. The HCLS exists in different cellular compartments including the nucleus, cytoplasm and mitochondria (Gravel and Narang 2005).

PC plays a role in the conversion of pyruvate to oxalate in the tricarboxylic acid cycle. Deficiency of this enzyme leads to lactic acidemia. PCC plays a role in the conversion of methylmalony-CoA to propionyl-CoA, which then becomes succinyl-CoA and enters into the tricarboxylic acid cycle. Deficiency of the latter enzyme leads to an increase in the excretion of the organic acids 3-hydroxypropionic acid and 2-methylcitric acid. MCC plays a role in the metabolism of leucine and deficiency of this enzyme leads to an increase in the urinary excretion of organic acids like 3-hydroxyisovaleric acid and 3-methylcrotonylglycine. ACC1 and ACC2 are involved in the generation of malonyl CoA.

In addition to the classical function of biotin described above, recent findings have indicated a role for the vitamin in histone modification (via biotinylation) (Gravel and Narang 2005; Ballard et al. 2002; Stanley et al. 2004). This function may be important for cell proliferation (Stanley et al. 2002) and DNA repair (Peteres et al. 2002) and may explain a number of findings on the effect of biotin on cell function. Both the HCLS and biotinidase (see latter) were suggested to play a role in histones biotinylation (Gravel and Narang 2005; Ballard et al. 2002; Stanley et al. 2004). Patients with HCLS deficiency have very low histone biotinylation level in their lymphocytes (Narang et al. 2004).

Evidence for a role for biotin in the regulation of expression of a variety of genes has also been reported. An excellent review of this subject by Rodriguez-Melendez and Zempleni has recently been published (Rodriguez-Melendez and Zempleni 2003). Briefly, biotin appears to stimulate the level of expression of the insulin receptor (De La Vega and Stockert 2000), glucokinase (a key enzyme in glycolysis) (Chauhan and Dakshinamurti 1991, Dakshinamurti and Cheah-Tan 1968), and that of the human thiamin transporter-2 (Vlasova et al. 2005), while it suppresses the level of expression of the hepatic phosphoenolpyruvate carboxykinase (a key enzyme in gluconeogenesis; Dashinamurti and Li 1994). In addition, expression of the genes of the holocarboxylase synthetase and of the biotin-dependent carboxylases is affected by the prevailing level of biotin (Rodriguez-Melendez et al. 2001; Solorzano-Vargas et al. 2002; Wiedmann et al. 2003). Furthermore, the level of expression of the human sodium-multivitamin transporter (hSMVT, a major biotin transporter in different tissues, see latter) has also been shown to be modulated (in a tissue specific manner) by the prevailing level of biotin (Rodriguez-Melendez and Zempleni 2003; Crisp et al. 2004; Reidling and Said 2006; Balamurugan et al. 2005). Evidence has also been forthcoming demonstrating an effect for biotin on expression of oncogenes like N-myc, c-myb, N-ras and raf (Scheerger and Zempleni 2003). A post-transcriptional regulation of the asialoglycoprotein expression in hepatocytes by biotin has also been reported and is believed to be due to improper cell surface targeting of the glycoprotein (Collins et al. 1988). The mechanisms through which biotin exerts its above effects are not fully understood but could involve activation of soluble guanylate cyclase (by biotinyl-AMP), translocation of NF-kB to the nucleolus, and histone modifications via biotinylation (Rodriguez-Melendez and Zempleni 2003; Gravel and Narang 2005; Ballard et al. 2002; Stanley et al. 2004; Solorzano-Vargas et al. 2002).

Biotin also appears to play a role in normal immune functions including the production of antibodies, normal macrophage function, and the differentiation of T and B lymphocytes (Rabin 1983; Pruzansky and Axelrod 1955; Petrelli et al. 1981; Kung et al. 1979; Kumar and Axelrod 1978; Baéz-Saldaña et al. 1998). The vitamin may also be important for the normal function of natural killer cells as biotin supplementation alleviates the level of suppression in the activity of these cells in patients with Crohn's disease (Okabe et al. 1988). A role for biotin in cell proliferation has also been reported (Dakshinamurti et al. 1985; Mathey et al. 2002).

1.4 Biotin Catabolism

Biotin appears in the intact form in the urine and feces with the output being always higher than the daily intake of the vitamin. Some biotin, however, undergoes a limited-degree of catabolism to biotin sulfoxide (a process that takes place mainly in the liver; McCormick 1975) and to bisnorbiotin (a process that takes place mainly in the mitochondria; McCormick and Wright 1971). These catabolic events appear to increase in pregnancy, with cigarette smoking, and following the use of anticonvulsant medications (Mock et al. 1997, 2002).

1.5 The Physiology of Biotin

The intestine and the kidney play important roles in regulating biotin body homeostasis via their involvement in the entry and the exit processes of the vitamin. Thus, significant attention has been paid toward the understanding of the mechanisms involved in biotin uptake by absorptive epithelial cells of these organs and how these processes are regulated at the cellular and molecular levels. In this section, we will describe in details the current knowledge of the mechanisms and regulation of the intestinal and renal transport process of biotin; more extensive reviews in the area also exist (Said 2004; Said et al. 2000; Said and Seetheram 2006). We will also provide a description of biotin transport into other tissues that are important in biotin metabolism/function.

1.5.1 Intestinal Absorption of Biotin

1.5.1.1 Digestion of Dietary Biotin

As mentioned earlier humans are exposed to two sources of biotin, the first being the dietary source, and the second being the bacterial source in the large intestine (Wrong et al. 1981). Dietary biotin exists in the free and protein-bound forms with the ratio of the two in a given dietary source being dependent on the type of that source (Lampen et al. 1942). Ingested protein-bound forms of biotin are first broken down by gastrointestinal proteases and peptidases to biocytin (biotinyl-L-lysine; Fig. 1.2) and biotin-oligopeptides (Wolf et al. 1984). These products are then further processed in the intestinal lumen (i.e., prior to absorption) to release the free biotin (Fig. 1.2). The latter process is performed enzymetically by the action of biotinidase (EC 3.5.1.12). The source of the involved intestinal biotinidase is believed to be the pancreas (Wolf et al. 1984). Clinical and experimental evidence have demonstrated the importance of the hydrolysis step of biocytin and biotin-oligopeptides to free biotin for the efficient absorption and optimal bioavailability of dietary biotin (Wolf et al. 1984; Said et al. 1993). The liberated biotin is then absorbed in the small intestine.

1.5.1.2 Mechanism of Intestinal Biotin Uptake

The mechanism of biotin uptake by the small intestine has been the subject of intense investigations and has been thoroughly reviewed in recent years (Said 2004; Said et al. 2000; Said and Seetheram 2006). Using a variety of intestinal in vitro and in vivo preparations, it has been well established that the intestinal biotin uptake

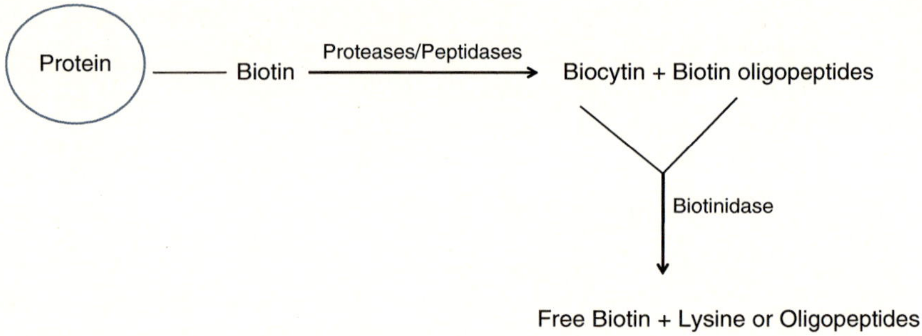

Fig. 1.2 Uptake of dietary biotin

process occurs via a Na+-dependent, carrier-mediated mechanism. This mechanism is inhibited by biotin structural analogues with a free carboxyl group at the valeric acid moiety of the biotin molecule (as in the case with desthiobiotin), but not by analogues with a blocked carboxyl group (as in the case of biocytin). Uptake of biotin in both human and animals is higher in the proximal compared to the distal part of the small intestine (Said et al. 1988; Said and Redha 1987). Functional and immunological studies have shown that the biotin Na+-dependent, carrier-mediated mechanism is expressed only at the apical membrane domain of the polarized enterocytes (Said et al. 1987, 1998; Said and Redha 1988b; Said and Derweesh 1991; Said 1991; Nabokina et al. 2003). This has been further confirmed by means of confocal imaging of living intestinal epithelial cells transfected with fluorescently tagged biotin transporter, SMVT (Fig. 1.4; Subramanian et al. 2006). The apical Na+-dependent biotin uptake system appears to be the rate-limiting step in the overall movement of biotin across the intestinal epithelial cells and is capable of transporting the substrate against a concentration gradient (Said et al. 1987; Said and Redha 1988a; Said and Derweesh 1991). The role of Na+ in biotin transport across the intestinal apical membrane domain is mediated via the inwardly directed Na+ gradient (which provides the needed energy for the transport of biotin against the concentration gradient) and not through the mere existence of Na+ in the incubation medium (Said et al. 1987) (Fig. 1.3). Internalized biotin then leaves the intestinal epithelial cells via the basolateral membrane by means of a Na+-independent, electrogenic, carrier-mediated mechanism (Said and Redha 1988c; Said 1991) (Fig. 1.3).

One of the interesting features of the Na+-dependent biotin uptake system is its ability to also transport two other functionally unrelated nutrients, namely pantothenic acid and lipoate (Fig. 1.1) (Said et al. 1998; Nabokina et al. 2003). Pantothenic acid is a member of the B family of water-soluble vitamins which is required for the biosynthesis of coenzyme A and acyl carrier proteins, and thus, it assumes important roles in carbohydrate, fat and protein metabolism. Lipoate is a potent intracellular and extracellular antioxidant in mammalian cells and is involved in the redox cycling of other antioxidants like vitamins C and E; it is also involved in regulating the intracellular level of glutathione. The ability of the intestinal biotin

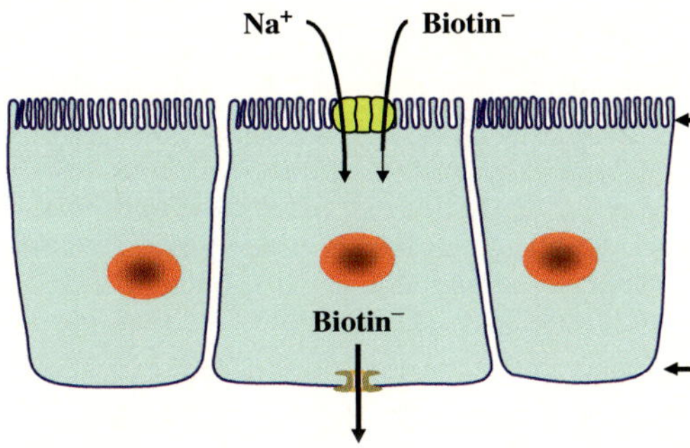

Fig. 1.3 Processing of dietary protein-bond biotin in the intestinal lumen

uptake system to also transport pantothenic acid and lipoate is not unique to the intestine but has also been observed in other cellular systems such as the brain, heart, placenta and kidney (Said 1991; Spector and Mock 1987; Beinlich et al. 1990; Grassl 1992). This ability increases the physiological and nutritional significance of the involved uptake system and was the basis for its naming as the sodium-dependent multivitamin transporter (SMVT).

With regards to the bacterially synthesized biotin, a substantial amount of this biotin exists in large intestinal lumen in the absorbable/unbound form (Wrong et al. 1981; Streit and Entcheva 2003). Also, in vivo studies in humans and animal models have shown that the large intestine is capable of absorbing luminal biotin (Barth et al. 1986; Brown and Rosenberg 1987; Sorrell et al. 1971). The mechanism involved in biotin uptake in the large intestine has also been the subject of investigation in recent years (Said et al. 1998). Using the human-derived colonic epithelial NCM460 cells as a model system for human colonocytes, studies have shown the existence of an efficient, Na+ -dependent, carrier-mediated mechanism for biotin uptake in these cells, which is again shared by pantothenic acid and lipoate (Said et al. 1998). These findings provided further evidence for the accessibility of the bacterially synthesized biotin into the human and its contribution toward the overall biotin pool, and especially that of the localized colonocytes.

1.5.1.3 Molecular Identity of the Intestinal Biotin Transport System

Cloning studies have determined the molecular identity of the intestinal SMVT of humans and of a number of animal models (Prasad et al. 1998; Wang et al. 1999; Chatterjee et al. 1999; and GenBank accession # AY572835). The human SMVT gene is located on chromosome 2p23 and consists of 17 exons (Wang et al. 1999). Significant sequence homology, at both the nucleotide and the amino acid levels, was found in the SMVT of the different mammalian species, and the polypeptide was predicted to have 12 trans-membrane domains with both of its ends (i.e., the N- and the C-terminal tails) oriented inwardly. The hSMVT polypeptide also appears to have a number of potential post-translational modification sites including sites for phosphorylation by protein kinases and glycosylation sites. When expressed in heterologus systems, the cloned SMVT displayed Na+-dependency transport specificity for biotin, pantothenic acid and lipoate, and showed similar kinetic parameters to those observed for biotin uptake in the native intestine. The SMVT transcripts were shown to be expressed at a markedly higher level in the differentiated epithelial cells of the intestinal villi compared to those of the immature and undifferentiated cells of the crypt, a finding that corresponds with the higher level of biotin uptake in the former compared to the latter cells (Chatterjee et al. 1999). Distribution of the SMVT transcripts along the longitudinal axis of the intestine has also been delineated and shown to be similar in the different regions of the gut (Prasad et al. 1998; Wang et al. 1999; Chatterjee et al. 1999). This is in contrast to the observations of a higher biotin uptake in the proximal compared to the distal part of the small intestine, and the colon (Said et al. 1988; Said and Redha 1987). These findings suggest the possible involvement of specific post-translational modification(s) that may regulate biotin transport activity in the different regions of the intestinal tract. Cellular localization of the SMVT protein in human intestinal epithelial cells have also been examined using confocal imaging approach with results showing exclusive expression of the protein the apical membrane of polarized cells (Fig. 1.4; Unpublished observation from one laboratory). Other studies, have reported significant heterogeneity in the 5′ un-translated region of the rat SMVT, with four distinct variants (I, II, III, IV) being identified (Chatterjee et al.

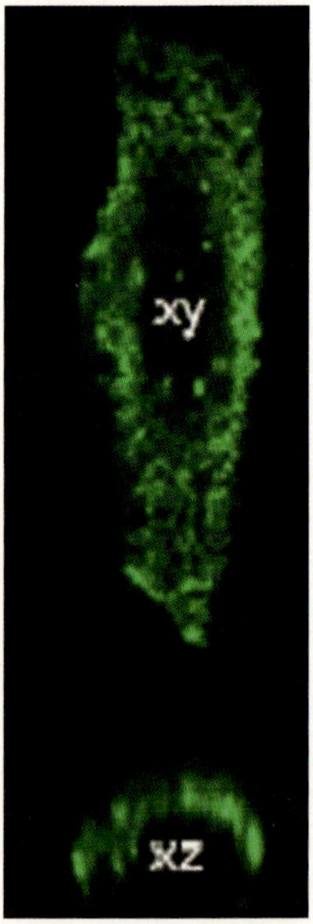

Fig. 1.4 Confocal imaging of live intestinal epithelial Caco-2 cells grown on filter showing exclusive expression of the human SMVT at the apical membrane domain. The XZ image shows exclusive expression of the human SMVT at the apical membrane domain of the polarized Caco-2 cells

1999); variant II was shown to be the predominant variant expressed in the intestinal tract (Chatterjee et al. 1999). A major role for the human SMVT system in intestinal carrier-mediated biotin uptake has been recently established with the use of a SMVT gene-specific siRNA approach (Balamurugan et al. 2003).

1.5.1.4 Regulation of the Intestinal Biotin Absorption Process

The intestinal biotin uptake process is regulated by a number of intracellular and extracellular factors. A role for intracellular protein kinase-C (PKC)- and Ca^{2+}/calmodulin-mediated pathways in the regulation of the intestinal biotin uptake process has been reported (Said 1999; Said et al. 1998). While each of these intracellular regulatory pathways was found to act via altering the activity (but not the affinity) of the SMVT system, they appeared to do so via different mechanisms (Said 1999; Said et al. 1998).

Extracellular biotin levels exert adaptive regulatory effects on the intestinal biotin uptake process in both humans and animal models (Reidling and Said 2006; Said et al. 1989a). Biotin deficiency leads to a specific and significant up-regulation in intestinal carrier-mediated biotin uptake, while biotin over-supplementation appears to have the opposite effect (ref). The up-regulation in intestinal biotin uptake observed in biotin deficiency occurs in association with a parallel increase in the level of SMVT protein and mRNA in intestinal epithelial cells with no changes in mRNA stability (Reidling and Said 2006). These findings suggest the involvement of transcriptional regulatory mechanism(s) in the observed adaptive regulatory effects in the intestinal biotin uptake process in biotin deficiency.

The intestinal biotin uptake process also undergoes developmental regulation via changes in the preferential site of biotin absorption and via changes in the kinetic parameters of the uptake process (Nabokina et al. 2003; Said and Redha 1988b). The latter changes involve the entry step of biotin across the apical membrane domain of the polarized intestinal epithelial cells and occurs in association with parallel changes in the level of SMVT protein and mRNA, as well as the transcription rate of the SMVT gene (Nabokina et al. 2003).

Recent studies using the human-derived intestinal epithelial Caco-2 cells as a model have also shown that the intestinal biotin uptake process is under differentiation-dependent regulation. Biotin uptake and the level of expression of hSMVT mRNA and protein as well as activity of the hSMVT promoter were found to be higher in post-confluent (differentiated) Caco-2 cells compared to pre-confluent (undifferentiated) cells (Reidling J and Said HM; unpublished observations). These findings clearly point to the possible involvement of transcriptional regulatory mechanism(s) in the differentiation-dependent regulation of the biotin uptake process.

Insight into the transcriptional regulation of the SMVT gene under basal and regulated conditions has also been forthcoming following the cloning and the characterization of the 5′-regulatory region of the human and rat SMVT genes (Chatterjee et al. 2001; Dey et al. 2002). In the case of the rat SMVT gene, three distinct promoters were identified (Chatterjee et al. 2001), while two promoters were identified in the case of the human SMVT gene (Dey et al. 2002). In both cases activities of the cloned promoters were demonstrated by fusing the promoter fragments with the Firefly luciferase reporter gene followed by expression of the constructs in the appropriate cellular systems. Human promoter I was found to be more active than promoter II and required functional GKLF and AP-2 cis-regulatory elements for its activity in intestinal epithelial cells (Reidling and Said 2006). Activity of the cloned human SMVT promoter has also been confirmed in vivo in transgenic mice (Reidling and Said 2006).

1.5.1.5 Effect of Anti-Epileptic Drugs and Alcohol on Intestinal Biotin Uptake

As mentioned earlier, long-term use of anticonvulsant drugs leads to impairment in normal biotin status (Krause et al. 1982a, b). While the mechanism(s) involved in causing this abnormality is not fully clear, competitive inhibition of intestinal biotin uptake by these agents (Said et al. 1989b; Prasad and Ganapathy 2000) as well as accelerated biotin catabolism (Mock et al. 1997, 1998) and impairment renal reclamation of the vitamin (Chauhan and Dakshinamurti 1988) have all been reported. Similarly, the reduced blood biotin levels observed in alcoholics (Fennelly et al. 1969; Bonjour 1980) is believed to be, at least in part mediated via impairment in intestinal uptake of biotin (Said et al. 1990b).

1.5.2 Renal Uptake of Biotin

Circulating biotin undergoes filtration in the renal glomeruli. The vitamin is then salvaged via reabsorption by renal proximal tubular epithelial cells. Studies on the mechanism of renal biotin uptake have shown the involvement of a concentrative Na-dependent, carrier-mediated mechanism localized at the apical membrane domain of the polarized renal epithelial cells (Balamurugan et al. 2005; Baur and Baumgartner 1993; Baur et al. 1990; Podevin and Barbarat 1986). The system involved in renal biotin uptake is also shared by pantothenic acid and lipoate (Balamurugan et al. 2005) and is inhibited by biotin structural analogues with a free carboxyl group in the valeric acid moiety (like desthiobiotin), but not by analogues with a blocked carboxyl group (e.g., biocyin and biotin sulfoxide) (Balamurugan et al. 2005; Baur and Baumgartner 1993; Baur et al. 1990; Podevin and Barbarat 1986). Exit of biotin from the renal epithelial cells occurs via a carrier-mediated mechanism that is Na-independent and electrogenic in nature (Podevin and Barbarat 1986).

Regulation of the renal biotin uptake process has also been examined using the human-derived renal proximal epithelial HK-2 cells as in vitro model to human renal epithelial cells (Balamurugan et al. 2005). The results showed the process to be under the regulation of intracellular PKC- and Ca^{2+}/calmodulin (CaM)-mediated pathways. The renal biotin uptake process is also adaptively regulated by extracellular biotin levels via a mechanism that appears to be transcriptionally mediated (Balamurugan et al. 2005). Other studies have shown that native human renal epithelial cells and the culture renal epithelial HK-2 cells to both express hSMVT at the protein and the mRNA levels (Balamurugan et al. 2005; Chatterjee et al. 1999). This hSMVT appears to be the main (if not the only) carrier-mediated mechanism for biotin uptake by renal epithelial cells as shown in studies utilizing gene silencing approach with gene-specific siRNA (Balamurugan et al. 2005).

An interesting finding with the renal biotin uptake process when compared to the intestinal uptake process is the finding that the former process does not appear to undergo developmental regulation (Nabokina et al. 2003). This conclusion is based on the observations that biotin uptake as well as the levels of the SMVT protein and mRNA is similar in suckling and adult rat kidney epithelial cells (Nabokina et al. 2003).

1.5.3 Biotin Uptake by the Liver

The liver plays an important role in normal biotin physiology and represents the major organ for biotin metabolism and utilization. While the liver contains the highest amount of biotin compared to other tissues, its capacity to store the vitamin is limited compared to other water-soluble vitamins like cobalamin, folate and riboflavin (Danford and Munro 1982). The liver extracts (from the portal circulation) a major portion of the newly absorbed biotin (Brown and Rosenberg 1987), and evidence exists that recycled biotin (i.e., the biotin that is generated as a result of degradation of holocarboxylase) is generated not in cells but in the extracellular compartment (Dakshinamurti and Chauhan 1988; Freytag and Utter 1983; Heard et al. 1985). Thus, the liver relies heavily on circulating and extracellular biotin for its needs, and on its transport across the hepatocyte basolateral membrane. Using isolated and cultured hepatocytes as well as purified liver basolateral membrane vesicle preparations of human and rat origin (Komro and McCormick 1985; Said et al. 1990a, 1992b, 1994), studies have shown the involvement of a concentrative, Na+-dependent carrier-mediated mechanism. This mechanism is sensitive to the effect of sulfhydryl group inhibitors and is inhibited by biotin structural analogues like desthiobiotin but not by biocytin. Human liver and the human-derived liver HepG2 cells both express the hSMVT at the protein and mRNA level (Balamurugan et al. 2003), and gene silencing studies with siRNA have shown that this system is the main biotin uptake system that operates in human hepatocytes (Balamurugan et al. 2003).

1.5.4 Biotin Transport Across the Blood Brain Barrier and the Placenta

Biotin transport across the blood brain barrier is carrier-mediated and is inhibited by probenecid and pantothenic acid but it is not affected by biocytin. This has been shown in vivo in the rat (Spector and Mock 1987) and in vitro using cultured calf brain microvessel endothelial cells (Baur and Baumgartner 2000). The recent description of patients with a novel biotin-responsive basal ganglia disease (Ozand et al. 1998; Zeng et al. 2005), and a patient with sudden onset biotin-responsive coma (Mardach et al. 2002), underscore the importance of biotin for brain function.

The concentration of biotin in the plasma of human fetuses is many folds higher than that in the plasma of the mother, clearly suggesting the involvement of an efficient transport system that transports biotin in the direction of the fetus. The existence of such a system has indeed been demonstrated using different placental preparations (Karl and Fisher 1992; Schenker et al. 1993; Hu et al. 1994). This system transports biotin by a Na-dependent, carrier-mediated mechanism and the system is again shared with pantothenic acid and lipoate. These are characteristics of SMVT, which is expressed at a high level in the human placenta (Balamurugan et al. 2003).

1.5.5 Intracellular Transport of Biotin: Transport into the Mitochondria

Studies have shown that free biotin can be transported into the mitochondria for utilization in biotinylation reactions of apo carboxylases (Ahmad and Ahmad 1991). The mechanism involved in this transport has been studied using isolated liver mitochondria and shown to occur by an acid pH-dependent, non-mediated simple diffusion process (Said et al. 1992). It is believed that biotin enters the intramitochondrial space in the protonated (neutral) form, dissociates into its anionic form (pKa = 4.5) at the alkaline pH of the mitochondria, and becomes trapped within (Said et al. 1992).

1.6 Disorders of Biotin Metabolism and Physiology

As mentioned earlier, biotin in humans acts as a coenzyme to four carboxylases, and that these enzymes exist in the inactive apo forms which are converted to the holo-active forms by the action of HCLS by means of biotinylation (McMahon 2002; Mock 2004; Sweetman and Nyhan 1986; Dakshinamurti and Chauhan 1988). At the end of their functional life, the holocarboxylases undergo proteolytic degradation (in the lysosomes) that lead to the generation of biocytin and biotin-short peptides. Free biotin is then released from biocytin and biotin-oligopeptides via the action of biotinidase. The freed biotin is then reutilized by cells, i.e., recycled, in what is sometimes referred to as the "biotin cycle."

There are two major genetic disorders in biotin metabolism in humans, both of which lead to multiple carboxylase deficiency (reviewed in McMahon 2002; Mock 2004; Sweetman and Nyhan 1986). The first metabolic disorder leads to a defect in the process of biotinylation due to deficiency in HCLS. The second metabolic disorder leads to a defect in biotin recycling (release of free biotin from its conjugated forms) due to deficiency in biotinidase. These two genetic disorders have also been classified according to the age at which their symptoms appear and are called early-onset (neonatal) multicarboxylase deficiency, and late-onset (juvenile) multiplecarboxylase deficiency syndromes, respectively.

HCLS deficiency is an autosomal recessive disorder. Patients with this disorder exhibit hypotonia, seizures, difficulties in breathing and in feeding, skin rash, and alopecia; in extreme conditions, patients may also exhibit developmental delay and coma. Metabolically, children affected by this disorder exhibit metabolic acidosis, organics aciduria and hyperammonemia. Multiple mutations in the HLCS gene have been identified and shown to lead to the generation of an enzyme with an impaired activity mainly due to decreased affinity. This condition can be diagnosed prenatally by determining the level of the relevant organic acids in the amniotic fluid, activity levels of the mitochondrial caboxylases in amniocytes, and by means of mutational analysis. This disorder can be treated with oral administration of pharmacological doses of biotin (10 mg/day), with most of the affected children demonstrating significant improvement.

Biotinidase deficiency is also an autosomal recessive disorder. The clinical symptoms associated with this disorder vary and include hypotonia, seizures, difficulty in breathing, ataxia, visual and hearing disturbances, skin rash, alopecia, immunological disturbances, and developmental delay. Metabolically, most of the affected children exhibit metabolic acidosis and organic aciduria. Multiple mutations in the biotinidase gene have been identified, which leads to the generation of an enzyme that is severely or partially incapable of recycling/re-utilization endogenous conjugated biotin; it is also incapable of releasing free biotin from its conjugated forms in the ingested food. Biotinidase deficiency can be diagnosed by measuring the biotinidase activity in the serum. Prenatally, the condition can also be diagnosed by means of mutational analysis. When diagnosed early all affected children can be successfully treated with oral administration of pharmacological doses of free biotin (5–20 mg).

Another inherited disorder related to biotin is the recently described biotin-responsive basal ganglia disease (BRGD). BBGD is a recessive disorder with childhood onset that exhibit subacute encephalopathy, confusion, dysarthria, disphagia with occasional external opthalmopelgia or supranuclear facial nerve palsy (Ozand et al. 1998; Zeng et al. 2005). The condition could progress to severe cogwheel rigidity, dystonia and quadriparesis, and even death if untreated. Patients with this disorder respond well to pharmacological doses of biotin (5–10 mg/kg/day), but the symptoms re-appear within a month if biotin supplementation is discontinued (Ozand et al. 1998; Zeng et al. 2005). Recent studies have shown that the cause of BRDG is mutations in the SLC19A3 gene, a member of the solute transporter family of genes (Zeng et al. 2005). More recent studies, however, have shown that the product of the SLC19A3 is not a biotin transporter but rather a specific membrane transporter for the water-soluble vitamin thiamin (Subramanian et al. 2006). Thus, it is still unclear at this stage how a disease like BRGD with mutations in a thiamin transporter responds to biotin over-supplementation, and further studies are needed to understand the connection(s).

A biotin-dependent sudden onset coma in an 18 months old child has also been recently described and is believed to be due to a defect in a biotin transport other than SMVT (Mardach et al. 2002). The patient in this case did not have holocarboxylase synthetase or biotinidase deficiencies, nor did the child appear to have nutritional biotin deficiency or accelerated biotin catabolism. The actual cause behind this case is under further investigation.

1.7 Concluding Remarks

Significant progress has occurred in recent years in our understanding of the physiology, biochemistry and nutritional roles of biotin. Despite that, there is a clear need for better understanding of the effects of suboptimal (marginal) biotin status on cellular metabolism and on human health and well-being. Additional effort is also needed for better appreciation of the degree of occurrence of suboptimal biotin status in the general population. There is also a need for better understanding of the nature of the molecular and biochemical defects in children who respond to biotin over-supplementation yet they are neither biotinidase nor holocarboxylase deficient. Finally a better understanding of the biotin physiology at the integrated whole animal level in vivo is required.

Acknowledgment We appreciate the kind support of the Department of Veterans Affairs and the National Institutes of Health (grants # DK 56061 and DK58057).

References

Ahmad PM, Ahmad F (1991) Mammalian pyruvate carboxylase: effect of biotin on the synthesis and translocation of apoenzyme into 3T3-L adipocyte mitochondria. FASEB J 5:2482–2485

Baéz-Saldaña A., Díaz G., Espinoza B et al (1998) Biotin deficiency induces changes in subpopulations of spleen lymphocytes in mice. Am J Clin Nutr 67:431–437

Balamurugan K, Ortiz A, Said HM (2003) Biotin uptake by human intestinal and liver epithelial cells: role of the SMVT system. Am J Physiol 285:G73–G77

Balamurugan K, Vaziri ND, Said HM (2005) Biotin uptake by human proximal tubular epithelial cells: cellular and molecular aspects. Am J Physiol 288:F23–F31

Ballard T, Wolff J, Griffin J et al (2002) Biotinidase catalyzes debiotinylation of histones. Eur J Nutr 41:78–84

Banares FF, Lacruz AA, Gine JJ et al (1989) Vitamin status in patients with inflammatory bound disease. Am J Gastroenterol 84:744–748

Barth CA, Frigg M, Hogemeister H (1986) Biotin absorption from the hindgut of the pig. J Anim Physiol Anim Nutr 55:128–134

Baur B, Baumgartner ER (1993) Na-dependent biotin transport into brush-border membrane vesicles from rat kidney. Am J Physiol 258:F840–F847

Baur B, Baumgartner R (2000) Biotin and biocytin uptake into cultured primary calf brain microvessel endothelial cells of the blood-brain barrier. Brain Res 858:348–355

Baur B, Wick H, Baumgartner ER (1990) Na-dependent biotin transport into brush-border membrane vesicles from rat kidney. Am J Physiol 258:F840–F847

Beinlich CJ, Naumovitz RD, Song WO et al. (1990) Myocardial metabolism of pantothenic acid in chronically diabetic rats. J Mol Cell Cardiol 22:323–332

Bonjour JP (1980) Vitamins and alcoholism. Int J Vitam Nutr Res 50:425–440

Brown BB, Rosenberg JH (1987) Biotin absorption by distal rat intestine. J Nutr 117:2121–2126

Chatterjee NS, Kumar CK, Ortiz A et al (1999) Molecular mechanism of the intestinal biotin transport process. Am J Physiol 277:C605–C613

Chatterjee NS, Rubin SA, Said HM (2001) Molecular characterization of the 5′ regulatory region of rat sodium-dependent multivitamin transporter gene. Am J Physiol 280:C548–C555

Chauhan J, Dakshinamurti K (1988) Role of human serum biotinidase as biotin-binding protein. Biochem J 256:265–270

Chauhan J, Dakshinamurti K (1991) Transcriptional regulation of the glucokinase gene by biotin in starved rats. J Biol Chem 266:10035–10038

Collins JC, Paietta E, Green R et al (1988) Biotin-dependent expression of the asialoglycoprotein receptor in HepG2. J Biol Chem 263:11280–11283

Combs GF (ed) (1992) Biotin. The vitamins: fundamental aspects in nutrition and health. Academic, San Diego, CA, pp 329–343

Cooper WA, Brown SO (1958) Tissue abnormalities in newborn rats from biotin-deficient mothers. Texas J Sci 10:60–68

Crisp S, Camporeale G, White BR et al (2004) Biotin supply affects rates of cell proliferation, biotinylation of carboxylases and histones and expression of the gene encoding the sodium-dependent multivitamin transporter in Jar choriocarcinoma cells. Eur J Nutr 43:23–31

Dakshinamurti K, Cheah-Tan C (1968) Biotin-mediated synthesis of hepatic glucokinase in the rat. Ach Biochem Biophys 127:17–21

Dakshinamurti K, Chauhan J (1988) Regulation of biotin enzymes. Annu Rev Nutr 8:211–233

Dakshinamurti K, Chalifour LE, Bhullar RJ (1985) Requirement for biotin and the function of biotin in cells in culture. In: Dakshinamurti K, Bhagavan HN (eds) Biotin. Academy of Science, New York, pp 38–55

Danford DE, Munro HN (1982) The liver in relation to the B-vitamins. In: Arias J, Pepper H, Schachter D, Shafritz DA (eds) The liver: biology and pathobiology. Raven Press, New York, pp 367–384

Dashinamurti K, Li W (1994) Transcriptional regulation of liver phosphoenolpyruvate carboxykinase by biotin in diabetic rats. Mol Cell Biochem 132:127–132

De La Vega L, Stockert RJ (2000) Regulation of the insulin and asialoglycoprotein receptors via cGMP-dependent protein kinase. Am J Physiol 279:C2037–C2042

Dey S, Subramanian VS, Chatterjee NS (2002) Characterization of the 5′ regulatory region of the human sodium-dependent multivitamin transporter, hSMVT. Biochim Biophys Acta 1574:187–192

Fennelly J, Frank O, Baker H et al (1969) Peripheral neuropathy of the alcoholics: I Aetiological role of aneurin and other B-complex vitamins. Br Med J 2:1290–1292

Forbes GM, Forbes A (1997) Micronutrient status in patients receiving home parenteral nutrition. Nutrition 13:941–944

Freytag SO, Utter MF (1983) Regulation of the synthesis and degradation of pyruvate carboxylases in 3T3-L1 cells. J Biol Chem 258:6307–6312

Grassl SM (1992) Human placental brush-border membrane Na+-pantothenate cotransport. J Biol Chem 267:22902–22906

Gravel R, Narang M (2005) Molecular genetics of biotin metabolism: old vitamin, new science. J Nutr Biochem 16:428–431

Green NM (1990) Avidin and streptavidin. In: Wilchek M, Bayer E (eds) Methods in enzymology, vol 186. Academic, New York, pp 51–67

Harding MG, Crooks H (1961) Lesser known vitamins in foods. J Am Diet Assoc 38:204

Heard GS, Grier RE, Weiner D et al (1985) Biotinidase-a possible mechanism for the recycling of biotin. Ann NY Acad Sci 447:400 (abst)

Hu ZQ, Henderson GI, Mock DM et al (1994) Biotin uptake by basolateral membrane of human placenta: normal characteristics and role ethanol. Proc Soc Biol Exp Med 206:404–408

Karl PI, Fisher SE (1992) Biotin transport in microvillous membrane vesicles, cultured trophblasts and the isolated perfused cotyledon of the human placenta. Am J Physiol 262:C302–C308

Komro DM, McCormick D (1985) Biotin uptake by isolated rat liver hepatocytes. Ann NY Acad Sci 447:350–358

Krause KH, Berlit P, Bonjour JP (1982a) Impaired biotin status in anticonvulsant therapy. Ann Neurol 12:485–486

Krause KH, Berlit P, Bonjour JP (1982b) Vitamin status in patients on chronic anticonvulsant therapy. Int J Vitam Nutr Res 52:375–385

Krause KH, Bonjour J, Berlit P et al (1985) Biotin status of epileptics. Ann New York Acad Sci 447:297–313

Kumar M, Axelrod AE (1978) Cellular antibody synthesis in thiamin, riboflavin, biotin and folic acid-deficient rats. Proc Soc Exp Biol Med 157:421–423

Kung JT, MacKenzie CG, Talmage DW (1979) The requirement for biotin and fatty acids in the cytotoxic T-cell response. Cell Immmunol 48:100–110

Lampen JO, Bahler GP, Peterson WH (1942) The occurrence of free and bound biotin. J Nutr 23:11–18

Mardach R, Zempleni J, Wolf B et al. (2002) Biotin dependency due to a defect in biotin transport. J Clin Invest 109:1617–1623

Mathey KC, Griffin JB, Zempleni J (2002) Biotin supply affects expression of biotin transporters, biotinylation of carboxylases, and metabolism of interleukin-2 in Jurkat cells. J Nutr 132:2316–2322

McCormick DB (1975) Biotin. Nutr Rev 33:97

McCormick DB, Wright LD (1971) The metabolism of biotin and analogues. In: Florkin M, Stolz EH (eds) Comprehensive biochemistry: metabolism of vitamins and trace elements. Elsevier, Amsterdam, pp. 81–99

McMahon RJ (2002) Biotin in metabolism and molecular biology. Annu Rev Nutr 22:221–239

Messaritakis J, Katlamis C, Karabula C et al. (1975) Generalized seborrhoeic dermatitis: clinical and therapeutic data of 25 patients. Arch Dis Child 50:871–874

Mock D (2004) Biotin: physiology, dietary sources and requirements. In: Caballero B, Allen L, Prentice A (eds) Encyclopedia of human nutrition, 2nd edn. Academic, London

Mock DM, Stadler DD, Stratton SL et al (1997) Biotin status assessed longitudinally in pregnant women. J Nutr 127:710–716

Mock DM, Mock NI, Lombard KA et al (1998) Distributions in biotin metabolism in children undergoing long-term anticonvulsant therapy. J Pediatr Gastroenterol Nutr 26:245–250

Mock DM, Quirk JG, Mock NI (2002) Marginal biotin deficiency during normal pregnancy. Am J Clin Nutr 75:295–299

Mock DM, Mock NI, Stewart CW (2003) Marginal biotin deficiency is teratogenic in ICR mice. J Nutr 133:2519–2525

Nabokina SM, Subramanian VS, Said HM (2003) Comparative analysis of ontogenic changes in renal and intestinal biotin transport in the rat. Am J Physiol 284:F737–F742

Narang MA, Dumas R, Ayer LM et al. (2004) Reduced histone biotinylation in multiple carboxylase deficiency patients: a nuclear role for holocarboxylase synthetase. Hum Mol Genet 13:15–23

National Research Council (1980) Recommended dietary allowances. National Academy of Sciences, Washington, DC

Nisenson A (1957) Seborrhoeic dermatitis of infants and Leiner's disease: a biotin deficiency. J Pediatr 51:537–548

Okabe N, Urabe K, Fujita K et al (1988) Biotin effects in Crohn's disease. Dig Dis Sci 33:1495–1496

Ozand PT, Gascon GG, Al Essa M et al (1998) Biotin-responsive basal ganglia disease: a novel entity. Brain 121:1267–1279

Paul PK (1978) Effects of nutrient toxicities in animals and man. CRC Press, Boca Raton, FL

Peteres DM, Griffin JB, Stanley JS et al (2002) Exposure to UV light causes increased biotinylation of histones in Jurkat cells. Am J Physiol 283:C878–C884

Petrelli F, Moretti P, Campanati G (1981) Studies on the relationships between biotin and the behavior of B and T lymphocytes in the guinea pig. Experientia 37:1204–1206

Podevin RA, Barbarat B (1986) Biotin uptake mechanism in brush-border and basolateral membrane vesicles isolated from rabbit leading cortex. Biochim Biophys Acta 856:471–481

Prasad PD, Ganapathy V (2000) Structure and function of mammalian sodium-dependent multivitamin transporter. Curr Opin Clin Nutr Metab Care 8:263–266

Prasad PD, Wang H, Kekuda R et al (1998) Cloning and functional expression of a cDNA encoding a mammalian sodium-dependent vitamin transporter mediating the uptake of pantothenate, biotin, and lipoate. J Biol Chem 273:7501–7506

Pruzansky J, Axelrod AE (1955) Antibody production to diphtheria toxoid in vitamin deficiency states. Proc Soc Exp Biol Med 89:323–325

Rabin BS (1983) Inhibition of experimentally induced autoimmunity in rats by biotin deficiency. J Nutr 113:2316–2322

Reidling J, Said HM (2006) In Vitro and in vivo characterization of the minimal promoter of the human intestinal biotin transporter HSMVT: adaptive regulation of the biotin uptake process. Gastroenterol 130:A68 (abst)

Rodriguez-Melendez R, Zempleni J (2003) Regulation of gene expression by biotin. J Nutr Biochem 14:680–690

Rodriguez-Melendez R, Cano S, Mendez ST et al. (2001) Biotin regulates the genetic expression of holocarboxylase synthetase and mitochondrial carboxylases in rats. J Nutr 131:1909–1913

Said HM (1991) Movement of biotin across the rat intestinal basolateral membrane: studies with membrane vesicles. Biochem J 279:671–674

Said HM (1999) Cellular uptake of biotin: mechanisms and regulation. J Nutr 129:490S–493S

Said HM (2004) Recent advances in carrier-mediated absorption of water-soluble vitamins. Ann Review Physiol 66:419–446

Said HM, Derweesh I (1991) A carrier-mediated mechanism for biotin transport in rabbit intestine: studies with brush border membrane vesicles. Am J Physiol 261:R94–R97

Said HM, Redha RN (1987) A carrier-mediated transport system for biotin in rat intestine in vitro. Am J Physiol 252:G52–G55

Said HM, Redha R (1988a) Biotin transport in basolateral membrane vesicles of human intestine. Gastroenterology 94:1157–1163

Said HM, Redha R (1988b) Biotin transport in brush border membrane vesicles of rat small intestine. Biochim Biophys Acta 945:195–201

Said HM, Redha R (1988c) Ontogenesis of the intestinal transport of biotin in the rat. Gastroenterology 94:68–72

Said HM, Redha R, Nylander W (1987) A carrier-mediated, Na+ gradient-dependent transport system for biotin in human intestinal brush border membrane vesicles. Am J Physiol 253:G631–G636

Said HM, Nylander W, Redha R (1988) Biotin transport in human intestine: site of maximum transport and effect of pH. Gastroenterology 95:1312–1317

Said HM, Mock DM, Collins J (1989a) Regulation of intestinal biotin transport in the rat: effect of biotin deficiency and supplementation. Am J Physiol 256:G306–G311

Said HM, Redha R, Nylander W (1989b) Biotin transport in the human intestine: inhibition by anticonvulsant drugs. Am J Clin Nutr 49:127–131

Said HM, Korchid S, Horne DW et al (1990a) Transport of biotin in basolateral membrane vesicles of rat liver. Am J Physiol 259:G865–G872

Said HM, Sharifian A, Bagherzadeh A et al. (1990b) Chronic ethanol feeding and acute ethanol exposure in vitro: effect on intestinal transport of biotin. Am J Clin Nutr 52:1083–1086

Said HM, Hoefs J, Mohammadkhani R et al (1992a) Biotin transport in human liver basolateral membrane vesicles: a carrier-mediated, Na+ gradient-dependent process. Gastroenterology 102:2120–2125

Said HM, Mc Alister-Henn L, Mohammmadkhani R et al (1992b) Uptake of biotin by isolated rat liver mitochondria. Am J Physiol 263:G81–G86

Said HM, Thuy LP, Sweetman L et al (1993) Transport of the biotin dietary derivative biocytin (N-biotinyl-L-lysine) in rat small intestine. Gastroenterology 104:75–79

Said HM, Ma TY, Kamanna VS (1994) Uptake of biotin by human hepatoma cell line, Hep G2: a carrier-mediated process similar to that of normal liver. J Cell Physiol 161:483–439

Said HM, Ortiz A, McCloud E et al. (1998) Biotin uptake by the human colonic epithelial cells NCM460: a carrier-mediated process shared with pantothenic acid. Am J Physiol 44:C1365–C1371

Said HM, Rose R, Seetharam B (2000) Intestinal absorption of water-soluble vitamins: cellular and molecular aspects. In: Barrett KE, Donowitz M (eds) Gastrointestinal transport: molecular physiology. Academic, San Diego, pp. 35–76

Said HM, Seetheram P (2006) Intestinal absorption pf water-soluble vitamins. In: Johnson L (ed) Physiology of the G.I. tract. San Diego, Elsevier, pp. 1791–1826

Scheerger SB, Zempleni J (2003) Expression of oncogenes depends on biotin in human small cell lung cancer cells NCI-H69. Int J Vitam Nutr Res 73:461–467

Schenker S, Hu Z, Johnson RF et al. (1993) Human placental Biotin transport: normal characteristics and effect of ethanol. Alcohol Clin Exp Res 17:566–575

Solorzano-Vargas RS, Pacheco-Alvarez D, Leon-Del-Rio A (2002) Holocarboxylase synthetase is an onligate participant in biotin-mediated regulation of its own expression and of biotin-dependent carboxylases mRNA levels in human cells. Proc Natl Acad Sci USA 99:5325–5330

Sorrell MF, Frank O, Thomson AD et al. (1971) Absorption of vitamins from the large intestine. Nutr Res Int 3:143–148

Spector R, Mock D (1987) Biotin transport through the blood brain barrier. J Neurochem 48:400–404

Stanley JS, Grifin JB, Zempleni J (2002) Biotinylation of histones in human cells. Effects of cell proliferation. Eur J Biochem 268:5424–5429

Stanley C, Hymes J, Wolf B (2004) Identification of alternatively spliced human biotinidase mRNAs and putative localization of endogenous biotinidase. Mol Genet Metab 81:300–312

Streit WR, Entcheva P (2003) Biotin in microbes, the genes involved in its biosynthesis, its biochemical role and perspectives for biotechnological production. Appl Microbiol Biotechnol 61: 21–31

Subramanian V, Marchand JS, Said HM (2006) Biotin-responsive basal ganglia disease-linked mutations inhibit thiamine transporter-2 (hTHTR-2): biotin is not a substrate for hTHTR-2. Am J Physiol Cell Physiol 291:C851–C859

Sweetman L, Nyhan WL (1986) Inheritable biotin-treatable disorders and associated phenomena. Ann Rev Nutr 6:314–343

Urabe K, Fujita K, Okabe N et al. (1986) Decreased plasma biotin levels in patients with Crohn's disease. Jpn J Gastroenterol 83:307–309

Vlasova TI, Stratton SL, Wells AM et al (2005) Biotin deficiency reduces expression of SLC19A3, a potential biotin transporter, in leukocytes from human blood. J Nutr 135:42–47

Wang H, Huang W, Fei YJ (1999) Human placental Na+-dependent multivitamin transporter. J Biol Chem 274:14875–14883

Watanabe T (1983) Teratogenic effect of biotin deficiency in mice. J Nutr 113:574–581

Watanabe T (1996) Morphological and biochemical effects of excessive amounts of biotin on the embryonic development in mice. Experientia 52:149–157

Wiedmann S, Eudy JD, Zempleni J (2003) Biotin supplementation causes increased expression of genes encoding interferon-γ, interleukin-1β, and 3-methylcrotonyl-CoA carboxylase, and causes decreased expression of the gene encoding interleukin-4 in human peripheral blood mononuclear cells. J Nutr 133:716–719

Wolf B (2001) Disorders of biotin metabolism. In: Scriver CR, Beaudet AL, Aly WS, Valle D, Childs B, Kinzler KW, Vogelstein B (eds) The metabolic and molecular bases of inherited disease. McGraw-Hill Medical Publishing Division, New York, pp 3935–3962

Wolf B, Heard GS, Mc Voy JR et al. (1984) Biotinidase deficiency: the possible role of biotinidase in the processing of dietary protein-bound biotin. J Inherited Metab Dis 7:121–122

Wrong OM., Edmonds CJ, Chadwich VS (1981) Vitamins. In: The large intestine: its role in mammalian nutrition and homeostasis. Wiley, New York, pp 157–166

Zempleni, J, Mock, DM (2000) Marginal biotin deficiency is teratogenic. Proc Soc Exp Biol Med 223:14–21

Zeng W, Al-Yamani E, Acierno J et al. (2005) Biotin-responsive basal ganglia disease maps to 2936.3 and is due to mutations in SLC19SA3. Am J Hum Genet 77:16–26

Chapter 2
Niacin Status and Genomic Instability in Bone Marrow Cells; Mechanisms Favoring the Progression of Leukemogenesis

James B. Kirkland

Abstract Niacin deficiency causes dramatic genomic instability in bone marrow cells in an in vivo rat model. The end result is seen in the increased incidence of sister chromatid exchanges, micronuclei, chromosomal aberrations and the eventual development of nitrosourea-induced leukemias. From a mechanistic perspective, niacin deficiency delays excision repair and causes double strand break accumulation, which in turn favor chromosome breaks and translocations. Niacin deficiency also impairs cell cycle arrest and apoptosis in response to DNA damage, which combine to encourage the survival of cells with leukemogenic potential. Niacin deficiency also enhances the level of oxidant damage found in cellular proteins and DNA, but not through depression of GSH levels. Pharmacological supplementation of niacin decreases the development of nitrosourea-induced leukemias, while short term effects of high niacin intake include a large increase in cellular NAD+ and poly(ADP-ribose) content and enhanced apoptosis. These results are important to cancer patients, which tend to be niacin deficient, are exposed to large doses of genotoxic drugs, and suffer short-term bone marrow suppression and long-term development of secondary leukemias. The data from our rat model suggest that niacin supplementation of cancer patients may decrease the severity of short and long-term side effects, and may also improve tumor cell killing through activation of poly(ADP-ribose)-dependent apoptosis pathways.

Keywords Niacin · NAD+ · Poly(ADP-ribose) · Genomic instability · Leukemogenesis

2.1 Niacin and Genomic Stability, Underlying Mechanisms

The human disease of niacin deficiency, pellagra, is characterized by diarrhea, dementia and sun-sensitive dermatitis, the last of which is suggestive of problems with DNA repair (Kirkland 2007). In the last few years there has been an explosion

---

J.B. Kirkland (✉)
Department of Human Health and Nutritional Sciences, University of Guelph, Guelph, ON, Canada N1G 2W1
e-mail: jkirklan@uoguelph.ca

O. Stanger (ed.), *Water Soluble Vitamins*, Subcellular Biochemistry 56,
DOI 10.1007/978-94-007-2199-9_2, © Springer Science+Business Media B.V. 2012

of knowledge in the area of niacin and its role in DNA function (Kirkland 2003, 2007; Kirkland et al. 2007). Niacin has been viewed traditionally as a central player in catabolic and synthetic pathways through a multitude of redox reactions based on NAD and NADP cofactors. A new appreciation of niacin function in DNA stability has come with the knowledge that NAD+ participates in a wide variety of ADP-ribosylation reactions. While redox reactions cycle pyridine nucleotides between oxidized and reduced forms, ADP-ribosylation reactions cleave NAD+, releasing nicotinamide and using the ADP-ribose moiety in a variety of interesting reactions. These reactions can be roughly divided into four categories, including; poly (ADPribose) synthesis, mono ADP-ribosylation, cyclic ADP-ribose formation and NAD-dependent deacetylation reactions (sirtuins).

2.1.1 Poly(ADP-Ribose) Polymerases

In the last few years rapid progress has been made in the study of poly(ADP-ribose) metabolism (Burkle 2006; Gagne et al. 2006; Schreiber et al. 2006). It is now recognized that 17 genes in humans have the potential to produce enzymes that synthesize poly(ADP-ribose), and catalytic activity has been proven in eight of these (Schreiber et al. 2006). PARP-1 has been recognized for several decades, and makes the majority of cellular poly(ADP-ribose). It is a nuclear protein that binds to and is catalytically activated by DNA strand breaks, creating one of the earliest metabolic signals at a site of DNA damage (Huber et al. 2004). PARP-2 has functional similarities to PARP-1; animals with either gene deleted experience genomic instability while deletion of both genes is embryonic lethal (Menissier et al. 2003). PARP-3 associates with the daughter centriole during cell division (Schreiber et al. 2006). Tankyrases one and two synthesize poly(ADP-ribose) at the telomere, opening the structure and allowing telomerase to maintain the length and stability of the chromosome terminus (Sbodio et al. 2002). vPARP is associated with vault particles and may play a role in multidrug resistance (Kickhoefer et al. 1999). PARP-10 interacts with c-Myc and inhibits cell transformation (Yu et al. 2005). tiPARP is induced by dioxin, with uncertain effects on metabolism (Ma et al. 2001).

While various individual functions have been described, it is also clear that PARP genes function together in the maintenance of genomic stability. PARP-1 and 2 have clear interactive roles in base excision repair of DNA, and likely function in other forms of DNA repair (Huber et al. 2004). PARP-1 and 3, tankyrase-1 and vPARP localize to centrosomes and mitotic spindle apparatus, while PARP-1 and 2 localize to centromeres, demonstrating an integrated role of poly(ADPribose) metabolism in chromosome segregation (Schreiber et al. 2006).

There is an extensive literature describing the complex mechanisms by which PARP enzymes accomplish these varied tasks (Burkle 2006; Gagne et al. 2006; Schreiber et al. 2006), and these will be described only briefly below:

(i) Poly(ADP-ribose) is negatively charged, as is DNA, and these two molecules repel one another. When histones are poly(ADP-ribosyl)ated they lose their affinity for DNA and this leads to a local chromatin relaxation. PARP-1 and 2 become automodified, and this eventually pushes them away from the site of damage, allowing repair to finish. Also, clouds of poly(ADPribose) around sites of DNA damage likely repel other free ends of DNA and prevent nonhomologous translocation events.

(ii) Proteins have specific high-affinity binding sites for poly(ADP-ribose), which draws them to sites of DNA damage, where they may participate in repair reactions or initiate signaling pathways for inflammatory or apoptotic responses (Malanga and Althaus 2005). This also contributes to local chromosome relaxation, as histones have high affinity sites, and are pulled out of chromatin onto the polymer bound to automodified PARP-1/2. Other proteins involved in DNA repair, cell cycle arrest, apoptosis and inflammation (topoisomerases, DNAPK, NF-kB, XRCC1, p53) are drawn to the site of injury through high affinity poly(ADP-ribose) binding (Malanga and Althaus 2005).

(iii) Covalent poly(ADP-ribosyl)ation of enzymes, transcription factors and signaling molecules can increase or decrease their activities, controlling the complex stress response of the cell, directing repair activities and survival decisions. Acceptor proteins for poly(ADP-ribose) addition include DNA topoisomerase, polymerase and ligase enzymes (D'Amours et al. 1999).

2.1.2 Sirtuins

There has been great interest in the sirtuin gene family since the finding that Sir2 (SirT in humans) is central to the extension of lifespan in response to caloric restriction, and that lifespan extension can be mimicked by the sirtuin activator resveratrol (Kruszewski and Szumiel 2005). Of interest, Sir2 is an NAD-dependent deacetylase, which removes acetyl groups from key proteins like histones and p53, leading to changes in chromatin structure and the cellular stress response. The deacetylation is an ADP-ribosylation reaction, producing acetyl-ADP-ribose as an end product. The sirtuins provide a mechanistic connection between cellular energy status, regulation of DNA structure and metabolism, and genomic stability and lifespan. Sir2 and PARP-1/2 may have close interactions, based on competition for NAD+ and inhibition of Sir2 by nicotinamide. It has been suggested that cell survival is dependent on the balance of PARP versus sirtuin activity, with PARP directing apoptotic signals and sirtuins promoting survival (Kruszewski and Szumiel 2005). It is clear that PARP and sirtuin enzymes play strong integrated roles in the maintenance of genomic stability in eukaryotes.

2.1.3 Mono ADP-Ribosyltransferases and ADP-Ribosyl cyclases

While PARP enzymes and sirtuins provide the most plausible links between niacin status and genomic stability, there are also potential effects of mono ADP-ribosylation reactions and cyclic ADP-ribose metabolism. Mono ADP-ribosylation is represents the post-translational modification of cellular proteins with a single ADP-ribose unit derived from NAD+. There are many different transferases and protein substrates involved, and diverse processes are controlled, from immune response, to cell adhesion and G-protein signaling (Di Girolamo et al. 2005). Cyclic ADP-ribose causes release of intracellular calcium stores, controlling functions in all cell types, from nerve conductance, to muscle contraction to insulin secretion (Guse 2005). There is potential for alterations in mono ADP-ribosylation and cyclic ADP-ribose metabolism to impact on genomic stability, though control of cell cycle and other signaling pathways. For example, the nuclear envelope contains calcium stores that are released by cyclic ADP-ribose, and another niacin based molecule, nicotinic acid adenine dinucleotide phosphate, exerting control over nuclear processes such as DNA replication and cell cycle checkpoints (Gerasimenko and Gerasimenko 2004). Basic redox metabolism could also be altered by niacin deficiency leading to genomic instability. One example of this would be the maintenance of reduced glutathione (GSH) levels using NADPH as a substrate.

2.2 Niacin Deficiency in a Whole Animal Model

Early experiments on NAD and genotoxic stress made use of cell culture models, which allowed for decreases in cellular NAD+ in excess of 90% (Durkacz et al. 1980; Jacobson et al. 1992). Of interest, normal cell growth and division were maintained in these models until DNA-damaging agents were added, suggesting that essential redox functions are relatively protected during niacin depletion. Competition for limiting NAD+ pools will occur at the subcellular level, but in whole animal models there will also be competition between tissues. The specific reactions that lose this tissue and subcellular competition for NAD+ will determine the pathologies that arise as the deficiency progresses. The end result may differ depending on the age, sex and species of animal model, and depending on the external stresses that are applied during deficiency. This was observed historically in human populations, where working class pellagrins would tend to present with sun-sensitive dermatitis, while wealthier, less sun-exposed patients often presented with dementia, and the full triad of dermatitis, dementia and diarrhea was observed less frequently (Kirkland 2007).

In our early work on niacin deficiency in rats, we found that liver and lung tissues have very high basal levels of NAD+, are relatively resistant to niacin depletion, and are not sensitized to genotoxic damage during niacin deficiency (Rawling et al. 1994, 1995, 1996). Noting that the symptoms of pellagra tend to appear in highly proliferative tissues like the skin and intestinal tract, we began to study bone marrow tissue. We have found that bone marrow is the most sensitive tissue that we have characterized in our model, with decreases in NAD+ of 80%, an almost complete loss of basal poly(ADP-ribose) and a large inhibition of nitrosourea-induced poly(ADP-ribose) formation (Fig. 2.1) (Boyonoski et al. 2002a). Bone marrow poly(ADP-ribose) is also very sensitive to high niacin intake,

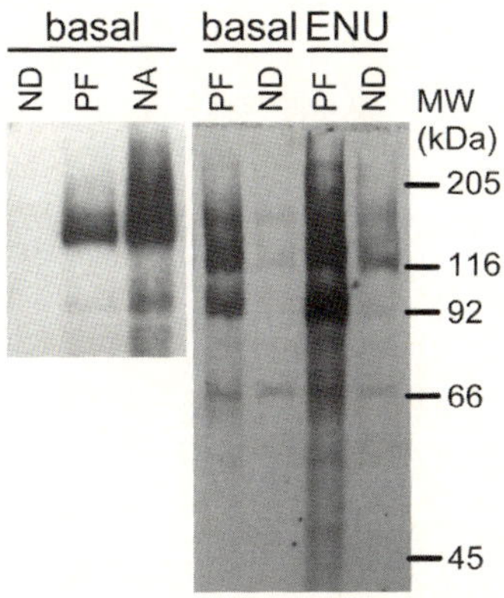

Fig. 2.1 The effect of niacin status on bone marrow poly(ADP-ribose) levels. ND, niacin deficient, 0 mg added dietary niacin; PF, pair-fed control, control diet with 30 mg nicotinic acid/kg diet, fed in identical quantities to ND; NA, high niacin diet, 4,000 mg nicotinic acid/kg diet, also pair-fed to ND. Ethylnitrosourea (ENU) is used to induce DNA damage. Cellular proteins are run using SDS PAGE, transferred and blotted with anti-poly(ADP-ribose) polyclonal Ab 96–10 (Dr. Guy Poirier)

as basal and damage induced poly(ADP-ribose) levels are much higher when rats are fed 4,000 mg nicotinic acid/kg diet compared to the 30 mg/kg diet found in the control diet (Fig. 2.1) (Boyonoski et al. 2002a). In total, bone marrow NAD+ and poly(ADP-ribose) concentrations vary 30- and 65-fold, respectively, going from niacin deficient to pharmacologically supplemented status (Boyonoski et al. 2002a). The majority of polymer appears to be bound to the 116 kDa PARP-1 and its 97 kDa apoptotic fragment (Boyonoski et al. 2002a).

The next question to be addressed was whether the diet-induced changes in NAD+ and poly(ADP-ribose) where associated with pathological endpoints associated with genomic instability. From a human clinical perspective, bone marrow progenitors experience the critical side effects of most types of cancer chemotherapy, leading to dose-limiting bone marrow suppression in the short term, and the development of secondary leukemias over time (Terada 2006). We set out to model the stress of chemotherapy on the bone marrow by treating Long-Evans rats with 12 doses of ethylnitrosourea (ENU), spread over a Langley et al. (2002) day period. Long-Evans rats tend to develop non-lymphocytic leukemias in response to alkylating agents (Shisa and Hiai 1985), similar to human cancer patients (Tucker 1993). Rats that were niacin deficient during ENU treatment displayed more severe short term bone marrow suppression (Boyonoski et al. 2000). We also collected long-term cancer data from these cohorts (Boyonoski et al. 2002a, b), but the results were compromised by excessively rapid appearance of cancers and differences in the basal diets between deficient and pharmacological models. We have recently repeated this experiment using 6 doses of ENU and have found that niacin status has a significant impact on nitrosourea-induced leukemogenesis (Fig. 2.2) (Bartleman et al. 2008). While non-leukemia cancers were not affected by diet, the incidence

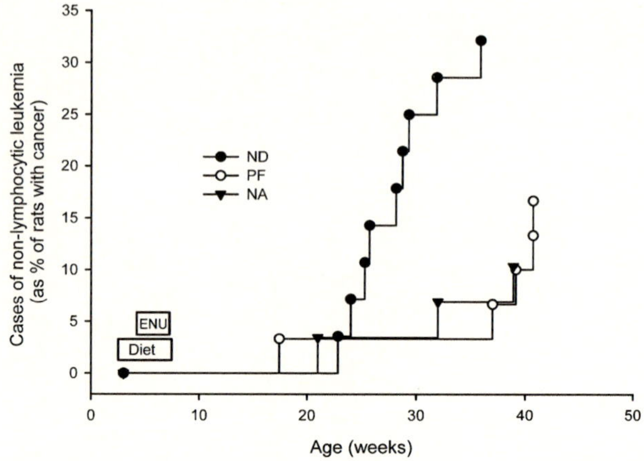

Fig. 2.2 The effect of niacin status during exposure to ENU on the long term development of non-lymphocytic leukemias in Long-Evans rats. ND, niacin deficient, 0 mg added dietary niacin; PF, pair-fed control, control diet with 30 mg nicotinic acid/kg diet, fed in identical quantities to ND; NA, high niacin diet, 4,000 mg nicotinic acid/kg diet, also pair-fed to ND. Ethylnitrosourea (ENU) is used to induce DNA damage. Rats consumed experimental diets and were exposed to ENU between 3 and 7 weeks of age, after which all three groups consumed control diet (AIN93M) and were observed for development of cancer

of non-lymphocytic leukemias was very sensitive to niacin status. While there were not any significant survival differences between the adequate and pharmacological niacin groups, the pharmacological niacin intake group did provide the greatest degree of protection, relative to niacin deficient rats, in the incidence of total cancers, total leukemias and non-lymphocytic leukemias (Bartleman et al. 2008).

Given the close connection between PARP-1/2 function and base excision repair (BER) (Malanga and Althaus 2005), we sought to characterize the impact of niacin deficiency on the repair of alkylation injury in our model of niacin and leukemogenesis. ENU was an ideal compound to initiate the damage response since it decomposes rapidly and spontaneously to produce reactive ethyl ions that damage DNA in an even pattern across tissues, independently of phase I or II xenobiotic metabolism. These small ethyl adducts are then mainly resolved by BER. DNA excision repair pathways involve incision, excision, repair synthesis and ligation (Seeberg et al. 1995). During base excision repair (BER), the removal of altered bases is initiated by DNA glycosylases, creating an abasic (AB) site, but not a strand break. The phosphate group 5′ to the AB site is incised by endonuclease, or the 3′ site is cleaved by lyase. This creates the strand break that is recognized by PARP-1 and 2, initiating the synthesis of poly(ADP-ribose), leading to changes in local chromatin structure, and the protein:protein and protein:poly(ADP-ribose) interactions that help to organize the repair process. The deoxyribose phosphate is excised by phophodiesterase, and DNA polymerases fill the resulting gap. The final nick is sealed by DNA ligase. At some point, the synthesis of poly(ADP-ribose) on PARP enzymes becomes extensive enough that anionic repulsion allows it to leave the strand break to allow completion of repair. The ability of catalytically inactive PARP-1 to block excision repair has been demonstrated through NAD+ removal (Durkacz et al. 1980; Satoh et al. 1993), competitive inhibitor treatment and mutagenesis of the catalytic domain (D'Amours et al. 1999). We examined repair kinetics using the alkaline comet assay during the 36 h following a single dose of ethylnitrosourea (ENU) (30 mg/kg bw). The comet assay uses carefully isolated bone marrow cells, which are suspended in soft agar and incubated at alkaline pH. The slides are then electrophoresed, causing loose strands of DNA to migrate out of the nucleus, forming a comet-like image. The current thinking is that single strand breaks cause a loss of supercoiling between sites of attachment to the nuclear matrix, allowing loops of DNA to migrate with the electrical current. The alkaline assay is fully sensitive to existing single strand breaks, but is also partially sensitive to AB sites and some other DNA lesions. The comet images are analyzed by computer, generating a proportionate index of tail versus nuclear DNA, referred to as mean tail moment (MTM). In our time course, there was no effect of ND on MTM before ENU treatment, or on the development of strand breaks between 0 and 8 h after ENU (Fig. 2.3) (Kostecki et al. 2007). This indicates that the ENU is causing similar levels of DNA alkylation, and that subsequent differences are due to changes in DNA repair processes. The peaks at 4 and 9 h likely represent formation of AB sites and strand breaks, respectively, through the action of glycosylases

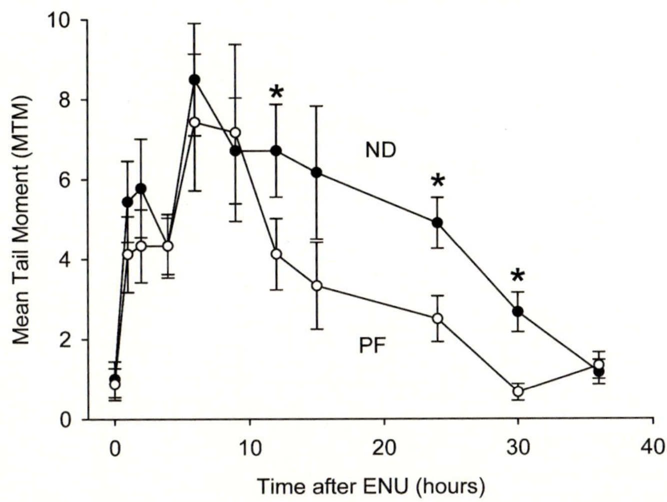

Fig. 2.3 The effect of niacin deficiency and ethylnitrosourea treatment on DNA strand breaks in bone marrow cells. Weanling Long-Evans rats were fed niacin deficient (ND) or pair fed niacin replete control diet (PF) for 3 weeks. Rats were gavaged with ethylnitrosourea (ENU, 30 mg/kg bw in pH 4 water) at 10:00 am. Time 0 rats were gavaged with pH 4 water and killed at the 3 h time point, to assess the influence of vehicle and procedures. ENU treated ND and PF rats were killed at the times indicated, and the bone marrow cells from the femurs were blindly analyzed by the alkaline comet assay. $n = 6$–14. *ND significantly greater than PF ($p < 0.05$, t-test)

and endonucleases. These processes are not affected by niacin deficiency. In contrast, repair kinetics between 12 and 30 h were significantly delayed by ND, with a doubling of area under the MTM curve during this period (Fig. 2.3).

It is logical to look at the lack of basal and damage-induced poly(ADP-ribose) synthesis in Fig. 2.1 and postulate that the observed delay in repair kinetics represents the effect of PARP-1 and/or 2 binding to strand breaks and blocking the completion of the repair process due to the lack of automodification. This mechanism has been demonstrated in many in vitro systems using NAD deprivation, competitive inhibitors or catalytically inactive mutants to produce dominant-negative forms of PARP-1 that block the later stages of BER (D'Amours et al. 1999). This may well be occurring in our in vivo model, but PARP enzymes play other roles in the BER process that should be considered. The local chromatin relaxation induced by poly(ADP-ribose) formation around strand breaks is likely decreased, and this may delay the completion of repair. PARP-1/2 activity draws XRCC1 to DNA strand breaks, where it acts as a scaffold for the repair machinery (Malanga and Althaus 2005). PARP-1 also attaches poly(ADP-ribose) covalently to DNA protein kinase, topoisomerases, polymerases and ligase (D'Amours et al. 1999), and these may all suffer functionally during niacin deficiency, leading to delayed excision repair.

O6-alkylguanine is a lesion that is not recognized by the glycosylases of BER. In recognition of its significant mutagenicity, there is a specific protein, O6-alkylguanine-DNAalkyltransferase (AGT), which is devoted to detecting and removing these lesions. AGT transfers the alkyl group to itself and is then degraded, at a metabolic cost of about 850 ATP for each lesion that is repaired. During niacin deficiency the removal of O6-ethylguanine is delayed, presumably by a lower availability of AGT, which may be due to a small drop in mRNA levels (Kostecki et al. 2007). There is no obvious reason for a drop in expression, although decreased ADPribosylation activity may be impairing the damage-induced signaling pathways that control AGT expression. Of interest, resveratrol has been shown to induce MGMT activity (Niture et al. 2007), and it is also known as an activator Sir2 catalytic activity (Wood et al. 2004), demonstrating one possible link between NAD+ pools and AGT expression. If O6-ethlyguanine persists through DNA replication, it may cause point mutations or it may be recognized by the mismatch repair system (MMS), which removes a long patch of the newly synthesized strand, and replaces it, without resolving the lesion. This continuing futile cycle of repair leads to double strand breaks which lead to chromosome breaks, translocations and other aberrations (Margison et al. 2002).

Chromosome breaks and translocations are key players in carcinogenesis, and they are dependent on the formation of double strand breaks in DNA. However, these lesions are increased by many forms of DNA damage which are known to cause single strand breaks, either directly (some forms of radiation) or indirectly through pathways like BER. In this model of simple alkylation damage, double strand breaks could form by several mechanisms, including multiple nearby single strand breaks, stalled replication forks (Cox et al. 2000), futile cycles of mismatch repair (as described above) and the blockage of strand break repair by catalytically-inactive PARP-1 or 2. Double strand breaks can be measured using the neutral comet assay. Neutral MTM was nearly doubled in ND bone marrow cells Langley et al. (2002) hours following ENU treatment, indicating an accumulation of double strand breaks (Kostecki et al. 2007). PARP-1 has been shown to be required for efficient resolution of stalled replication forks (Yang et al. 2004). We have also shown that niacin deficiency impairs damage-induced cell cycle arrest in our model (Spronck et al. 2007). Poorly resolved replication forks proceeding into DNA replication are likely to produce double strand breaks (Cox et al. 2000). Double strand breaks may also accumulate due to a lack of repair, and PARP enzymes have been sown to participate in various forms of double strand break repair (Audebert et al. 2004; Frosina et al. 1996).

The traditional analysis of metaphase spreads for chromosome aberrations should reveal differences if there are more double strand breaks in niacin deficient bone marrow cells, and this is what we have observed (Kostecki et al. 2007). Figure 2.4a shows examples of metaphase spreads from normal and niacin deficient bone marrow cells. Figure 2.4b shows the results of blinded analysis of metaphase arrests for chromosome breaks, which are breaks that appear in both sister chromatids. There was a dramatic (4-fold) increase in chromosome breaks caused by niacin deficiency alone. Niacin deficiency enhanced the actual number of additional breaks caused by ENU, although not the proportion.

Other short-term biomarkers of genomic instability are also increased in this model. We found that sister chromatid exchanges and micronuclei were increased by 3- and 6-fold, respectively, by niacin deficiency alone (Spronck and Kirkland

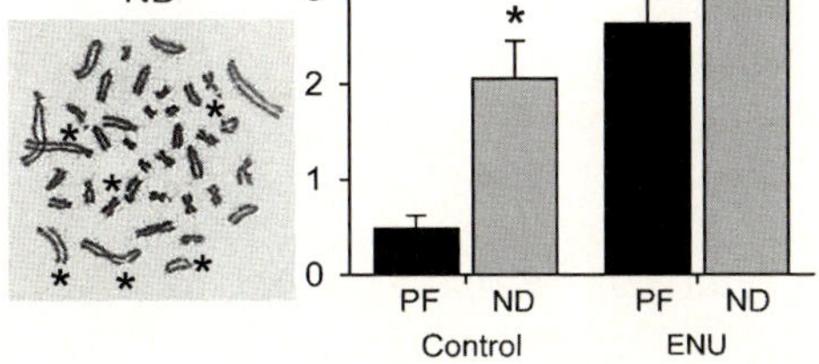

Fig. 2.4 The effect of niacin deficiency and ENU treatment on Chromosome Breaks. Weanling Long-Evans rats were fed niacin deficient (ND) or pair fed niacin replete control diet (PF) for 3 weeks. Rats were gavaged with ethylnitrosourea (ENU, 30 mg/kg bw in pH 4 water) at 10:00 am. Control rats were gavaged with pH 4 water to assess the influence of vehicle and procedures. Four hours following the ENU treatment, rats were injected with colcemid (0.6 mg/kg bw, i.p.) and killed 2 h later. (**a**) Metaphase arrests from PF and ND bone marrow cells (*site of chromatid or chromosome break), and (**b**) Blinded analysis of chromosome breaks. $n = 9$–11, 50 spreads per animal. *significant effect of diet, **significant effect of ENU ($p < 0.05$, t-test)

2002). Pharmacological niacin intake offered no additional benefit using these endpoints (Spronck and Kirkland 2002). Sister chromatid exchanges represent homologous recombination events that do not cause damage per se, but they generally correlate with increased non-homologous recombination events, which underlie chromosomal translocations. Essentially all models of PARP disruption have caused increased sister chromatid exchanges. The bone marrow micronucleus assay is a simple measure of genomic instability, and one that is very sensitive to niacin deficiency. The micronucleus is easily observed in polychromatic erythrocytes, which have recently extruded their nucleus. Micronuclei may arise from clastogenic events, which cause a chromosome fragment lacking a centromere. Conversely, an aneugenic micronucleus is complete and has been left behind due to some failure of the sorting mechanisms, either the physical microtubular structures, or the checkpoint signals that ensure all chromosomes are attached. Thus the micronucleus assay is sensitive to many defects in DNA repair, checkpoint control or chromosome sorting. In looking at total numbers of micronuclei, niacin deficiency alone caused a large increase in micronuclei, and further increased the response to ENU treatment (Fig. 2.5) (Kostecki et al. 2007). It would be quite interesting, from a mechanistic perspective, to know which type of micronuclei are being produced. I would speculate that the numerous micronuclei caused by niacin deficiency alone are predominantly aneugenic. This is based on the fact that we do not see evidence that

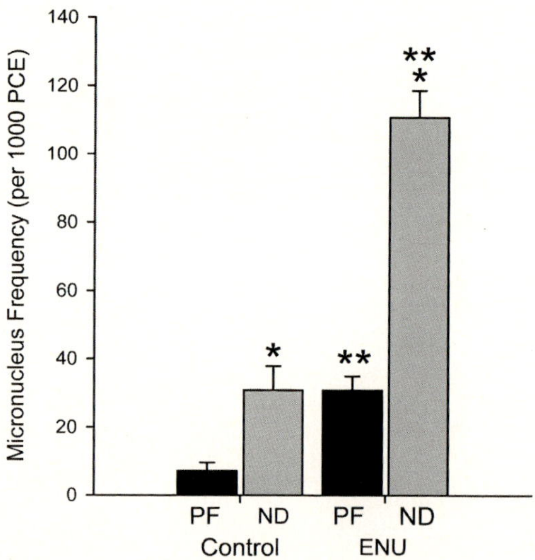

Fig. 2.5 The effect of niacin deficiency and ENU treatment on the frequency of micronuclei (MN) in bone marrow polychromatic erythrocytes (PCE). Weanling Long-Evans rats were fed niacin deficient (ND) or pair fed niacin replete control diet (PF) for 3 weeks. Rats were gavaged with ethylnitrosourea (ENU, 30 mg/kg bw in pH 4 water) at 10:00 am. Control rats were gavaged with pH 4 water to assess the influence of vehicle and procedures. Twenty four hours after ENU treatment bone marrows cells were processed to view MN in PCE. 2000 PCE were evaluated for each sample ($n = 8$–13). *significant effect of diet, **significant effect of ENU ($p < 0.05$, t-test)

niacin deficiency alone causes a significant excess of strand breaks using the comet assay (Fig. 2.3, time 0) or chromatid gaps as end points (Kostecki et al. 2007). Conversely, ENU-induced micronuclei are likely to be clastogenic in nature since they are being induced by a DNA-breaking compound. It seems probable that niacin deficiency is favoring the formation of micronuclei through changes in both DNA repair and chromosome sorting.

Tankyrases I and II are PARP enzymes that associate with telomeres, which are repetitive sequences at the ends of chromosomes. Normal DNA replication causes a small loss of telomeric DNA with each cell division, and the erosion of these areas plays an important role in cell senescence. Telomerase is an enzyme that extends the telomere to allow continued cell division. Tankyrase enzymes use NAD+ to add poly(ADP-ribose) to certain telomere binding proteins, causing them to dissociate and enable telomerase activity. This represents another connection between cellular NAD+ status and chromosome function and stability. When the telomeres erode, they tend to become instable and participate in translocation events. During niacin deficiency one would predict that telomere erosion would occur, and we have preliminary results that suggest this is true (unpublished data).

DNA damage and genomic instability can also be induced through oxidant stress. Oxidant stress results from an imbalance between the formation and protective metabolism of reactive oxygen species. Reactive oxygen is an unavoidable aspect of aerobic metabolism, and also plays specialized roles in metabolism and cell signaling. An excess of reactive oxygen species leads to DNA damage through the formation of adducts such as 8-oxo-deoxyguanosine (8-oxodG). Niacin deficiency could lead to oxidant stress through decreased levels of NADPH, limiting the regeneration of GSH, and thus an inhibition of glutathione peroxidase activity. We have found that niacin deficient bone marrow cells display increased levels of oxidant damage to both cellular proteins and DNA, as reflected by 8-oxodG and protein carbonyl formation (Tang et al. 2008). However, there was no effect of niacin status on NADPH or GSH levels, indicating that the activity of glutathione peroxidase was not being limited directly by niacin deficiency. Inflammation can also cause oxidant stress, but cDNA microarray analysis did not reveal any changes in gene expression that were suggestive of enhanced inflammatory signaling (Tang et al. 2008). Not all increases in reactive oxygen represent cellular stress or injury. NAD(P)H, or mitogenic, oxidases produce reactive oxygen species that act as signals for cell division, adhesion and survival (through inhibition of apoptosis) (Terada 2006). This is similar to what we observe in niacin deficient bone marrow cells. Mitochondria use reactive oxidants as signals to control cellular energetics (Terada 2006), and this process could be disrupted directly or indirectly by changes in niacin status. The causes and impact of oxidant damage during niacin deficiency are uncertain and require further research.

There are many aspects to genomic stability beyond the correct replication and repair of DNA and the sorting of chromosomes. These processes can be disrupted indirectly by improper control of the cell cycle. A failure to maintain genomic stability is also mitigated by the process of apoptosis, and defects in these pathways allow damaged cells to survive and progress to leukemias. One protein that plays a key role in coordinating DNA repair, cell cycle arrest and apoptosis is p53. This protein plays a strong role in tumor suppression and is mutated or downregulated in most human cancers. p53 is regulated by several NAD-dependent processes. It has been reported as a covalent acceptor of poly(ADP-ribose), via PARP-1 activation (Won et al. 2006). It also has a high affinity binding site for non-covalent interactions with poly(ADP-ribose) (Malanga and Althaus 2005) and it is deacetylated by Sir2 (Langley et al. 2002). The end result of these interactions is uncertain and there is a lack of consensus on the role of ADP-ribosylation in the regulation of p53 function. We have examined p53 expression in bone marrow cells in our model, and have found that niacin deficiency causes accumulation of two slow mobility forms of p53 (Spronck et al. 2007). We have determined that they are not due to alternate splicing of the mRNA (Spronck et al. 2007). The change in mobility is suggestive of one or two ubiquitination sites, and may indicate a defect in proteasome function (Won et al. 2006), but further work is needed to characterize these changes, and determine if they occur in other species.

If p53 function is disrupted there can be impairment of cell cycle arrest and apoptotic signaling. We found that niacin deficient bone marrow cells displayed an altered response to etoposide-induced DNA damage, with a blunting of the G1 arrest, and a greater portion of cells staying in S-phase (active DNA synthesis). In addition, there was an impaired apoptotic response in niacin deficient cells. Of interest, this is one of the few biomarkers that we have found to be altered by pharmacological supplementation of niacin relative to adequate niacin intake. The progressive increase in apoptotic frequency from niacin deficient through adequate to pharmacological intake correlates with cellular NAD+ and poly(ADP-ribose) content in these groups (Spronck et al. 2007). This relationship is also reflected by the eventual development of ENU-induced leukemias (Bartleman et al. 2008). Essentially all of the other intermediate endpoints that we have examined have not shown additional benefit of pharmacological supplementation, suggesting that regulation of apoptosis may be a key aspect of the impact of dietary niacin on leukemogenesis. There are a number of possible connections between niacin status and apoptosis, but the most compelling of these involves the activation of apoptosis-inducing factor (AIF) (Porter and Urbano 2006). AIF is a 67 kDa flavoprotein that appears to play important roles in mitochondrial metabolism of normal cells and in apoptosis of injured cells (similar to cytochrome c). Decreased AIF levels are associated with impaired mitochondrial oxidative phosphorylation and increased neuronal cell death in Harlequin mice (Porter and Urbano 2006). Conversely, following cell injury, AIF may be released from the mitochondria after PARP-1 is activated in the nucleus (Yu et al. 2006). It can then translocate to the nucleus and initiate or contribute to apoptotic cell death.

Most of this is not well understood, especially the mechanisms by which nuclear formation of poly(ADP-ribose) leads to AIF release from mitochondria. It does provide an interesting explanation for the impact of niacin status on apoptosis in bone marrow cells. The fact that cell cycle arrest and p53 expression patterns are not changed between control and pharmacological niacin intake, while poly(ADP-ribose) and apoptosis are changed, suggests that high polymer levels during niacin megadosing help to favor AIF-induced apoptosis.

For the cancer patient, the critical relationship is between toxicity and efficacy of the chemotherapy treatment. Many side effects, especially those that may happen in the long term, are tolerated to achieve treatment efficacy. Correcting niacin status may not be clinically beneficial if tumor cells are also protected from the toxicity of the chemotherapy drugs. Optimally, niacin supplementation could protect the bone marrow and increase tumor cell killing, and there are several indications that this could be true. Tumors evolve protective mechanisms against endogenous controls over cell division by downregulation of apoptosis pathways, and these changes also protect them from chemotherapy drugs. When a cancer patient is niacin deficient, it is likely that the tumor cells will have decreased NAD levels and experience some of the physiological changes that we observe in the bone marrow, including depression of apoptosis, decreased control over the cell cycle and increased genomic instability. These are all effects that increase the malignant potential of a tumor, and/or make it more resistant to chemotherapy. One intriguing possibility is that pharmacological niacin intake could increase poly(ADP-ribose) synthesis in tumor cells, activating the AIF-dependent apoptotic pathway and increasing treatment efficacy. Tumors tend to display decreased mitochondrial metabolism and become more dependent on anaerobic glycolysis (Bonnet et al. 2007). This has been shown to inhibit apoptosis and contribute to chemoresistance. Dichloroacetate (DCA) activates pyruvate dehydrogenase, increasing mitochondrial metabolism and sensitizing cancer cells and tumors in vivo to chemotherapy (Bonnet et al. 2007). We have recently measured mitochondrial NAD+ in bone marrow cells, and found that they are not protected during niacin deficiency, depleting to the same extent as whole cell pools (unpublished data). Niacin supplementation may work in a similar fashion to DCA treatment, and may provide additive or synergistic increases in tumor cell killing, by supporting the increase in mitochondrial metabolism and driving up the poly(ADP-ribose) dependent activation of AIF.

2.3 Summary

Niacin deficiency decreases bone marrow NAD+ and poly(ADP-ribose) levels, disrupts cell cycle regulation, depresses apoptosis and enhances multiple forms of genomic instability, leading to bone marrow suppression and leukemogenesis in response to alkylating agents. Pharmacological supplementation with niacin increases bone marrow NAD+ and poly(ADPribose) levels, enhances apoptosis and decreases leukemogenesis. The next stage of research in this area should be to characterize the effect of dietary niacin status on the response of implanted tumors to chemotherapy in various animal models. This will allow a decision to be made on whether aggressive supplements of niacin during therapy are likely to benefit the balance between efficacy and toxicity in human cancer patients.

References

Audebert, M, Salles B, Calsou P (2004) Involvement of poly(ADP-ribose) polymerase1 and XRCC1/DNA ligase III in an alternative route for DNA double-strand breaks rejoining. J Biol Chem 279:55117–55126

Bartleman AP, Jacob RA, Kirkland JB (2008) Niacin supplementation decreases the incidence of alkylation-induced non-lymphocytic leukemia in Long-Evans rats. Nutr Cancer 60:251–258

Bonnet S, Archer SL, Allalunis-Turner J, Haromy A, Beaulieu C, Thompson R, Lee CT, Lopaschuk GD, Puttagunta L, Bonnet S, Harry G, Hashimoto K, Porter CJ, Andrade MA, Thebaud B, Michelakis ED (2007) A mitochondria-K+ channel axis is suppressed in cancer and its normalization promotes apoptosis and inhibits cancer growth. Cancer Cell 11:37–51

Boyonoski AC, Gallacher LM, ApSimon MM, Jacobs RM, Shah GM, Poirier GG, Kirkland JB (2000) Niacin deficiency in rats increases the severity of ethylnitrosourea-induced anemia and leukopenia. J Nutr 130:1102–1107

Boyonoski AC, Spronck JC, Gallacher LM, Jacobs RM, Shah GM, Poirier GG, Kirkland JB (2002a) Niacin deficiency decreases bone marrow poly(ADP-ribose) and the latency of ethylnitrosourea-induced carcinogenesis in rats. J Nutr 132:108–114

Boyonoski AC, Spronck JC, Jacobs RM, Shah GM, Poirier GG, Kirkland JB (2002b) Pharmacological intakes of niacin increase bone marrow poly(ADP-ribose) and the latency of ethylnitrosourea-induced carcinogenesis in rats. J Nutr 132:115–120

Burkle A (2006) DNA repair and PARP in aging. Free Radic Res 40:1295–1302

Cox MM, Goodman MF, Kreuzer KN, Sherratt DJ, Sandler SJ, Marians KJ (2000) The importance of repairing stalled replication forks. Nature 404:37–41

D'Amours D, Desnoyers S, D'Silva I, Poirier GG (1999) Poly(ADP-ribosyl)ation reactions in the regulation of nuclear functions. Biochem J 342(Pt 2):249–268

Di Girolamo M, Dani N, Stilla A, Corda D (2005) Physiological relevance of the endogenous mono(ADP-ribosyl)ation of cellular proteins. FEBS J 272:4565–4575

Durkacz BW, Omidiji O, Gray DA, Shall S (1980) (ADP-ribose)n participates in DNA excision repair. Nature 283:593–596

Frosina G, Fortini P, Rossi O, Carrozzino F, Raspaglio G, Cox LS, Lane DP, Abbondandolo A, Dogliotti E (1996) Two pathways for base excision repair in mammalian cells. J Biol Chem 271:9573–9578

Gagne JP, Hendzel MJ, Droit A, Poirier GG (2006) The expanding role of poly(ADP-ribose) metabolism: current challenges and new perspectives. Curr Opin Cell Biol 18:145–151

Gerasimenko O, Gerasimenko J (2004) New aspects of nuclear calcium signalling. J Cell Sci 117: 3087–3094

Guse AH (2005) Second messenger function and the structure-activity relationship of cyclic adenosine diphosphoribose (cADPR). FEBS J 272:4590–4597

Huber A, Bai P, de Murcia JM, de Murcia G (2004) PARP-1, PARP-2 and ATM in the DNA damage response: functional synergy in mouse development. DNA Repair (Amst) 3:1103–1108

Jacobson EL, Nunbhakdi-Craig V, Smith DG, Chen HY, Wasson BL, Jacobson MK (1992) ADP-ribose polymer metabolism: implications for human nutrition. In: Poirier GG, Moreau P (eds) ADP-ribosylation reactions. Springer, New York, pp 153–162

Kickhoefer VA, Siva AC, Kedersha NL, Inman EM, Ruland C, Streuli M, Rome LH (1999) The 193-kD vault protein, VPARP, is a novel poly(ADP-ribose) polymerase. J Cell Biol 146:917–928

Kirkland JB (2003) Niacin and carcinogenesis. Nutr Cancer 46:110–118

Kirkland JB (2007) Niacin. In: Rucker R, Zempleni J, Suttie JW, McCormick DB (eds) Handbook of vitamins, 4th edn. Taylor and Francis, Boca Raton, pp 191–232

Kirkland JB, Zempleni J, Buckles LK, Christman JK (2007) Vitamin-dependent modifications of chromatin: epigenetic events and genomic stability. In: Rucker RB, Zempleni J, Suttie JW, McCormick DB (eds) Handbook of vitamins, 4th edn. Taylor and Francis, pp 521–544

Kostecki LM, Thomas M, Linford G, Lizottee M, Toxopeus L, Bartleman AP, Kirkland JB (2007) Niacin deficiency delays DNA excision repair and increases spontaneous and nitrosourea-induced chromosomal instability in rat bone marrow. Mutat Res 625:50–61

Kruszewski M, Szumiel I (2005) Sirtuins (histone deacetylases III) in the cellular response to DNA damage–facts and hypotheses. DNA Repair (Amst) 4:1306–1313

Langley E, Pearson M, Faretta M, Bauer UM, Frye RA, Minucci S, Pelicci PG, Kouzarides T (2002) Human SIR2 deacetylates p53 and antagonizes PML/p53induced cellular senescence. EMBO J 21:2383–2396

Ma Q, Baldwin KT, Renzelli AJ, McDaniel A, Dong L (2001) TCDD-inducible poly(ADP-ribose) polymerase: a novel response to 2,3,7,8-tetrachlorodibenzo-pdioxin. Biochem Biophys Res Commun 289:499–506

Malanga M, Althaus FR (2005) The role of poly(ADP-ribose) in the DNA damage signaling network. Biochem Cell Biol 83:354–364

Margison GP, Santibanez Koref MF, Povey AC (2002) Mechanisms of carcinogenicity/chemotherapy by O6-methylguanine. Mutagenesis 17:483–487

Menissier DM, Ricoul M, Tartier L, Niedergang C, Huber A, Dantzer F, Schreiber V, Ame JC, Dierich A, LeMeur M, Sabatier L, Chambon P, de Murcia G (2003) Functional interaction between PARP-1 and PARP-2 in chromosome stability and embryonic development in mouse. EMBO J 22:2255–2263

Niture SK, Velu CS, Smith QR, Bhat GJ, Srivenugopal KS (2007) Increased expression of the MGMT repair protein mediated by cysteine prodrugs and chemopreventative natural products in human lymphocytes and tumor cell lines. Carcinogenesis 28:378–389

Porter AG, Urbano AG (2006) Does apoptosis-inducing factor (AIF) have both life and death functions in cells? Bioessays 28:834–843

Rawling JM, Jackson TM, Driscoll ER, Kirkland JB (1994) Dietary niacin deficiency lowers tissue poly(ADP-ribose) and NAD+ concentrations in Fischer-344 rats. J Nutr 124:1597–1603

Rawling JM, Jackson TM, Roebuck BD, Poirier GG, Kirkland JB (1995) The effect of niacin deficiency on diethylnitrosamine-induced hepatic poly(ADP-ribose) levels and altered hepatic foci in the Fischer-344 rat. Nutr Cancer 24:111–119

Rawling JM, ApSimon MM, Kirkland JB (1996) Lung poly(ADP-ribose) and NAD+ concentrations during hyperoxia and niacin deficiency in the Fischer-344 rat. Free Radic Biol Med 20:865–871

Satoh MS, Poirier GG, Lindahl T (1993) NAD(+)-dependent repair of damaged DNA by human cell extracts. J Biol Chem 268:5480–5487

Sbodio JI, Lodish HF, Chi NW (2002) Tankyrase-2 oligomerizes with tankyrase-1 and binds to both TRF1 (telomere-repeat-binding factor 1) and IRAP (insulinresponsive aminopeptidase). Biochem J 361:451–459

Schreiber V, Dantzer F, Ame JC, de Murcia G (2006) Poly(ADP-ribose): novel functions for an old molecule. Nat Rev Mol Cell Biol 7:517–528

Seeberg E, Eide L, Bjoras M (1995) The base excision repair pathway. Trends Biochem Sci 20:391–397

Shisa H, Hiai H (1985) Genetically determined susceptibility of Fischer 344 rats to propylnitrosourea-induced thymic lymphomas. Cancer Res 45:1483–1487

Spronck JC, Kirkland JB (2002) Niacin deficiency increases spontaneous and etoposide-induced chromosomal instability in rat bone marrow cells in vivo. Mutat Res 508:83–97

Spronck JC, Nickerson JL, Kirkland JB (2007) Niacin deficiency alters p53 expression and impairs etoposide-induced cell cycle arrest and apoptosis in rat bone marrow cells. Nutr Cancer 57:88–89

Tang K, Sham H, Hui E, Kirkland JB (2008) Niacin deficiency causes oxidative stress in rat bone marrow cells but not through decreased nadph or glutathione status. J Nutr Biochem 19:746–753

Terada LS (2006) Specificity in reactive oxidant signaling: think globally, act locally. J Cell Biol 174:615–623

Tucker MA (1993) Secondary cancers. In: DeVita VT, Hellman S, Rosenberg SA (eds) Cancer: principles and practice of oncology. JB Lippincott Co., Philadelphia, pp 2407–2419

Won J, Chung SY, Kim SB, Byun BH, Yoon YS, Joe CO (2006) Dose-dependent UV stabilization of p53 in cultured human cells undergoing apoptosis is mediated by poly(ADP-ribosyl)ation. Mol Cells 21:218–223

Wood JG, Rogina B, Lavu S, Howitz K, Helfand SL, Tatar M, Sinclair D (2004) Sirtuin activators mimic caloric restriction and delay ageing in metazoans. Nature 430:686–689

Yang YG, Cortes U, Patnaik S, Jasin M, Wang ZQ (2004) Ablation of PARP-1 does not interfere with the repair of DNA double-strand breaks, but compromises the reactivation of stalled replication forks. Oncogene 23:3872–3882

Yu M, Schreek S, Cerni C, Schamberger C, Lesniewicz K, Poreba E, Vervoorts J, Walsemann G, Grotzinger J, Kremmer E, Mehraein Y, Mertsching J, Kraft R, Austen M, Luscher-Firzlaff J, Luscher B (2005) PARP-10, a novel Mycinteracting protein with poly(ADP-ribose) polymerase activity, inhibits transformation. Oncogene 24:1982–1993

Yu SW, Andrabi SA, Wang H, Kim NS, Poirier GG, Dawson TM, Dawson VL (2006) Apoptosis-inducing factor mediates poly(ADP-ribose) (PAR) polymerinduced cell death. Proc Natl Acad Sci USA 103:18314–18319

Chapter 3
Niacin: Vitamin and Antidyslipidemic Drug

Elaine L. Jacobson, H. Kim, M. Kim, and M.K. Jacobson

Abstract Niacin is defined collectively as nicotinamide and nicotinic acid, both of which fulfill the vitamin functions of niacin carried out by the bioactive forms NAD(P). In the last few decades numerous new enzymes that consume NAD(P) as substrates have been identified. The functions of these enzymes are emerging as exciting paradigm shifts, even though they are in early stages of discovery. The recent identification of the nicotinic acid receptor has allowed distinction of the drug-like roles of nicotinic acid from its vitamin functions, specifically in modulating blood lipid levels and undesirable side effects such as skin vasodilation and the more rare hepatic toxicities. This information has led to a new strategy for drug delivery for niacin, which, if successful, could have a major impact on human health through decreasing risk for cardiovascular disease. Understanding the many other effects of niacin has much broader potential for disease intervention and treatment in numerous diseases including cancer.

Keywords Cardiovascular diseases · Clinical trial · HDL · LDL · Modulating blood lipids · Niacin · Nicotinamide · Nicotinic acid · Risk for triglycerides · Vasodilation · Topical delivery strategy

3.1 Niacin's Multifaceted Metabolism

Nicotinic acid and nicotinamide, collectively referred to as niacin, were identified as compounds that prevent and cure the dietary deficiency disease, pellagra (Elvehjem et al. 1938). Since then the identification of functions of niacin and it's active metabolites, NAD(H) and NADP(H), in cell metabolism constitute a large metabolome in and of itself [Fig. 3.1, reviewed in (Hassa et al. 2006; Ziegler 2000)].

Both nicotinic acid and nicotinamide function as vitamins, supporting the biosynthesis of NAD(P)(H) (Fig. 3.1). The de novo pathway from tryptophan to NAD is not functional in most human tissues and may not be significant in

E.L. Jacobson (✉)
Division of Medicinal Chemistry, Department of Pharmacology and Toxicology, College of Pharmacy and Arizona Cancer Center, University of Arizona, Tucson, AZ, USA

Niadyne Development, Inc., Tucson, AZ, USA
e-mail: jacobse@pharmacy.arizona.edu; elaine.Jacobson@pharmacy.arizona.edu

O. Stanger (ed.), *Water Soluble Vitamins*, Subcellular Biochemistry 56,
DOI 10.1007/978-94-007-2199-9_3,

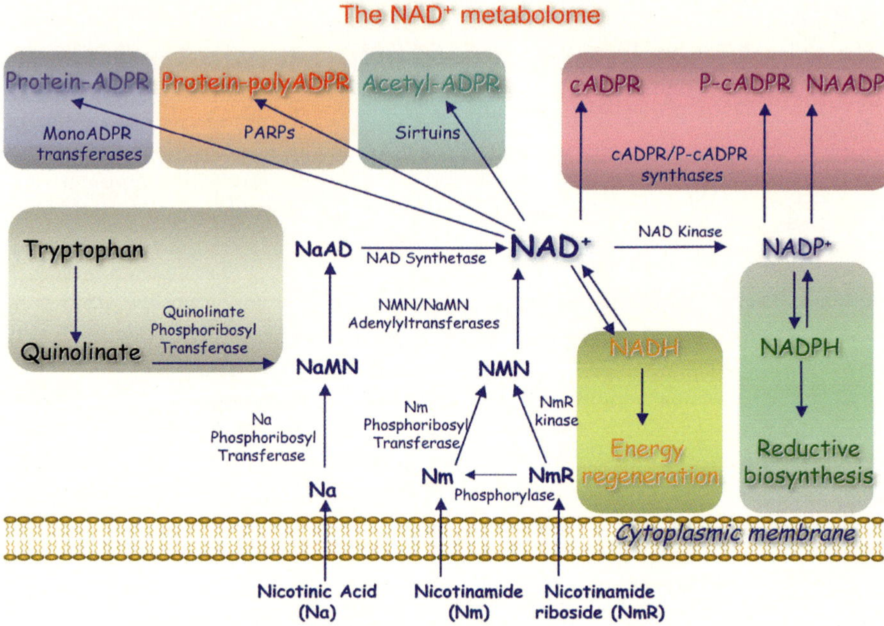

Fig. 3.1 The vitamin functions of niacin

humans, at least under partial dietary restriction (Fu et al. 1989). NAD+/NADH play crucial roles in energy generation and in regulating numerous enzymes of energy metabolism, while NADPH is essential for reductive biosynthesis and maintenance of the antioxidant status of the cell. In more recent decades, numerous enzyme activities have been identified that consume NAD+ or NADP+ as a substrate for post-translational modification of proteins and to synthesize small signaling molecules. Multiple classes of mono-ADP-ribosyl transferases have been identified that have specificity for unique proteins at a specific amino acid side chain, e.g., arginine, cysteine, histidine, etc. While the chemistry and biochemistry of these modifications have evolved, understanding the biological impact of these modifications is an early science. Multiple poly (ADP-ribosyl) polymerases (PARPs) have been discovered and shown to consume NAD+ to regulate many cellular functions that involve assembly and disassembly of protein-nucleic acid complexes, including DNA repair complexes, mitotic spindles, telomeres, etc. NAD+ dependent protein deacetylases (SIRTs) have emerged more recently with seven putative gene products identified to date in human tissues. These enzymes have been implicated in many cellular regulatory events from effecting longevity to regulating specific enzymes of energy metabolism such as glutamate dehydrogenase (Haigis et al. 2006). Finally, ADP-ribosyl transferases that catalyze an internal transfer within the ADP-ribose moiety to form the small signaling molecules, cyclic ADP-ribose (cADPR) and phospho-cyclic-ADP-ribose (Pc-ADPR) appear to effect numerous regulatory roles, many of them involving calcium signaling, however the functions of these enzymes also are poorly understood. The enzymes/pathways mentioned here and noted in

Fig. 3.1 have been briefly mentioned only to illustrate the large number of functions niacin effects as a vitamin. It is important to note that the recent discovery of the nicotinic acid receptor separates the vitamin functions of these two forms of niacin from drug-like functions, since only nicotinic acid can bind the receptor [reviewed in (Offermanns 2006)].

3.2 Niacin's Potential and Pitfalls as an Antidyslipidemic Drug

Blood levels of different lipoprotein classes are strongly associated with the risk of cardiovascular diseases leading to death by heart attack and stroke. Thus, the management of blood lipid levels is an area of enormous significance to public health. As an example, Fig. 3.2 shows data from the Framingham Heart Study documenting that elevated high density lipoprotein cholesterol (LDL) and insufficient high density lipoprotein cholesterol (HDL) are independent risk factors for coronary artery disease (CAD). Most public health messages over the past 15 years have focused on the link between LDL and cardiovascular disease, but the medical community is increasingly realizing that both LDL reduction and HDL elevation are critical for reducing the risk of cardiovascular disease.

The importance of HDL to public health is illustrated further by the fact that nearly 40% of males and 15% of females in the United States have HDL values below 40 mg/dL. These data along with the findings that a 6% increase in HDL can translate to 22–29% decreases in coronary events and deaths illustrate the enormous public health significance of identifying a safe and effective therapy to elevate HDL. The importance of blood lipids to public health has led to the development of intensive therapy to modify blood lipids. Table 3.1 provides an overview of agents used

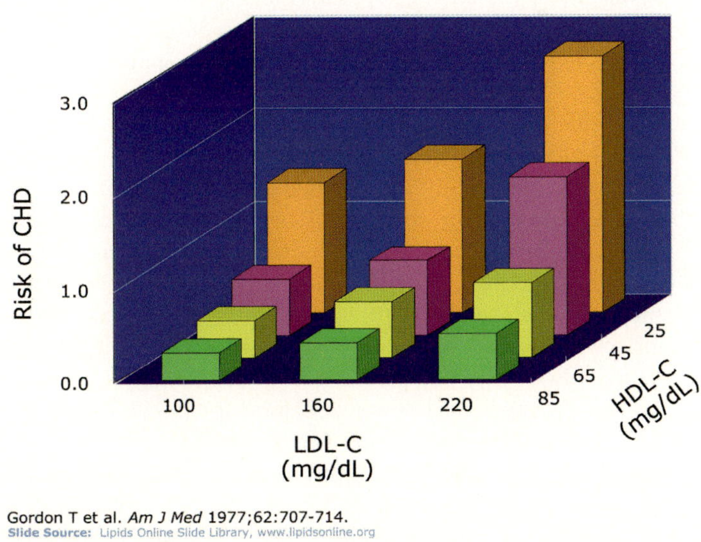

Fig. 3.2 Risk of CHD as a function of blood cholesterol

Table 3.1 Drugs for the management of blood lipids

Drug class	% ↑ in HDL	% ↓ in LDL	% ↓ in TG
Statins	5–6	25–28	11–15
Bile acid sequestrants	3–4	9–18	5–10
Cholesterol absorption inhibitors	1	15–18	5–10
Fibrates	6–18	4–11	21–35
Nicotinic acid	26–32	17–18	27–35

and the range of effects of different agents that have been compiled from many different clinical studies. Nicotinic acid is the oldest drug known to modify serum lipids and it is the most effective drug currently available for elevation of HDL. The statin drugs are effective in reducing LDL but they do not effectively raise HDL. In addition to its ability to raise HDL, nicotinic acid also is effective in reducing cardiovascular risk by lowering serum LDL and triglycerides (TG), which also are a risk factor in CAD. So why is nicotinic acid not the most widely used drug? It is because compliance is poor due to its side effect profile of skin flushing and associated puritis. Furthermore, many individuals with hepatic dysfunction are contraindicated for oral niacin therapy and other oral lipid modification therapies including statin therapy. Standard protocols indicate that nicotinic acid, combined with diet and exercise, should be the first agent used in an attempt to lower LDL, lower TG, and raise HDL. However, the side effect of flushing frequently drives patients with low HDL to seek alternatives.

Currently, there are two sources of nicotinic acid available to elevate HDL levels. First, it can be obtained as over the counter (OTC) formulations in a variety of different forms. The second source is a slow release formulation approved by the US Food and Drug Administration as a once per day dose under the trade name Niaspan. The Niaspan formulation of nicotinic acid decreases vasodilation but still has a very significant side effect profile including skin flushing. Niacin also became available in 2002 as Advicor, a combination therapy that contains a statin, lovastatin, along with Niaspan.

The issue of a side effect profile that limits compliance with therapy is particularly significant for the treatment of a risk factor such as low HDL because most patients on therapy are outwardly healthy and thus an individual who feels fine is reluctant to take a medication that makes one feel uncomfortable. A recently published study (LaRosa and LaRosa 2000) estimates the discontinuation rate of patients placed on oral nicotinic acid therapy at 46%. While the number of patients currently on nicotinic acid therapy is difficult to estimate, it is reasonable to assume that the side effect profile of oral nicotinic acid limits use to far less than 10% of the number of individuals who could benefit from this therapy. We describe below the strategy for development of a new technology for the elevation of HDL and reduction of LDL and TG by transdermal delivery using a nicotinic acid prodrug, lauryl nicotinate. This technology is designed to achieve the stated endpoints without the limiting side effect profile and to provide lipid modulation therapy to those whose hepatic dysfunction contraindicates oral therapies, because transdermal administration greatly reduces the liver first pass effect.

3.3 Niacin's Unique Effects on Blood Lipids

Nicotinic acid is known to inhibit lipolysis in adipose tissue leading to decreases in plasma levels of free fatty acids, decreases in lipoprotein Lp(a), very low-density lipoprotein cholesterol (VLDL or TG) levels, and LDL, while it increases HDL up to 35%. However, it is not completely clear how nicotinic acid increases HDL. A growing interest in the ability of nicotinic acid to inhibit cAMP in adipose tissue due to the Gi-mediated inhibition of adenlyl cyclase (Aktories et al. 1980) and the ensuing hypothesis that nicotinic acid acts as an agonist of a Gi-coupled receptor ultimately resulted in the discovery of the nicotinic acid receptor (Soga et al. 2003; Tunaru et al. 2003; Wise et al. 2003). The potential mechanisms by which nicotinic acid may function through this receptor as noted in the model of Fig. 3.3 include a decrease in TG (VLDL) through decreased cAMP effects on lipolysis (yellow adipocyte, Fig. 3.3), leading to reduced cholesterol ester transfer protein (CETP) mediated exchange of TG and cholesterol esters between Apo-B containing lipoproteins (VLDL, LDL) and HDL, with the end result being increased HDL (grey box, Fig. 3.3) [reviewed in (Offermanns 2006)].

Recent studies demonstrating that nicotinic acid modulates leptin and adiponectin release and down stream signaling as depicted in Fig. 3.3 support alternate mechanisms for affecting HDL metabolism to provide cardioprotective effects. Here, leptin signals via the leptin receptor (LR), a member of the Type 1 cytokine receptor family that mediates intracellular signaling via the activation of associated

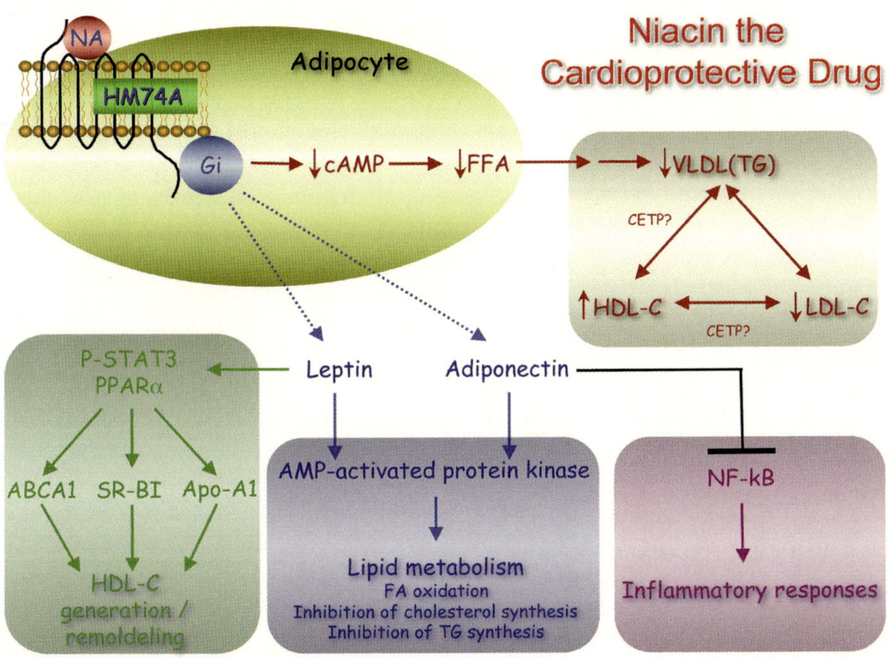

Fig. 3.3 Niacin receptor mediated signaling

Jak family tyrosine kinases (Myers 2004). The LR recruits the signal transducer and activator of transcription 3 (STAT3, green box, Fig. 3.3), which in turn induces important positive effectors of leptin action. The peroxisome proliferator-activated receptor alpha (PPAR-..) has been shown to be a key proximal mediator of the lipopenic action of leptin (Lee et al. 2002), mediating pleiotropic effects such as stimulation of lipid oxidation, alterations in lipoprotein metabolism, and inhibition of vascular inflammation. PPAR-.. binds the hypolipidemic fibrates to increase plasma HDL dependent on expression of apoA-I (Fruchart et al. 1999). Furthermore, PPAR- activation induces cholesterol removal from human macrophage foam cells through stimulation of the ATP binding cassette transporter 1 (ABCA1) pathway (Chinetti-Gbaguidi et al. 2005). Functional interactions between apoA-I and ABCA1 are necessary for the initial lipidation of apoA-I. The expression of hepatic scavenger receptor class BI (SR-BI) is modulated also by leptin (Lundasen et al. 2003). Thus, leptin appears to be important in integrating the transport of cholesterol from extrahepatic tissue to the liver since HDL lipidation, mediated by ABCA1, and delipidation, mediated by SR-BI, are modulated by leptin. Functional SR-BI is critical to the lipid and apolipoprotein composition of HDL and its absence is associated with increased susceptibility to atherosclerosis. It also has been suggested that decreased hepatic uptake and catabolism of HDL-ApoA-I may contribute to increased HDL, however this mechanism is not likely to be mediated via the nicotinic acid receptor as it is not expressed in the liver (Soga et al. 2003; Tunaru et al. 2003; Wise et al. 2003). In addition to modulation of HDL, Fig. 3.3 also shows additional cardioprotective effects mediated by the niacin receptor on NF-kB and the blocking of inflammatory responses known to be very important in CAD (purple box, Fig. 3.3) and effects via leptin and adiponectin illustrated in the blue box of Fig. 3.3. Many studies of the mechanisms that cause diseases of the artery now indicate that inflammation leading to oxidative stress is a major factor in atherosclerosis and that the ability of HDL to prevent oxidative damage to LDL is crucial to the ability of HDL to decrease the risk of arterial diseases (Kontush et al. 2003). HDL is comprised of multiple fractions classified as HDL2 and HDL3, each of which can be further separated into sub fractions designated as HDL3a, b, c, etc.

Many studies now suggest that it is the HDLC3 subfraction that reduces the risk of arterial disease progression and protects LDL from oxidation. Taken together, these observations provide compelling evidence that the goal of therapy to raise HDL should include a focus on the HDL3. Here we overview the evidence for this conclusion. Recent clinical studies have evaluated the relative risk of progression of coronary artery disease to HDL subfractions concluding that risk reduction is associated with the HDL3 and not with HDL2. Yu et al. (2003) examined the risk of incident coronary heart disease in more than 1,700 patients over a 9 year follow up in the Caerphilly coronary heart disease study. The odds ratios demonstrate that HDL3 shows a strong association with risk reduction while HDL2 did not. The study of Syvanne et al. that was derived from the Lopid coronary angiography trial (Syvanne et al. 1998) evaluated the progression of atherosclerosis in 372 subjects. Like the Caerphilly study, this study demonstrated a strong risk reduction associated with HDL3. But in contrast to the Caerphilly study, this study actually detected an increased risk of disease progression associated with HDL2. Additional evidence supporting the role of HDL in antioxidant functions is derived from recent biochemical studies that have specifically implicated HDL3 as an important protective factor by which HDL achieves risk reduction through prevention of accumulation of oxidized LDL particles. In this case, the small, dense HDL3 particles conferred the greatest protection against LDL oxidation. The order of protection and size from smallest to largest is HDL3c > HDL3b > HDL3a > HDL2a > HDL2b (Yoshikawa et al. 1997) (Kontush et al. 2003). Studies of the fibrate class of drugs provide further evidence implicating HDL3 as being crucial to CAD risk reduction. These drugs reduce the risk of CAD, and they also have been shown to raise HDL3 but not HDL2 (Sasaki et al. 2002). Finally, evidence supporting a role of HDL3 in modulating risk relates to promoting efflux of excess cholesterol from cells in the arterial wall, transporting it back to the liver for elimination, a process known as 'reverse cholesterol transport (RCT)'. Nascent apoA-I containing HDL particles interact with peripheral cells and acquire lipid through a transport process facilitated by ABCA1. RCT has been implicated in atheroprotective role of HDL but its efficacy had not been demonstrated in acute coronary syndromes. However, a recent study has reported a rapid reduction of atherosclerosis with intravenous administration of a recombinant apoA-I Milano (apoA-IM) (Nissen et al. 2003). ApoA-IM is a variant of the naturally occurring apoA-I that was discovered in residents of an Italian village characterized by low rates of cardiovascular disease and longer life spans. ApoA-IM differs from naturally occurring apoA-I in that it has a different amino acid composition. Interestingly, carriers of apoA-IM are characterized by predominant, heterogeneous HDL3 and a marked reduction of HDL2, indicating that apoA-IM has a major effect in HDL particle interconversion and that a shift in HDL subfractions from HDL2 to the smaller, more dense HDL3 confers cardioprotection (Kim et al. 2004).

3.4 Making a Good Old Drug Better

With the goal of enhancing the potential and ameliorating the pitfalls of niacin, we have synthesized, characterized, and screened niacin derivatives as potential prodrugs for transdermal delivery. The objectives in searching for effective transdermal prodrugs were to control the rate of delivery in order to keep niacin concentration below the threshold for vasodilation and to decrease the dose required by avoiding first pass metabolism. An overview of the strategy to deliver nicotinic acid transdermally to the blood circulation is shown in Fig. 3.4. The relative lipophilicity of a number of niacin derivatives was determined using the octanol/water partition coefficients. Partition coefficients were then related to the rates of partitioning of the derivatives from the stratum corneum to the epidermis following topical application using intracellular NAD as a biomarker of delivery. Prodrug candidates from methyl nicotinate through octyl nicotinate failed screening tests because they partitioned at a rate rapid enough to cause vasodilation at the site of application, since following conversion to nicotinic acid they exceeded the threshold for vasodilation. Decyl

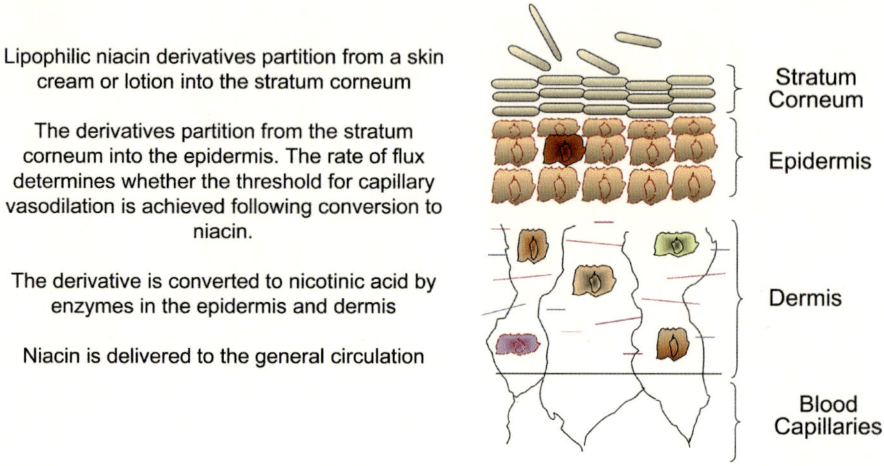

Fig. 3.4 Overview of transdermal delivery strategy using niacin derivatives

nicotinate showed variable effects on vasodilation. Finally, lauryl nicotinate was chosen as the lead candidate because it did not cause vasodilation, yet demonstrated systemic delivery of nicotinic acid in animal models as evidenced by achieving tissue saturation at a skin site distal to the site of application.

The feasibility of modulating blood lipids by transdermal delivery of niacin was then demonstrated in an animal model. Wild type mice are severely limited as an animal model for lipoprotein metabolism as their lipoprotein profile is more than 80% HDL. Transgenic mice carrying human lipoprotein metabolism genes have been used to overcome this limitation for the study of lipoprotein metabolism. For proof of principle studies, transgenic mice carrying the human genes for CETP and apoB 100 were studied. An oral dose of 0.75% nicotinic acid in the drinking water was compared to topical application of lauryl nicotinate doses that corresponded in nicotinic acid equivalents to 12.5 and 25% of the oral dose. Decreases in LDL and TG observed in both topical dosages and in the orally dosed animals were nearly identical, demonstrating that transdermal delivery of nicotinic acid to modify blood lipids is feasible. This led to completion of preclinical toxicology and pharmacology studies whose outcomes supported further drug development.

3.5 Progression to Clinical Trials

A Phase I clinical trial was conducted in a double blind, randomized, placebo controlled dose escalation design. Healthy males ranging in age from 18 to 72 years of age were recruited. Subjects enrolled in the trial were required to have HDL levels below 40 mg/dL, which represent the low end of normal HDL ranges. Each treatment group had nine subjects, six receiving a cream containing 20% lauryl nicotinate, three receiving a placebo control cream. The daily dosages for topical application were 1.5, 3.0, 6.0, and 9.0 ml of cream that corresponded to approximately 150, 300, 600, and 900 mg of nicotinic acid equivalents. Fasting blood samples were collected at each treatment visit for lipid analysis and safety monitoring. After 1 week of application of placebo cream, dosing was initiated with the lowest dosing group and moved to the higher dose groups based on the lack of any adverse events that would dictate not initiating treatment of the higher dose group. Treatment was for a period of 8 weeks, which is about 50% of the treatment period required to see maximum pharmacodynamic effects on blood lipids when niacin is administered orally.

The safety of lauryl nicotinate was assessed by the reporting of adverse events and by analysis of blood chemistry parameters. There were no instances of systemic skin flushing or puritis in the study. There were no severe adverse events related to drug. Blood chemistry, hematology, and urine parameters also were monitored to assess safety. Oral nicotinic acid therapy has been associated with increased fasting levels of blood glucose, uric acid, and the liver enzymes aspartate aminotransferase (AST) and alanine aminotransferase (ALT). In this study, lauryl nicotinate treated subjects did not show a statistically significant elevation in blood glucose, uric acid, AST, or ALT. In addition, there were no statistically significant different values between lauryl nicotinate and placebo subjects in any of the other blood chemistry or hematology parameters that included alkaline phosphatase, total bilirubin, inorganic phosphate, hemoglobin, hematocrit, platelet count, WBC count, and RBC count. In summary, the data indicate that lauryl nicotinate is both safe and well tolerated with no side effects that would limit compliance with the therapy.

The primary endpoint of the Phase I clinical trial was assessment of safety and any side effects not related to safety that might limit compliance with therapy. However, blood lipid profiles were determined to provide preliminary pharmacodynamic information that could prove useful for follow up studies. The statistical power of the pharmacodynamic information obtained in this study was limited by several factors that included (i) lack of information on an optimal dosing schedule, (ii) a limited number of subjects in the study, and (iii) inherent variability of the laboratory determination for HDL. Despite these limitations, useful information was obtained. Blood lipid parameters were determined at baseline and following 4 and 8 weeks of treatment. The data demonstrate a statistically significant increase in both total HDL of approximately 10% and in the ratio of HDL to total cholesterol of approximately 16%. As discussed above, compelling evidence is accumulating that indicates that a desired goal of nicotinic acid therapy is elevation of the HDL3c subfraction. Thus, we Fig. 3.5: HDL subfraction distributions following 8 weeks treatment analyzed HDL subfractions. The data are shown in Fig. 3.5 for the lowest (1.5 ml) dosing group and demonstrate that lauryl nicotinate treatment preferentially increases the HDL3 subfraction. Indeed, the smallest and most potent antioxidant subfraction, HDL3c is elevated an average of 47%, and the next most potent antioxidant subfraction HDL3b is elevated by an average of 19% relative to placebo treatment. The results of lauryl nicotinate treatment reported here are in sharp contrast to results from studies with oral niacin, which increases the HDL2 subfractions (Morgan et al. 2003). An important quantitative note regarding

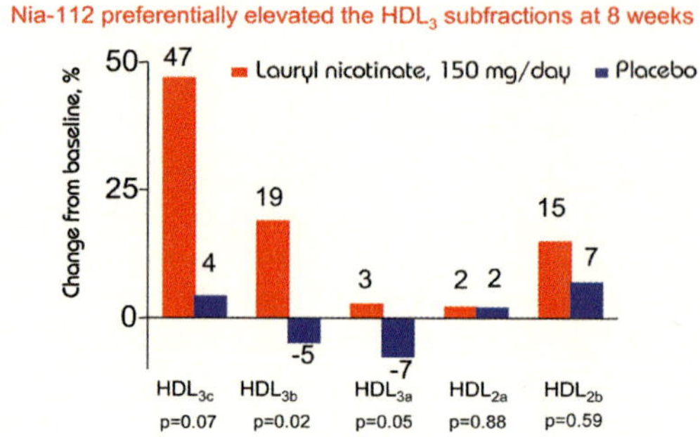

Fig. 3.5 HDL-C subfraction distributions following 8 weeks treatment

percentage changes in cholesterol fractions should be made here. Since HDL is comprised of the smallest particles containing the least cholesterol, it is likely that a 10% increase relative to the total cholesterol pool represents much more than a 10% increase in total number of HDL particles. Reporting changes in percent cholesterol within HDL in this case may represent an under estimate of efficacy, especially if the smaller particles of the HDL fraction are the most efficacious. In summary, this phase 1 clinical trial demonstrated that application of a cream containing lauryl nicotinate was well tolerated and at doses 10% of effective oral doses resulted in statistically significant elevation of HDL and the ratio of HDL to total cholesterol in 8 weeks.

Clearly, the functions of niacin are many (Fig. 3.1). The recent identification of the niacin receptor introduces a new phase of discovery regarding the effects of this vitamin and drug. The potential positive impact on human health of modulating niacin dependent metabolism is immense if its metabolic processes can be further understood to unleash its potential in a selective manner without side effects. Such efforts could have a major effect on both cardiovascular diseases as well as others.

Acknowledgments The research described herein was supported in part by the National Institutes of Health and Niadyne, Inc. ELJ and MKJ are principles in Niadyne, Inc. and conflict of interest management is conducted by the University of Arizona Board of Regents.

References

Aktories K, Jakobs KH, Schultz G (1980) Nicotinic acid inhibits adipocyte adenylate cyclase in a hormone – like manner. FEBS Lett 115:11–14

Chinetti-Gbaguidi G, Rigamonti E, Helin L, Mutka AL, Lepore M, Fruchart JC, Clavey V, Ikonen E, Lestavel S, Staels B (2005) Peroxisome proliferator-activated receptor alpha controls cellular cholesterol trafficking in macrophages. J Lipid Res 46:2717–2725

Elvehjem CA, Madden RJ, Strong FM, Woolley DW (1938) The isolation and identification of the anti-black tongue factor. J Biol Chem 123:137–149

Fruchart JC, Duriez P, Staels B (1999) Peroxisome proliferator-activated receptor-alpha activators regulate genes governing lipoprotein metabolism, vascular inflammation and atherosclerosis. Curr Opin Lipidol 10:245–257

Fu CS, Swendseid ME, Jacob RA, McKee RW (1989) Biochemical markers for assessment of niacin status in young men: levels of erythrocyte niacin coenzymes and plasma tryptophan. J Nutr 119:1949–1955

Haigis MC, Mostoslavsky R, Haigis KM, Fahie K, Christodoulou DC, Murphy AJ, Valenzuela DM, Yancopoulos GD, Karow M, Blander G, Wolberger C, Prolla TA, Weindruch R, Alt FW, Guarente L (2006) SIRT4 inhibits glutamate dehydrogenase and opposes the effects of calorie restriction in pancreatic beta cells. Cell 126:941–954

Hassa PO, Haenni SS, Elser M, Hottiger MO (2006) Nuclear ADP-ribosylation reactions in mammalian cells: where are we today and where are we going? Microbiol Mol Biol Rev 70:789–829

Kim H, Jacobson EL, Jacobson MK, Lacko AG (2004) Apoa-1 milano and regression of atherosclerosis. JAMA 291:1319

Kontush A, Chantepie S, Chapman MJ (2003) Small, dense HDL particles exert potent protection of atherogenic LDL against oxidative stress. Arterioscler Thromb Vasc Biol 23:1881–1888

LaRosa JH, LaRosa JC (2000) Enhancing drug compliance in lipid-lowering treatment. Arch Fam Med 9:1169–1175

Lee Y, Yu X, Gonzales F, Mangelsdorf DJ, Wang MY, Richardson C, Witters LA, Unger RH (2002) PPAR alpha is necessary for the lipopenic action of hyperleptinemia on white adipose and liver tissue. Proc Natl Acad Sci USA 99:11848–11853

Lundasen T, Liao W, Angelin B, Rudling M (2003) Leptin induces the hepatic high density lipoprotein receptor scavenger receptor B type I (SR-BI) but not cholesterol 7alpha-hydroxylase (cyp7a1) in leptin-deficient (ob/ob) mice. J Biol Chem 278:43224–43228

Morgan JM, Capuzzi DM, Baksh RI, Intenzo C, Carey CM, Reese D, Walker K (2003) Effects of extended-release niacin on lipoprotein subclass distribution. Am J Cardiol 91:1432–1436

Myers MG Jr (2004) Leptin receptor signaling and the regulation of mammalian physiology. Recent Prog Horm Res 59:287–304

Nissen SE, Tsunoda T, Tuzcu EM, Schoenhagen P, Cooper CJ, Yasin M, Eaton GM, Lauer MA, Sheldon WS, Grines CL, Halpern S, Crowe T, Blankenship JC, Kerensky R (2003) Effect of recombinant apoA-I milano on coronary atherosclerosis in patients with acute coronary syndromes: a randomized controlled trial. JAMA 290:2292–2300

Offermanns S (2006) The nicotinic acid receptor GPR109A (HM74A or PUMA-G) as a new therapeutic target. Trends Pharmacol Sci 27:384–390

Sasaki J, Yamamoto K, Ageta M (2002) Effects of fenofibrate on high-density lipoprotein particle size in patients with hyperlipidemia: a randomized, double-blind, placebo-controlled, multicenter, crossover study. Clin Ther 24:1614–1626

Soga T, Kamohara M, Takasaki J, Matsumoto S, Saito T, Ohishi T, Hiyama H, Matsuo A, Matsushime H, Furuichi K (2003) Molecular identification of nicotinic acid receptor. Biochem Biophys Res Commun 303:364–369

Syvanne M, Nieminen MS, Frick MH, Kauma H, Majahalme S, Virtanen V, Kesaniemi YA, Pasternack A, Ehnholm C, Taskinen MR (1998) Associations between lipoproteins and the progression of coronary and vein-graft atherosclerosis in a controlled trial with gemfibrozil in men with low baseline levels of HDL cholesterol. Circulation 98:1993–1999

Tunaru S, Kero J, Schaub A, Wufka C, Blaukat A, Pfeffer K, Offermanns S (2003) PUMA-G and HM74 are receptors for nicotinic acid and mediate its anti-lipolytic effect. Nat Med 9:352–355

Wise A, Foord SM, Fraser NJ, Barnes AA, Elshourbagy N, Eilert M, Ignar DM, Murdock PR, Steplewski K, Green A, Brown AJ, Dowell SJ, Szekeres PG, Hassall DG, Marshall FH, Wilson S, Pike NB (2003) Molecular identification of high and low affinity receptors for nicotinic acid. J Biol Chem 278:9869–9874

Yoshikawa M, Sakuma N, Hibino T, Sato T, Fujinami T (1997) HDL3 exerts more powerful anti-oxidative, protective effects against copper-catalyzed LDL oxidation than HDL2. Clin Biochem 30:221–225

Yu S, Yarnell JW, Sweetnam P, Bolton CH (2003) High density lipoprotein subfractions and the risk of coronary heart disease: 9-years follow-up in the caerphilly study. Atherosclerosis 166:331–338

Ziegler M (2000) New functions of a long-known molecule. Emerging roles of NAD in cellular signaling. Eur J Biochem 267:1550–1564

Chapter 4
Beyond the Antioxidant: The Double Life of Vitamin C

Mario C. De Tullio

Abstract When considering the history of vitamin C, and the names given to this molecule in early days, the Latin proverb nomen est omen suddenly comes to mind. Around 1920, when Casimir Funk introduced the term Vitamin C to indicate the nutritional factor necessary to prevent the pathological state known as scurvy, the nature of the active molecule was still unknown (Davies MB, Austin J, Partridge DA (1991) Vitamin C: Its chemistry and biochemistry. The Royal Society of Chemistry, Cambridge UK). Almost in the same years, Albert Szent-Giörgyi was striving to identify a new 6-carbon sugar he had obtained in crystal form from oranges, lemons, cabbage and adrenal glands. As humorously described by Szent-Giörgyi himself (Szent-Giörgyi A (1963) Lost in the twentieth century. Annu Rev Biochem 36:1–15), he intended to name this yet unknown carbohydrate "ignose". When this name was rejected by Sir Arthur Harden, editor of the Biochemical Journal, he suggested to name it "godnose", meaning that only God could know the real identity of the molecule. Obviously, also this choice was considered inappropriate by Harden, who suggested the plain name "hexuronic acid". Only later, when the structure of "hexuronic acid" had been completely elucidated, and biological tests performed by Swirbely identified this molecule as the anti-scurvy factor vitamin C, Szent-Giörgyi and Walter Norman Haworth decided to eventually name it ascorbic acid (Szent-Giörgyi A (1963) Lost in the twentieth century. Annu Rev Biochem 36:1–15). "Ascorbic" literally means "against scurvy", and scurvy is known to be mainly due to the inactivation of some important dioxygenases involved in the synthesis of a few key molecules, including different collagen forms (De Tullio MC (2004) How does ascorbic acid prevent scurvy? A survey of the nonantioxidant functions of vitamin C. In: Asard H, May J, Smirnoff N (eds) Vitamin C, its functions and biochemistry in animals and plants. Bios Scientific Publishers, Oxford, UK, pp. 159–172). All this has very little to do with the celebrated role of ascorbic acid (ASC) as an antioxidant. So, if the fate of ASC had to be found in its name, its role in the prevention of scurvy (i.e. beyond the antioxidant function) should be considered its main feature. But, in spite of more than 80 years of extensive research (34,424 hits in a PubMed query on January 6 2007), an unprecedented popularity among the general public, an estimated market of several billion dollars (Hancock

M.C. De Tullio (✉)
Dipartimento di Biologia e Patologia Vegetale, Universita' di Bari, 70125 Bari, Italia
e-mail: detullio@botanica.uniba.it

O. Stanger (ed.), *Water Soluble Vitamins*, Subcellular Biochemistry 56,
DOI 10.1007/978-94-007-2199-9_4, © Springer Science+Business Media B.V. 2012

RD, Viola R (2005) Improving the nutritional value of crops through enhancement of l-ascorbic acid (vitamin C) content: Rationale and biotechnological opportunities. J Agr Food Chem 53:5248–5257), we should honestly conclude that the fate of vitamin C is still in the first name it received, many years ago: we still ignore much of its actual relevance in cell metabolism, although we are progressively getting aware of the many facets of this fascinating molecule, and its direct involvement in the regulation of apparently unrelated pathways (Arrigoni O, De Tullio MC (2002) Ascorbic acid, much more than just an antioxidant. Biochim Biophys Acta 1569:1–9; De Tullio MC, Arrigoni O (2004) Hopes, disillusions and more hopes from vitamin C. Cell Mol Life Sci 61:209–219; Duarte TL, Lunec J (2005) When is an antioxidant not an antioxidant? A review of novel actions and reactions of vitamin C. Free Rad Res 39:671–686). Recent data on ASC involvement in cell signalling and gene expression open new perspectives, that will be presented and discussed in this chapter.

Keywords Vitamin C · Ascorbic acid · Scurvy · HIF1-α · Antioxidome

4.1 The Biochemical Role of Vitamin C

The accumulation of mutations in the gene coding for l-guolonolactone oxidase, last enzyme in ASC biosynthetic pathway, made primates (including humans), guinea pigs and a few other animal species unable to synthesize ASC (Nishikimi and Yagi 1991; Nishikimi et al. 1994). The evolutionary reason for this loss has been debated. According to some authors, the loss of ASC synthesis could have increased the rate of mutation and consequent genetic variability, boosting further evolution (Challem and Taylor 1998). More realistically, other authors suggest that dietary intake of vitamin C present in fruit and vegetables was enough to compensate for the lack of synthesis, and that, on the contrary, this event could have been an advantage, since hydrogen peroxide is generated in the reaction catalyzed by l-guolonolactone oxidase (Halliwell 2001). Whatever the reason for the loss of synthesis, humans can be considered as a knock-out experimental model, and scurvy as the most compelling evidence available about actual ASC functions.

Scurvy is a pathological state caused by ASC deficiency. Symptoms of scurvy include lassitude, vision problems, and neurological disorders, but the most evident signs of the disease are caused by alteration of the extracellular matrix in blood vessels, bones, skin, gums and tendons, which result in spots, haemorrhages and bone fragility (Fain 2005). All symptoms can be reversed by ASC administration (Johnston et al. 1985). Many, if not all, symptoms of scurvy can be explained in terms of the inactivation of a few ASC-dependent enzymes, all belonging to the large class of 2-oxoglutarate dependent dioxygenases (2-ODDs), widespread in the cells of all organisms (De Tullio 2004). 2-ODDs catalyze the incorporation of O_2 into an organic substrate. In general, these enzymes share a catalytic mechanism specifically requiring Fe^{2+}, 2-oxoglutarate and ASC as co-substrates (Tuderman et al. 1977). In some cases, however, differences in co-substrate requirement may occur (Schofield and Zhang 1999; Clifton et al. 2006).

Different 2-ODDs catalyze reactions of hydroxylation, desaturation, oxidative ring-closure or expansion (Fig. 4.1). Such reactions, leading to the incorporation of oxygen into a given substrate, are critical steps in a surprisingly large number of biochemical pathways. It has been suggested that the role of ASC in the reaction is just related to keeping iron in the Fe^{2+} state, but this is contradicted by the fact that many other electron donors cannot replace ASC in maintaining 2-ODDs in their active form. Detailed studies on the catalytic mechanism of peptidyl-prolyl-4-hydroxylase (P4H), the enzyme responsible for post-translational hydroxylation at carbon 4 of proline residues incorporated into polypeptide chains, showed the complexity of the reaction. Molecular oxygen is split by the enzyme and used for both oxidative decarboxylation of 2-oxoglutarate and oxidation of peptidyl proline. The latter is obtained via the generation of an iron-oxygen-atom complex, the ferryl ion, which hydroxylates an appropriate proline residue (Hanauske-Abel and Günzler 1982; Wu et al. 2000). The generation of the ferryl ion can proceed without subsequent hydroxylation in so-called uncoupled reaction cycles (Myllyla et al. 1984). ASC is utilized as a specific alternative acceptor of the ferryl oxygen in these reaction cycles. In the absence of ASC, P4H is rapidly inactivated by self oxidation (Tschank et al. 1994; Kivirikko and Pihlajaniemi 1998). From a chemical point of view, managing the highly reactive ferryl ion is not easy, and highly specialized organization of the enzyme is required (Hoffart et al. 2006). It is very likely that ASC and 2-ODDs underwent molecular co-evolution thus making such complex reaction possible, as can be deduced by the high specificity for ASC in the reactions catalyzed by those enzymes (Majamaa et al. 1986).

Inactivation of P4H by ASC deficiency is the first identified cause of scurvy (Stone and Meister 1962). Hydroxylation of peptidyl-proline is strictly required to ensure proper collagen folding (Pihlajaniemi et al. 1991). Mature collagen is present as chains wound in tight triple helices organized into fibrils of great tensile strength and thermal stability (Holmgren et al. 1998). If ASC is not available, proline

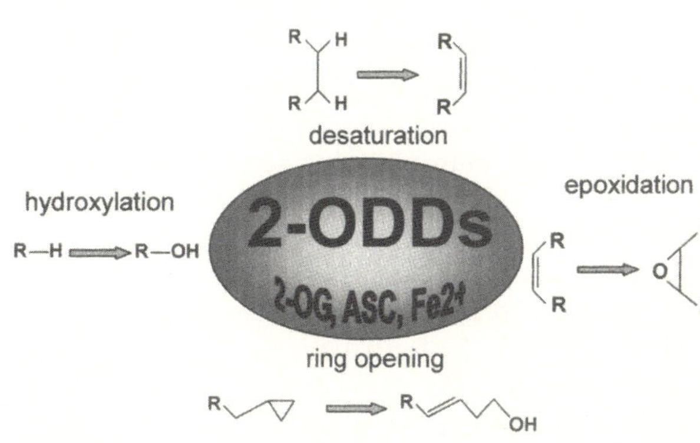

Fig. 4.1 Partial list of reactions catalyzed by 2-oxoglutarate-dependent dioxygenases

residues of collagen are not properly hydroxylated and collagen trimers cannot form. Human unhydroxylated collagen (obtained by heterologous expression in transgenic plants) showed increased flexibility as well as a reduced melting temperature, and this recombinant unhydroxylated collagen did not self-assemble into banded fibrils in physiological conditions (Perrett et al. 2001). In humans, underhydroxylated collagen is degraded in the endoplasmic reticulum (Hosokawa and Nagata 2000).

Also involved in post-translational collagen stabilization is lysyl hydroxylase, another 2-ODD sharing the same reaction mechanism of P4H, but highly specific for lysine residues and present in humans with different development-specific isoforms (Salo et al. 2005). Low expression of lysyl hydroxylase caused marked collagen alterations in a knock-out mouse experimental model (Takaluoma et al. 2006). It is noteworthy that both ASC deficiency (scurvy) and low expression of prolyl hydroxylase due to inherited defects (Prockop and Kivirikko 1995; Myllyharju and Kivirikko 2001) or gene manipulation, converge to the same result.

Many different collagen types have been described, with specific functions in maintaining the integrity of skin, tendons, cartilage, bones, teeth, cornea, muscles, blood vessels and other tissues and organs (Boot-Handford et al. 2003; Jenkins et al. 2005). In some cases, the link between ASC deficiency and altered structure of specific collagen types has been directly investigated. For instance, guinea pigs fed an ASC-free diet showed lower expression of type IV collagen and elastin mRNAs in blood vessels, as well as lower amount of hydroxyproline in the protein. This may contribute to defects observed in blood vessels during scurvy (Mahmoodian and Peterkofsky 1999).

Collagen being the most abundant protein in animals, its hydroxylation conceivably consumes large ASC amounts. This also explains why collagen malfunctioning is the most evident sign of scurvy. However, ASC deficiency also affects other important targets. Both ASC and its oxidized form dehydroascorbate are involved in protein folding in the endoplasmic reticulum (Arrigoni and De Tullio 2002). Recent studies showed that persistent ascorbate deficiency in guinea pigs leads to ER stress, unfolded protein response, and apoptosis in the liver, suggesting that insufficient protein processing participates in the pathology of scurvy (Margittai et al. 2005). In addition, other enzymes sharing ASC dependency with collagen hydroxylases are also affected by ASC deficiency, and their inactivation can lead to less dramatic, but relevant consequences.

Carnitine is an essential metabolite having an indispensable role in energy metabolism, since it enables activated fatty acids to enter the mitochondria, where they are broken down via β-oxidation. Two ASC-dependent 2-ODDs, namely, ε-trimethyl-lysine dioxygenase and γ-butyrobetaine dioxygenase are necessary for the synthesis of carnitine (Vaz and Wanders 2002). ASC deficiency affects carnitine synthesis (Thoma and Henderson 1984).

The basis of neurological disorders observed in scurvy are probably related to low activity of other ASC-dependent enzymes. Indoleamine 2,3-dioxygenase, the first and rate-limiting enzyme in human tryptophan metabolism, has been implicated in the pathogenesis of many diseases (Littlejohn et al. 2000). This enzyme is necessary to form kynurenine (Widner et al. 2000). The kynurenine pathway regulates the metabolism of tryptophan to neuroactive compounds, and also seems to be a key factor in the communication between the nervous and immune systems (Stone and Darlington 2002). The conversion of tyrosine to 3,4-hydroxyphenyl alanine (DOPA) is catalyzed by tyrosine hydroxylase. In turn, DOPA is further modified by dopamine-β-hydroxylase, yielding the neurotransmitter noradrenalin (Diliberto et al. 1991). DOPA and noradrenalin metabolism is apparently correlated with depression (Racagni and Brunello 1999).

Synthesis of vitamin A requires the activity of β,β-carotene-15,15′-oxygenase, an ASC-dependent enzyme catalyzing the cleavage of β-carotene into two molecules of retinal (von Lintig and Vogt 2000; von Lintig et al. 2005).

Peptidylglycine R-amidating monooxygenase (PAM) is responsible for the activation of many peptide hormones and neuropeptides that require amidation of their C terminus for biological activity. PAM uses separate enzymatic domains to catalyze this amidation reaction in two steps: (i) hydroxylation of the CR of a C-terminal glycine and (ii) disproportionation of the R-hydroxyglycine. The first step is catalyzed by the peptidylglycine R-hydroxylating monooxygenase (PHM), a two-copper, ascorbate-dependent enzyme (Crespo et al. 2006).

Recently, a new group of ASC-dependent dioxygenases has been identified, with the unique property of catalyzing the repair of methylated bases (1-methyladenine and 3-methylcytosine) in DNA sequences (Duncan et al. 2002). E. coli AlkB and human ABH1-3 proteins catalyze oxidation of the relevant methyl groups and release them as formaldehyde (Koivisto et al. 2004).

The biological role of the many ASC-dependent plant dioxygenases involved in the synthesis of hormones such as gibberellins and ethylene and other signalling molecules was discussed elsewhere (Arrigoni and De Tullio 2000).

All those findings not only prove that the general reaction mechanism of dioxygenases is common to many different pathways and functions, but also that novel dioxygenases with surprising functions are likely to be discovered in the future.

4.2 ASC in HIF Hydroxylation Signaling

The reaction catalyzed by dioxygenases obviously cannot take place in the absence of oxygen. Therefore, a complex and elegant mechanism evolved in animal cells, coupling the capability of dioxygenases to sense oxygen with the orchestration of cellular defenses against oxygen deprivation (hypoxia). In 2001, two independent groups discovered that hydroxylation of two proline residues present in a specific sequence of the transcription factor hypoxia-inducible factor 1α (HIF1α) underlies the mechanism of oxygen sensing in mammalian cells (Ivan et al. 2001; Jaakkola et al. 2001; Epstein et al. 2001). HIF1, the first characterized member of the HIF family, transcriptionally activates hundreds of genes associated with angiogenesis, energy metabolism, nutrient transport; and cell migration (Chun et al. 2002; Greijer et al. 2005). It is a heterodimer, comprised of α and β subunits. In human cells, 2 proline residues (Pro402 and Pro564) of HIF1α are hydroxylated by 3 different prolyl hydroxylases (HIF-P4H) which are not directly related to collagen hydroxylases mentioned above, although sharing the same catalytic mechanism (Tuckerman et al. 2004). HIF-P4H are in the cytosol, whereas collagen hydroxylases are typically in the endoplasmic reticulum. In HIF-P4H, the Km values for O_2 are slightly above its atmospheric concentration, indicating that the HIF-P4Hs are effective oxygen sensors (Hirsila et al. 2003). The 3 HIF-P4H have different roles in the hydroxylation of the 2 proline residues of HIF1$\tilde{\alpha}$ (Berra et al. 2003; Appelhoff et al. 2004).

When oxygen is available (normoxic conditions), proline residues of HIF1α are hydroxylated (Fig. 4.2). Hydroxylation of HIF 1α is necessary for the binding of a multiprotein complex containing the von Hippel-Lindau tumor suppression protein (pVHL). In turn, the formation of this multiprotein complex leads to rapid targeting of HIF1α to proteasome-mediated degradation (Schofield and Ratcliffe 2004). When oxygen availability lowers (hypoxia) or no oxygen at all is available (anoxia), hydroxylation cannot take place, HIF1α cannot bind pVHL, is not degraded, and migrates to the nucleus, where it binds HIF1β. The dimer binds a specific sequence (the Hypoxia Responsive Element) in the promoter of an array of hypoxia-induced genes, including those for the synthesis of erythropoietin, transferrin, ceruloplasmin, vascular endothelial growth factor, nitric oxide synthase, carbonic anhydrase9, insulin-like growth factor, glyceraldehyde-3-phosphate dehydrogenase, and many more (Schofield and Ratcliffe 2005). The activation of these genes is of paramount importance in many important processes, including angiogenesis in cancer (Harris 2002; Hickey and Simon 2006) and ischemia (Kido et al. 2005). The potential

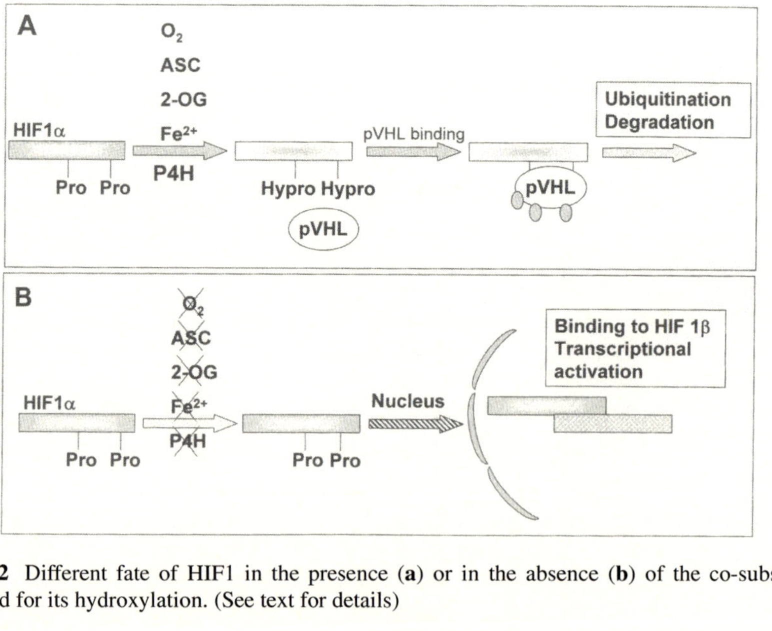

Fig. 4.2 Different fate of HIF1 in the presence (**a**) or in the absence (**b**) of the co-substrates required for its hydroxylation. (See text for details)

therapeutic fallout of those findings is currently under investigation (Hewitson and Schofield 2004).

Beside proline hydroxylation, HIF signaling is also regulated by hydroxylation of an asparagine residue (Asn803 in the human HIF1α) by asparaginyl hydroxylase, an enzyme similar to P4H, but whose ASC dependency is still uncertain (Knowles et al. 2003; Schofield and Ratcliffe 2005). Hydroxylation of this residue prevents binding of HIF1α to a co-activator (p300) required for transcriptional activation (Schofield and Ratcliffe 2004).

When the mechanism of oxygen sensing and activation of hypoxic response by hydroxylases was identified, the role of ASC in the regulation of this process was underestimated. Only later, a few studies directly addressed the role of ASC in controlling the expression and activation of HIF1α expression. Knowles et al. (2003) reported that, under normoxic and mild hypoxic conditions, ASC addition in physiological concentrations decreased the amount of HIF1 protein, by increasing the rate of its degradation. This means that lower ASC content could activate molecular responses to hypoxia, thus being beneficial in case of ischemia. This could possibly explain the contradictory results obtained when attempting to correlate plasma ASC concentrations with the risk of myocardial infarction (Padayatty and Levine 2000a).

Kinetic analyses of recombinant HIF hydroxylases have given apparent Km values of 140–180 μM, a value in a similar range to those cited for procollagen prolyl hydroxylase (Hirsila et al. 2003). Under certain conditions, ASC availability may be limiting for hydroxylase activity. ASC addition has been shown to reduce the normoxic accumulation of HIF-α that is observed in certain rapidly growing tissue culture cells (Knowles et al. 2003; Lu et al. 2005). Such effects were not seen in cells treated with 2-oxoglutarate analogues that block HIF hydroxylase activity, and are not seen in VHL deficient cells, supporting the interpretation that they arise from the promotion of hydroxylase activity by ASC, and not by its generic "antioxidant function". In another study, ASC was shown to promote HIF hydroxylation in cells deficient in the expression of junD, a member of the AP-1 family of transcription factors (Gerald et al. 2004).

A recent study providing a computational model of oxygen sensing by HIF1α (Qutub and Popel 2006) shows that low ASC concentration markedly affects the rate of hydroxylation of HIF1α protein, slowing down its degradation. According to the model, ASC has a dual role in the reaction: reactivating the enzyme by reducing the accumulating Fe^{3+} and binding independently to the saturable enzyme complex formed by the binding of the hydroxylase with Fe^{2+}, 2-oxoglutarate and O_2 (Majamaa et al. 1986). A third function of ASC in the hydroxylation reaction, i.e. its role as an acceptor of the ferryl ion in uncoupled decarboxylation (Myllyla et al. 1984), was not considered in the model calculated by Qutub and Popel (2006), but is likely to significantly affect the activity of HIF-hydroxylating enzymes.

Additional studies on HIF signalling have demonstrated that HIF1α not only accumulates in hypoxic conditions, but also in response to other nonhypoxic stimuli, comprising the Krebs cycle intermediates pyruvate and oxaloacetate. ASC proved a major controller of such nonhypoxic activation of HIF signalling (Lu et al.

2005). Enhancement of HIF-1 by glucose metabolites has been proposed as a novel feed-forward signalling mechanism involved in malignant progression (Lu et al. 2005). This leads to conclusion that hydroxylase-mediated signalling not only acts in oxygen sensing, but functions in a broader context than previously appreciated, and that ASC has a key role in its regulation.

A fierce debate on ASC in cancer treatment has been going on over more than 3 decades, following the high hopes raised by the controversial work started in 1971 by Linus Pauling and Ewan Cameron (Pauling and Cameron 1979). However, contrasting results have been obtained on this important issue (Moertel et al. 1985; Riordan et al. 1995; Padayatty and Levine 2000b). Angiogenesis, a process stimulated by HIF-1α, has a fundamental role in cancer development (Harris 2002) and HIF-1α, is constitutively highly expressed in cancer cells (Zhong et al. 1998). ASC deficiency has been observed in cancer patients (Mayland et al. 2005). All in all, it seems now possible to find a rationale explanation for the use of ASC in cancer therapy. Jones et al. (2006) successfully explored this possibility by combining the use of monoclonal antibodies against transferrin (resulting in the inhibition of tumor cell proliferation, but in the induction of angiogenetic responses via HIF signalling) with the use of physiologic concentration of ASC (which causes HIF down-regulation).

The involvement of HIF1 in immune reactions has been recently identified (Hellwig-Burgel et al. 2005). Also in this case, this could explain the postulated, but never really proved, function of ASC in improving immune defences (Hemila 1997). There is little doubt that those very promising fields of research shall be further explored in the next years.

The discovery of the role of peptidyl-proline hydroxylation in the control of the stability of HIF1α and in oxygen sensing induced more scientists to investigate whether the same mechanism could be involved in the control of the turnover of other proteins. Kuznetsova et al. (2003) reported that ubiquitination and degradation of the large subunit of RNA polymerase II follows a mechanism requiring hydroxylation of proline residues in the protein, followed by binding of pVHL. In this case, degradation is regulated by UV radiation and not by oxygen availability. A striking difference between the mechanism of ubiquitination of HIF1α and that of RNA polymerase II is that in the former, degradation is the default process occurring under normal conditions (normoxia), whereas in the latter, hydroxylation, binding of pVHL and ubiquitination occur under stress conditions (UV). Hydroxylation of proline residues represents therefore a dynamic and efficient tool for the control of protein turnover. A similar system has recently been invoked for the degradation of the Iron Regulatory Protein 2 (IRP2), a posttranscriptional regulator of iron metabolism (Wang and Pantopoulos 2005). It can be easily predicted that in the near future more proteins sharing this regulative mechanism will be identified.

Notably, the activities of other ASC-dependent 2-ODDs, besides P4H, are apparently involved in important signal transduction pathways: this is the case of tyrosine hydroxylase (Paulding et al. 2002) and aspartyl hydroxylase (Maeda et al. 2003). ASC is also required for the hydroxylation, by aspartyl-β-hydroxylase (AAH), of aspartyl and asparaginyl residues in epidermal growth factor (EGF)-like domains

(Gronke et al. 1990) present in different signaling proteins, involved in protein-protein interaction (Stenflo et al. 2000). Some of these target proteins (Notch and Notch homologs) have roles in cell migration and development (Selkoe and Kopan 2003; Lubman et al. 2007).

4.3 ASC and Gene Expression

A large wealth of literature demonstrates ASC involvement in the transcription of many different genes.

Early observations suggested that vitamin C could be involved in gene transcription (Price 1966). Many recent reports further substantiate this hypothesis. In some cases, vitamin C seems to stabilize specific mRNAs (Arrigoni and De Tullio 2002 and references therein). For example, ASC-dependent stabilization of collagen transcript and destabilization of elastin transcript in smooth muscle cells and skin fibroblasts was observed (Davidson et al. 1997). Transcription of the gene coding for tyrosine hydroxylase is enhanced by ASC (Seitz et al. 1998).

A special role of ASC in inducing differentiation of several mesenchymal cell types was observed (Duarte and Lunec 2005, and references therein). Differentiation of chondrocytes (Dozin et al. 1992), cardiomyocytes (Sato et al. 2006) and osteoblasts (Harada et al. 1991) is apparently induced by ASC via collagen synthesis. ASC is specifically required for the expression of osteocalcin, a calcium-binding protein made by osteoblasts. ASC treatment increases osteocalcin gene transcription by way of a complex mechanism that is blocked by the inhibition of collagen hydroxylation, and that requires a specific sequence present in the promoter of the osteocalcin gene (Xiao et al. 1997). ASC was also successfully used to induce myelination in a mouse model of Charcot Marie Tooth syndrome, a severe neuropathy (Passage et al. 2004), although the underlying biochemical mechanism is not clear.

It has been suggested that, due to its effects in cell differentiation, ASC might be useful in the in vitro production of specialized cells/tissues from embryonic stem cells for future clinical treatment (Duarte and Lunec 2005).

The number of microarray studies investigating ASC-dependent gene expression is increasing. Catani et al. (2001) demonstrated that ASC administration induces transcription of the fra-1 gene, which codes for a transcription factor of the Fos family, and down-regulates the activator protein-1 (AP-1)-target genes. Expression profiles of human mesenchymal stem cells during the mineralization process were analyzed in the presence of ASC and other osteogenic supplements. ASC and other agents were necessary for the mineralization process and for transcriptional stimulation of 55 genes and repression of 82 genes among more than 20,000 examined (Doi et al. 2002). In an attempt to identify the mechanism of ASC-induced differentiation of embryonic stem cells into mineralised osteoblasts (Harada et al. 1991), Carinci et al. (2005) analysed the trascriptome of pre-osteoblasts treated with ASC, and identified 97 up-regulated and 373 down-regulated genes involved in cell growth, metabolism, morphogenesis, cell death, and cell communication.

The picture emerging from these data suggests that ASC controls gene expression by still unclear mechanisms. ASC seems to operate in transcription; RNA stabilization; post-translational modification of proteins; hydroxylation, epoxidation and desaturation of many different substrates. Although the complete picture is far from being complete, this is enough to get an idea of the many still ignored functions of ASC.

4.4 Antioxidant vs. Non-antioxidant Functions of ASC: A False Problem?

The free radical theory of aging and disease is one of the most fortunate scientific hypotheses of the last century (Halliwell and Gutteridge 1984). The possibility to explain in a single, unifying theory the etiology of almost all pathological conditions (including ageing!) in all organisms was, and still is, very seductive (Marx 1985). In addition, at least in its early formulation, the theory was based on a simple dichotomy between "bad" oxidants and "good" antioxidants (Azzi et al. 2004). This simple, schematic concept rapidly became very popular also among non-scientists.

The antioxidative properties of already known and newly discovered molecules became the focus of much research. Provided that all in vitro antioxidants actually have biological activities (Azzi et al. 2004), different organisms have apparently evolved a different blend of antioxidant molecules and enzymes (introducing a new term this could be called "antioxidome"), which is in part genetically determined, and in part responsive to environmental clues. The great relevance of ASC in the antioxidome of both plants and animals has been underlined in much scientific literature, due to its abundant presence in almost all biological systems (Frei et al. 1989; Rose and Bode 1993; Smirnoff 2005).

There is an increasing number of reports also pointing at "non-antioxidant" functions of vitamin E (Zingg and Azzi 2004) and vitamin C (De Tullio and Arrigoni 2004; Duarte and Lunec 2005), but a contraposition between "oxidant" and "non-antioxidant" activities runs the risk of becoming a new, worthless scientific dispute. It is imperative to find a reasonable explanation accounting for this "double life" of vitamin C.

The involvement of ASC in cellular responses to nickel toxicity (Salnikow et al. 2004; Salnikow and Kasprzak 2005; Karaczyn et al. 2006) possibly opens new perspectives to our understanding of actual ASC function in biological systems.

Exposure of cells to carcinogenic compounds of nickel and cobalt caused hydroxylase-dependent activation of the HIF-1 transcription factor and of its transcriptional activity under normoxic conditions. As mentioned above, hydroxylation of HIF1α cannot take place if any of the co-substrates in the reaction catalyzed by prolyl hydroxylase (2-oxoglutarate, Fe^{2+}, oxygen, and ASC) is missing (Fig. 4.2). Nickel or cobalt did not affect the level of intracellular iron, but greatly depleted intracellular ASC. Co-exposure of cells to metals and ASC reversed both metal-induced stabilization of HIF-1α and HIF-1-dependent gene transcription

(Salnikow et al. 2004). Studying the expression of carbonic anhydrase IX (CA IX) and NDRG1/Cap43, both known as intrinsic hypoxia markers and cancer-associated genes, Karaczyn et al. (2006) observed a clear ASC-dependent regulation via HIF signalling in fibroblasts exposed to nickel or grown to high cell density without medium change, but with adequate oxygen supply. The loss of intracellular ASC triggered the induction of both tumour markers. The involvement of an ASC-dependent, but HIF-independent signalling pathway was also observed, suggesting that other transcription factors beside HIF are ASC-regulated. Karaczyn et al. (2006) concluded that, in addition to low oxygenation, insufficient ASC supply can contribute to the induction of hypoxia-associated proteins via both HIF-dependent and independent mechanisms.

Information so far presented and discussed in this chapter can be summarized in three sentences:

(i) ASC acts as an antioxidant
(ii) ASC is specifically required for the activity of dioxygenases (non-antioxidant role)
(iii) ASC affects gene expression

A model of ASC-dependent gene regulation was recently proposed (Duarte and Lunec 2005), suggesting that ASC can influence gene expression by means of a "redox regulation" mechanism. However, as in many other papers, this redox regulation is ill-defined, and largely remains a big black box in which hiding anything that cannot be explained in terms of molecules and mechanisms (Azzi et al. 2004).

Are both the antioxidant and the non-antioxidant role of ASC just the two sides of the same coin? Do they co-operate in the regulation of gene expression? The identification of ASC as an important regulator of HIF signalling induces to hypothesize that HIF is not just an oxygen sensor, as initially considered, but a general mechanism to perceive and integrate different stimuli, directly connecting them to an array of different cellular response via immediate transcription of "first aid" genes. This can eventually explain why such complex mechanism evolved for the activation of the HIF (and possibly other) transcription factor: why going through a double round of hydroxylation? Probably because all co-substrates of the hydroxylation reaction (oxygen, 2-oxoglutarate, iron and ASC) can bring useful information about cellular health conditions. So the role of HIF signalling is not only oxygen sensing: the Krebs cycle intermediate 2-oxoglutarate can act as an indicator of cellular metabolism (hence the down-regulation of HIF by other Krebs cycle intermediates; Lu et al. 2005). In this context, ASC would have a key role in perceiving oxidative conditions (antioxidant role) and in transducing this signal via the activation of hydroxylases (non-antoxidant role) in a bona fide two-component signalling system operated by the same molecule (Fig. 4.3). This model explains both the prominent role of ASC in the antioxidome of almost all organisms, and its molecular co-evolution with 2-ODDs in their complex catalytic mechanism.

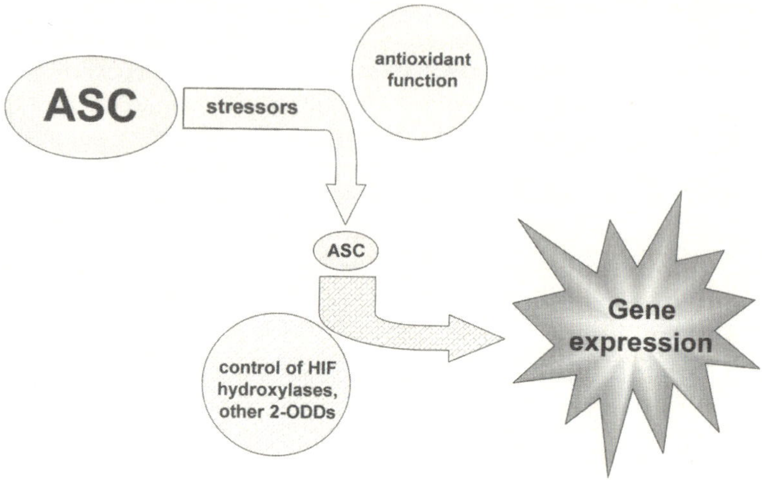

Fig. 4.3 A model of ASC-dependent gene regulation mediated by oxidative stress and dioxygenase activity

4.5 Conclusions

Is it time to change once more the name of vitamin C? Should we call it "ignose" again? Maybe this is not necessary. We are now learning more and more about this tricky molecule and its decisive role in cellular processes, and hopefully next years' research will allow us to overcome present limitations. On the other hand, it is important to avoid sterile discussion on the relative importance of the antioxidant and the non-antioxidant roles, as they are likely to be the Yin and Yang of ASC reactivity. Time (and hard bench work!) will tell.

References

Appelhoff RJ, Tian YM, Raval RR, Turley H, Harris AL, Pugh CW, Ratcliffe PJ, Gleadle JM (2004) Differential function of the prolyl hydroxylases PHD1, PHD2, and PHD3 in the regulation of hypoxia-inducible factor. J Biol Chem 279:38458–38465

Arrigoni O, De Tullio MC (2000) The role of ascorbic acid in plant metabolism: Between gene-directed functions and unpredictable chemical reactions. J Plant Physiol 157:481–488

Arrigoni O, De Tullio MC (2002) Ascorbic acid, much more than just an antioxidant. Biochim Biophys Acta 1569:1–9

Azzi A, Davies KJ, Kelly F (2004) Free radical biology – terminology and critical thinking. FEBS Lett 558:3–6

Berra E, Benizri E, Ginouves A, Volmat V, Roux D, Pouyssegur J (2003) HIF prolyl-hydroxylase 2 is the key oxygen sensor setting low steady-state levels of HIF-1alpha in normoxia. EMBO J 22:4082–4090

Boot-Handford RP, Tuckwell DS, Plumb DA, Rock CF, Poulsom R (2003) A novel and highly conserved collagen (pro(alpha)1(XXVII)) with a unique expression pattern and unusual molecular characteristics establishes a new clade within the vertebrate fibrillar collagen family. J Biol Chem 278:31067–31077.

Carinci F, Pezzetti F, Spina AM, Calmieri A, Laino G, De Rosa A, Farina E, Illiano F, Stabellini G, Pernotti V, Piattelli A (2005) Effect of Vitamin C on pre-osteoblast gene expression. Arch Oral Biol 50:481–496

Catani MV, Rossi A, Costanzo A, Sabatini S, Levrero M, Melino G, Avigliano L (2001) Induction of gene expression via activator protein-1 in the ascorbate protection against UV-induced damage. Biochem J 356:77–85

Challem JJ, Taylor EW (1998) Retroviruses, ascorbate, and mutations, in the evolution of Homo sapiens. Free Rad Biol Med 25:130–132

Chun YS, Kim MS, Park JW (2002) Oxygen-dependent and -independent regulation of HIF1α. J Kor Med Sci 17:581–588

Clifton IJ, McDonough MA, Ehrismann D, Kershaw NJ, Granatino N, Schofield CJ (2006) Structural studies on 2-oxoglutarate oxygenases and related double-stranded beta-helix fold proteins. J Inorg Biochem 100:644–669

Crespo, A, Marti MA, Roitberg AE, Amzel LM, Estrin DA (2006) The catalytic mechanism of peptidylglycine alpha-hydroxylating monooxygenase investigated by computer simulation. J Am Chem Soc 128:12817–12828

Davidson JM, LuValle PA, Zoia O, Quaglino D Jr, Giro M (1997) Ascorbate differentially regulates elastin and collagen biosynthesis in vascular smooth muscle cells and skin fibroblasts by pretranslational mechanisms. J Biol Chem 272:345–352

De Tullio MC (2004) How does ascorbic acid prevent scurvy? A survey of the nonantioxidant functions of vitamin C. In: Asard H, May J, Smirnoff N (eds) Vitamin C, its functions and biochemistry in animals and plants. Bios Scientific Publishers, Oxford, UK, pp 159–172

De Tullio MC, Arrigoni O (2004) Hopes, disillusions and more hopes from vitamin C. Cell Mol Life Sci 61:209–219

Diliberto EJ Jr, Daniels AJ, Viveros OH (1991) Multicompartmental secretion of ascorbate and its dual role in dopamine beta-hydroxylation. Am J Clin Nutr 54:1163S–1172S

Doi M, Nagano A, Nakamura Y (2002) Genome-wide screening by cDNA microarray of genes associated with matrix mineralization by human mesenchymal stem cells in vitro. Biochem Biophys Res Commun 290:381–390

Dozin B, Quarto R, Campanile G, Cancedda R (1992) In vitro differentiation of mouse embryo chondrocytes: requirement for ascorbic acid. Eur J Cell Biol 58:390–394

Duarte TL, Lunec J (2005) When is an antioxidant not an antioxidant? A review of novel actions and reactions of vitamin C. Free Rad Res 39:671–686

Duncan T, Trewick SC, Koivisto P, Bates PA, Lindahl T, Sedgwick B (2002) Reversal of DNA alkylation damage by two human dioxygenases. Proc Natl Acad Sci USA 99:16660–16665

Epstein AC, Gleadle JM, McNeill LA, Hewitson KS, O'Rourke J, Mole DR, Mukherji M, Metzen E, Wilson MI, Dhanda A, Tian YM, Masson N, Hamilton DL, Jaakkola P, Barstead R, Hodgkin J, Maxwell PH, Pugh CW, Schofield CJ, Ratcliffe PJ (2001) C. elegans EGL-9 and mammalian homologs define a family of dioxygenases that regulate HIF by prolyl hydroxylation. Cell 107:43–54

Fain O (2005) Musculoskeletal manifestations of scurvy. Joint Bone Spine 72:124–128

Frei B, England L, Ames B (1989) Ascorbate is an outstanding antioxidant in human blood plasma. Proc Natl Acad Sci USA 86:6377–6381

Gerald D, Berra E, Frapart YM, Chan DA, Giaccia AJ, Mansuy D, Pouyssegur J, Yaniv M, Mechta-Grigoriou F (2004) JunD reduces tumor angiogenesis by protecting cells from oxidative stress. Cell 118:781–794

Greijer AE, van der Groep P, Kemming D, Shvarts A, Semenza GL, Meijer GA, van de Wiel MA, Belien JA, van Diest PJ, van der Wall E (2005) Up-regulation of gene expression by hypoxia is mediated predominantly by hypoxia-inducible factor 1 (HIF-1). J Pathol 206:291–304

Gronke RS, Welsch DJ, VanDusen WJ, Garsky VM, Sardana MK, Stern AM, Friedman PA (1990) Partial purification and characterization of bovine liver aspartyl-β-hydroxylase. J Biol Chem 265: 8558–8565

Halliwell B (2001) Vitamin C and genomic stability. Mutat Res 475:29–35

Halliwell B, Gutteridge JMC (1984) Oxygen toxicity, oxygen radicals, transition metals and disease. Biochem J 219:1–14

Hanauske-Abel HM, Günzler V (1982) A stereochemical concept for the catalytic mechanism of prolyl-hydroxylase: applicability to classification and design of inhibitors. J Theor Biol 94:421–455

Harada S, Matsumoto T, Ogata E (1991) Role of ascorbic acid in the regulation of proliferation in osteoblast-like MC3T3-E1 cells. J Bone Miner Res 6:903–908

Harris AL (2002) Hypoxia – a key regulatory factor in tumour growth. Nat Rev Cancer 2:38–47

Hellwig-Burgel T, Stiehl DP, Wagner AE, Metzen E, Jelkmann W (2005) Review: hypoxia-inducible factor-1 (HIF-1): a novel transcription factor in immune reactions. J Interferon Cytokine Res 25:297–310

Hemila H (1997) Vitamin C supplementation and the common cold – was Linus Pauling right or wrong? Int J Vitam Nutr Res 67:329–335

Hewitson KS, Schofield CJ (2004) The HIF pathway as a therapeutic target. Drug Discov Today 9:704–711

Hickey MM, Simon MC (2006) Regulation of angiogenesis by hypoxia and hypoxia-inducible factors. Curr Top Dev Biol 76:217–257

Hirsila M, Koivunen P, Gunzler V, Kivirikko KI, Myllyharju J (2003) Characterization of the human prolyl 4-hydroxylases that modify the hypoxia-inducible factor. J Biol Chem 278:30772–30780

Hoffart LM, Barr EW, Guyer RB, Bollinger JM Jr, Krebs C (2006) Direct spectroscopic detection of a C-H-cleaving high-spin Fe(IV) complex in a prolyl-4-hydroxylase. Proc Natl Acad Sci USA 103:14738–14743

Holmgren SK, Taylor KM, Bretscher LE, Raines RT (1998) Code for collagen's stability deciphered. Nature 392:666–667

Hosokawa N, Nagata K (2000) Procollagen binds to both prolyl 4-hydroxylase/protein disulfide isomerase and HSP47 within the endoplasmic reticulum in the absence of ascorbate. FEBS Lett 466:19–25

Ivan M, Kondo K, Yang H, Kim W, Valiando J, Ohh M, Salic A, Asara JM, Lane WS, Kaelin WG Jr (2001) HIFα targeted for VHL-mediated destruction by proline hydroxylation: implications for O_2 sensing. Science 292:464–468

Jaakkola P, Mole DR, Tian YM, Wilson MI, Gielbert J, Gaskell SJ, Kriegsheim A, Hebestreit HF, Mukherji M, Schofield CJ, Maxwell PH, Pugh CW, Ratcliffe PJ (2001) Targeting of HIF-alpha to the von Hippel-Lindau ubiquitylation complex by O_2-regulated prolyl hydroxylation. Science 292:468–472

Jenkins E, Moss JB, Pace JM, Bridgewater LC (2005) The new collagen gene COL27A1 contains SOX9-responsive enhancer elements. Matrix Biol 24:177–84

Johnston CS, Cartee GD, Haskell BE (1985) Effect of ascorbic acid nutriture on protein-bound hydroxyproline in guinea pig plasma. J Nutr 115:1089–1093

Jones DT, Trowbridge IS, Harris AL (2006) Effects of transferrin receptor blockade on cancer cell proliferation and hypoxia-inducible factor function and their differential regulation by ascorbate. Cancer Res 66:2749–2756

Karaczyn A, Ivanov S, Reynolds M, Zhitkovich A, Kasprzak KS, Salnikow K (2006) Ascorbate depletion mediates up-regulation of hypoxia-associated proteins by cell density and nickel. J Cell Biochem 97:1025–1035

Kido M, Du L, Sullivan CC, Li X, Deutsch R, Jamieson SW, Thistlethwaite PA (2005) Hypoxia-inducible factor 1-alpha reduces infarction and attenuates progression of cardiac dysfunction after myocardial infarction in the mouse. J Am Coll Cardiol 46:2116–2124

Kivirikko KI, Pihlajaniemi T (1998) Collagen hydroxylases and the protein disulfide isomerase subunit of prolyl 4-hydroxylases. Adv Enzymol Relat Areas Mol Biol 72: 325–398
Knowles HJ, Raval RR, Harris AL, Ratcliffe PJ (2003) Effect of ascorbate on the activity of hypoxia-inducible factor in cancer cells. Cancer Res 63:1764–1768
Koivisto P, Robins P, Lindahl T, Sedgwick B (2004) Demethylation of 3-methylthymine in DNA by bacterial and human DNA dioxygenases. J Biol Chem 279:40470–40474
Kuznetsova AV, Meller J, Schnell PO, Nash JA, Ignacak ML, Sanchez Y, Conaway JW, Conaway RC, Czyzyk-Krzeska MF (2003) von Hippel-Lindau protein binds hyperphosphorylated large subunit of RNA polymerase II through a proline hydroxylation motif and targets it for ubiquitination. Proc Natl Acad Sci USA 100:2706–2711
Littlejohn TK, Takikawa O, Skylas D, Jamie JF, Walker, MJ, Truscott RJ (2000) Expression and purification of recombinant human indoleamine 2, 3-dioxygenase. Protein Expr Purif 19:22–29
Lu H, Dalgard CL, Mohyeldin A, McFate T, Tait AS, Verma A (2005) Reversible inactivation of HIF-1 prolyl hydroxylases allows cell metabolism to control basal HIF-1. J Biol Chem 280:41928–41939
Lubman OY, Ilagan MX, Kopan R, Barrick D (2007) Quantitative dissection of the Notch:CSL interaction: insights into the Notch-mediated transcriptional switch. J Mol Biol 365:577–589
Maeda T, Sepe P, Lahousse S, Tamaki S, Enjoji M, Wands JR, de la Monte SM (2003) Antisense oligodeoxynucleotides directed against aspartyl (asparaginyl) beta-hydroxylase suppress migration of cholangiocarcinoma cells. J Hepatol May 38(5):615–622
Mahmoodian F, Peterkofsky B (1999) Vitamin C deficiency in guinea pigs differentially affects the expression of type IV collagen, laminin, and elastin in blood vessels. J Nutr 129:83–91
Majamaa K, Gunzler V, Hanauske-Abel HM, Myllyla R, Kivirikko KI (1986) Partial identity of the 2-oxoglutarate and ascorbate binding sites of prolyl 4-hydroxylase. J Biol Chem 261:7819–7823
Margittai E, Banhegyi G, Kiss A, Nagy G, Mandl J, Schaff Z, Csala M (2005) Scurvy leads to endoplasmic reticulum stress and apoptosis in the liver of Guinea pigs. J Nutr 135:2530–2534
Marx JL (1985) Oxygen free radicals linked to many diseases. Science 235:529–531
Mayland CR, Bennett MI, Allan K (2005) Vitamin C deficiency in cancer patients. Palliat Med 19:17–20
Moertel CG, Fleming TR, Creagan ET, Rubin J, O'Connell MJ, Ames MM (1985) High-dose vitamin C versus placebo in the treatment of patients with advanced cancer who have had no prior chemotherapy. A randomized double-blind comparison. N Engl J Med 312:137–141
Myllyharju J, Kivirikko KI (2001) Collagen and collagen-related diseases. Ann Med 33:7–21
Myllyla R, Majamaa K, Günzler V, Hanauske-Abel HM, Kivirikko KI (1984) Ascorbate is consumed stoichiometrically in the uncoupled reactions catalyzed by prolyl 4-hydroxylase and lysyl hydroxylase. J Biol Chem 259:5403–5405
Nishikimi M, Yagi K (1991) Molecular basis for the deficiency in humans of gulonolactone oxidase, a key enzyme for ascorbic acid biosynthesis. Am J Clin Nutr 54:1203S–1208S.
Nishikimi M, Fukuyama R, Minoshima S, Shimizu N, Yagi K (1994) Cloning and chromosomal mapping of the human nonfunctional gene for L-gulono-gamma-lactone oxidase, the enzyme for L-ascorbic acid biosynthesis missing in man. J Biol Chem 269:13685–13688
Padayatty SJ, Levine M (2000a) Vitamin C and myocardial infarction: the heart of the matter. Am J Clin Nutr 71:1027–1028
Padayatty SJ, Levine M (2000b) Reevaluation of ascorbate in cancer treatment: emerging evidence, open minds and serendipity. J Am Coll Nutr 19:423–425
Passage E, Norreel JC, Noack-Fraissignes P, Sanguedolce V, Pizant J, Thirion X, Robaglia-Schlupp A, Pellissier JF, Fontes M (2004) Ascorbic acid treatment corrects the phenotype of a mouse model of Charcot-Marie-Tooth disease. Nat Med 10:396–401
Paulding WR, Schnell PO, Bauer AL, Striet JB, Nash JA, Kuznetsova AV, Czyzyk-Krzeska MF (2002) Regulation of gene expression for neurotransmitters during adaptation to hypoxia in oxygen-sensitive neuroendocrine cells. Microsc Res Tech 59:178–187
Pauling L, Cameron E (1979) Cancer and vitamin C. Warner Books, New York

Perrett S, Merle C, Bernocco S, Berland P, Garrone R, Hulmes DJS, Theisen M, Ruggiero F (2001) Unhydroxylated triple helical collagen I produced in transgenic plants provides new clues on the role of hydroxyproline in collagen folding and fibril formation. J Biol Chem 276:43693–43698

Pihlajaniemi T, Myllyla R, Kivirikko KI (1991) Prolyl-4-hydroxylase and its role in collagen synthesis. J Hepatol 13(Suppl 3):S2–S7

Price CE (1966) Ascorbate stimulation of RNA synthesis. Nature 212:481

Prockop DJ, Kivirikko KI (1995) Collagens: molecular biology, diseases, and potentials for therapy. Annu Rev Biochem 64:403–434

Qutub AA, Popel AS (2006) A computational model of intracellular oxygen sensing by hypoxia-inducible factor HIF1α. J Cell Sci 119:3467–3480

Racagni G, Brunello N (1999) Physiology to functionality: The brain and neurotransmitter activity. Int Clin Psychopharmacol 14:S3–S7

Riordan NH, Riordan HD, Meng X, Li Y, Jackson JA (1995) Intravenous ascorbate as a tumor cytotoxic hemoterapeutic agent. Med Hypoth 44:207–213

Rose AM, Bode AM (1993) Antioxidants: an evaluation of ascorbate. FASEB J 7:1135–1142

Salnikow K, Kasprzak KS (2005) Ascorbate depletion: A critical step in nickel carcinogenesis? Environ Health Perspect 113:577–584

Salnikow K, Donald SP, Bruick RK, Zhitkovich A, Phang JM, Kasprzak KS (2004) Depletion of intracellular ascorbate by the carcinogenic metals nickel and cobalt results in the induction of hypoxic stress. J Biol Chem 279:40337–40344

Salo AM, Sipila L, Sormunen R, Ruotsalainen H, Vainio S, Myllyla R (2005) The lysyl hydroxylase isoforms are widely expressed during mouse embryogenesis, but obtain tissue- and cell-specific patterns in the adult. Matrix Biol 25:475–483

Sato H, Takahashi M, Ise H, Yamada A, Hirose S, Tagawa Y, Morimoto H, Izawa A, Ikeda U (2006) Collagen synthesis is required for ascorbic acid-enhanced differentiation of mouse embryonic stem cells into cardiomyocytes. Biochem Biophys Res Commun 342:107–112

Schofield CJ, Ratcliffe PJ (2004) Oxygen sensing by HIF hydroxylases. Nat Rev Mol Cell Biol 5:343–354

Schofield CJ, Ratcliffe PJ (2005) Signalling hypoxia by HIF hydroxylases. Biochem Biophys Res Comm 338:617–626

Schofield CJ, Zhang Z (1999) Structural and mechanistic studies on 2-oxoglutarate-dependent oxygenases and related enzymes. Curr Opin Struct Biol 9:722–731

Seitz G, Gebhardt S, Beck JF, Bohm W, Lode HN, Niethammer D, Bruchelt G (1998) Ascorbic acid stimulates DOPA synthesis and tyrosine hydroxylase gene expression in the human neuroblastoma cell line SK-N-SH. Neurosci Lett 244:33–36

Selkoe D, Kopan R (2003) Notch and presenilin: Regulated intramembrane proteolysis links development and degeneration. Annu Rev Neurosci 26:565–597

Smirnoff N (ed) (2005) Antioxidants and reactive oxygen species in plants. Blackwell, Oxford

Stenflo J, Stenberg Y, Muranyi A (2000) Calcium-binding EGF-like modules in coagulation proteinases: function of the calcium ion in module interactions. Biochim Biophys Acta 1477:51–63

Stone TW, Darlington LG (2002) Endogenous kynurenines as targets for drug discovery and development. Nat Rev Drug Discov 1:609–620

Stone N, Meister A (1962) Function of ascorbic acid in the conversion of proline to collagen hydroxyproline. Nature 194:555–557

Takaluoma K, Hyry M, Lantto J, Sormunen R, Bank RA, Kivirikko KI, Myllyharju J, Soininen R (2006) Tissue-specific changes in the hydroxylysine content and crosslinks of collagens and alterations in fibril morphology in lysyl hydroxylase 1 knock-out mice. J Biol Chem 2006 Dec 29; [Epub ahead of print]

Thoma WJ, Henderson LM (1984) Effect of vitamin C deficiency on hydroxylation of trimethylaminobutyrate to carnitine in the guinea pig. Biochim Biophys Acta 797:136–139

Tschank G, Sanders J, Baringhaus KH, Dallacker F, Kivirikko K, Günzler V (1994) Structural requirements for the utilisation of ascorbate analogues in the prolyl 4-hydroxylase reaction. Biochem J 300:75–79

Tuckerman JR, Zhao Y, Hewitson KS, Tian YM, Pugh CW, Ratcliffe PJ, Mole DR (2004) Determination and comparison of specific activity of the HIF-prolyl hydroxylases. FEBS Lett 576:145–150

Tuderman L, Myllyla R, Kivirikko KI (1977) Mechanism of the prolyl hydroxylase reaction. 1. Role of co-substrates. Eur J Biochem 80:341–348

Vaz FM, Wanders RJA (2002) Carnitine biosynthesis in mammals. Biochem J 361:417–429

von Lintig J, Vogt K (2000) Filling the gap in vitamin A research. Molecular identification of an enzyme cleaving beta-carotene to retinal. J Biol Chem 275:11915–11920

von Lintig J, Hessel S, Isken A, Kiefer C, Lampert JM, Voolstra O, Vogt K (2005) Towards a better understanding of carotenoid metabolism in animals. Biochim Biophys Acta 1740:122–131

Wang J, Pantopoulos K (2005) The pathway for IRP2 degradation involving 2-oxoglutarate-dependent oxygenase(s) does not require the E3 ubiquitin ligase activity of pVHL. Biochim Biophys Acta 1743:79–85

Widner B, Ledochowski M, Fuchs D (2000) Interferon-gamma-induced tryptophan degradation: neuropsychiatric and immunological consequences. Curr Drug Metab 1:193–204

Wu M, Moon HS, Pirskanen A, Myllyharju J, Kivirikko KI, Begley TP (2000) Mechanistic studies on prolyl-4-hydroxylase: the vitamin C requiring uncoupled oxidation. Bioorg Med Chem Lett 10:1511–1514

Xiao G, Cui Y, Ducy P, Karsenty G, Franceschi RT (1997) Ascorbic acid-dependent activation of the osteocalcin promoter in MC3T3-E1 preosteoblasts: Requirement for collagen matrix synthesis and the presence of an intact OSE2 sequence. Mol Endocrin 11:1103–1113

Zhong H, Agani F, Baccala AA, Laughner E, Rioseco-Camacho N, Isaacs WB, Simons JW, Semenza G (1998) Increased expression of hypoxia-inducible factor-1α in rat and human prostate cancer. Cancer Res 58:5280–5284

Zingg JM, Azzi A (2004) Non-antioxidant activities of vitamin E. Curr Med Chem 11:1113–1133

Chapter 5
Vitamin C in Sepsis

John X. Wilson and F. Wu

Abstract Bacterial bloodstream infection causes septic syndromes that range from systemic inflammatory response syndrome (SIRS) and encephalopathy to severe sepsis and septic shock. Microvascular dysfunction, comprising impaired capillary blood flow and arteriolar responsiveness, precedes multiple organ failure. Vitamin C (ascorbate) levels are low in critically ill patients. The impact of ascorbate administered orally is moderate because of its limited bioavailability. However, intravenous injection of ascorbate raises plasma and tissue concentrations of the vitamin and may decrease morbidity. In animal models of polymicrobial sepsis, intravenous ascorbate injection restores microvascular function and increases survival. The protection of capillary blood flow and arteriolar responsiveness by ascorbate may be mediated by inhibition of oxidative stress, modulation of intracellular signaling pathways, and maintenance of homeostatic levels of nitric oxide. Ascorbate scavenges reactive oxygen species (ROS) and also inhibits the NADPH oxidase that synthesizes superoxide in microvascular endothelial cells. The resulting changes in redox-sensitive signaling pathways may diminish endothelial expression of inducible nitric oxide synthase (iNOS), tissue factor and adhesion molecules. Ascorbate also regulates nitric oxide concentration by releasing nitric oxide from adducts and by acting through tetrahydrobiopterin (BH4) to stimulate endothelial nitric oxide synthase (eNOS). Therefore, it may be possible to improve microvascular function in sepsis by using intravenous vitamin C as an adjunct therapy.

Keywords Adjuvant therapy · Ascorbic acid · Bacteremia · Blood pressure · Brain · Capillary blood flow · Critical illness · Dehydroascorbic acid · Endothelial cells · Endotoxin · Infection · Inflammation · Microvascular function · Sepsis · Vitamin C

J.X. Wilson (✉)
Department of Exercise and Nutrition Sciences, University at Buffalo, Buffalo, NY 14214-8028, USA
e-mail: jxwilson@buffalo.edu

O. Stanger (ed.), *Water Soluble Vitamins*, Subcellular Biochemistry 56,
DOI 10.1007/978-94-007-2199-9_5,

5.1 Causes of Morbidity and Mortality in Sepsis

Sepsis is defined as suspected or proven infection plus systemic inflammatory response syndrome (SIRS; e.g., fever, tachycardia, tachypnea, leukocytosis). Severe sepsis is sepsis with organ dysfunction (hypotension, hypoxemia, oliguria, metabolic acidosis, thrombocytopenia, or obtundation). The mainstay of emergency management is goal-directed therapy plus ventilation, antibiotics and possibly activated protein C (Rivers et al. 2001; Russell 2006). Nevertheless, despite best management, mortality remains high. One third to half of all patients with severe sepsis die before leaving the hospital (Rice and Bernard 2005).

Bacterial bloodstream infection, trauma and burn injury are common causes of SIRS. But the pathophysiological progression of septic syndromes has been studied most extensively in the case of Gram-negative organisms. All clinically significant aerobic Gram-negative bacteria contain a macromolecule, endotoxin (e.g. Escherichia coli lipopolysaccharide, LPS), embedded in the outer layer of their cell walls. In patients, LPS unleashes a proinflammatory cascade involving production of cytokines (e.g., interferon-gamma, IFNgamma). This cascade induces septic syndromes that range from SIRS and encephalopathy to severe sepsis and septic shock (Russell 2006; Sharshar et al. 2005; Wilson and Young 2003).

Sepsis is associated with maldistribution of blood flow within organs and loss of microvascular control of tissue oxygenation, as well as with arteriolar hyporesponsiveness to vasoconstrictors and vasodilators (Sakr et al. 2004; Spronk et al. 2002; Tyml et al. 2005) (Fig. 5.1). Red blood cell flow in the microcirculation is altered by sepsis even after fluid resuscitation, as evidenced by the decreased density of perfused capillaries and increased proportion of non-perfused capillaries observed in fluid-resuscitated septic patients with adequate arterial blood oxygenation and cardiac output (Sakr et al. 2004). This increased heterogeneity of capillary blood flow leads to tissue hypoxia because the diffusion distance for oxygen from perfused capillaries to parenchymal cells becomes longer (Goldman et al. 2004). Tissue hypoxia may precipitate organ failure. Indeed, microvascular dysfunction is a strong predictor of death and one-third of severe sepsis patients die of organ failure even when shock is prevented (Rivers et al. 2001, Sakr et al. 2004). Capillary blood flow improves in survivors but fails to improve in non-survivors (Rivers et al. 2001, Sakr et al. 2004). Therefore, septic patients may benefit from adjunctive therapy that targets microvascular dysfunction acutely.

Sepsis and tissue hypoxia are also associated with increased production of reactive oxygen species (ROS) and peroxynitrite that deplete antioxidant molecules and cause oxidative stress (Wilson and Gelb 2002; Wilson and Young 2003). The latter involves oxidization of the body's proteins, lipids and nucleic acids. Furthermore, ROS modulate redox-sensitive intracellular signaling pathways that control the expression of genes critical to the initiation and perpetuation of sepsis; these genes encode proteins that synthesize inflammatory cytokines, increase blood coagulation, and alter endothelial cells' regulation of blood pressure and capillary blood flow (Fig. 5.1). Therefore, patients with sepsis may benefit from adjuvant therapy that counters ROS and peroxynitrite by increasing antioxidant levels acutely.

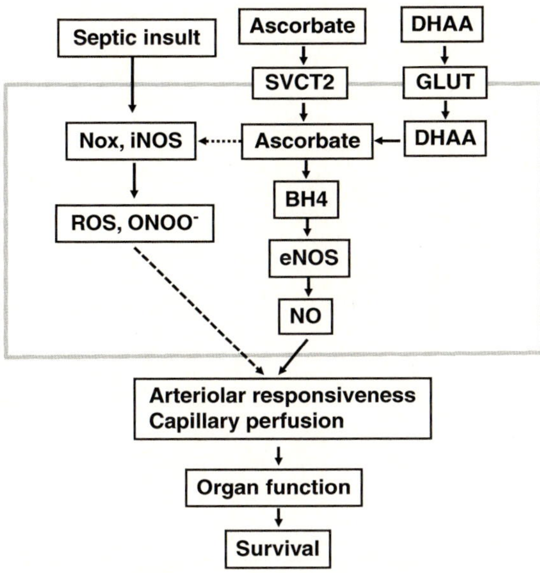

Fig. 5.1 Septic insult and vitamin C alter microvascular function. The *largest rectangle* represents a microvascular endothelial cell. *Arrows with solid lines* indicate stimulation and those with *dotted lines* indicate inhibition. Septic insult increases NADPH oxidase (Nox) and inducible nitric oxide synthase (iNOS) activities, which elevate reactive oxygen species (ROS) and peroxynitrite (ONOO-) production. The ROS and ONOO- impair capillary blood flow and arteriolar responsiveness to vasoconstrictors and vasodilators. Septic insult may also inhibit transport of ascorbate into cells by SVCT2. Dehydroascorbic acid (DHAA) enters the cells through GLUT and then is reduced to ascorbate. Intracellular ascorbate scavenges ROS and ONOO-, while preventing induction of iNOS and Nox enzymes. Ascorbate also acts through tetrahydrobiopterin (BH4), endothelial nitric oxide synthase (eNOS) and nitric oxide adducts, to increase the local concentration of nitric oxide (NO). Therefore, vitamin C may maintain ROS and NO at homeostatic levels that promote normal organ function and survival

5.2 Vitamin C Deficiency in Sepsis

Ascorbate (the predominant form of reduced vitamin C at physiological pH) is an antioxidant that defends against excess ROS. In so doing, it becomes oxidized to ascorbyl free radical (AFR) and dehydroascorbic acid (DHAA). Ascorbate concentrations are subnormal in plasma and leukocytes, whereas plasma AFR concentration in plasma is elevated, in critically ill patients (Borrelli et al. 1996; Galley et al. 1996; Long et al. 2003; Metnitz et al. 1999; Rumelin et al. 2005; Schorah et al. 1996). Ascorbate deficiency may have serious consequences, since prior depletion of ascorbate decreases survival in mice injected with pathogenic bacteria (Gaut et al. 2006).

In most cell types that have been studied, the uptake and reduction of DHAA to ascorbate is not impaired by LPS. DHAA is carried into cells through facilitative hexose transporters such as GLUT1 (Wilson and Dragan 2005) (Fig. 5.1). LPS stimulates DHAA uptake through the 55 kDa isoform of GLUT1 in macrophages (May et al. 2005) and also upregulates this transporter in endothelial cells (Spolarics et al. 1996). Exposure to LPS in vivo accelerates DHAA reduction in mouse liver (Kuo et al. 2005). Similarly, preincubation with LPS and IFNgamma (LPS+IFNgamma) increases DHAA reduction in astrocytes permeabilized with saponin, although this septic insult decreases the activity of the 45 kDa isoform of GLUT1 in astrocytes with intact plasma membrane (i.e., absent saponin; Wilson and Dragan 2005). Taken together, these results suggest that septic insult may accelerate the rate at which extracellular DHAA is taken up and reduced to ascorbate in cells that express 55 kDa GLUT1. Indeed, injection of a sublethal dose of LPS increases ascorbate concentration in the liver, kidney, adrenal gland, and heart of mice (Kuo et al. 2005).

One cause of ascorbate deficiency in patients may be hospital food that contains too little vitamin C (McErlain et al. 2001). Even when patients are nourished intravenously, the parenteral nutrition solutions may generate hydrogen peroxide that degrades ascorbate (Dupertuis et al. 2005; Knafo et al. 2005; Laborie et al. 2002). A second cause of vitamin C deficiency in critically ill patients is an increased requirement for ascorbate (Baines and Shenkin 2002). The basis for this increased requirement may be oxidation of ascorbate by excess ROS. However, changes in cellular mechanisms of vitamin C transport and metabolism may also be important.

Inflammation may decrease the transport of ascorbate into cells. The underlying mechanism involves the ascorbate-specific transporters, SVCT1 and SVCT2, through which ascorbate enters most cell types (Wilson and Dragan 2005) (Fig. 5.1). Inflammatory cytokines (tumor necrosis factor-alpha, interleukin-1beta) inhibit ascorbate uptake in human endothelial cells that spontaneously express SVCT2 (Seno et al. 2004). Moreover, incubation with LPS+IFNgamma inhibits SVCT2-mediated ascorbate uptake in cultured cerebral astrocytes, thereby decreasing the rate of intracellular ascorbate accumulation from extracellular vitamin C in this abundant brain cell type (Korcok et al. 2002).

5.3 Adjuvant Therapy with Intravenous Ascorbate

Ingestion of ascorbic acid, ascorbate or its reversibly oxidized metabolite, DHAA, raises the plasma concentration of ascorbate in normal human subjects. The impact of oral vitamin C on plasma ascorbate concentration in these subjects is moderate, however, because of its limited bioavailability (Wilson 2005). In contrast, intravenous ascorbate can raise plasma ascorbate concentration markedly (Padayatty et al. 2004). Clinical consequences arise from the limited bioavailability of oral ascorbate. For example, when oral and intravenous routes of ascorbate administration (500 mg/day for 30 days) were compared in sedentary men, only intravenous ascorbate improved endothelial function as indicated by flow-mediated vasodilatation (Eskurza et al. 2004).

There is evidence that ascorbate may improve outcome in critically ill patients, who either are or are at risk of becoming septic. In a randomized, prospective, double-blind, placebo-controlled trial with 216 critically ill patients, 28-day mortality was decreased in the patients who received combined ascorbate and vitamin E by enteral feeding compared to those who did not (Crimi et al. 2004). A randomized, prospective clinical trial of 595 critically ill surgical patients found that a treatment combining ascorbate (1,000 mg q8h by intravenous injection) and vitamin E (1,000 IU q8h by naso- or orogastric tube) decreased relative risk of pulmonary morbidity and multiple organ failure and also shortened the duration of mechanical ventilation and intensive care unit stay (Nathens et al. 2002). Yet another randomized, prospective study reported decreased morbidity for severely burned patients who were infused with high-dose ascorbate (Tanaka et al. 2000).

A great deal has been learned about ascorbate therapy in animal and cell culture models of septic syndromes. Ascorbate prevents hypotension and lung injury in LPS-injected animals (Dwenger et al. 1994; Shen et al. 2005); it prolongs survival in experimental bacteremia (Gaut et al. 2006); and it improves capillary blood flow, arteriolar responsiveness, arterial blood pressure and survival in polymicrobial sepsis models (Armour et al. 2001; Tyml et al. 2005; Wu et al. 2003, 2004).

Some of the most clinically relevant models of polymicrobial sepsis involve cecal ligation and puncture (CLP). Similar to the changes observed in septic patients, CLP increases oxidative stress markers and decreases ascorbate concentration in plasma and tissue (Armour et al. 2001; Tyml et al. 2005; Wu et al. 2003). Injection of ascorbate bolus (200 mg/kg i.v.) increases survival in CLP mice (Wu et al. 2004). Survival rates at 24 h post-CLP are 9 and 65%, respectively, in vehicle-injected and ascorbate-injected mice. The protective effect of the vitamin does not appear to be due to inhibition of bacterial replication at the site of infection, because the number of bacterial colony forming units in peritoneal lavage fluid at 6 h post-CLP does not differ between vehicle- and ascorbate-injected mice (Wu et al. 2004).

5.4 Effects of Vitamin C on Capillary Blood Flow

CLP causes tissue hypoxia due to microvascular dysfunction (Goldman et al. 2004; Miyaji et al. 2003). Similar to what was reported for septic patients, capillary blood flow is altered by CLP even after volume resuscitation, as evidenced by the decreased density of perfused capillaries and increased proportion of nonperfused capillaries in skeletal muscle (Armour et al. 2001; Tyml et al. 2005). Maldistribution of capillary blood flow at 24 h post-CLP is prevented by ascorbate even when intravenous injection of the vitamin is delayed until 6 h post-CLP. Moreover, the increased proportion of nonperfused capillaries at 48 h post-CLP is prevented even when the ascorbate injection is delayed until 24 h post-CLP (Tyml et al. 2005). These results show that ascorbate can reverse microvascular dysfunction after the onset of sepsis.

The restoration of capillary blood flow by a single intravenous ascorbate bolus (76 mg/kg) lasts at least 18–24 h in CLP rats (Tyml et al. 2005). Microvascular endothelial cells in culture can accumulate a high level of intracellular ascorbate (e.g., 4–16 mmol/L) that persists longer than does extracellular ascorbate (Wilson et al. 1996; Wu et al. 2007). Taken together, these findings suggests that prophylactic, intravenous injection of vitamin C may elevate ascorbate concentration within endothelial cells long after plasma ascorbate concentration has returned to pre-injection levels.

High concentrations of intracellular ascorbate affect nitric oxide through several mechanisms. First, millimolar concentrations of ascorbate prevent consumption of nitric oxide by superoxide (Jackson et al. 1998) (Fig. 5.2). Secondly, ascorbate releases nitric oxide from adducts (e.g., by ascorbate's reduction of S-nitrosothiols; May 2000). Thirdly, ascorbate in endothelial cells acts through (6R)-5,6,7,8-tetrahydro-L-biopterin (BH4) to stimulate nitric oxide synthesis by nitric oxide synthases (NOS) (Kim et al. 2006; Smith et al. 2002) (Fig. 5.1). This

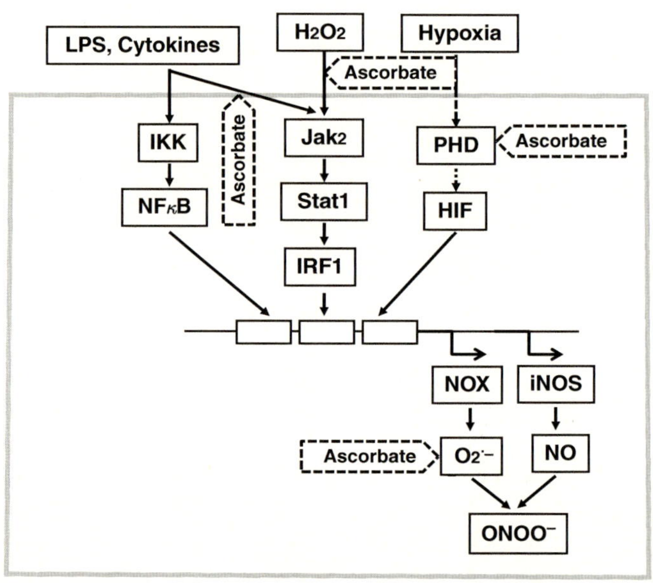

Fig. 5.2 Redox-sensitive signaling pathways by which septic insult and ascorbate may alter inflammatory responses in microvascular endothelial cells. The *largest rectangle* represents an endothelial cell. *Arrows with solid lines* indicate stimulation and those with *dotted lines* indicate inhibition. Lipopolysaccharide (LPS), cytokines and hydrogen peroxide (H_2O_2) stimulate activation of signaling pathways that trigger expression of inducible nitric oxide synthase (iNOS), NADPH oxidase (Nox) and other inflammatory mediators. Activation of the Jak2-Stat1-IRF1 pathway is inhibited by ascorbate. Hypoxia increases hypoxia inducible factor 1 (HIF-1), by inhibiting HIF-1 proly-hydroxylase (PHD), and thereby induces expression of sepsis-associated genes. Ascorbate increases the activity of PHD. Further, ascorbate acts as a scavenger of reactive oxygen species (e.g. superoxide, O_2.-) to decrease peroxynitrite (ONOO-) concentration and regulate local nitric oxide (NO) at homeostatic levels

action distinguishes ascorbate from other low molecular weight antioxidants such as N-acetylcysteine (Schneider et al. 2005). As described later in this section, homeostatic levels of nitric oxide prevent platelet and leukocyte adhesion to the microvascular endothelium, and thereby improve capillary blood flow in sepsis.

Septic insult increases the activity and expression levels of NADPH oxidases in vascular cells, as has been shown in blood vessels of CLP animals and in endothelial cell cultures (Wu et al. 2005, 2007; Yu et al. 2006). Endothelial NADPH oxidases are composed of several combinations of regulatory subunits (p22phox, p40phox, p47phox, p67phox) and catalytic subunits (Nox1, Nox2, Nox4) that synthesize intracellular superoxide (Li and Shah 2004).

Endothelial NADPH oxidases synthesize intracellular superoxide, which reacts to form other ROS (e.g., dismutation of superoxide produces hydrogen peroxide) that induce prolonged redox signaling effects (Ardanaz and Pagano 2006; Cai 2005; Griendling et al. 2000). Increases in intracellular ROS may stimulate expression of NADPH oxidase subunits that subsequently assemble and produce more ROS in endothelial cells (Jacobi et al. 2005) (Fig. 5.2). By scavenging the signal ROS that are produced in microvascular endothelial cells exposed to septic insult (Wu et al. 2002), intracellular ascorbate may interrupt this feed-forward mechanism and thereby prevent expression of NADPH oxidase subunits. Indeed, Wu et al. (2007) observed that preincubation of microvascular endothelial cells with high concentrations of ascorbate (500 and 1,000 umol/L) prevented the induction by LPS+IFNgamma or hydrogen peroxide of endothelial NADPH oxidase activity and p47phox expression. The effect on p47phox expression was mediated by the Jak2/Stat1/IRF1 signaling pathway; ascorbate prevented activation of this pathway by LPS+IFNgamma and hydrogen peroxide (Wu et al. 2007) (Fig. 5.2).

Endothelial NADPH oxidase-derived ROS increase the expression of inducible nitric oxide synthase (iNOS) in microvascular endothelial cells exposed to septic insult (Wu et al. 2001, 2002, 2005). Ascorbate acts within endothelial cells to prevent induction of iNOS by CLP, bacterial endotoxin and inflammatory cytokines (Wu et al. 2002, 2003, 2004). A preliminary report suggests that ascorbate's suppression of NADPH oxidase activity mediates this inhibition of iNOS expression (Wu et al. 2005). Ascorbate injection also prevents induction by CLP of iNOS in liver and this effect may not be confined to endothelial cells (Kim and Lee 2004).

Clearly, ascorbate modifies inflammatory responses. The way this is accomplished, in addition to the mechanisms described above, may involve suppression of gene expression that is dependent on the transcription factor, hypoxia inducible factor-1 (HIF-1) (Fig. 5.2). HIF-1 is an alpha-beta heterodimer that stimulates expression of iNOS, GLUT1 and other inflammatory mediators (Wenger 2002; Vissers et al. 2007). Normally the HIF-1alpha subunit has a short half-life because it is covalently modified by HIF-1 prolyl-hydroxlase (PHD), which is a ferrous iron (Fe^{2+})-dependent enzyme that targets HIF-1alpha for proteolysis by the ubiquitin-proteosome system. NADPH oxidase-derived ROS inhibit PHD activity (Goyal et al. 2004). Inhibition by ascorbate of endothelial NADPH oxidase (Wu et al. 2007) may thereby preserve PHD activity. Further, experiments in cells exposed to oxidative stress indicate that intracellular ascorbate reduces ferric iron (Fe^{3+}) to

Fe^{2+}, which then activates PHD to modify HIF-1alpha for proteolysis (Knowles et al. 2003). In various cell types, including human umbilical vein endothelial cells, ascorbate has been shown to decrease HIF-1 levels and inhibit HIF-1alpha–dependent expression of inflammatory mediators (Lu et al. 2005; Vissers et al. 2007).

NADPH oxidase-derived superoxide avidly interacts with NOS-derived nitric oxide to form peroxynitrite. Further, peroxynitrite oxidizes the essential NOS cofactor. BH4. The lack of BH4 makes each NOS isoform a superoxide-producing (i.e., uncoupled) enzyme. Therefore, septic activation of NADPH oxidase may lead to a local deficiency of nitric oxide and excesses of peroxynitrite and ROS in endothelial cells. These intracellular changes may cause plugging of microvessels that subsequently slows capillary blood flow, as is described below.

The plugging of microvessels may be an important factor in septic impairment of capillary flow, because adhesion of leukocytes, platelets and red blood cells to the endothelium is elevated in experimental sepsis (Cerwinka et al. 2003; Merx et al. 2005; Ondiveeran and Fox-Robichaud 2004; Secor et al. 2010; Singer et al. 2006; Sundrani et al. 2000). Local application of exogenous nitric oxide prevents adhesion of blood cells to the venular endothelium during experimental sepsis (Cerwinka et al. 2002). Similarly, ascorbate prevents the increase in leukocyte adherence that is induced by LPS in mice (De la Fuente and Victor 2001). The mechanism underlying this effect on cell adhesion may involve ascorbate inhibiting NADPH oxidase activity, p47phox expression and iNOS expression, thereby preventing the intracellular scarcity of nitric oxide and excess of ROS and peroxynitrite.

The constitutive NOS isoform in endothelial cells is endothelial nitric oxide synthase (eNOS). During ascorbate therapy in sepsis, this enzyme may play a role in rapid restoration of capillary blood flow, which may precede de novo synthesis of p47phox and iNOS (Secor et al. 2010). Moreover, ascorbate increases synthesis of nitric oxide by eNOS in endothelial cells and the underlying mechanism, at least in part, appears to involve enhancing BH4 levels to prevent eNOS uncoupling (Forstermann and Munzel 2006; Kim et al. 2006; Mittermayer et al. 2005).

Superoxide stimulates expression of cell surface intercellular adhesion molecule 1 (ICAM-1) in microvascular endothelial cells (Fan et al. 2002; Li et al. 2005). ICAM-1 mediates adhesion of leukocytes to the endothelium and may thereby impair the microcirculation. Since ascorbate inhibits superoxide production in microvascular endothelial cells exposed to septic insult (Wu et al. 2002, 2007), further research is required to determine if this action suppresses ICAM-1 expression and leukocyte plugging of venules.

Sepsis decreases the deformability of red blood cells (Bateman et al. 2001) and this may contribute to plugging of microvessels. There is evidence that antioxidants (e.g., tirilazad) can maintain red blood cell deformability in stressed animals (Chmiel et al. 2001). Further, it has been shown that aminoguanidine treatment, which probably lowers the concentration of nitric oxide metabolites (e.g., peroxynitrite), prevents the decrease in red blood cell deformability caused by CLP (Bateman et al. 2001). Ascorbate prevents overproduction of nitric oxide metabolites in vivo (Wu et al. 2003, 2004). Thus, it is plausible that ascorbate may maintain red blood cell deformability, although further research is necessary to test this hypothesis.

Coagulability is another potential cause of capillary blood flow impairment that may be modulated by ascorbate therapy. ROS stimulate induction of tissue factor in endothelial cells (Salvemini and Cuzzocrea 2002). Tissue factor (formerly known as thromboplastin) is a transmembrane glycoprotein that functions as the primary cellular initiator of blood coagulation in vivo (Steffel et al. 2006). Because endothelial cells and monocytes do not express tissue factor under physiological conditions, there is no appreciable contact of cellular tissue factor with the circulating blood. However, these cells do express tissue factor when exposed to inflammatory cytokines and this contributes to coagulopathy in sepsis (Steffel et al. 2006). Coagulation is activated rapidly after septic insult and results in diffuse microvascular clot formation that may disrupt blood flow in capillary beds (Laudes et al. 2002; Salvemini and Cuzzocrea 2002). Similar to the effect observed after injection of ascorbate (Wu et al. 2004), injection of tissue factor pathway inhibitor increased survival in a CLP model of sepsis (Opal et al. 2001).

In critically ill patients, vasodilators transiently increase the proportion of perfused capillaries (Spronk et al. 2002). Although ascorbate enhances arteriolar vasodilatation in some patient populations (Schneider et al. 2005; Tousoulis et al. 2005), whether vasodilatation is a direct cause of ascorbate's restoration of capillary blood flow in sepsis is unknown.

5.5 Effects of Vitamin C on Arteriolar Responsiveness and Hypotension

Infusion of ascorbate reverses arteriolar hyporesponsiveness to vasoconstrictors (norepinephrine, angiotensin, vasopressin) in patients who have inflammatory disease or have been injected with LPS (Ferlitsch et al. 2005; Pleiner et al. 2003). Infusion of ascorbate or BH4 also prevents inhibition by LPS of endothelium-dependent vasodilatation responses (assessed as changes in forearm blood flow) to acetylcholine in human subjects, and this effect of ascorbate is associated with a 32% increase in plasma BH4 concentration (Mittermayer et al. 2005). It is likely ascorbate raises BH4 levels within the endothelial cells of arterioles and thereby enhances acetylcholine-stimulated synthesis by eNOS of the vasodilator, nitric oxide.

Bacterial bloodstream infection can lead to septic shock in patients (Stoclet et al. 1999). Similarly, in animals made septic by CLP, increased heterogeneity of capillary blood flow is followed by the development of arterial hypotension (Tyml et al. 2005). The hypotension may be caused by impairment of myocardial function (Piper et al. 1999) and arteriolar responsiveness to vasoconstrictors (Wu et al. 2003, 2004). Arteriolar responsiveness and arterial blood pressure are higher in CLP rats injected intravenously with ascorbate, compared to those injected with vehicle, when those parameters are measured at 18–24 h post-injection (Armour et al. 2001,

Tyml et al. 2005). However, ascorbate improves only arteriolar responsiveness and not hypotension in CLP mice at 3 h post-injection, indicating that the initial target of ascorbate protection is arteriolar rather than myocardial function (Wu et al. 2004). Arteriolar vasoconstriction and arterial blood pressure pressor responses to norepinephrine and angiotensin II are inhibited in mice at 6 h post-CLP (Wu et al. 2003, 2004). Intravenous ascorbate and iNOS gene deficiency (iNOS$^{-/-}$ mice) are equally effective at preventing this CLP-induced impairment of arteriolar responsiveness. iNOS synthesizes abundant superoxide (when uncoupled) or nitric oxide (when coupled) that reacts with superoxide to form the strong oxidant, peroxynitrite. This process may lead to impairment of vascular function. For example, iNOS expression in omental arteries of septic patients is associated with a subnormal vasoconstrictor response to norepinephrine that is normalized by the NOS inhibitor, N(G)-methyl-l-arginine (L-NMMA) (Stoclet et al. 1999). A phase III clinical trial of L-NMMA in patients with septic shock was terminated early because of increased mortality, but post hoc analysis indicated an overall survival benefit for low-dose L-NMMA (Lopez et al. 2004). The adverse effect of high-dose L-NMMA may be due to inhibition of endothelial cells' eNOS that is necessary for vascular homeostasis (Wang et al. 2004).

5.6 Vitamin C in Sepsis-Associated Encephalopathy

Most hospital patients with bacteremia show neurologic symptoms ranging from lethargy to coma (Wilson and Young 2003). Microcirculatory dysfunction in the brain may contribute to sepsis-associated encephalopathy, since impairment of cerebral activation flow coupling precedes changes in evoked potentials in LPS-injected rats (Rosengarten et al. 2006). Arteriolar responsiveness is depressed in the cerebral circulation of patients with severe sepsis (Terborg et al. 2001). Therefore, the hypothesis that the effects of ascorbate on capillaries and arterioles (reviewed above) may ameliorate encephalopathy in sepsis is plausible although untested.

Sepsis is also associated with ascorbate depletion, excessive inflammation (i.e., NADPH oxidase and NOS activation), and glutamatergic excitoxicity in brain parenchymal cells (Sharshar et al. 2005; Wilson and Young 2003; Wilson and Dragan 2005). The depletion of ascorbate may contribute to the inflammation and excitoxicity, since lowering intracellular ascorbate concentration in cerebral astrocytes increases iNOS expression and decreases glutamate clearance by these cells (Korcok et al. 2002).

Whereas normal neurotransmission requires high ascorbate and low DHAA concentrations in brain extracellular fluid, there is evidence that septic insult inhibits the clearance of extracellular DHAA and export of ascorbate by astrocytes (Wilson and Dragan 2005). Normally, glutamate released from active neurons stimulates astrocytes to export ascorbate, which the astrocytes synthesize by taking up and reducing DHAA (Wilson 2005). The ascorbate that enters the extracellular fluid is then accumulated by neurons and may protect them from the toxic effects of excess glutamate. Additionally, the ascorbate released from glutamate-stimulated astrocytes may potentiate nitric oxide-dependent vasodilatation and contribute to activation flow coupling. In other words, a demand-driven component of ascorbate export from astrocytes may occur just in time and in the right place to reach active neurons and also redirect blood flow to those neurons. However, cell culture experiments have shown that septic insult inhibits the replacement of extracellular DHAA with ascorbate by astrocytes (Wilson and Dragan 2005). The underlying mechanism involves decreased activities of astrocytes' high-affinity glutamate transporters and 45-kDa GLUT1 transporter (Wilson and Dragan 2005); the latter transporter mediates DHAA uptake in astrocytes and is distinct from the 55-kDa isoform that is common in other cell types (Korcok et al. 2003). The decrease in extracellular ascorbate concentration in septic brain may result from this derangement of astrocyte function. Therefore, interventions that restore the ability of astrocytes to replace extracellular DHAA with ascorbate should be investigated as adjunctive treatments to the current antimicrobial and supportive therapy of sepsis.

5.7 Other Issues Concerning Vitamin C in Sepsis

There have been no studies published that identified which redox state of injectable vitamin C (i.e., ascorbate or DHAA) is most effective for treating sepsis. The maximal transport capacity for DHAA uptake exceeds that for ascorbate uptake in most mammalian cell types. Therefore, when present in the extracellular fluid at identical concentrations and under physiological conditions, DHAA increases intracellular ascorbate concentration more rapidly than does ascorbate (Wilson and Dragan 2005). It is not known if intravenous DHAA is superior to ascorbate for raising intracellular ascorbate in sepsis, but it is in experimental ischemia-reperfusion of heart and liver (De Tata et al. 2005; Guaiquil et al. 2004). Furthermore, injection of DHAA improves neurologic outcome in mice subjected to cerebral ischemia-reperfusion, whereas equal doses of ascorbate do not (Huang et al. 2001). Intravenous injection of DHAA increases cerebral ascorbate concentration (Mack et al. 2006) and the underlying mechanism likely involves uptake and reduction of DHAA to ascorbate by brain cells.

Maintenance of neutrophils' antimicrobial function is an important consideration during infection. It appears that ascorbate does not impair the respiratory burst in neutrophils. Bolus intravenous injection of 2 g ascorbate in human subjects does not alter superoxide production by neutrophils (Ellis et al. 2000). Similarly, incubation of neutrophils with either ascorbate or DHAA, to raise intracellular ascorbate concentration, does not inhibit the production of extracellular ROS elicited when these cells are subsequently challenged by stimuli such as E. coli (Carr and Frei 2002; Sharma et al. 2004). The preservation of neutrophils' ability to produce ROS during exposure to high extracellular ascorbate levels may allow these cells to destroy pathological microorganisms and thus defend against infection.

Finally, the safety of high-dose vitamin C therapy is controversial. For vitamin C administered orally, doses up to 2,000 mg/day appear to be safe for the general adult population (Hathcock et al. 2005). But less is known about the safety of vitamin C injections. Pharmacokinetic modeling indicates that bolus intravenous injection of 10 g ascorbate raises plasma ascorbate to a peak of 5,000 umol/L and keeps it above 2,000 umol/L for 2 h in human subjects (Padayatty et al. 2004). In vitro experiments showed that these concentrations of ascorbate are not lethal to normal human cells (Chen et al. 2005). Similarly, repeated intravenous injections of 750 mg/day or 7,500 mg/day of ascorbate for 6 days in healthy volunteers do not induce a pro-oxidant change in plasma markers of oxidative stress (Muhlhofer et al. 2004). However, injection of ascorbate 2 h before major surgery increases oxidative modification of plasma lipids in venous blood during the ischemic phase of surgery, and the consequences of this lipid modification are unknown (Bailey et al. 2006).

References

Ardanaz N, Pagano PJ (2006) Hydrogen peroxide as a paracrine vascular mediator: Regulation and signaling leading to dysfunction. Exp Biol Med (Maywood) 231:237–251

Armour J, Tyml K, Lidington D, Wilson JX (2001) Ascorbate prevents microvascular dysfunction in the skeletal muscle of the septic rat. J Appl Physiol 90:795–803

Bailey DM, Raman S, McEneny J, Young IS, Parham KL, Hullin DA, Davies B, McKeeman G, McCord JM, Lewis MH (2006) Vitamin C prophylaxis promotes oxidative lipid damage during surgical ischemia-reperfusion. Free Radic Biol Med 40:591–600

Baines M, Shenkin A (2002) Lack of effectiveness of short-term intravenous micronutrient nutrition in restoring plasma antioxidant status after surgery. Clin Nutr 21:145–150

Bateman RM, Jagger JE, Sharpe MD, Ellsworth ML, Mehta S, Ellis CG (2001) Erythrocyte deformability is a nitric oxide-mediated factor in decreased capillary density during sepsis. Am J Physiol Heart Circ Physiol 280:H2848–H2856

Borrelli E, Roux-Lombard P, Grau GE, Girardin E, Ricou B, Dayer J, Suter PM (1996) Plasma concentrations of cytokines, their soluble receptors, and antioxidant vitamins can predict the development of multiple organ failure in patients at risk. Crit Care Med 24:392–397

Cai H (2005) Hydrogen peroxide regulation of endothelial function: origins, mechanisms, and consequences. Cardiovasc Res 68:26–36

Carr AC, Frei B (2002) Human neutrophils oxidize low-density lipoprotein by a hypochlorous acid-dependent mechanism: the role of vitamin C. J Biol Chem 383:627–636

Cerwinka WH, Cooper D, Krieglstein CF, Feelisch M, Granger DN (2002) Nitric oxide modulates endotoxin-induced platelet-endothelial cell adhesion in intestinal venules. Am J Physiol Heart Circ Physiol 282:H1111–H1117

Cerwinka WH, Cooper D, Krieglstein CF, Ross CR, McCord JM, Granger DN (2003) Superoxide mediates endotoxin-induced platelet-endothelial cell adhesion in intestinal venules. Am J Physiol Heart Circ Physiol 284:H535–H541

Chen Q, Espey MG, Krishna MC, Mitchell JB, Corpe CP, Buettner GR, Shacter E, Levine M (2005) Pharmacologic ascorbic acid concentrations selectively kill cancer cells: action as a pro-drug to deliver hydrogen peroxide to tissues. Proc Natl Acad Sci USA 102:13604–13609

Chmiel B, Grabowska-Bochenek R, Piskorska D (2001) Tirilazad mesylate improve water immersion stress induced decrease in erythrocyte deformability in the rats. Med Sci Monit 7:38–41

Crimi E, Liguori A, Condorelli M, Cioffi M, Astuto M, Bontempo P, Pignalosa O, Vietri MT, Molinari AM, Sica V, Della Corte F, Napoli C (2004) The beneficial effects of antioxidant supplementation in enteral feeding in critically ill patients: a prospective, randomized, double-blind, placebo-controlled trial. Anesth Analg 99:857–863

De la Fuente M, Victor VM (2001) Ascorbic acid and N-acetylcysteine improve in vitro the function of lymphocytes from mice with endotoxin-induced oxidative stress. Free Radic Res 35:73–84

De Tata V, Brizzi S, Saviozzi M, Lazzarotti A, Fierabracci V, Malvaldi G, Casini A (2005) Protective role of dehydroascorbate in rat liver ischemia-reperfusion injury. J Surg Res 123:215–221

Dupertuis YM, Ramseyer S, Fathi M, Pichard C (2005) Assessment of ascorbic acid stability in different multilayered parenteral nutrition bags: critical influence of the bag wall material. JPEN J Parenter Enteral Nutr 29:125–130

Dwenger A, Pape HC, Bantel C, Schweitzer G, Krumm K, Grotz M, Lueken B, Funck M, Regel G (1994) Ascorbic acid reduces the endotoxin-induced lung injury in awake sheep. Eur J Clin Invest 24:229–235

Ellis GR, Anderson RA, Lang D, Blackman DJ, Morris RH, Morris-Thurgood J, McDowell IF, Jackson SK, Lewis MJ, Frenneaux MP (2000) Neutrophil superoxide anion–generating capacity, endothelial function and oxidative stress in chronic heart failure: effects of short- and long-term vitamin C therapy. J Am Coll Cardiol 36:1474–1482

Eskurza I, Monahan KD, Robinson JA, Seals DR (2004) Effect of acute and chronic ascorbic acid on flow-mediated dilatation with sedentary and physically active human ageing. J Physiol 556:315–324

Fan J, Frey RS, Rahman A, Malik AB (2002) Role of neutrophil NADPH oxidase in the mechanism of tumor necrosis factor-alpha -induced NF-kappa B activation and intercellular adhesion molecule-1 expression in endothelial cells. J Biol Chem 277:3404–3411

Ferlitsch A, Pleiner J, Mittermayer F, Schaller G, Homoncik M, Peck-Radosavljevic M, Wolzt M (2005) Vasoconstrictor hyporeactivity can be reversed by antioxidants in patients with advanced alcoholic cirrhosis of the liver and ascites. Crit Care Med 33:2028–2033

Forstermann U, Munzel T (2006) Endothelial nitric oxide synthase in vascular disease: from marvel to menace. Circulation 113:1708–1714

Galley HF, Davies MJ, Webster NR (1996) Xanthine oxidase activity and free radical generation in patients with sepsis syndrome. Crit Care Med 24:1649–1653

Gaut JP, Belaaouaj A, Byun J, Roberts LJ 2nd, Maeda N, Frei B, Heinecke JW (2006) Vitamin C fails to protect amino acids and lipids from oxidation during acute inflammation. Free Radic Biol Med 40:1494–1501

Goldman D, Bateman RM, Ellis CG (2004) Effect of sepsis on skeletal muscle oxygen consumption and tissue oxygenation: interpreting capillary oxygen transport data using a mathematical model. Am J Physiol Heart Circ Physiol 287:H2535–H2544

Goyal P, Weissmann N, Grimminger F, Hegel C, Bader L, Rose F, Fink L, Ghofrani HA, Schermuly RT, Schmidt HH, Seeger W, Hanze J (2004) Upregulation of NAD(P)H oxidase 1 in hypoxia activates hypoxia-inducible factor 1 via increase in reactive oxygen species. Free Radic Biol Med 36:1279–1288

Griendling KK, Sorescu D, Lassegue B, Ushio-Fukai M (2000) Modulation of protein kinase activity and gene expression by reactive oxygen species and their role in vascular physiology and pathophysiology. Arterioscler Thromb Vasc Biol 20:2175–2183

Guaiquil VH, Golde DW, Beckles DL, Mascareno EJ, Siddiqui MA (2004) Vitamin C inhibits hypoxia-induced damage and apoptotic signaling pathways in cardiomyocytes and ischemic hearts. Free Radic Biol Med 37:1419–1429

Hathcock JN, Azzi A, Blumberg J, Bray T, Dickinson A, Frei B, Jialal I, Johnston CS, Kelly FJ, Kraemer K, Packer L, Parthasarathy S, Sies H, Traber MG (2005) Vitamins E and C are safe across a broad range of intakes. Am J Clin Nutr 81:736–745

Huang J, Agus DB, Winfree CJ, Kiss S, Mack WJ, McTaggart RA, Choudhri TF, Kim LJ, Mocco J, Pinsky DJ, Fox WD, Israel RJ, Boyd TA, Golde DW, Connolly ES Jr (2001) Dehydroascorbic acid, a blood-brain barrier transportable form of vitamin C, mediates potent cerebroprotection in experimental stroke. Proc Natl Acad Sci USA 98:11720–11724

Jackson TS, Xu A, Vita JA, Keaney JF Jr (1998) Ascorbate prevents the interaction of superoxide and nitric oxide only at very high physiological concentrations. Circ Res 83:916–922

Jacobi J, Kristal B, Chezar J, Shaul SM, Sela S (2005) Exogenous superoxide mediates pro-oxidative, proinflammatory, and procoagulatory changes in primary endothelial cell cultures. Free Radic Biol Med 39:1238–1248

Kim JY, Lee SM (2004) Effect of ascorbic acid on hepatic vasoregulatory gene expression during polymicrobial sepsis. Life Sci 75:2015–2026

Kim HJ, Lee SI, Lee DH, Smith D, Jo H, Schellhorn HE, Boo YC (2006) Ascorbic acid synthesis due to L-gulono-1,4-lactone oxidase expression enhances NO production in endothelial cells. Biochem Biophys Res Commun 345:1657–1662

Knafo L, Chessex P, Rouleau T, Lavoie JC (2005) Association between hydrogen peroxide-dependent byproducts of ascorbic acid and increased hepatic acetyl-CoA carboxylase activity. Clin Chem 51:1462–1471

Knowles HJ, Raval RR, Harris AL, Ratcliffe PJ (2003) Effect of ascorbate on the activity of hypoxia-inducible factor in cancer cells. Cancer Res 63:1764–1768

Korcok J, Wu F, Tyml K, Hammond RR, Wilson JX (2002) Sepsis inhibits reduction of dehydroascorbic acid and accumulation of ascorbate in astroglial cultures: Intracellular ascorbate depletion increases nitric oxide synthase induction and glutamate uptake inhibition. J Neurochem 81:185–193

Korcok J, Dixon SJ, Lo TC, Wilson JX (2003) Differential effects of glucose on dehydroascorbic acid transport and intracellular ascorbate accumulation in astrocytes and skeletal myocytes. Brain Res 993:201–207

Kuo SM, Tan CH, Dragan M, Wilson JX (2005) Endotoxin increases ascorbate recycling and concentration in mouse liver. J Nutr 135:2411–2416

Laborie S, Lavoie JC, Rouleau T, Chessex P (2002) Multivitamin solutions for enteral supplementation: A source of peroxides. Nutrition 18:470–473

Laudes IJ, Chu JC, Sikranth S, Huber-Lang M, Guo RF, Riedemann N, Sarma JV, Schmaier AH, Ward PA (2002) Anti-c5a ameliorates coagulation/fibrinolytic protein changes in a rat model of sepsis. Am J Pathol 160:1867–1875

Li JM, Shah AM. (2004) Endothelial cell superoxide generation: Regulation and relevance for cardiovascular pathophysiology. Am J Physiol Regul Integr Comp Physiol 287:R1014–R1030

Li JM, Fan LM, Christie MR, Shah AM (2005) Acute tumor necrosis factor alpha signaling via NADPH oxidase in microvascular endothelial cells: Role of p47phox phosphorylation and binding to TRAF4. Mol Cell Biol 25:2320–2330

Long CL, Maull KI, Krishnan RS, Laws HL, Geiger JW, Borghesi L, Franks W, Lawson TC, Sauberlich HE (2003) Ascorbic acid dynamics in the seriously ill and injured. J Surg Res 109:144–148

Lopez A, Lorente JA, Steingrub J, Bakker J, McLuckie A, Willatts S, Brockway M, Anzueto A, Holzapfel L, Breen D, Silverman MS, Takala J, Donaldson J, Arneson C, Grove G, Grossman S, Grover R (2004) Multiple-center, randomized, placebo-controlled, double-blind study of the nitric oxide synthase inhibitor 546C88: effect on survival in patients with septic shock. Crit Care Med 32:21–30

Lu H, Dalgard CL, Mohyeldin A, McFate T, Tait AS, Verma A (2005) Reversible inactivation of HIF-1 prolyl hydroxylases allows cell metabolism to control basal HIF-1. J Biol Chem 280:41928–41939

Mack WJ, Mocco J, Ducruet AF, Laufer I, King RG, Zhang Y, Guo W, Pinsky DJ, Connolly ES Jr (2006) A cerebroprotective dose of intravenous citrate/sorbitol-stabilized dehydroascorbic acid is correlated with increased cerebral ascorbic acid and inhibited lipid peroxidation after murine reperfused stroke. Neurosurgery 59:383–388

May JM (2000) How does ascorbic acid prevent endothelial dysfunction? Free Radic Biol Med 28:1421–1429

May JM, Huang J, Qu ZC (2005) Macrophage uptake and recycling of ascorbic acid: response to activation by lipopolysaccharide. Free Radic Biol Med 39:1449–1459

McErlain L, Marson H, Ainsworth P, Burnett SA (2001) Ascorbic acid loss in vegetables: adequacy of a hospital cook-chill system. Int J Food Sci Nutr 52:205–211

Merx MW, Liehn EA, Graf J, van de Sandt A, Schaltenbrand M, Schrader J, Hanrath P, Weber C (2005) Statin treatment after onset of sepsis in a murine model improves survival. Circulation 112:117–124

Metnitz PG, Bartens C, Fischer M, Fridrich P, Steltzer H, Druml W (1999) Antioxidant status in patients with acute respiratory distress syndrome. Intensive Care Med 25:180–185

Mittermayer F, Pleiner J, Schaller G, Zorn S, Namiranian K, Kapiotis S, Bartel G, Wolfrum M, Brugel M, Thiery J, Macallister RJ, Wolzt M (2005) Tetrahydrobiopterin corrects Escherichia coli endotoxin-induced endothelial dysfunction. Am J Physiol Heart Circ Physiol 289:H1752–H1757

Miyaji T, Hu X, Yuen PS, Muramatsu Y, Iyer S, Hewitt SM, Star RA (2003) Ethyl pyruvate decreases sepsis-induced acute renal failure and multiple organ damage in aged mice. Kidney Int 64:1620–1631

Muhlhofer A, Mrosek S, Schlegel B, Trommer W, Rozario F, Böhles H, Schremmer D, Zoller WG, Biesalski HK (2004) High-dose intravenous vitamin C is not associated with an increase of pro-oxidative biomarkers. Eur J Clin Nutr 58:1151–1158

Nathens AB, Neff MJ, Jurkovich GJ, Klotz P, Farver K, Ruzinski JT, Radella F, Garcia I, Maier RV (2002) Randomized, prospective trial of antioxidant supplementation in critically ill surgical patients. Ann Surg 236:814–822

Ondiveeran HK, Fox-Robichaud AE (2004) Pentastarch in a balanced solution reduces hepatic leukocyte recruitment in early sepsis. Microcirculation 11:679–687

Opal SM, Palardy JE, Parejo NA, Creasey AA (2001) The activity of tissue factor pathway inhibitor in experimental models of superantigen-induced shock and polymicrobial intra-abdominal sepsis. Crit Care Med 29:13–17

Padayatty SJ, Sun H, Wang Y, Riordan HD, Hewitt SM, Katz A, Wesley RA, Levine M (2004) Vitamin C pharmacokinetics: implications for oral and intravenous use. Ann Intern Med 140:533–537

Piper RD, Li FY, Myers ML, Sibbald WJ (1999) Effect of isoproterenol on myocardial structure and function in septic rats. J Appl Physiol 86:993–1001

Pleiner J, Mittermayer F, Schaller G, Marsik C, MacAllister RJ, Wolzt M (2003) Inflammation-induced vasoconstrictor hyporeactivity is caused by oxidative stress. J Am Coll Cardiol 42:1656–1662

Rice TW, Bernard GR (2005) Therapeutic intervention and targets for sepsis. Annu Rev Med 56:225–248

Rivers E, Nguyen B, Havstad S, Ressler J, Muzzin A, Knoblich B, Peterson E, Tomlanovich M (2001) Early goal-directed therapy collaborative group. Early goal-directed therapy in the treatment of severe sepsis and septic shock. N Engl J Med 345:1368–1377

Rosengarten B, Hecht M, Auch D, Ghofrani HA, Schermuly RT, Grimminger F, Kaps M (2006) Microcirculatory dysfunction in the brain precedes changes in evoked potentials in endotoxin-induced sepsis syndrome in rats. Cerebrovasc Dis 23:140–147

Rumelin A, Humbert T, Luhker O, Drescher A, Fauth U (2005) Metabolic clearance of the antioxidant ascorbic acid in surgical patients. J Surg Res 129:46–51

Russell JA (2006) Management of sepsis. N Engl J Med 355:1699–16713

Sakr Y, Dubois MJ, De Backer D, Creteur J, Vincent JL (2004) Persistent microcirculatory alterations are associated with organ failure and death in patients with septic shock. Crit Care Med 32:1825–1831

Salvemini D, Cuzzocrea S (2002) Oxidative stress in septic shock and disseminated intravascular coagulation. Free Radic Biol Med 33:1173–1185

Schneider MP, Delles C, Schmidt BM, Oehmer S, Schwarz TK, Schmieder RE, John S (2005) Superoxide scavenging effects of N-acetylcysteine and vitamin C in subjects with essential hypertension. Am J Hypertens 18:1111–1117

Schorah CJ, Downing C, Piripitsi A, Gallivan L, Al-Hazaa AH, Sanderson MJ, Bodenham A (1996) Total vitamin C, ascorbic acid, and dehydroascorbic acid concentrations in plasma of critically ill patients. Am J Clin Nutr 63:760–765

Secor D, Li F, Ellis CG, Sharpe MD, Gross PL, Wilson JX, Tyml K (2010) Impaired microvascular perfusion in sepsis requires activated coagulation and P-selectin-mediated platelet adhesion in capillaries. Intensive Care Med 36:1928–1934

Seno T, Inoue N, Matsui K, Ejiri J, Hirata KI, Kawashima S, Yokoyama M (2004) Functional expression of sodium-dependent vitamin C transporter 2 in human endothelial cells. J Vasc Res 41:345–351

Sharma P, Raghavan SA, Saini R, Dikshit M (2004) Ascorbate-mediated enhancement of reactive oxygen species generation from polymorphonuclear leukocytes: modulatory effect of nitric oxide. J Leukoc Biol 75:1070–1078

Sharshar T, Hopkinson NS, Orlikowski D, Annane D (2005) Science review: the brain in sepsis – culprit and victim. Crit Care 9:37–44

Shen KP, Lo YC, Yang RC, Liu HW, Chen IJ, Wu BN (2005) Antioxidant eugenosedin-A protects against lipopolysaccharide-induced hypotension, hyperglycaemia and cytokine immunoreactivity in rats and mice. J Pharm Pharmacol 57:117–125

Singer G, Urakami H, Specian RD, Stokes KY, Granger DN (2006) Platelet recruitment in the murine hepatic microvasculature during experimental sepsis: role of neutrophils. Microcirculation 13:89–97

Smith AR, Visioli F, Hagen TM (2002) Vitamin C matters: increased oxidative stress in cultured human aortic endothelial cells without supplemental ascorbic acid. FASEB J 16:1102–4110

Spolarics Z, Stein DS, Garcia ZC (1996) Endotoxin stimulates hydrogen peroxide detoxifying activity in rat hepatic endothelial cells. Hepatology 24:691–696

Spronk PE, Ince C, Gardien MJ, Mathura KR, Oudemans-van Straaten HM, Zandstra DF (2002) Nitroglycerin in septic shock after intravascular volume resuscitation. Lancet 360:1395–1396

Steffel J, Luscher TF, Tanner FC (2006) Tissue factor in cardiovascular diseases: molecular mechanisms and clinical implications. Circulation 113:722–731

Stoclet JC, Martinez MC, Ohlmann P, Chasserot S, Schott C, Kleschyov AL, Schneider F, Andriantsitohaina R (1999) Induction of nitric oxide synthase and dual effects of nitric oxide and cyclooxygenase products in regulation of arterial contraction in human septic shock. Circulation 100:107–112

Sundrani R, Easington CR, Mattoo A, Parrillo JE, Hollenberg SM (2000) Nitric oxide synthase inhibition increases venular leukocyte rolling and adhesion in septic rats. Crit Care Med 28:2898–2903

Tanaka H, Matsuda T, Miyagantani Y, Yukioka T, Matsuda H, Shimazaki S (2000) Reduction of resuscitation fluid volumes in severely burned patients using ascorbic acid administration: a randomized, prospective study. Arch Surg 135:326–331

Terborg C, Schummer W, Albrecht M, Reinhart K, Weiller C, Rother J (2001) Dysfunction of vasomotor reactivity in severe sepsis and septic shock. Intensive Care Med 27:1231–1234

Tousoulis D, Xenakis C, Tentolouris C, Davies G, Antoniades C, Crake T, Stefanadis C (2005) Effects of vitamin C on intracoronary L-arginine dependent coronary vasodilatation in patients with stable angina. Heart 91:1319–1323

Tyml K, Li F, Wilson JX (2005) Delayed ascorbate bolus protects against maldistribution of microvascular blood flow in a rat model of sepsis. Crit Care Med 33:1823–1828

Vissers MCM, Gunningham SP, Morrison MJ, Dachs GU, Currie MJ (2007) Modulation of hypoxia-inducible factor-1 alpha in cultured primary cells by intracellular ascorbate. Free Radic Biol Med, doi: 10.1016/j.freeradbiomed.2006.11.023

Wang W, Mitra A, Poole B, Falk S, Lucia MS, Tayal S, Schrier R (2004) Endothelial nitric oxide synthase-deficient mice exhibit increased susceptibility to endotoxin-induced acute renal failure. Am J Physiol Renal Physiol 287:F1044–F1048

Wenger RH (2002) Cellular adaptation to hypoxia: O_2 sensing protein hydroxylases, hypoxia-inducible transcription factors, and O_2 regulated gene expression. FASEB J 16:1151–1162

Wilson JX (2005) Regulation of vitamin C transport. Annu Rev Nutr 25:105–125

Wilson JX, Dragan M (2005) Sepsis inhibits recycling and glutamate-stimulated export of ascorbate by astrocytes. Free Radic Biol Med 39:990–998

Wilson JX, Gelb AW (2002) Free radicals, antioxidants, and neurologic injury: Possible relationship to cerebral protection by anesthetics. J Neurosurg Anesthesiol 14:66–79

Wilson JX, Young GB (2003) Progress in clinical neurosciences: Sepsis-associated encephbreakalopathy: Evolving concepts. Can J Neurol Sci 30:98–105

Wilson JX, Dixon SJ, Yu J, Nees S, Tyml K (1996) Ascorbate uptake by microvascular endothelial cells of rat skeletal muscle. Microcirculation 3:211–221

Wu F, Cepinskas G, Wilson JX, Tyml K (2001) Nitric oxide attenuates but superoxide enhances iNOS expression in endotoxin- and IFNgamma-stimulated skeletal muscle endothelial cells. Microcirculation 8:415–425

Wu F, Tyml K, Wilson JX (2002) Ascorbate inhibits iNOS expression in endotoxin- and IFN gamma-stimulated rat skeletal muscle endothelial cells. FEBS Lett 520:122–126

Wu F, Wilson JX, Tyml K (2003) Ascorbate inhibits iNOS expression and preserves vasoconstrictor responsiveness in skeletal muscle of septic mice. Am J Physiol Regul Integr Comp Physiol 285:R50–R56

Wu F, Wilson JX, Tyml K (2004) Ascorbate protects against impaired arteriolar constriction in sepsis by inhibiting inducible nitric oxide synthase expression. Free Radic Biol Med 37:1282–1289

Wu F, Tyml K, Wilson JX (2005) Ascorbate inhibits iNOS induction in microvascular endothelial cells through inhibiting NADPH oxidase activity. FASEB J 19:A507

Wu F, Schuster DP, Tyml K, Wilson JX (2007) Ascorbate inhibits NADPH oxidase subunit p47phox expression in microvascular endothelial cells. Free Radic Biol Med 42:124–131

Yu HP, Lui PW, Hwang TL, Yen CH, Lau YT (2006) Propofol improves endothelial dysfunction and attenuates vascular superoxide production in septic rats. Crit Care Med 34:453–460

Chapter 6
Vitamin C Transport and Its Role in the Central Nervous System

James M. May

Abstract Vitamin C, or ascorbic acid, is important as an antioxidant and participates in numerous cellular functions. Although it circulates in plasma in micromolar concentrations, it reaches millimolar concentrations in most tissues. These high ascorbate cellular concentrations are thought to be generated and maintained by the SVCT2 (Slc23a2), a specific transporter for ascorbate. The vitamin is also readily recycled from its oxidized forms inside cells. Neurons in the central nervous system (CNS) contain some of the highest ascorbic acid concentrations of mammalian tissues. Intracellular ascorbate serves several functions in the CNS, including antioxidant protection, peptide amidation, myelin formation, synaptic potentiation, and protection against glutamate toxicity. The importance of the SVCT2 for CNS function is supported by the finding that its targeted deletion in mice causes widespread cerebral hemorrhage and death on post-natal day 1. Neuronal ascorbate content as maintained by this protein also has relevance for human disease, since ascorbate supplements decrease infarct size in ischemia-reperfusion injury models of stroke, and since ascorbate may protect neurons from the oxidant damage associated with neurodegenerative diseases such as Alzheimer's, Parkinson's, and Huntington's. The aim of this review is to assess the role of the SVCT2 in regulating neuronal ascorbate homeostasis and the extent to which ascorbate affects brain function and antioxidant defenses in the CNS.

Keywords Brain · SVCT2 · Ascorbate transport · Dehydroascorbate · Neurons · Glutamate

6.1 Ascorbic Acid Physiology and Recycling

Ascorbic acid is a vitamin because humans, higher primates, and a few other vertebrates such as guinea pigs have lost the ability to synthesize it from glucose. These species lack the enzyme L-gulono-1,4-lactone oxidase (EC 1.1.3.8), which

J.M. May (✉)
Departments of Medicine and Molecular Physiology and Biophysics, Vanderbilt University School of Medicine, Nashville, TN 37232, USA
e-mail: james.may@vanderbilt.edu

O. Stanger (ed.), *Water Soluble Vitamins*, Subcellular Biochemistry 56,
DOI 10.1007/978-94-007-2199-9_6, © Springer Science+Business Media B.V. 2012

carries out the last step in ascorbate biosynthesis (Chatterjee et al. 1960). Species that cannot synthesize ascorbate must therefore have efficient mechanisms for both absorption of the vitamin and for its recycling. Absorption of vitamin C occurs largely in the distal ileum of humans and guinea pigs, where uptake is mediated by a saturable, sodium- and energy-dependent transporter (Stevenson and Brush 1969; Bianchi et al. 1986) that is located in the apical brush border membrane (Bianchi et al. 1986; Rose 1988). This transporter has been cloned (Tsukaguchi et al. 1999) and is termed SVCT1 (sodium-dependent vitamin C transporter-1, Slc23a1). The SVCT1 helps to generate plasma ascorbate concentrations of 40–60 μM in humans (Table 6.1), which can double with typical levels of dietary ascorbate supplementation (e.g., 250–1,000 mg daily) (Levine et al. 1996). In contrast to plasma ascorbate concentrations, tissue concentrations are typically 1 mM or higher, as exemplified by leukocytes and platelets in human blood (Table 6.1). High intracellular ascorbate concentrations are achieved by two mechanisms. The most important is probably unidirectional transport of the vitamin into cells on a transporter very closely related to the SVCT1, termed the SVCT2 (Slc23a2) (Tsukaguchi et al. 1999). These transporters share 65% sequence homology, substrate saturability, high affinity for ascorbate (20–60 μM), as well as energy- and sodium-dependence. Intracellular ascorbate tends to exit cells only slowly (May et al. 1998a), probably because of its hydrophilic structure and negative charge at physiologic pH. The second route by which ascorbate can enter cells is by uptake of its two-electron-oxidized form, dehydroascorbic acid (DHA, Fig. 6.1). DHA is taken up on the ubiquitous

GLUT-type glucose transporters (Vera et al. 1993). Although DHA, a triketone with a strained lactone ring, has little resemblance to glucose, it forms a hydrated bicyclic hemiketal (Pastore et al. 2001) for which these transporters have affinity (Bigley and Stankova 1974). Since the GLUTs mediate facilitated diffusion, transport of DHA by this mechanism is bi-directional. This also means that any DHA formed within cells by ascorbate oxidation will rapidly efflux and be lost. This is prevented by efficient cellular mechanisms of DHA reduction or recycling to ascorbate.

Table 6.1 Ascorbate concentrations in plasma and blood cells from normal humans

Tissue/fluid	Ages; # of subjects	Diet (mg/day)	Concentration (mM)	References
Plasma	>60 years; 226	< 120	0.057 ± 0.003	Jacques et al. (1995)
Plasma	73 years; 161	257	0.072 ± 0.027	Taylor et al. (1997)
Aqueous humor	73 years; 161	257	1.64 ± 0.58	Taylor et al. (1997)
Lens	73 years; 161	257	3.74 ± 1.4	Taylor et al. (1997)
Neutrophils	19–27 years; 13	100	1.17 ± 0.17	Levine et al. (2001)
Monocytes	17–55 years; 41	Unknown	3.8	Evans et al. (1982)
Platelets	17–55 years; 41	Unknown	1.9	Evans et al. (1982)
Erythrocytes	17–55 years; 41	Unknown	0.043	Evans et al. (1982)
Plasma	17–55 years; 41	Unknown	0.045	Evans et al. (1982)

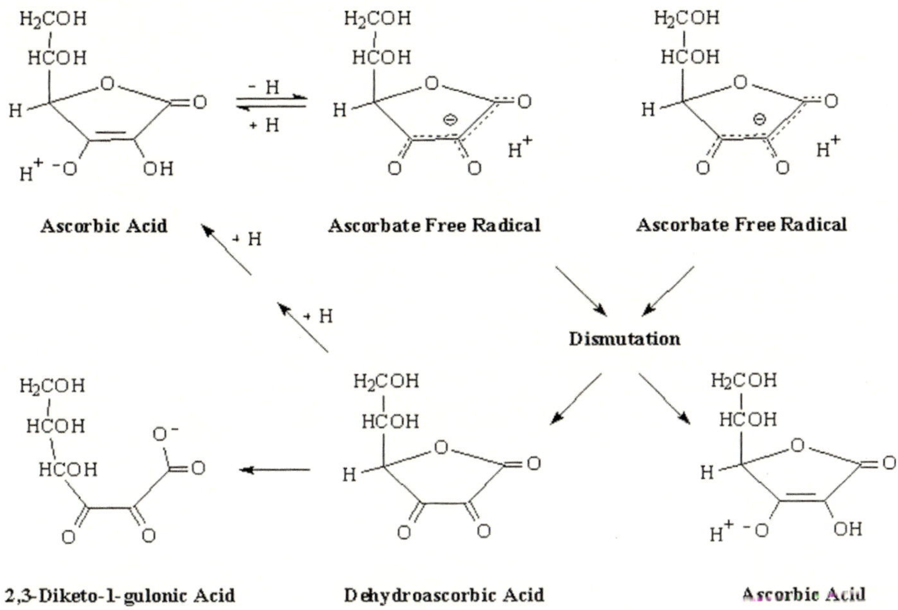

Fig. 6.1 Ascorbate oxidation and recycling

Ascorbate recycling from both of its oxidized forms occurs largely within cells, although cell-surface reduction of its radical one electron-oxidized form has been described (May et al. 2000, VanDuijn et al. 2000). As shown in Fig. 6.1, ascorbate is oxidized when it donates a hydrogen atom (proton + electron) to enzyme reactions or to scavenge free radicals. The one-electron oxidized form of ascorbate, commonly termed the ascorbate free radical (AFR) or monodehydroascorbate, is surprisingly stable and can be detected at 10 nM concentrations in biological fluids by EPR (Mehlhorn 1991, Coassin et al. 1991, Buettner and Jurkiewicz 1993). Although the AFR has a lower mid-point reduction potential than ascorbate (Buettner 1993), it is not very reactive. Instead of donation of another hydrogen and electron, the preferred reaction shown in Fig. 6.1 is for two molecules of the AFR to dismutate, forming ascorbate and dehydroascorbate (Bielski et al. 1981). DHA is unstable at physiologic pH, with a half-life of about 6 min (Drake et al. 1942; Winkler 1987) (Tolbert and Ward 1982). Hydrolysis of the lactone ring irreversibly converts DHA to 2,3-diketo-1-gulonic acid (Fig. 6.1) (Chatterjee 1970; Bode et al. 1990). Ascorbate loss by DHA decomposition is obviously wasteful of the vitamin, and DHA reduction to ascorbate is rapid and efficient within cells.

Mechanisms of GSH-dependent DHA reduction include direct chemical reduction (Winkler et al. 1994) and enzyme-dependent reduction with GSH serving as electron donor (Maellaro et al. 1994; Winkler et al. 1994; Washburn and Wells 1999). DHA is also reduced to ascorbate by NADPH-dependent enzymes, including thioredoxin reductase (May et al. 1997) and a liver 3α-hydroxysteroid dehydrogenase (Del Bello et al. 1994). The AFR is also reduced to ascorbate with high affinity by both NADH-dependent reductases (Nazemi and Staudinger 1968; Ito et al. 1981; Villalba et al. 1993) and by the NADPH-dependent enzyme thioredoxin reductase (May et al. 1998a). As reviewed previously (May and Asard 2004), these cellular systems provide redundant and efficient mechanisms for recycling of ascorbate from its oxidized forms.

6.2 Functions of Ascorbate

As noted above, ascorbate provides electrons for crucial enzyme reactions in cells and acts as a primary antioxidant capable of scavenging radicals generated within cells or plasma. The chapter by Dr. deTulio details the non-antioxidant functions of ascorbate, which are crucial in preventing scurvy and for maintaining cell integrity and health. The main antioxidant function of the vitamin involves direct scavenging of radical species before they can damage DNA, proteins, or lipids. Another mechanism by which ascorbate functions as an antioxidant is to recycle other antioxidants. For example, ascorbate can reduce the α-tocopheroxyl radical at the surface of biological membranes, thus contributing to the ability of α-tocopherol to break the chain of lipid peroxidation in lipid bilayers (Buettner 1993). Another example relates to the finding that ascorbate spares tetrahydrobiopterin in cultured endothelial cells, thus allowing continued action of endothelial nitric oxide synthase (Heller et al. 2001; Baker et al. 2001). As with α-tocopherol, the mechanism of this effect also appears to be due to reduction of the tetrahydrobiopterin radical by ascorbate (Patel et al. 2002). Consideration of both non-antioxidant and antioxidant functions of the vitamin shows that it has multiple sites of action in cells. Although it may not be absolutely required for any specific action in a cell, ascorbate is clearly necessary for proper function of many organs. Perhaps most important is its requirement in the brain and central nervous system (CNS), as discussed next.

6.3 Importance of Vitamin C in the CNS

Ascorbic acid is an essential micronutrient in the CNS. Whereas whole brain ascorbate concentrations are one to two mM, intracellular neuronal concentrations have been calculated to be much higher (Rice and Russo-Menna 1998; Rice 2000), as discussed below. Neurons are also especially sensitive to ascorbate deficiency, perhaps because they have 10-fold higher rates of oxidative metabolism than supporting glia (Wilson 1997; Hediger 2002). This heightened neuronal sensitivity is most apparent in states of ascorbate deficiency and in conditions in which there is excess oxidant stress. A neuroprotective role for ascorbate is also suggested by the existence of homeostatic mechanisms that maintain high concentrations of ascorbate in cerebrospinal fluid (CSF) and in neurons. A key feature in this regard is the ability to sustain steep ascorbate concentration gradients: (a) from plasma to the CSF across

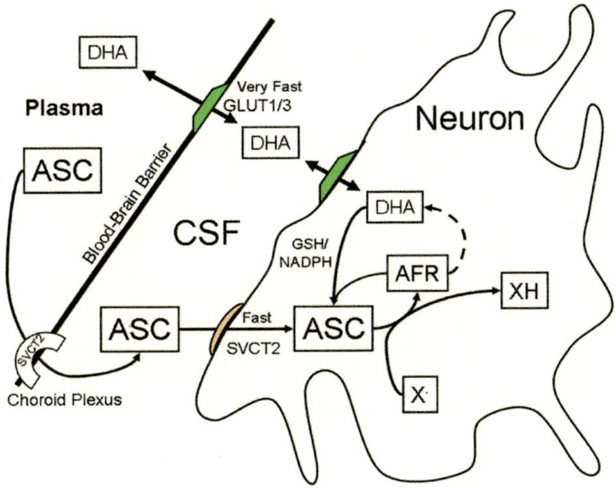

Fig. 6.2 Ascorbate uptake and metabolism in the CNS. ASC, ascorbate; AFR, ascorbate free radical; DHA, dehydroascorbate; X• oxidizing free radical species. AFR generated by X• dismutates to form ascorbate and dehydroascorbate. Both the AFR and DHA are recycled back to ascorbate by cellular metabolism

the choroid plexus and (b) from the CSF and interstitium into neurons (Fig. 6.2). In both instances, this gradient is generated for the most part by the SVCT2 protein. As discussed next, intracellular ascorbate and thus the SVCT2 protects the CNS in syndromes of oxidant stress.

Scurvy causes severe lassitude and asthenia in humans. Although the disease has been associated with paraparesis in humans, death appears to be due more to complications of systemic collagen dysfunction and not to a distinct neurologic syndrome (Hirschmann and Raugi 1999). This likely relates to the fact that ascorbate is avidly retained by the CNS during ascorbate deficiency (Hornig 1975). For example, 14 days after removal of ascorbate from their diet, guinea pig brains still had 24% of their original ascorbate content, compared to only 4% for adrenal gland and 3% for spleen (Hughes et al. 1971). Whereas this could indicate that decreases in CNS ascorbate do not play a major role in generalized scurvy, the argument can also be made that the strong retention of ascorbate in the CNS reflects its importance to neuronal function. This hypothesis is supported by our own studies in guinea pigs with moderate vitamin E deficiency in which an acute deficiency of ascorbate was superimposed (Hill et al. 2003). After 2 weeks of vitamin E deprivation in male guinea pigs, plasma and brain α-tocopherol concentrations were modestly decreased by 65% and 32%, respectively. These animals gained weight and appeared completely normal without evidence of abnormal gait or ataxia (signs of vitamin E deficiency). However, within 5–6 days of removing vitamin C from their diet, the animals typically developed a progressive ascending paralysis and died within 24 h. No neurologic signs were apparent in animals with single deficiencies of either vitamin C or E alone. Death occurred in the doubly deficient animals despite only 37% and 67% decreases in brain and spinal cord ascorbate concentrations, respectively. Despite over 95% decreases in plasma and liver ascorbate in the animals on the vitamin C-deficient diet, neither the C-deficient nor the doubly deficient animals had signs of scurvy (no skin or hair changes, no hemarthroses). The doubly deficient animals had only small increases in liver and brain F2-isoprostanes (a marker of oxidant stress). Standard hematoxylin and eosin staining of the brains and spinal cords showed no specific abnormalities. However, subsequent studies using Nissl and silver degeneration stains revealed widespread neuronal loss and degeneration in the pons and long tracts of the spinal cord (Burk et al. 2006). Thus, even a modest decrease in CNS ascorbate accelerated signs of vitamin E deficiency in this model, indicative of a crucial role for ascorbate in protecting the brain against oxidant stress.

If the SVCT2 transporter is responsible for maintaining high CSF and neuronal ascorbate concentrations, what happens when the protein is knocked out in the mouse? Targeted deletion of the SVCT2 protein results in homozygotes that die shortly after birth: although they appear to develop normally in utero, they never take a breath and have diffuse cerebral hemorrhage (Sotiriou et al. 2002). Whereas the latter likely reflects increased capillary fragility due to defects in basement membrane type IV collagen (Gould et al. 2005), the animals do not show hemorrhage elsewhere and lack signs of generalized scurvy. Further, hydroxyproline levels in skin are normal (Sotiriou et al. 2002). The cause of death is unknown. Catecholamine synthesis is decreased, but this is apparently not the cause of death (Bornstein et al. 2003). Although the lungs fail to expand, their architecture is normal, as are surfactant levels (Sotiriou et al. 2002). Brain ascorbate concentrations are decreased 30–50% in the heterozygotes, and are very low to undetectable in the homozygous mice (Sotiriou et al. 2002). It seems likely that diffuse cortical capillary hemorrhage along with neuronal dysfunction contributed significantly to the death of these mice, although further studies are needed to assess the role of the lung and metabolic defects.

Perhaps the most dramatic acute oxidant stress in the CNS is the ischemia-reperfusion injury that occurs with stroke. It is generally accepted that ischemia depletes intracellular GSH and ascorbate in brain, and that subsequent reperfusion generates reactive oxygen species (ROS) that further deplete these two antioxidants and extend tissue damage to areas with decreased oxidant defenses (Rice et al. 1995). Ascorbate supplements decrease infarct size in animal studies. For example, in monkeys given 1 g/day of ascorbate parenterally for 6 days before middle cerebral artery occlusion, brain ascorbate was increased by 50%. This indicates that ascorbate had access to the CNS. Further, infarct size was decreased by 50% in the ascorbate-treated group compared to the control group (Ranjan et al. 1993). Similarly, in rats with middle cerebral artery occlusion, high dose dehydroascorbate given by intraperitoneal injection just before, 15 min after, or three h after occlusion markedly decreased infarct volume, mortality, and neurological deficits in mice (Huang et al. 2001). In contrast, ascorbate itself showed no protective effect in these studies. As shown in Fig. 6.2, dehydroascorbate is rapidly transported across the blood-brain barrier by glucose transporters (Agus et al. 1997; Huang et al. 2001; Mack et al. 2006), taken up by glucose transporters in neurons, and reduced to ascorbate (May and Asard 2004). The inability of acute treatment with ascorbate to prevent neuronal damage in the rat model is supported by other evidence that ascorbate does not cross the blood-brain barrier in its fully reduced form (Agus et al.

1997; Hosoya et al. 2004). This appears to be due to lack of SVCT2 in endothelial cells of the blood-brain barrier, as shown by in situ hybridization (Berger et al. 2003; García et al. 2005) and more recently by immunostaining for the SVCT2 protein (Mun et al. 2006).

The ability of ascorbate to decrease infarct size following ischemia-reperfusion likely relates to scavenging of ROS, both within and around cells. Ascorbate at the concentrations present in CSF and neurons in vivo (i.e., >100 μM) will effectively scavenge superoxide (Jackson et al. 1998) and spare or recycl-tocopherol in the lipid bilayer (Niki et al. 1995; May et al. 1998b). For example, an increase in intracellular (but not extracellular) ascorbate content of (rat) brain slices decreased the swelling induced by oxidant stress (Brahma et al. 2000). We reported that ascorbate prevented loss of α-tocopherol and lipid peroxidation induced by culture in oxygenated medium (Li et al. 2003). Finally, in our recent studies, primary cultures of hippocampal neurons prepared from homozygous SVCT2 knockout mice showed increased susceptibility to cell death induced by N-methyl-D-aspartate (NMDA) and H_2O_2 compared to cultures from wild-type mice (Qiu et al. 2007). The improved survival of neurons containing the SVCT2 was likely due to small amounts of endogenous ascorbate (0.1–0.2 mM) retained by the cells after 2 weeks in culture, since the culture medium did not contain ascorbate.

Regarding extracellular ascorbate, it has long been known that ischemia releases large amounts of ascorbate from brain cells (Hillered et al. 1988). Since this release was associated with glutamate uptake by neurons and glia (Miele et al. 1994; Rice 2000), these two processes were thought to be linked in mechanism involving heteroexchange across one or more transporters (Grünewald 1993; Rebec and Pierce 1994; Yusa 2001). Removal of extracellular glutamate by such heteroexchange would decrease excitotoxicity caused by activation of cell surface and synaptic glutamate receptors (Grünewald 1993; Rebec and Pierce 1994; Yusa 2001). In addition, intracellular ascorbate would be expected to scavenge ROS generated during to glutamate-induced excitotoxicity. In this regard, ascorbate added to the culture medium was found to prevent glutamate-induced cell damage and death in cultured cerebellar granule cells (Ciani et al. 1996; Atlante et al. 1997). Ascorbate also has been shown to protect neurons from excitotoxicity induced by activation of the NMDA receptor (MacGregor et al. 1996; Majewska et al. 1990; Majewska and Bell 1990). Whereas intracellular ascorbate protects cells against glutamate toxicity, recent evidence suggests that it does not do so via a heteroexchange mechanism. Wilson and colleagues confirmed that glutamate increases ascorbate efflux from cerebral astrocytes, but found no effect of intracellular ascorbate on glutamate uptake (Wilson et al. 2000). The release of ascorbate was attributed to glutamate-induced cell swelling followed by opening of organic ion channels through which ascorbate can pass (Wilson et al. 2000). We found the analogous results in SH-SY5Y neuroblastoma cells: glutamate caused release of intracellular ascorbate, but neither intra- nor extracellular ascorbate affected glutamate uptake (May et al. 2006). Together, the results of these studies strongly support a protective role for ascorbate and possibly for the SVCT2 in brain during acute ROS generation.

6.4 Ascorbate Function in the CNS

Whereas ascorbate is well known to be involved in neuronal biochemistry (e.g., peptide amidation, myelination, and catecholamine synthesis), evidence for a direct role in neuronal maturation and function has only recently been forthcoming. Culture of embryonic cortical precursor cells in medium containing ascorbate was first shown to induce their maturation into neurons and astrocytes by Lee and colleagues (Lee et al. 2003). We confirmed and extended these findings in primary cultures of hippocampal neurons (Qiu et al. 2007). Key to these experiments was the use of cells prepared from late-stage embryos (gestational day 18–19) that either expressed or completely lacked the SVCT2. After 10–14 days in culture in serum-free medium in the absence of ascorbate, no ascorbate was present in cells prepared from Slc23a2 (–/–) mice, whereas the ascorbate content of cells from wild-type littermates was 0.1–0.2 mM. The SVCT2 was not detected by immunostaining in cells from the Slc23a2 (–/–) mice, whereas it was clearly present in cells from wild-type mice. Although the SVCT2 was previously reported to be restricted to neuronal cell bodies in sections prepared from adult rat brain (Mun et al. 2006), we found that it was present in punctate pattern almost exclusively in neuronal axons and not in dendrites or cell bodies. The presence of the SVCT2 enhanced neuronal maturation in culture. Compared to neurons from wild-type mice, neurons from Slc23a2 (–/–) mice showed slower neuronal growth, as indicated by fewer dendritic branches and reduced total dendritic length. In addition, the numbers of α-amino-5-hydroxy-3-methyl-4-isoxazole propionic acid (AMPA) receptor subunit GluR1clusters were also significantly reduced in cells lacking the SVCT2. The mechanism for reduced GluR1 clustering was not established, but it may be due to ascorbate-dependent abnormalities in neuronal activity, since neuronal activity can induce neurotransmitter receptor clustering (Rao et al. 2000). Perhaps related to the abnormalities in AMPA receptor subunit clustering, cells lacking the SVCT2 also had decreased neuronal activity. In patch-clamp experiments, cells from Slc23a2 (–/–) mice showed both decreased amplitude and frequency of miniature excitatory post-synaptic currents. It seems likely that these findings are due to the ability of cells containing the SVCT2 to retain intracellular ascorbate in culture, rather than to the presence of the SVCT2 per se.

6.5 Ascorbate Accumulation and Maintenance in the CNS: Role of SVCT2

As shown in Fig. 6.2, there are two mechanisms by which ascorbate can enter neurons and glia: transport of ascorbate on the SVCT2 and uptake and reduction of dehydroascorbate. The SVCT2 mediates transport of ascorbate both from plasma across the choroid plexus to the CSF and across the neuronal cell plasma membrane. The plasma-CSF ascorbate gradient is about 4-fold in humans (Reiber et al. 1993; Lönnrot et al. 1996) and animals (Spector and Lorenzo 1973; Spector 1977), resulting in CSF ascorbate concentrations of about 200 μM, compared to plasma concentrations of 50 μM or less (Table 6.1). Although DHA enters the CNS more rapidly than ascorbate, the latter does readily penetrate the central nervous system after oral administration (Kontush et al. 2001). For example, supplements of one g vitamin C a day to volunteers already replete with vitamin C increased plasma levels from 50 to 75 μM, and increased CSF ascorbate from 200 to 250 μM (Lönnrot et al. 1996). Neuronal ascorbate uptake further accentuates the ascorbate concentration gradient from plasma to neurons, and brain neurons are said to have the highest ascorbate concentrations in the body (Rose 1988). In this regard, neuronal ascorbate concentrations in young rats have been calculated to be as high as 10 mM, whereas glial ascorbate concentrations are only 0.9 mM (Rice and Russo-Menna 1998; Rice 2000). Such high ascorbate concentrations in young rats agree with the finding that SVCT2 mRNA is increased in fetal mouse brain compared to brain from adult animals (Castro et al. 2001).

It also appears that much of the ascorbate in brain is in neurons. This was first shown over four decades ago in histochemical studies (Shimizu et al. 1960). More recent investigations have localized high ascorbate concentrations to neuron-rich areas of the hippocampus and neocortex in normal human brain, where the ascorbate content is as much as two-fold higher in than in other brain regions (Mefford et al. 1981; Milby et al. 1982). The higher ascorbate contents of neurons in comparison to glia, as alluded to above, may well account for varying ascorbate contents of different brain regions. In post-natal rat brain, for example, areas rich in glia have ascorbate levels only about 10–20% those of areas with high neuronal density (Rice and Russo-Menna 1998). Indeed, whereas GSH is the major low molecular weight cytosolic reducing agent in most cells (e.g., in glia the ascorbate/GSH ratio is 0.24), in neurons ascorbate may take that role, since neuron-rich brain areas have an ascorbate/GSH ratio of 4.0 (Rice and Russo-Menna 1998). This differential in ascorbate content is probably due to the SVCT2, since in situ hybridization studies in rat brain show that SVCT2 mRNA is present only in neurons and not in astrocytes (Castro et al. 2001; Berger et al. 2003). More recent studies show that SVCT2 mRNA is expressed in certain glial elements in the hypothalamus (García et al. 2005).

Although astrocytes may not express the SVCT2 in vivo, oxidant stress can induce SVCT2 expression in astrocytes and increase expression in neurons. For example, ischemia-reperfusion injury following middle cerebral artery occlusion in rats increased SVCT2 mRNA over several hours in the peri-infarct penumbra, both in neurons and in glia (Berger et al. 2003). Also, astrocytes in culture develop high affinity ascorbate transport and SVCT2 mRNA (Wilson et al. 1990), an effect that has been attributed to exposure to high oxygen concentrations in culture (Berger et al. 2003).

As noted previously, the second route by which ascorbate can enter the CNS is via transport of dehydroascorbate across the blood-brain barrier on the ubiquitous GLUT1 glucose transporter. Any DHA generated in the brain interstitium would also be taken up by GLUT1 and GLUT3 in glia and neurons (Fig. 6.2). Since dehydroascorbate concentrations in plasma are usually only 0–2% of ascorbate concentrations (Dhariwal et al. 1991), this route of entry across the blood-brain barrier may not contribute substantially to total brain ascorbate concentrations under normal conditions. However, during oxidant stress in the CNS, ascorbate in the CSF and interstitial space will be oxidized to the AFR, which will then dismutate to form ascorbate and DHA (Fig. 6.1). The latter can then enter both glia and neurons on glucose transporters in substantial amounts, where it will be rapidly reduced to ascorbate. This likely serves as a back-up mechanism to recover oxidized ascorbate before it is irretrievably lost with degradation of DHA. A similar mechanism in cultured HL-60 cells has recently been termed the "bystander" effect (Nualart et al. 2003).

The SVCT2 thus appears to be crucial in maintaining ascorbate homeostasis in the CNS and in neurons. However, little known about SVCT2 structure as it relates to function or about specific mechanisms by which SVCT2 expression or function are regulated.

6.6 Structure-Function and Regulation of the SVCT2 in Neurons

The SVCT2 transporter was one of two SVCT-type transporters cloned in 1999 by Hediger's group (Tsukaguchi et al. 1999), both of which turned out to be members of the solute carrier (Slc) family of transporters. As noted previously, the two SVCT proteins are quite homologous, with only minor differences between species (Takanaga et al. 2004). There has been confusion in the literature regarding official designations of the ascorbate transporter genes, but currently accepted usage is that the Slc23a1 gene encodes the SVCT1 protein, and that Slc23a2 encodes the SVCT2 protein (Takanaga et al. 2004). The distributions of the two SVCT proteins differ in that mRNA for the SVCT1 is found in epithelial tissues involved in ascorbate absorption (intestine) and re-absorption (kidney). The SVCT2 (Fig. 6.3) is the major isoform in most other tissues in the mouse, with highest expression in brain and in areas of high neuronal density in particular (Tsukaguchi et al. 1999).

Neuronal and most other cells have long been known to accumulate ascorbate against a concentration gradient via a saturable, sodium-dependent mechanism (Wilson 2004). With the cloning of the protein, it became apparent that this sodium- and energy-dependent transport is mediated largely by SVCT2 (Tsukaguchi et al. 1999). The SVCT2 has a high affinity for ascorbate (Km = 20–40 μM) that corresponds well to plasma ascorbate concentrations (30–50 μM). In neuronal cells the apparent Km for ascorbate appears to be somewhat higher (113 μM (May et al. 2006)), but even with this modest decrease in affinity, the SVCT2 will still be nearly saturated at the ascorbate concentrations present in CSF (200–400 μM). If the SVCT2 is saturated, then acute regulation of its function by changes in substrate affinity will be minimal in contrast to changes in transporter Vmax or transporter expression in the membrane. It seems likely that transporter number and ability to function rather than affinity are the major mechanisms involved in regulation of the SVCT2.

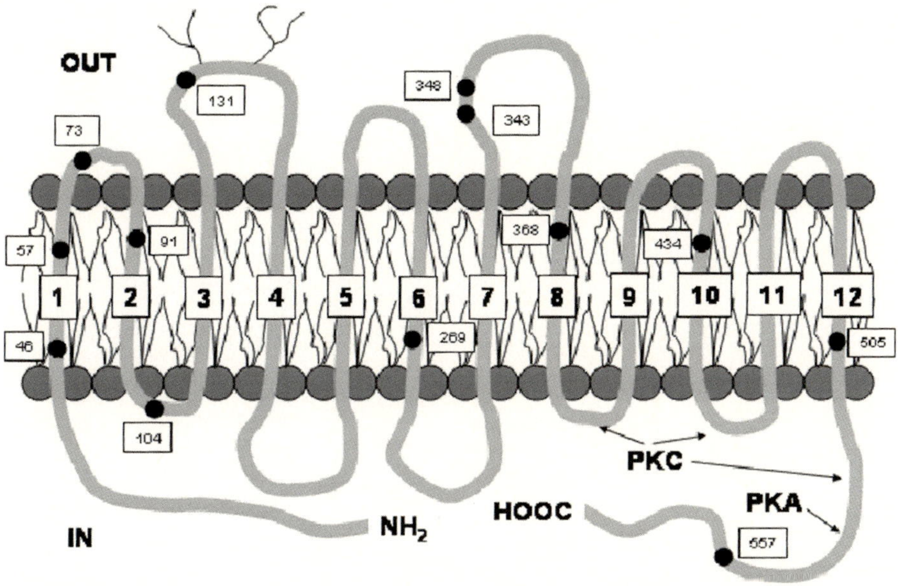

Fig. 6.3 Trans-membrane structure of rat brain SVCT2 (Tsukaguchi et al. 1999). *Numbered dots* show the location of the 13 cysteines in the sequence

Rat brain SVCT2 cDNA codes for a 647-amino acid protein that hydropathy analysis shows to cross the cell membrane 12 times, with both the N- and C-termini in the cytoplasm (Fig. 6.3) (Tsukaguchi et al. 1999). Assuming an average amino acid molecular weight of 121, this predicts that it will migrate as a 78 kDa protein on electrophoretic gels. This electrophoretic migration on Western blotting has been reported by two laboratories (Lutsenko et al. 2004; García et al. 2005). Although there have been reports of the SVCT2 from various tissues migrating in the 50–65 kDa region on Western blots using detection by polyclonal antibodies (Li et al. 2003; Wu et al. 2003; Jin et al. 2005; Mun et al. 2006), these results could be explained by proteolysis during sample preparation, by expression of short forms of the transporter (Lutsenko et al. 2004), or by variable glycosylation of the protein. Regarding the latter, the SVCT2 does have potential N-glycosylation sites on an exofacial loop between trans-membrane segments three and four (Tsukaguchi et al. 1999). Confirmation that the protein is glycosylated was provided by Liang and co-workers (Liang et al. 2002), who showed that endoglycosidase digestion of extracted protein decreased its apparent migration on electrophoretic gels from about 80 to 65 kDa. This finding adds support to the trans-membrane loop topology originally projected by hydropathy analysis (Tsukaguchi et al. 1999). Moreover, the expression of non-glycosylated transporter precursors could explain the presence of bands migrating at lower apparent molecular weights on Western blots in the studies noted above. Shortening of the protein will also generate lower molecular weight bands. Lutsenko et al. (2004) showed that a truncated form of the SVCT2 was present in cultured cells by Western blotting and in normal human brain by PCR. They hypothesized that this truncated SVCT2 corresponds to a non-functional variant that is missing 115 amino acids from domains five and six (Fig. 6.3). Whereas a 75–80 kDa protein found on Western blots likely represents the SVCT2, shorter forms may also be present as well.

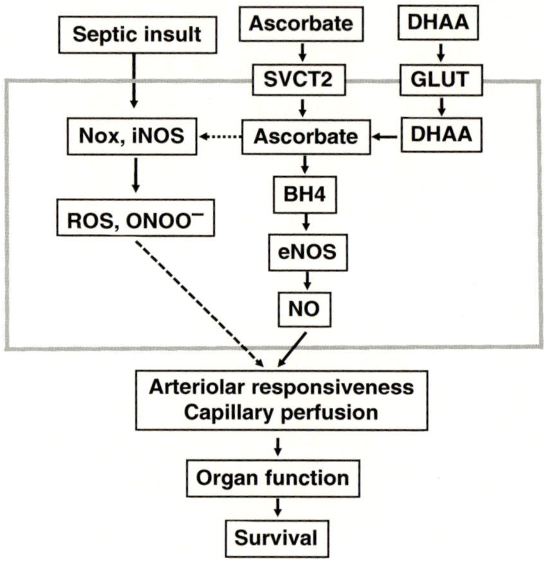

Fig. 6.4 Ascorbate preservation of nitric oxide sustains capillary perfusion

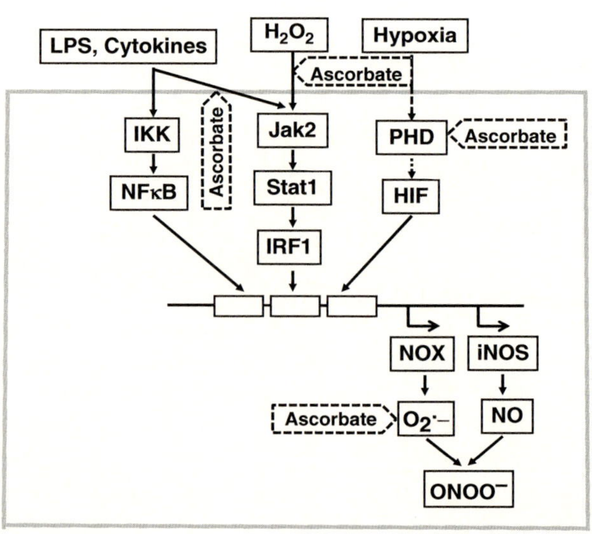

Fig. 6.5 Sites of cellular protection by ascorbate against oxidative stress

There are several plausible mechanisms for regulation of SVCT2 function. First, the SVCT2 protein has putative sites for protein kinase-A and protein kinase-C regulation on the cytoplasmic portions (Tsukaguchi et al. 1999) (Fig. 6.3) Studies in cultured cells support the notion that these are functional in acute regulation of the protein, since ascorbate transport Vmax is decreased by both dibutyryl cyclic AMP (Siushansian and Wilson 1995; Korcok et al. 2000) and by protein kinase-C agonists (Daruwala et al. 1999; Liang et al. 2002) (Figs. 6.4 and 6.5). Intracellular ascorbate may also regulate SVCT2 expression, based on the findings of Wilson et al., that there is a time-dependent up-regulation of ascorbate transport in astrocytes depleted of the vitamin in culture (Wilson et al. 1990). A recent study (Seno et al. 2004) has also shown that function of the SVCT2 expressed in human umbilical vein endothelial cells is inhibited by tumor necrosis factor-α and by interleukin-1β. Since these inhibitor effects were maximal after 5 or more hours of exposure in culture, it seems likely that they were due to changes in protein expression rather than affinity. If so, these studies bring up the notion that the SVCT2 may be regulated by inflammatory cytokines and thus possibly by the level of oxidant stress experienced by the cells.

In addition to regulating SVCT2 transporter expression, oxidant stress could directly modify transporter function by oxidizing key amino acids in the protein. This would most likely involve formation of mixed disulfides or oxidation of cysteine sulfhydryls or methionine, thus causing changes in cell surface or protein redox status. For example, we found in cultured endothelial cells (May and Qu 2004) that rates of ascorbate transport on the SVCT2 are very sensitive to sulfhydryl reagents. Transport was inhibited by agents restricted to the extracellular space, suggesting that one or more of the four cysteines on the exofacial transporter (Fig. 6.2) are involved in the transport mechanism. The presence of one or more exofacial sulfhydryls on the SVCT2 mirrors the situation with the GLUT1 glucose transporter. On that protein, modification of a single sulfhydryl (Cys-429) decreased affected transporter affinity for glucose (May 1988; May et al. 1990), although cysteine-scanning mutagenesis showed that its substitution with serine did not affect transporter function in Xenopus oocytes (Due et al. 1995). SVCT2 expression or function may also be regulated by changes in intracellular redox stress. For example, ascorbate transport in cultured endothelial cells is decreased in parallel with intracellular GSH concentrations in cultured endothelial cells (May and Qu 2004), which suggests that it is susceptible to redox regulation.

The presence of a non-functional truncated SVCT2 may also modify transporter function. In cultured cells truncated SVCT2 was present on the cell surface and expressed inversely in proportion to rates of ascorbate transport (Lutsenko et al. 2004). Further, transfection of the cells with truncated SVCT2 cDNA inhibited ascorbate transport. This dominant-negative inhibition was thought due to protein-protein interaction between functional and non-functional transporters. decreased ascorbate transport (Lutsenko et al. 2004).

Whether any of these potential mechanisms for SVCT2 regulation are important for the CNS protein remains to be determined. Other considerations regarding how neuronal ascorbate concentrations are regulated are the extent to which entry of ascorbate as DHA on the GLUT transporters supplies neuronal ascorbate, whether neurons and glia interact to maintain CNS ascorbate, and the extent to which glutamate-induced ascorbate efflux from neurons modulates intracellular ascorbate concentrations.

6.7 Conclusion

Lethality of the SVCT2 knockout on day 1 of life could reflect the need for ascorbate in the lung or the CNS. Indeed, any damage to the CNS in utero may simply show that ascorbate is required for neuronal maturation, not for maintenance and function of mature neurons. Nonetheless, differential retention of brain ascorbate in scurvy and the novel two-step concentrating mechanism used by the CNS to get ascorbate into neurons suggests that the importance of the vitamin for neuronal repair, myelination, and function may have been underestimated.

Acknowledgement This work was supported by NIH grant AG023138.

References

Agus DB, Gambhir SS, Pardridge WM, Speilholz C, Baselga J, Vera JC (1997) Vitamin C crosses the blood-brain barrier in the oxidized form through the glucose transporters. J Clin Invest 100:2842–2848

Atlante A, Gagliardi S, Minervini GM, Ciotti MT, Marra E, Calissano P (1997) Glutamate neurotoxicity in rat cerebellar granule cells: a major role for xanthine oxidase in oxygen radical formation. J Neurochem 68:2038–2045

Baker RA, Milstien S, Katusic ZS (2001) Effect of vitamin C on the availability of tetrahydrobiopterin in human endothelial cells. J Cardiovasc Pharmacol 37:333–338

Berger UV, Lu XC, Liu W, Tang Z, Slusher BS, Hediger MA (2003) Effect of middle cerebral artery occlusion on mRNA expression for the sodium-coupled vitamin C transporter SVCT2 in rat brain. J Neurochem 86:896–906

Bianchi J, Wilson FA, Rose RC (1986) Dehydroascorbic acid and ascorbic acid transport systems in the quinea pig ileum. Am J Physiol 250:G461–G468

Bielski BH, Allen AO, Schwarz HA (1981) Mechanism of disproportionation of ascorbate radicals. J Am Chem Soc 103:3516–3518

Bigley RH, Stankova L (1974) Uptake and reduction of oxidized and reduced ascorbate by human leukocytes. J Exp Med 139:1084–1092

Bode AM, Cunningham L, Rose RC (1990) Spontaneous decay of oxidized ascorbic acid (dehydro-L-ascorbic acid) evaluated by high-pressure liquid chromatography. Clin Chem 36:1807–1809

Bornstein SR, Yoshida-Hiroi M, Sotiriou S, Levine M, Hartwig HG, Nussbaum RL, Eisenhofer G (2003) Impaired adrenal catecholamine system function in mice with deficiency of the ascorbic acid transporter (SVCT2). FASEB J 17:1928–1930

Brahma B, Forman RE, Stewart EE, Nicholson C, Rice ME (2000) Ascorbate inhibits edema in brain slices. J Neurochem 74:1263–1270

Buettner GR (1993) The pecking order of free radicals and antioxidants: lipid peroxidation, alpha-tocopherol, and ascorbate. Arch Biochem Biophys 300:535–543

Buettner GR, Jurkiewicz BA (1993) Ascorbate free radical as a marker of oxidative stress: an EPR study. Free Radic Biol Med 14:49–55

Burk RF, Christensen JM, Maguire MJ, Austin LM, Whetsell WO Jr, May JM, Hill KE, Ebner FF (2006) A combined deficiency of vitamins E and C causes severe central nervous system damage in guinea pigs. J Nutr 136:1576–1581

Castro M, Caprile T, Astuya A, Millán C, Reinicke K, Vera JC, Vásquez O, Aguayo LG, Nualart F (2001) High-affinity sodium-vitamin C co-transporters (SVCT) expression in embryonic mouse neurons. J Neurochem 78:815–823

Chatterjee IB (1970) Biosynthesis of L-ascorbate in animals. Methods Enzymol 18:28–34

Chatterjee IB, Chatterjee GC, Ghosh NC, Guha BC (1960) Biological synthesis of L-ascorbic acid in animal tissues: conversion of L-gulonolactone into L-ascorbic acid. Biochem J 74:193–203

Ciani E, Groneng L, Voltattorni M, Rolseth V, Contestabile A, Paulsen RE (1996) Inhibition of free radical production or free radical scavenging protects from the excitotoxic cell death mediated by glutamate in cultures of cerebellar granule neurons. Brain Res 728:1–6

Coassin M, Tomasi A, Vannini V, Ursini F (1991) Enzymatic recycling of oxidized ascorbate in pig heart: one-electron vs two-electron pathway. Arch Biochem Biophys 290:458–462

Daruwala R, Song J, Koh WS, Rumsey SC, Levine M (1999) Cloning and functional characterization of the human sodium-dependent vitamin C transporters hSVCT1 and hSVCT2. FEBS Lett 460:480–484

Del Bello B, Maellaro E, Sugherini L, Santucci A, Comporti M, Casini AF (1994) Purification of NADPH-dependent dehydroascorbate reductase from rat liver and its identification with 3a-hydroxysteroid dehydrogenase. Biochem J 304:385–390

Dhariwal KR, Hartzell WO, Levine M (1991) Ascorbic acid and dehydroascorbic acid measurements in human plasma and serum. Am J Clin Nutr 54:712–716

Drake BB, Smythe CV, King CG (1942) Complexes of dehydroascorbic acid with three sulfhydryl compounds. J Biol Chem 143:89–98

Due AD, Cook JA, Fletcher SJ, Zhi-Chao Q, Powers AC, May JM (1995) A "cysteineless" GLUT1 glucose transporter has normal function when expressed in Xenopus oocytes. Biochem Biophys Res Commun 208:590–596

Evans RM, Currie L, Campbell A (1982) The distribution of ascorbic acid between various cellular components of blood, in normal individuals, and its relation to the plasma concentration. Br J Nutr 47:473–482

García ML, Salazar K, Millán C, Rodríguez F, Montecinos H, Caprile T, Silva C, Cortes C, Reinicke K, Vera JC, Aguayo LG, Olate J, Molina B, Nualart F (2005) Sodium vitamin C cotransporter SVCT2 is expressed in hypothalamic glial cells. Glia 50:32–47

Gould DB, Phalan FC, Breedveld GJ, van Mil SE, Smith RS, Schimenti JC, Aguglia U, van der Knaap MS, Heutink P, John SW (2005) Mutations in Col4a1 cause perinatal cerebral hemorrhage and porencephaly. Science 308:1167–1171

Grünewald RA (1993) Ascorbic acid in the brain. Brain Res Rev 18:123–133

Hediger MA (2002) New view at C. Nat Med 8:445–446

Heller R, Unbehaun A, Schellenberg B, Mayer B, Werner-Felmayer G, Werner ER (2001) L-ascorbic acid potentiates endothelial nitric oxide synthesis via a chemical stabilization of tetrahydrobiopterin. J Biol Chem 276:40–47

Hill KE, Montine TJ, Motley AK, Li X, May JM, Burk RF (2003) Combined deficiency of vitamins E and C causes paralysis and death in guinea pigs. Am J Clin Nutr 77:1484–1488

Hillered L, Persson L, Bolander HG, Hallstrom A, Ungerstedt U (1988) Increased extracellular levels of ascorbate in the striatum after middle cerebral artery occlusion in the rat monitored by intracerebral microdialysis. Neurosci Lett 95:286–290

Hirschmann JV, Raugi GJ (1999) Adult scurvy. J Am Acad Dermatol 41:895–906

Hornig D (1975) Distributin of ascorbic acid, metabolites and analogues in man and animals. Ann NY Acad Sci 258:103–118

Hosoya K, Minamizono A, Katayama K, Terasaki T, Tomi M (2004) Vitamin C transport in oxidized form across the rat blood-retinal barrier. Invest Ophthalmol Vis Sci 45:1232–1239

Huang J, Agus DB, Winfree CJ, Kiss S, Mack WJ, McTaggart RA, Choudhri TF, Kim LJ, Mocco J, Pinsky DJ, Fox WD, Israel RJ, Boyd TA, Golde DW, Connolly ES Jr (2001) Dehydroascorbic

acid, a blood-brain barrier transportable form of vitamin C, mediates potent cerebroprotection in experimental stroke. Proc Natl Acad Sci USA 98:11720–11724
Hughes RE, Hurley RJ, Jones PR (1971) The retention of ascorbic acid by guinea-pig tissues. Br J Nutr 26:433–438
Ito A, Hayashi S, Yoshida T (1981) Participation of a cytochrome b5-like hemoprotein of outer mitochondrial membrane (OM cytochrome b) in NADH- semidehydroascorbic acid reductase activity of rat liver. Biochem Biophys Res Commun 101:591–598
Jackson TS, Xu AM, Vita JA, Keaney JF Jr (1998) Ascorbate prevents the interaction of superoxide and nitric oxide only at very high physiological concentrations. Circ Res 83:916–922
Jacques PF, Halpner AD, Blumberg JB (1995) Influence of combined antioxidant nutrient intakes on their plasma concentrations in an elderly population. Am J Clin Nutr 62:1228–1233
Jin SN, Mun GH, Lee JH, Oh CS, Kim J, Chung YH, Kang JS, Kim JG, Hwang DH, Hwang YI, Shin DH, Lee WJ (2005) Immunohistochemical study on the distribution of sodium-dependent vitamin C transporters in the respiratory system of adult rat. Microsc Res Tech 68:360–367
Kontush A, Mann U, Arlt S, Ujeyl A, Lührs C, Müller-Thomsen T, Beisiegel U (2001) Influence of vitamin E and C supplementation on lipoprotein oxidation in patients with Alzheimer's disease. Free Radic Biol Med 31:345–354
Korcok J, Yan R, Siushansian R, Dixon SJ, Wilson JX (2000) Sodium-ascorbate cotransport controls intracellular ascorbate concentration in primary astrocyte cultures expressing the SVCT2 transporter. Brain Res 881:144–151
Lee JY, Chang MY, Park CH, Kim HY, Kim JH, Son H, Lee YS, Lee SH (2003) Ascorbate-induced differentiation of embryonic cortical precursors into neurons and astrocytes. J Neurosci Res 73:156–165
Levine M, Conry-Cantilena C, Wang YH, Welch RW, Washko PW, Dhariwal KR, Park JB, Lazarev A, Graumlich JF, King J, Cantilena LR (1996) Vitamin C pharmacokinetics in healthy volunteers: evidence for a recommended dietary allowance. Proc Natl Acad Sci USA 93:3704–3709
Levine M, Wang YH, Padayatty SJ, Morrow J (2001) A new recommended dietary allowance of vitamin C for healthy young women. Proc Natl Acad Sci USA 98:9842–9846
Li X, Huang J, May JM (2003) Ascorbic acid spares alpha-tocopherol and decreases lipid peroxidation in neuronal cells. Biochem Biophys Res Commun 305:656–661
Liang WJ, Johnson D, Ma LS, Jarvis SM (2002) Regulation of the human vitamin C transporters and expressed in COS-1 cells by protein kinase C. Am J Physiol Cell Physiol 283, C1696–C1704
Lönnrot K, Metsä-Ketelä T, Molnár G, Ahonen JP, Latvala M, Peltola J, Pietilä T, Alho H (1996) The effect of ascorbate and ubiquinone supplementation on plasma and CSF total antioxidant capacity. Free Radic Biol Med 21:211–217
Lutsenko EA, Carcamo JM, Golde DW (2004) A human sodium-dependent vitamin C transporter 2 isoform acts as a dominant-negative inhibitor of ascorbic acid transport. Mol Cell Biol 24:3150–3156
MacGregor DG, Higgins MJ, Jones PA, Maxwell WL, Watson MW, Graham DI, Stone TW (1996) Ascorbate attenuates the systemic kainate-induced neurotoxicity in the rat hippocampus. Brain Res 727:133–144
Mack WJ, Mocco J, Ducruet AF, Laufer I, King RG, Zhang Y, Guo W, Pinsky DJ, Connolly ES Jr (2006) A cerebroprotective dose of intravenous citrate/sorbitol-stabilized dehydroascorbic acid is correlated with increased cerebral ascorbic acid and inhibited lipid peroxidation after murine reperfused stroke. Neurosurgery 59:383–388
Maellaro E, Del Bello B, Sugherini L, Santucci A, Comporti M, Casini AF (1994) Purification and characterization of glutathione-dependent dehydroascorbate reductase from rat liver. Biochem J 301:471–476
Majewska MD, Bell JA (1990) Ascorbic acid protects neurons from injury induced by glutamate and NMDA. Neuroreport 1:194–196

Majewska MD, Bell JA, London ED (1990) Regulation of the NMDA receptor by redox phenomena: inhibitory role of ascorbate. Brain Res 537:328–332

May JM (1988) Inhibition of hexose transport and labelling of the hexose carrier in human erythrocytes by an impermeant maleimide derivative of maltose. Biochem J 254:329–336

May JM, Asard H (2004) Ascorbate recycling. In: Asard H, May JM, Smirnoff N (eds) Vitamin C. Functions and biochemistry in animals and plants. Bios Scientific Publishers, London, pp 139–158

May JM, Qu ZC (2004) Redox regulation of ascorbic acid transport: role of transporter and intracellular sulfhydryls. Biofactors 20:199–211

May JM, Buchs A, Carter-Su C (1990) Localization of a reactive exofacial sulfhydryl on the glucose carrier of human erythrocytes. Biochemistry 29:10393–10398

May JM, Mendiratta S, Hill KE, Burk RF (1997) Reduction of dehydroascorbate to ascorbate by the selenoenzyme thioredoxin reductase. J Biol Chem 272:22607–22610

May JM, Cobb CE, Mendiratta S, Hill KE, Burk RF (1998a) Reduction of the ascorbyl free radical to ascorbate by thioredoxin reductase. J Biol Chem 273:23039–23045

May JM, Qu Z-C, Mendiratta S (1998b) Protection and recycling of a-tocopherol in human erythrocytes by intracellular ascorbic acid. Arch Biochem Biophys 349:281–289

May JM, Qu ZC, Cobb CE (2000) Extracellular reduction of the ascorbate free radical by human erythrocytes. Biochem Biophys Res Commun 267:118–123

May JM, Li L, Hayslett K, Qu ZC (2006) Ascorbate transport and recycling by SH-SY5Y neuroblastoma cells: response to glutamate toxicity. Neurochem Res 31:785–794

Mefford IN, Oke AF, Adams RN (1981) Regional distribution of ascorbate in human brain. Brain Res 212:223–226

Mehlhorn RJ (1991) Ascorbate- and dehydroascorbic acid-mediated reduction of free radicals in the human erythrocyte. J Biol Chem 266:2724–2731

Miele M, Boutelle MG, Fillenz M (1994) The physiologically induced release of ascorbate in rat brain is dependent on impulse traffic, calcium influx and glutamate uptake. Neuroscience 62:87–91

Milby K, Oke A, Adams RN (1982) Detailed mapping of ascorbate distribution in rat brain. Neurosci Lett 28:15–20

Mun GH, Kim MJ, Lee JH, Kim HJ, Chung YH, Chung YB, Kang JS, Hwang YI, Oh SH, Kim JG, Hwang DH, Shin DH, Lee WJ (2006) Immunohistochemical study of the distribution of sodium-dependent vitamin C transporters in adult rat brain. J Neurosci Res 83:919–928

Nazemi M, Staudinger H (1968) Kinetische Untersuchungen zur mitochondrialen NADH-semidehydroascorbate- oxydoreduktase (EC 1.6.5.4) in der Rattenleber. Hoppe Seylers Z Physiol Chem 349:345–348

Niki E, Noguchi N, Tsuchihashi H, Gotoh N (1995) Interaction among vitamin C, vitamin E, and b-carotene. Am J Clin Nutr 62(Suppl):1322S–1326S

Nualart FJ, Rivas CI, Montecinos VP, Godoy AS, Guaiquil VH, Golde DW, Vera JC (2003) Recycling of vitamin C by a bystander effect. J Biol Chem 278:10128–10133

Pastore P, Rizzetto T, Curcuruto O, Cin MD, Zaramella A, Marton D (2001) Characterization of dehydroascorbic acid solutions by liquid chromatography/mass spectrometry. Rapid Commun Mass Spectrom 15:2051–2057

Patel KB, Stratford MRL, Wardman P, Everett SA (2002) Oxidation of tetrahydrobiopterin by biological radicals and scavenging of the trihydrobiopterin radical by ascorbate. Free Radic Biol Med 32:203–211

Qiu S, Li L, Weeber EJ, May JM (2007) Ascorbate transport `by primary cultured neurons and its role in neuronal function and protection against excitotoxicity. J Neurosci Res 85:1046–1056

Ranjan A, Theodore D, Haran RP, Chandy MJ (1993) Ascorbic acid and focal cerebral ischaemia in a primate model. Acta Neurochir (Wien) 123:87–91

Rao A, Cha EM, Craig AM (2000) Mismatched appositions of presynaptic and postsynaptic components in isolated hippocampal neurons. J Neurosci 20:8344–8353

Rebec GV, Pierce RC (1994) A vitamin as neuromodulator: ascorbate release into the extracellular fluid of the brain regulates dopaminergic and glutamatergic transmission. Prog Neurobiol 43:537–565

Reiber H, Ruff M, Uhr M (1993) Ascorbate concentration in human cerebrospinal fluid (CSF) and serum. Intrathecal accumulation and CSF flow rate. Clin Chim Acta 217:163–173

Rice ME (2000) Ascorbate regulation and its neuroprotective role in the brain. Trends Neurosci 23:209–216

Rice ME, Russo-Menna I (1998) Differential compartmentalization of brain ascorbate and glutathione between neurons and glia. Neuroscience 82:1213–1223

Rice ME, Lee EJ, Choy Y (1995) High levels of ascorbic acid, not glutathione, in the CNS of anoxia-tolerant reptiles contrasted with levels in anoxia-intolerant species. J Neurochem 64:1790–1799

Rose RC (1988) Transport of ascorbic acid and other water-soluble vitamins. Biochim. Biophys. Acta 947:335–366

Seno T, Inoue N, Matsui K, Ejiri J, Hirata K, Kawashima S, Yokoyama M (2004) Functional expression of sodium-dependent vitamin C transporter 2 in human endothelial cells. J Vasc Res 41:345–351

Shimizu N, Matsunami T, Onishi S (1960) Histochemical demonstration of ascorbic acid in the locus coeruleus of the mammalian brain. Nature 186:479–480

Siushansian R, Wilson JX (1995) Ascorbate transport and intracellular concentration in cerebral astrocytes. J Neurochem 65:41–49

Sotiriou S, Gispert S, Cheng J, Wang YH, Chen A, Hoogstraten-Miller S, Miller GF, Kwon O, Levine M, Guttentag SH, Nussbaum RL (2002) Ascorbic-acid transporter Slc23a1 is essential for vitamin C transport into the brain and for perinatal survival. Nature Med 8:514–517

Spector R (1977) Vitamin homeostasis in the central nervous system. N Engl J Med 296:1393–1398

Spector R, Lorenzo AV (1973) Ascorbic acid homeostasis in the central nervous system. Am J Physiol 225:757–763

Stevenson NR, Brush MK (1969) Existence and characteristics of Na+-dependent active transport of ascorbic acid in guinea pig. Am J Clin Nutr 22:318–326

Takanaga H, Mackenzie B, Hediger MA (2004) Sodium-dependent ascorbic acid transporter family SLC23. Pflugers Arch 447:677–682

Taylor A, Jacques PF, Nowell T, Perrone G, Blumberg J, Handelman G, Jozwiak B, Nadler D (1997) Vitamin C in human and guinea pig aqueous, lens and plasma in relation to intake. Curr Eye Res 16:857–864

Tolbert BM, Ward JB (1982) Dehydroascorbic acid. In: Seib PA, Tolbert BM (eds) Ascorbic acid: chemistry, metabolism, and uses. American Chemical Society, Washington, DC, pp 101–123

Tsukaguchi H, Tokui T, Mackenzie B, Berger UV, Chen X-Z, Wang YX, Brubaker RF, Hediger MA (1999) A family of mammalian Na+-dependent L-ascorbic acid transporters. Nature 399:70–75

VanDuijn MM, Tijssen K, VanSteveninck J, van den Broek PJA, Van der Zee J (2000) Erythrocytes reduce extracellular ascorbate free radicals using intracellular ascorbate as an electron donor. J Biol Chem 275:27720–27725

Vera JC, Rivas CI, Fischbarg J, Golde DW (1993) Mammalian facilitative hexose transporters mediate the transport of dehydroascorbic acid. Nature 364:79–82

Villalba JM, Canalejo A, Rodríguez-Aguilera JC, Burón MI, Moore DJ, Navas P (1993) NADH-ascorbate free radical and -ferricyanide reductase activities represent different levels of plasma membrane electron transport. J Bioenerg Biomembr 25:411–417

Washburn MP, Wells WW (1999) Identification of the dehydroascorbic acid reductase and thioltransferase (glutaredoxin) activities of bovine erythrocyte glutathione peroxidase. Biochem Biophys Res Commun 257:567–571

Wilson JX (1997) Antioxidant defense of the brain: a role for astrocytes. Can J Physiol Pharmacol 75:1149–1163

Wilson JX (2004) Vitamin C transport in animals and plants. In: Asard H, May JM, Smirnoff N (eds) Vitamin C. Functions and biochemistry in animals and plants. Bios Scientific Publishers, London, pp 97–113

Wilson JX, Jaworski EM, Kulaga A, Dixon SJ (1990) Substrate regulation of ascorbate transport activity in astrocytes. Neurochem. Res 15:1037–1043

Wilson JX, Peters CE, Sitar SM, Daoust P, Gelb AW (2000) Glutamate stimulates ascorbate transport by astrocytes. Brain Res 858:61–66

Winkler BS (1987) In vitro oxidation of ascorbic acid and its prevention by GSH. Biochim Biophys Acta 925:258–264

Winkler BS, Orselli SM, Rex TS (1994) The redox couple between glutathione and ascorbic acid: a chemical and physiological perspective. Free Radic Biol Med 17:333–349

Wu X, Itoh N, Taniguchi T, Nakanishi T, Tanaka K (2003) Requirement of calcium and phosphate ions in expression of sodium-dependent vitamin C transporter 2 and osteopontin in MC3T3-E1 osteoblastic cells. Biochim Biophys Acta 1641:65–70

Yusa T (2001) Increased extracellular ascorbate release reflects glutamate re-uptake during the early stage of reperfusion after forebrain ischemia in rats. Brain Res 897:104–113

Chapter 7
Genetic Aspects of Folate Metabolism

Anne M. Molloy

Abstract The vitamin folate functions within the cell as a carrier of one-carbon units. The requirement for one-carbon transfers is ubiquitous and all mammalian cells carry out folate dependent reactions. In recent years, low folate status has been linked to risk of numerous adverse health conditions throughout life from birth defects and complications of pregnancy to cardiovascular disease, cancer and cognitive dysfunction in the elderly. In many instances inadequate intake of folate seems to be the primary contributor but there is also evidence that an underlying genetic susceptibility can play a modest role by causing subtle alterations in the availability, metabolism or distribution of intermediates in folate related pathways. Folate linked one-carbon units are essential for DNA synthesis and repair and as a source of methyl groups for biological methylation reactions. The notion of common genetic variants being linked to risk of disease was relatively novel in 1995 when the first functional folate-related polymorphism was discovered. Numerous polymorphisms have now been identified in folate related genes and have been tested for functionality either as a modifier of folate status or as being associated with risk of disease. Moreover, there is increasing research into the importance of folate-derived one-carbon units for DNA and histone methylation reactions, which exert crucial epigenetic control over cellular protein synthesis. It is thus becoming clear that genetic aspects of folate metabolism are wide-ranging and may touch on events as disparate as prenatal imprinting to cancer susceptibility. This chapter will review the current knowledge in this area.

Keywords Folate · Vitamin B12 · Homocysteine · Methylenetetrahydrofolate reductase (MTHFR)

7.1 Introduction

The availability of one-carbon units is critical to cell proliferation and metabolism. The vitamin folate acts as a transitional molecule in the transfer of one-carbon units from donor molecules to various intermediates in the synthesis of DNA and in

---

A.M. Molloy (✉)
School of Medicine, Trinity College Dublin, Dublin 2, Ireland
e-mail: amolloy@tcd.ie

O. Stanger (ed.), *Water Soluble Vitamins*, Subcellular Biochemistry 56,
DOI 10.1007/978-94-007-2199-9_7,

biological methylation reactions, including the methylation of DNA, proteins, phospholipids, neurotransmitters, etc. Accordingly, maintenance of folate homeostasis is essential to growth and development. The effects of clinical folate deficiency on rapidly proliferating tissues such as those of the erythropoietic system are long known and well described (Lindenbaum and Allen 1995). More recent studies go beyond this knowledge by implicating a mild to moderately disrupted folate homeostasis in increased risk of numerous adverse health conditions throughout an individual's lifetime, including birth defects and other unfavorable developmental outcomes, cardiovascular disease (CVD), some cancers and cognitive dysfunction (Choi and Mason 2002; Herrmann and Knapp 2002; Mangoni and Jackson 2002; Molloy et al. 2008; Rampersaud et al. 2003; Ravaglia et al. 2005; Ueland et al. 2000; Vollset et al. 2000; Xu and Chen 2009). These diseases had long been known to have complex etiologies, involving genetic and environmental components.

The close association of plasma total homocysteine (tHcy) with folate status has generated more research interest than any other aspect of folate metabolism and has been a key analysis in determining the functionality of newly discovered polymorphisms in folate related genes. The first suggestions that common genetic diversity in folate related genes might influence risk of disease came from work by Kang et al. (1988a) who studied tHcy concentrations in subjects with CVD and identified a subset of patients with mild hyperhomocysteinemia who appeared to have a thermolabile form of the folate metabolizing enzyme 5,10-methylenetetrahydrofolate reductase (MTHFR). Around the same time, a growing body of evidence culminated in clinical trials that conclusively showed that periconceptional supplementation with folic acid could prevent the majority of neural tube defects (NTDs) (Czeizel and Dudas 1992; MRC 1991). Women who had given birth to children with NTDs were not clinically deficient in folate but they did tend to have lower blood folate concentrations (Daly et al. 1995), suggesting an interaction of folate nutrition and genetic factors that was negatively affecting some aspect of folate related pathways at the time of neural tube closure. Current knowledge of the genetic aspects of folate metabolism has been driven primarily by the search for polymorphisms in candidate genes conferring NTD risk (Molloy et al. 2009) but this has expanded logically to examine functionality in other respects by searching for alternative clinical or biochemical phenotypes such as an association of such polymorphisms with CVD or cancer and an effect on blood indices of folate status, mainly serum folate, red cell folate and tHcy. Numerous polymorphisms have now been identified in folate pathway genes and have been examined for such functionality. Perhaps not surprisingly, apart from the 677C→T polymorphism in the MTHFR gene (discussed in more detail later), the outcomes of these association studies have been inconsistent and often confusing, due largely to the fact that such common genetic diversity has a low penetrance in relation to disease and biochemical phenotype. Because of this low penetrance, studies are often limited by factors such as low sample size and inadequate power to detect small differences between groups; inappropriate control selection for such type of study; variability of genotype prevalence in different ethnic groups and complex interactions between nutrient status and genotype prevalence in different populations (Amorim et al. 2007; Gueant-Rodriguez et al. 2006; Mitchell et al. 2004).

An alternative line of research has focused on the links between folate and metabolic processes that control gene expression. These studies address the important question of whether low folate status can alter the quantity and pattern of methylated DNA or methylated histone in specific tissues, including DNA and histone methylations that may influence early carcinogenesis, or methylations in utero that may form part of the fetal programming of genes for long-term expression and silencing (Antony 2007; Duthie et al. 2004; Esteller 2008; Lillycrop et al. 2008; Sinclair et al. 2007; Xu and Chen 2009). Although the technology to carry out this type of research with adequate sensitivity is relatively new and is still being developed, impressive evidence is already accumulating to confirm a fundamental role of folate dependent one-carbon metabolism in epigenetic processes and a susceptibility of these processes to changes in folate status.

7.2 Folate Cofactor Interactions

The biochemical role of the folate cofactors in one-carbon metabolism has been documented in numerous reviews (Bailey and Gregory 1999; Mason 2003; Scott and Weir 1998; Selhub 2002; Stover 2004). Tetrahydrofolate (THF) is the biologically active parent molecule and consists of a fully reduced pteridine ring linked to a para-aminobenzoyl unit and thence to a glutamic acid residue (Fig. 7.1). Within the cell, up to eight additional glutamic acid residues are added to the molecule in order to ensure its retention in the cell and increase its affinity for folate dependent enzymes (Shane 1989). THF carries one-carbon units as methyl (CH3), formyl (CHO), methylene (CH2), methenyl (CH=) or formimino (CHNH) derivatives on carbon 5, or carbon 10 of the pteridine ring, or as a 5,10 adduct of the pteroyl-para-aminobenzoyl moiety.

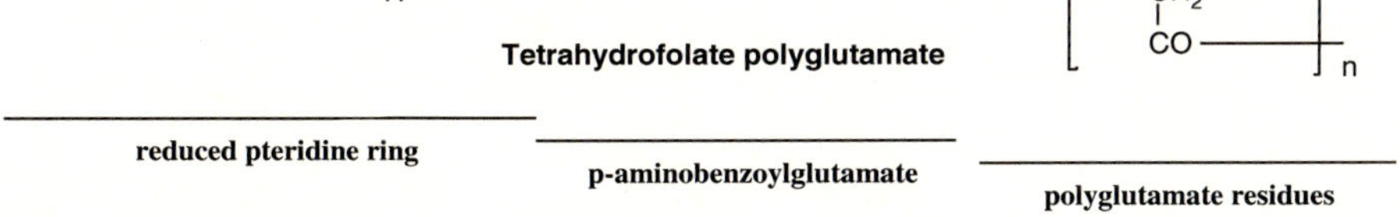

Fig. 7.1 Tetrahydrofolate (THF) polyglutamate, showing the reduction of the pteridine ring at positions 5,6,7 and 8 of the pteridine ring; the attachment of the pteridine to p-aminobenzoylglutamate at the C9–N10 bond; and the presence of a number of polyglutamate residues. One-carbon units are added to this molecule at positions 5 and 10 or as a 5,10 adduct of the pteroyl-para-aminobenzoyl moiety

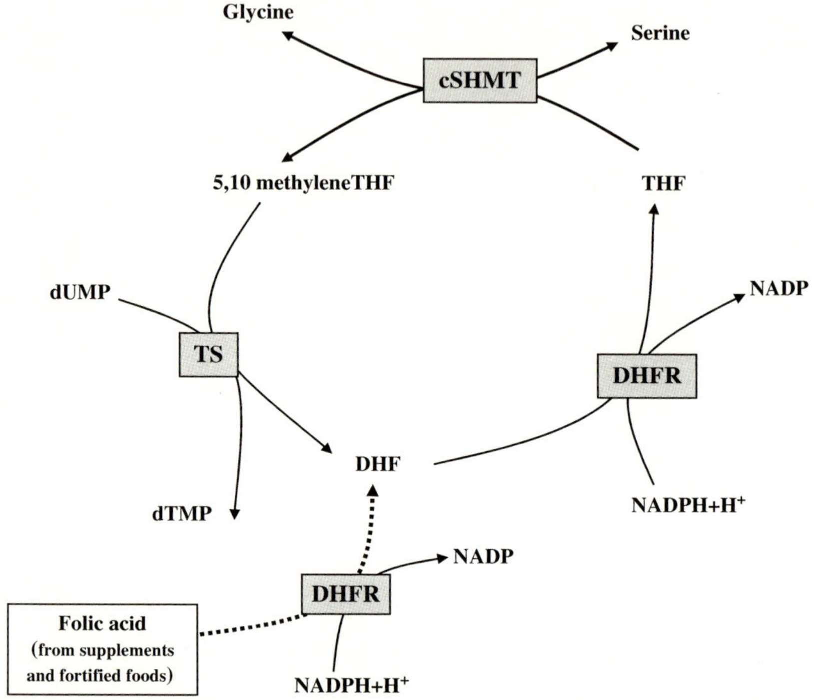

Fig. 7.2 Folate dependent thymidylate synthesis: the enzymes in this pathway are thymidylate synthase (TS); dihydrofolate reductase (DHFR); and cytosolic serine hydroxymethyltransferase (cSHMT). The metabolites are deoxyuridine monophosphate (dUMP); deoxythymidine monophosphate (dTMP); dihydrofolate (DHF); tetrahydrofolate (THF). Synthesis of thymidylate occurs in the nuclear fraction of the cell

One-carbon units are obtained from several sources, most importantly from mitochondrially generated formate through the trifunctional C1-synthase enzyme encoded by the MTHFD1 gene and from serine, through cytosolic and mitochondrial isoforms of serine hydroxymethyltransferase (SHMT). As folate derivatives, these one-carbon units are transferred to specific intermediates in the de-novo synthesis of purines and thymidylate and in the synthesis of methionine from homocysteine. The biological importance of folate lies in the crucial downstream functions of these molecules, as summarized below.

7.3 Folate in DNA Synthesis

Two metabolic systems in the folate network transfer folate cofactors to intermediates in DNA synthesis. In one system, deoxyuridine monophosphate (dUMP) is converted to deoxythymidine monophosphate (dTMP) by the addition of a methyl group from 5,10-methylene THF. Thymidylate synthase (TS) catalyzes this reaction and the co-product dihydrofolate (DHF) is then converted to THF via dihydrofolate reductase (DHFR). The cycle is completed with THF accepting another one-carbon unit and regenerating 5,10 methylene THF (principally via cSHMT). Recent work demonstrates that this thymidylate synthesis system is under a high degree of metabolic control and the three enzymes involved in de-novo thymidylate synthesis become localised to the nucleus or the nuclear periphery for activity (Anderson et al. 2007). A common non-synonymous single nucleotide polymorphism (SNP):1420C→T (L474F), in the cSHMT gene has been shown to weaken the interaction of SHMT with other components in this process (Woeller et al. 2007). Apart from the obvious importance that the thymidylate synthesis system has in relation to cell proliferation, there has been considerable recent research on its role in the ongoing maintenance and repair of DNA. Impaired folate status

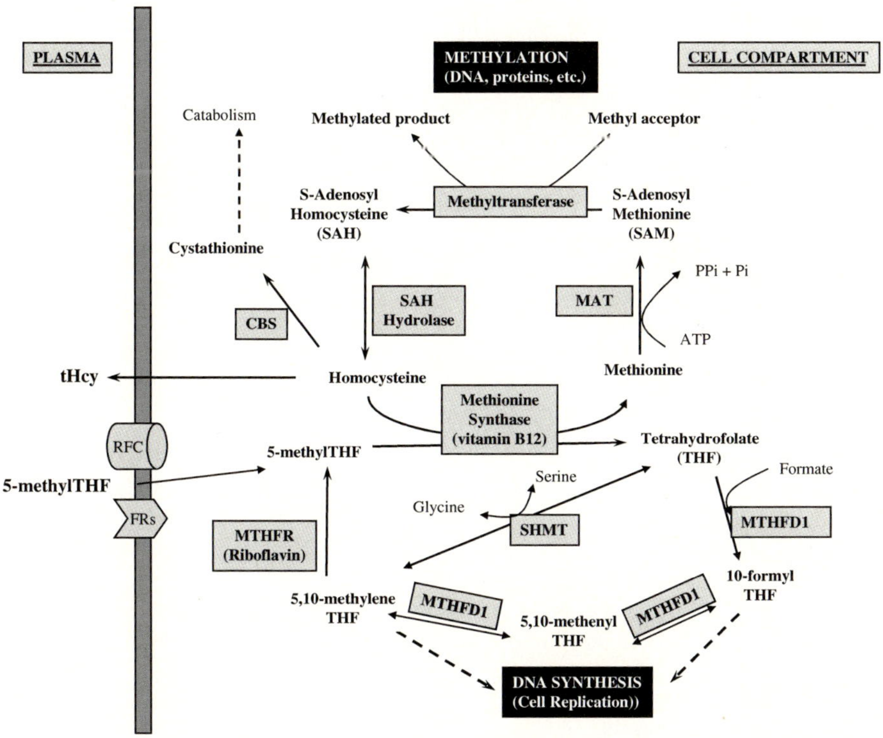

Fig. 7.3 The flux of one-carbon units through folate cofactors: the enzymes annotated in this figure are 5,10-methylenetetrahydrofolate reductase (MTHFR); methionine adenosyltransferase (MAT); S-adenosylhomocysteine hydrolase (SAH hydrolase); cystathionine-β-synthase (CBS); serine hydroxymethyltransferase (SHMT); the trifunctional C1-synthase enzyme incorporating the activities of formyl-THF synthetase, cyclohydrolase and dehydrogenase activities (MTHFD1, the gene encoding these enzyme activities). The transport of reduced folates into the cell is facilitated primarily by the reduced folate carrier (RFC). Folate receptors (FRs) also transport folates into cells. Plasma total homocysteine is denoted by tHcy

in the microenvironment of a specific tissue can result in abnormalities of DNA synthesis, such as uracil misincorporation into DNA, leading to replication abnormalities, strand breaks, increased micronucleus formation and genomic instability (Duthie et al. 2004; Fenech 2001). It is now accepted that these effects occur at sub-optimal folate concentrations above the clinically deficient cut-off level. Persistent uracil misincorporation into DNA has been one of the primary hypotheses for explaining the role of folate in the development of certain cancers (Choi and Mason 2002; Xu and Chen 2009).

Several studies have examined the SHMT 1420C→T polymorphism for association with disease risk. No significant association with NTD was observed (Heil et al. 2001; Relton et al. 2004) but a positive link was found between the occurrence of CVD in persons with the T allele, particularly in those who also carried the MTHFR 677 TT genotype (Lim et al. 2005). Common polymorphisms in TS and DHFR have also been studied, in particular a polymorphic 28 bp double or triple repeat in the TS promoter enhancer region (TSER) and a 19 bp insertion/deletion polymorphism in DHFR. In general, reported association studies of these polymorphisms with NTDs and with certain forms of cancer have been inconsistent (Kalmbach et al. 2008; Parle-McDermott et al. 2007; Skibola et al. 2002; Ulrich et al. 2005; Volcik et al. 2003) although it appears that an interaction between low dietary folate status and the TSER polymorphism may provide a decreased risk of some forms of cancer (Skibola et al. 2002; Ulrich et al. 2005). Interestingly, recent work suggests that the DHFR 19 bp polymorphism might influence an individual's ability to convert folic acid, the folate precursor found in vitamin supplements and fortified foods, to biologically active THF (Kalmbach et al. 2008). This is a biologically plausible effect since reduction to DHF and thence to THF via the enzyme DHFR is the mechanism whereby folic acid enters the cellular folate pool.

In the second DNA synthesis system, one-carbon units from mitochondrially-generated formate enter the tri-functional C1-synthase complex as 10-formyl THF via the enzyme's 10-formylTHF synthetase domain. These folate cofactors then participate in the formation of intermediates in de-novo purine synthesis (Anguera et al. 2006; Christensen and MacKenzie 2006; Fowler et al. 1997) or undergo rearrangement to 5,10 methyleneTHF for use in thymidylate synthesis or for channelling through MTHFR to 5-methylTHF. A polymorphism in the synthetase domain of MTHFD1 (R653Q) has been associated with NTDs and other pregnancy complications (Brody et al. 2002; De Marco et al. 2006; Parle-McDermott et al. 2005a, b). Consistent with the deleterious nature of the R653Q polymorphism, mice lacking this synthetase domain have been found not to survive the embryonic period (MacFarlane et al. 2009), but little else is known of the phenotype.

7.4 Folate in Methylation Reactions

The third fate of folate derived one-carbon units is in the remethylation of homocysteine to methionine, catalyzed by the vitamin B12 (cobalamin) dependent enzyme, methionine synthase. This reaction is crucial to maintain a supply of methyl groups through S-adenosylmethionine (SAM) for the methylation of DNA, proteins, neurotransmitters, etc. Methionine synthase, encoded by the MTR gene, is one of only two enzymes in mammalian systems that are known to require vitamin B12 as a cofactor, in this case as methylcobalamin. This cofactor is prone to oxidation and the enzyme therefore requires a reductive methylation system involving the flavoprotein methionine synthase reductase and SAM as the methyl donor to re-prime the enzyme every 200–1,000 turnovers (Bandarian et al. 2003; Yamada et al. 2006). Because of this, adequate functioning of both methionine synthase and methionine synthase reductase is required to maintain the methylation of homocysteine and the provision of methyl groups to SAM mediated methyltransferase reactions. Genetic mutations in these two enzymes are responsible for two complementation classes of vitamin B12 deficiency, CblG and CblE (Leclerc et al. 1996; Watkins et al. 2002; Wilson et al. 1999a). Complete ablation of the MTR gene in mice is embryonically lethal and mice lacking a functional MTRR gene (that encodes methionine synthase reductase) also show evidence of reproductive failure (Deng et al. 2008; Elmore et al. 2007). Common polymorphisms have been described in the genes for both enzymes. The most widely studied of these are the 2756A→G (D919G) variant in MTR, and the 66A→G (I22M) variant in MTRR. These polymorphisms have been evaluated in numerous studies for metabolic and disease risk effects. Modest impacts on vitamin B12 status, plasma tHcy concentration and on birth defect, CVD and cancer risk have been reported but the results are not consistent between studies and several large highly-powered negative studies have tended to cast doubt on the reliability of smaller positive studies (Bosco et al. 2003; Chen et al. 1997:1998, 2001; Christensen et al. 1999; Dekou et al. 2001; Klerk et al. 2003; Tsai et al. 1999a; Ulrich et al. 2005; Ulvik et al. 2004; Wilson et al. 1999b; Yang et al. 2008).

As indicated above, SAM is the universal donor of methyl groups for almost all biological methyltransferase reactions, involving in excess of 40 known enzymes in mammalian systems. SAM is synthesized from methionine and ATP via methionine adenosyltransferase (MAT). Liver expresses a MATIA isoform, which can produce high SAM concentrations as the intracellular methionine level increases (Finkelstein 2007; Molloy et al. 1990). Most other tissues express the MAT II isoform, which is inhibited by high tissue levels of SAM, ensuring that utilization and synthesis of SAM are tightly balanced (Kotb and Kredich 1990; LeGros et al. 2001). Severe genetic mutations in the MATIA gene lead to hyper-methioninemia without hyperhomocysteinemia and severe deficiency can lead to neurological symptoms similar to those seen in deficiency of MTHFR but little is known of genetic mutations in the MATII gene or of disease risk or biochemical effects associated with common polymorphisms in these genes. (Chamberlin et al. 2000; Couce et al. 2008; Perez Mato et al. 2001; Surtees et al. 1991; Ubagai et al. 1995).

The simultaneous conversion of SAM to S-adenosylhomocysteine (SAH) in methyltransferase reactions, followed by the hydrolysis of the SAH adenosyl moiety via SAH hydrolase, regenerates homocysteine and completes this methylation cycle whereby methyl groups from the folate one-carbon pool are incorporated into a multitude of biological compounds. Along with folate, homocysteine is an essential carrier of methyl groups in this process since folate derived methyl groups are provided to the methionine cycle via the constant recycling of homocysteine to form methionine. Methylation is a crucial regulatory modification that affects a wide spectrum of housekeeping functions within the cell. A decrease in availability of methyl groups from the folate cofactor pool leads to impaired recycling of homocysteine, and inhibition of methylation reactions, including those that regulate gene expression and other epigenetic or post-translational processes. The inhibition of methyltransferase reactions is, to a large extent, due to the equilibrium of the SAH hydrolase reaction, which favours synthesis of SAH rather than hydrolysis and leads to a build-up of intracellular SAH, a potent competitive inhibitor of all methyltransferase reactions (Caudill et al. 2001; Finkelstein 2007; McKeever et al. 1995; Molloy et al. 1992; Yi et al. 2000). Individuals with genetic mutations in SAH hydrolase show markedly elevated plasma SAH concentrations (Baric et al. 2004; Buist et al. 2006).

The influence of folate status on the methylation cycle has been difficult to demonstrate although many current studies are addressing this issue by examining the effect of folate deficiency on total and site specific genome DNA methylation patterns (Basten et al. 2006; Jacob et al. 1998; Shelnutt et al. 2004; Wasson et al. 2006; Yi et al. 2000). Histone methylation, which provides the next layer of epigenetic control (Esteller 2008; Lee and Mahadevan 2009; Mathers and McKay 2009) is also potentially susceptible to folate status. It is now thought that in conjunction with uracil misincorporation into DNA, as noted earlier, changes in DNA or histone methylation, resulting from persistent marginally low folate status, may be an important early mechanism predisposing to cancer development (Duthie et al. 2002; Fenech 2001; Friso and Choi 2002; Jang et al. 2005; Mason 2003). Furthermore, a line of current research is exploring the notion that maternal prenatal folate status may, through its effects on embryonic methylation reactions, influence the expression and repression of developmental genes and may culminate in permanent modifications to the expression of certain genes in the offspring (Burdge et al. 2009; Lillycrop et al. 2008; Sinclair et al. 2007; Zeisel 2009). Such changes may provide mechanistic support to the theory that there are developmental origins to diseases in later life (Mathers and McKay 2009; Zeisel 2009).

7.5 Homocysteine

Unlike folate, homocysteine is a highly reactive and toxic metabolite whose level must be carefully balanced within the cell. The ongoing requirement for methylation of DNA, proteins, lipids, neurotransmittors and other biological intermediates within the cell generates a persistent supply of homocysteine that either must be recycled to methionine or catabolized via cystathionine β-synthase, a vitamin B6 dependent enzyme that is only expressed at high levels in liver and kidney (Finkelstein 2000). The gene encoding cystathionine β-synthase has been well described and a number of severe genetic mutations in this gene have been shown to cause the classical clinical condition of homocystinuria (Kraus 1998; Linnebank et al. 2004; Miles and Kraus 2004; Vyletal et al. 2007). A relatively common 68 bp insertion polymorphism (844ins68) has been studied for its effect on plasma tHcy concentrations (Tsai et al. 1996, 1999b). Some studies suggest that this polymorphism interacts with polymorphisms in MTHFR and MTRR to modulate the plasma tHcy concentration, but the phenotypic effect is small (Dekou et al. 2001; Kluijtmans et al. 2003; Silaste et al. 2001; Summers et al. 2008; Tsai et al. 1999b).

Liver and kidney also contain high levels of a second enzyme, betaine-homocysteine methyltransferase (BHMT) that is capable of remethylating homocysteine independently of folate (Finkelstein 2007; Sunden et al. 1997; Ueland et al. 2005). The activity of this enzyme has an important sparing effect on folate in these tissues and the substrate, betaine, that provides the methyl group in the reaction, is useful as an alternative clinical therapy for lowering plasma tHcy in individuals with abnormally high tHcy concentrations (Holm et al. 2007; Melse-Boonstra et al. 2005; Schwab et al. 2006; Schwahn et al. 2007; Ueland et al. 2005). A common polymorphism in the BHMT gene (742G→A; R239Q) has been studied for disease or functional associations but no strong or consistent trends have been reported to date (Ananth et al. 2007; Boyles et al. 2006; da Costa et al. 2006; Fredriksen et al. 2007; Weisberg et al. 2003; Zhu et al. 2005).

7.6 Methylenetetrahydrofolate Reductase

One enzyme in particular, MTHFR, exerts central control over the partitioning of folate cofactors between methylation and DNA synthesis reactions. MTHFR catalyzes the essentially irreversible conversion of 5,10-methyleneTHF to 5-methylTHF. This enzyme therefore channels one-carbon units away from purine and pyrimidine synthesis and into the provision of methyl groups for SAM mediated methylation reactions. MTHFR is a flavoprotein containing loosely bound flavin adenine dinucleotide (FAD) that accepts reducing equivalents from NADP(H). The enzyme exhibits strong allosteric inhibition by SAM, the downstream product of the MTHFR-methionine synthase metabolic locus (Daubner and Matthews 1982; Kutzbach and Stokstad 1971). In mammals, the enzyme exists as a homo-dimer of about 77 kd, containing a catalytic domain and a SAM regulatory domain (Matthews et al. 1984). Several features are remarkable about the MTHFR-methionine synthase locus. Firstly, the reduction of 5,10-methyleneTHF via MTHFR is the only biological process known to generate 5-methylTHF. Secondly, newly absorbed dietary folates that are not retained within the enteric or hepatic folate pools are released into the circulation as 5-methylTHF. This therefore is the predominant form of the vitamin found in the circulation and presented to the cell. 5-methylTHF must be de-methylated through methionine synthase in order to be incorporated into the intracellular folate pool, therefore optimal activities of both MTHFR and methionine synthase (along with adequate availability of three vitamins; folate, vitamin B12 and riboflavin) are required to remethylate homocysteine and ensure constant functioning of methylation processes and also to incorporate circulating folate into the cell. Recent research has shown that MTHFR essentially competes with cSHMT for available 5,10-methyleneTHF (Herbig et al. 2002; Oppenheim et al. 2001; Woeller et al. 2007) indicating that there are several layers of control in maintaining the balance of folate cofactors between DNA synthesis and methylation reactions.

7.7 The MTHFR677C→T Polymorphism

Genetic inborn errors in MTHFR that give rise to less efficient function and animal models that diminish MTHFR activity reduce the supply of methyl groups to the methylation cycle. These conditions are characterised by abnormally high plasma tHcy concentrations and display a variety of neurological and vascular symptoms (Baumgartner et al. 1985; Chwatko et al. 2007; Erbe 1979; Goyette et al. 1996; Li et al. 2008; Rosenblatt 1995; Schwahn et al. 2007; Sibani et al. 2003). In the 1980s, Kang and co-workers (Kang et al. 1988a, b) discovered a 'thermolabile' version of MTHFR in patients with coronary artery disease, which they predicted was due to a genetic mutation in MTHFR. When the cDNA for the human MTHFR gene was isolated, (Goyette et al. 1994) a common polymorphism was discovered (677C→T) that resulted in an amino acid change in the protein from alanine to valine (A222V) (Frosst et al. 1995). The variant enzyme was thermolabile in vitro; i.e. lymphocyte cultures from homozygote individuals displayed reduced total MTHFR activity and significantly lower residual activity after heating to 42°C (Frosst et al. 1995). The observation that the polymorphism was also a cause of mild hyper-homocysteinemia helped to convince researchers that this was the mutation predicted by Kang and co-workers.

The polymorphism has been studied in great detail in the past decade and is somewhat unique among folate polymorphisms as having known functionality in terms of molecular mechanism, biochemical phenotype, effect modification by nutrient status and impact on risk of disease. Studies in vitro using E. coli constructs (the homologous mutant being A177V) (Guenther et al. 1999; Pejchal et al. 2006) or recombinant human enzyme (Yamada et al. 2001) demonstrated weaker binding affinity for the flavin cofactor when the protein contained valine instead of alanine at this position. These workers showed that in vitro the valine containing enzyme easily loses its flavin cofactor in conditions of low riboflavin or low folate, causing the biologically active tetrameric form of the enzyme to break down into inactive monomers, but that the presence of high folate concentrations could stabilize FAD binding and preserve enzyme functionality. Moreover, these effects could be modulated by SAM, indicating that riboflavin, folate, and methionine would all be important for maintenance of MTHFR activity in individuals with the variant enzyme (Yamada et al. 2001). These conclusions have now been abundantly supported by cell culture and human studies. Individuals homozygous for the TT genotype tend to have higher plasma tHcy (Frosst et al. 1995; Harmon et al. 1996; Hustad et al. 2007), and lower plasma and red cell folate concentrations, (Kluijtmans et al. 2003; Molloy et al. 1997; Yang et al. 2008) suggesting an increased requirement for folate. There is an altered distribution of folate cofactors within the cell (Bagley and Selhub 1998; Smulders et al. 2007) and riboflavin status appears to be critical to maintaining adequate enzyme function in TT homozygotes (Hustad et al. 2007; Jacques et al. 2002; Lathrop Stern et al. 2003; McNulty et al. 2002:2006). Several studies have shown that increased folate and riboflavin intake can lower the plasma tHcy in TT individuals (Ashfield-Watt et al. 2002; Chuang et al. 2006; de Bree et al. 2003; Jacques et al. 2002; McNulty et al. 2006; Yang et al. 2008). Recently, it was suggested that low vitamin B12 status might also modulate the effect of the TT genotype on plasma tHcy (Bailey et al. 2002; D'Angelo et al. 2000; Hustad et al. 2007).

The distribution of the polymorphism varies considerably worldwide. Mexican and Southern European populations have the highest frequency with some studies reporting more than 30% of individuals homozygous for the TT genotype. This contrasts with a TT homozygous frequency of about 12% in Northern Europeans and less than 1% among African groups (Botto and Yang 2000; Esfahani et al. 2003; Gueant-Rodriguez et al. 2006; Kirke et al. 2004; Mutchinick et al. 1999; Pepe et al. 1998; Wilcken et al. 2003). Such heterogeneity in prevalence has contributed to the difficulty in establishing the polymorphism as a risk factor for disease even though many studies on this topic have been reported (Ueland et al. 2001). To date, there is good evidence that the MTHFR 677C→T polymorphism confers increased risk of neural tube defects (Botto and Yang 2000; de Franchis et al. 2002; Kirke et al. 2004; Shields et al. 1999) and it may also be associated with other congenital defects (Botto and Yang 2000; Brouns et al. 2008; Mills et al. 2008). Many research groups have examined the prevalence of TT versus CC homozygotes among individuals with various forms of CVD, because of the association of the polymorphism with elevated plasma tHcy concentrations, an established risk factor for CVD. Two important meta-analyses of these studies (Klerk et al. 2002; Wald et al. 2002) confirmed a higher risk of vascular disease among TT individuals, although wide variations in risk were observed between populations, highlighting the difficulties in carrying out disease risk studies for genetic variants that can be modified by diet or environment. The polymorphism has also been studied as a risk factor for several forms of cancer (Boccia et al. 2008; Chen et al. 1998; Choi and Mason 2002; Ma et al. 1997; Robien and Ulrich 2003; Ryan et al. 2001; Sharp and Little 2004; Suzuki et al. 2008; Ulvik et al. 2004). Again there is considerable disagreement between studies, but in general, there are trends towards an interaction between folate status and risk, suggesting that when folate status is above average, TT status may be associated with decreased risk. Such an effect is consistent with a proportionally higher intracellular content of folate cofactors in the form available for maintenance and repair of DNA.

In summary of the above, human cross-sectional and intervention studies on the 677C→T polymorphism have led researchers to conclude that MTHFR is at the centre of a 'functional metabolic network' that may have important effects on nutritional status and disease risk (Hustad et al. 2007; Yang et al. 2008). It is important to note that a second polymorphism in the MTHFR gene:1298A→C, has also been described and tested for association with risk of several disease conditions and pregnancy complications. This SNP is in strong linkage disequilibrium with the 677C-T variant (Stegmann et al. 1999). Most studies find no discernable effect of the 1298C variant when considered independently of the 677C→T polymorphism, and in those studies that have found effects, the results tend to conflict (Devlin et al. 2006; Fredriksen et al. 2007; Nurk et al. 2004; Parle-McDermott et al. 2003, 2006; Stegmann et al. 1999; Ulvik et al. 2007).

7.8 Folate Absorption and Uptake into Cells

Because folate is not synthesized in mammalian tissues, the bioavailability of ingested dietary folate and the function of enzymes and transporters that move folates across intestinal and tissue membranes are an important aspect in the regulation of folate homeostasis (Gregory 1995; Hamid et al. 2009; McNulty and Pentieva 2004; Molloy 2002). All of the monoglutamate folate forms are extremely hydrophilic and have a low capacity to diffuse across biological membranes, therefore several carrier-mediated systems are responsible for the absorption of folate from the intestinal lumen and its transport across other cell membranes. These include the reduced folate carrier (RFC), folate receptors (FRs) and a recently discovered proton coupled folate transporter (PCFT). Several excellent reviews describe the different functions of these transporters (Hamid et al. 2009; Low and Antony 2004; Matherly and Goldman 2003; Zhao et al. 2009). Genetic factors that may affect these systems are of special interest and have been studied in some depth, including with the use of animal gene-knock-out models (Ma et al. 2005; Piedrahita et al. 1999) In particular, studies have examined the possibility that genetic variability in these binding proteins and enzymes can influence an individual's achievable folate status, since such an effect might have implications for risk of NTDs and other chronic disease states. Some of these are described below.

Before being absorbed, dietary folate polyglutamates within the food matrix must be hydrolysed to the monoglutamate form (Hannon-Fletcher et al. 2004; McKillop et al. 2006). This is catalysed by a brush border folylpolyglutamate hydrolase, encoded by the glutamate carboxypeptidase II (GCPII) gene (Halsted et al. 1998). A 1561C→T polymorphism in the GCPII gene was initially thought to be associated with lower folate status (Devlin et al. 2000) but this was not confirmed (Devlin et al. 2006; Vargas-Martinez et al. 2002) and other studies found a small but significant association of this polymorphism with higher plasma folate and lower tHcy concentrations (Afman et al. 2003; Fodinger et al. 2003; Halsted et al. 2007; Lievers et al. 2002; Winkelmayer et al. 2003). Studies to date have not shown an association of this polymorphism with risk of NTD (Afman et al. 2003; Morin et al. 2003).

Several of the folate transporters have also been examined for effect on status and risk of NTD. The RFC gene is a member of the SLC19 family of transporters and is ubiquitously expressed in normal tissues. A common polymorphism (80G→A) in this gene has been examined for association with blood folate levels and disease risk including NTDs, orofacial clefts and vascular disease (Devlin et al. 2006;

Fredriksen et al. 2007; O'Leary et al. 2006; Vieira et al. 2005; Winkelmayer et al. 2003) but no strong associations have been observed. Folate receptors (FRs) internalize folate molecules via endocytosis. Four isoforms of FRs are now known to exist (α,β,γ, and δ) (Low and Antony 2004; Matherly and Goldman 2003; Walker 2007; Yamaguchi et al. 2007) The high concentration of some of these receptor isoforms in placental tissue and fetal cells makes genetic variation in the genes for these receptors attractive candidates for conferring risk of NTD and other birth defects. Embryonic lethality or severe embryonic defects are an important feature in mouse studies where the genes for the α, or β FRs are inactivated, and in some instances these effects can be modulated or prevented by folic acid supplementation (Piedrahita et al. 1999; Spiegelstein et al. 2004; Tang and Finnell 2003; Taparia et al. 2007; Zhu et al. 2007). Nevertheless, human studies to date have not found significant associations between polymorphisms in folate transport genes and risk of NTDs (Barber et al. 1998:2000; Finnell et al. 2008; O'Leary et al. 2003). One reason for this may be the conserved nature of these genes. Genetic analysis of the gene for the α-FR in humans found that polymorphic nucleotides were extremely rare (Barber et al. 1998), but recent work suggests that the promoter region might have a higher level of genetic variability (Finnell et al. 2008) Finally, mutations in the recently discovered PCFT have been identified as the causative factor in congenital folate malabsorption (Min et al. 2008; Qiu et al. 2006; Zhao et al. 2007). All of these receptors remain an important focus for future study.

7.9 Summary

Folate cofactor metabolism involves an intricate network of at least 50 genes encoding products such as receptors, binding proteins and enzymes (Molloy 2004). When downstream factors such as methylation reactions that rely on folate derived metabolites are added to that list, the interplay of folate with genetic and epigenetic factors becomes very complex indeed. Although the phenotypic consequences of individual factors may be very small, the combined effect of small alterations in the function of many proteins may have an important influence on the availability, flux and final distribution of one-carbon units, leading to innate differential risk of disease between individuals. Therefore the impact of genetic variation needs to be considered in aiming to establish optimal protection against risk of diseases that are linked to folate status. In the past few years, epigenetic analysis techniques have become more sophisticated and high-throughput multiplex genotyping systems are widely available. The strategy for studying genetic variation has evolved to the extent where micro-satellite analysis of all the genes in a metabolic axis can be scanned for association with disease or phenotype and where genome-wide association studies are emerging for many diseases and biochemical phenotypes. This new strategy is being applied to the folate metabolic network and such analyses are beginning to emerge (Meyer et al. 2004; Molloy et al. 2009; Shaw et al. 2009; Souto et al. 2005; Tanaka et al. 2009; Vermeulen et al. 2006). Future studies on large cohorts will provide a clearer insight into the genetic influences involved at this important nutrient locus.

References

Afman LA, Trijbels FJ, Blom HJ (2003) The H475Y polymorphism in the glutamate carboxypeptidase II gene increases plasma folate without affecting the risk for neural tube defects in humans. J Nutr 133:75–77

Amorim MR, Lima MA, Castilla EE, Orioli IM (2007) Non-Latin European descent could be a requirement for association of NTDs and MTHFR variant 677C > T: a meta-analysis. Am J Med Genet A 143A:1726–1732

Ananth CV, Elsasser DA, Kinzler WL, Peltier MR, Getahun D, Leclerc D, Rozen RR (2007) Polymorphisms in methionine synthase reductase and betaine-homocysteine S-methyltransferase genes: risk of placental abruption. Mol Genet Metab 91:104–110

Anderson DD, Woeller CF, Stover PJ (2007) Small ubiquitin-like modifier-1 (SUMO-1) modification of thymidylate synthase and dihydrofolate reductase. Clin Chem Lab Med 45:1760–1763

Anguera MC, Field MS, Perry C, Ghandour H, Chiang EP, Selhub J, Shane B, Stover PJ (2006) Regulation of folate-mediated one-carbon metabolism by 10-formyltetrahydrofolate dehydrogenase. J Biol Chem 281:18335–18342

Antony AC (2007) In utero physiology: role of folic acid in nutrient delivery and fetal development. Am J Clin Nutr 85:598S–603S

Ashfield-Watt PA, Pullin CH, Whiting JM, Clark ZE, Moat SJ, Newcombe RG, Burr ML, Lewis MJ, Powers HJ, McDowell IF (2002) Methylenetetrahydrofolate reductase 677C→T genotype modulates homocysteine responses to a folate-rich diet or a low-dose folic acid supplement: a randomized controlled trial. Am J Clin Nutr 76:180–186

Bagley PJ, Selhub J (1998) A common mutation in the methylenetetrahydrofolate reductase gene is associated with an accumulation of formylated tetrahydrofolates in red blood cells. Proc Natl Acad Sci USA 95:13217–13220

Bailey LB, Gregory JF 3rd (1999) Folate metabolism and requirements. J Nutr 129:779–782

Bailey LB, Duhaney RL, Maneval DR, Kauwell GP, Quinlivan EP, Davis SR, Cuadras A, Hutson AD, Gregory JF 3rd (2002) Vitamin B-12 status is inversely associated with plasma homocysteine in young women with C677T and/or A(1298)C methylenetetrahydrofolate reductase polymorphisms. J Nutr 132:1872–1878

Bandarian V, Ludwig ML, Matthews RG (2003) Factors modulating conformational equilibria in large modular proteins: a case study with cobalamin-dependent methionine synthase. Proc Natl Acad Sci USA 100:8156–8163

Barber RC, Shaw GM, Lammer EJ, Greer KA, Biela TA, Lacey SW, Wasserman CR, Finnell RH (1998) Lack of association between mutations in the folate receptor-alpha gene and spina bifida. Am J Med Genet 76:310–317

Barber R, Shalat S, Hendricks K, Joggerst B, Larsen R, Suarez L, Finnell R (2000) Investigation of folate pathway gene polymorphisms and the incidence of neural tube defects in a Texas hispanic population. Mol Genet Metab 70:45–52

Baric I, Fumic K, Glenn B, Cuk M, Schulze A, Finkelstein JD, James SJ, Mejaski-Bosnjak V, Pazanin L, Pogribny IP, Rados M, Sarnavka V, Scukanec-Spoljar M, Allen RH, Stabler S, Uzelac L, Vugrek O, Wagner C, Zeisel S, Mudd SH (2004) S-adenosylhomocysteine hydrolase deficiency in a human: A genetic disorder of methionine metabolism. Proc Natl Acad Sci USA 101:4234–4239

Basten GP, Duthie SJ, Pirie L, Vaughan N, Hill MH, Powers HJ (2006) Sensitivity of markers of DNA stability and DNA repair activity to folate supplementation in healthy volunteers. Br J Cancer 94:1942–1947

Baumgartner ER, Stokstad EL, Wick SH, Watson JE, Kusano G (1985) Comparison of folic acid coenzyme distribution patterns in patients with methylenetetrahydrofolate reductase and methionine synthetase deficiencies. Pediatr Res 19:1288–1292

Boccia S, Hung R, Ricciardi G, Gianfagna F, Ebert MP, Fang JY, Gao CM, Gotze T, Graziano F, Lacasana-Navarro M, Lin D, Lopez-Carrillo L, Qiao YL, Shen H, Stolzenberg-Solomon R, Takezaki T, Weng YR, Zhang FF, van Duijn CM, Boffetta P, Taioli E (2008) Meta- and pooled analyses of the methylenetetrahydrofolate reductase C677T and A(1298)C polymorphisms and gastric cancer risk: a huge-GSEC review. Am J Epidemiol 167:505–516

Bosco P, Gueant-Rodriguez RM, Anello G, Barone C, Namour F, Caraci F, Romano A, Romano, C, Gueant JL (2003) Methionine synthase (MTR) (2756) (A→G) polymorphism, double heterozygosity methionine synthase (2756) AG/methionine synthase reductase (MTRR) 66 AG, and elevated homocysteinemia are three risk factors for having a child with Down syndrome. Am J Med Genet A 121A:219–224

Botto LD, Yang Q (2000) 5,10-Methylenetetrahydrofolate reductase gene variants and congenital anomalies: a HuGE review. Am J Epidemiol 151:862–877

Boyles AL, Billups AV, Deak KL, Siegel DG, Mehltretter L, Slifer SH, Bassuk AG, Kessler JA, Reed MC, Nijhout HF, George TM, Enterline DS, Gilbert JR, Speer MC (2006) Neural tube defects and folate pathway genes: family-based association tests of gene-gene and gene-environment interactions. Environ Health Perspect 114:1547–1552

Brody LC, Conley M, Cox C, Kirke PN, McKeever MP, Mills JL, Molloy AM, O'Leary VB, Parle-McDermott A, Scott JM, Swanson DA (2002) A polymorphism, R653Q, in the trifunctional enzyme methylenetetrahydrofolate dehydrogenase/methenyltetrahydrofolate cyclohydrolase/formyltetrahydrofolate synthetase is a maternal genetic risk factor for neural tube defects: report of the Birth Defects Research Group. Am J Hum Genet 71:1207–1215

Brouns R, Ursem N, Lindemans J, Hop W, Pluijm S, Steegers E, Steegers-Theunissen R (2008) Polymorphisms in genes related to folate and cobalamin metabolism and the associations with complex birth defects. Prenat Diagn 28:485–493

Buist NR, Glenn B, Vugrek O, Wagner C, Stabler S, Allen RH, Pogribny I, Schulze A, Zeisel SH, Baric I, Mudd SH (2006) S-adenosylhomocysteine hydrolase deficiency in a 26-year-old man. J Inherit Metab Dis 29:538–545

Burdge GC, Lillycrop KA, Phillips ES, Slater-Jefferies JL, Jackson AA, Hanson MA (2009) Folic acid supplementation during the juvenile-pubertal period in rats modifies the phenotype and epigenotype induced by prenatal nutrition. J Nutr 139:1054–1060

Caudill MA, Wang JC, Melnyk S, Pogribny IP, Jernigan S, Collins MD, Santos-Guzman J, Swendseid ME, Cogger EA, James SJ (2001) Intracellular S-adenosylhomocysteine concentrations predict global DNA hypomethylation in tissues of methyl-deficient cystathionine beta-synthase heterozygous mice. J Nutr 131:2811–2818

Chamberlin ME, Ubagai T, Mudd SH, Thomas J, Pao VY, Nguyen TK, Levy HL, Greene C, Freehauf C, Chou JY (2000) Methionine adenosyltransferase I/III deficiency: novel mutations and clinical variations. Am J Hum Genet 66:347–355

Chen LH, Liu ML, Hwang HY, Chen LS, Korenberg J, Shane B (1997) Human methionine synthase. cDNA cloning, gene localization, and expression. J Biol Chem 272:3628–3634

Chen J, Giovannucci E, Hankinson SE, Ma J, Willett WC, Spiegelman D, Kelsey KT, Hunter DJ (1998) A prospective study of methylenetetrahydrofolate reductase and methionine synthase gene polymorphisms, and risk of colorectal adenoma. Carcinogenesis 19:2129–2132

Chen J, Stampfer MJ, Ma J, Selhub J, Malinow MR, Hennekens CH, Hunter DJ (2001) Influence of a methionine synthase (D919G) polymorphism on plasma homocysteine and folate levels and relation to risk of myocardial infarction. Atherosclerosis 154:667–672

Choi SW, Mason JB (2002) Folate status: effects on pathways of colorectal carcinogenesis. J Nutr 132:2413S–2418S

Christensen KE, MacKenzie RE (2006) Mitochondrial one-carbon metabolism is adapted to the specific needs of yeast, plants and mammals. Bioessays 28:595–605

Christensen B, Arbour L, Tran P, Leclerc D, Sabbaghian N, Platt R, Gilfix BM, Rosenblatt DS, Gravel RA, Forbes P, Rozen R (1999) Genetic polymorphisms in methylenetetrahydrofolate reductase and methionine synthase, folate levels in red blood cells, and risk of neural tube defects. Am J Med Genet 84:151–157

Chuang CZ, Boyles A, Legardeur B, Su J, Japa S, Lopez SA (2006) Effects of riboflavin and folic acid supplementation on plasma homocysteine levels in healthy subjects. Am J Med Sci 331:65–71

Chwatko G, Boers GH, Strauss KA, Shih DM, Jakubowski H (2007) Mutations in methylenetetrahydrofolate reductase or cystathionine beta-synthase gene, or a high-methionine diet, increase homocysteine thiolactone levels in humans and mice. FASEB J 21:1707–1713

Couce ML, Boveda MD, Castineiras DE, Corrales FJ, Mora MI, Fraga JM, Mudd SH (2008) Hypermethioninaemia due to methionine adenosyltransferase I/III (MAT I/III) deficiency: diagnosis in an expanded neonatal screening programme. J Inherit Metab Dis 31(Suppl 2):233–239

Czeizel AE, Dudas I (1992) Prevention of the first occurrence of neural-tube defects by periconceptional vitamin supplementation. N Engl J Med 327:1832–1835

da Costa KA, Kozyreva OG, Song J, Galanko JA, Fischer LM, Zeisel SH (2006) Common genetic polymorphisms affect the human requirement for the nutrient choline. FASEB J 20:1336–1344

Daly LE, Kirke PN, Molloy A, Weir DG, Scott JM (1995) Folate levels and neural tube defects. Implications for prevention. JAMA 274:1698–1702

D'Angelo A, Coppola A, Madonna P, Fermo I, Pagano A, Mazzola G, Galli L, Cerbone AM (2000) The role of vitamin B12 in fasting hyperhomocysteinemia and its interaction with the homozygous C677T mutation of the methylenetetrahydrofolate reductase (MTHFR) gene. A case-control study of patients with early-onset thrombotic events. Thromb Haemost 83:563–570

Daubner SC, Matthews RG (1982) Purification and properties of methylenetetrahydrofolate reductase from pig liver. J Biol Chem 257:140–145

de Bree A, Verschuren WM, Bjorke-Monsen AL, van der Put NM, Heil SG, Trijbels FJ, Blom HJ (2003) Effect of the methylenetetrahydrofolate reductase 677C→T mutation on the relations among folate intake and plasma folate and homocysteine concentrations in a general population sample. Am J Clin Nutr 77:687–693

de Franchis R, Botto LD, Sebastio G, Ricci R, Iolascon A, Capra V, Andria G, Mastroiacovo P (2002) Spina bifida and folate-related genes: a study of gene-gene interactions. Genet Med 4:126–130

Dekou V, Gudnason V, Hawe E, Miller GJ, Stansbie D, Humphries SE (2001) Gene-environment and gene-gene interaction in the determination of plasma homocysteine levels in healthy middle-aged men. Thromb Haemost 85:67–74

De Marco P, Merello E, Calevo MG, Mascelli S, Raso A, Cama A, Capra V (2006) Evaluation of a methylenetetrahydrofolate-dehydrogenase (1958)G>A polymorphism for neural tube defect risk. J Hum Genet 51:98–103

Deng L, Elmore CL, Lawrance AK, Matthews RG, Rozen R (2008) Methionine synthase reductase deficiency results in adverse reproductive outcomes and congenital heart defects in mice. Mol Genet Metab 94:336–342

Devlin AM, Ling EH, Peerson JM, Fernando S, Clarke R, Smith AD, Halsted CH (2000) Glutamate carboxypeptidase II: A polymorphism associated with lower levels of serum folate and hyperhomocysteinemia. Hum Mol Genet 9:2837–2844

Devlin AM, Clarke R, Birks J, Evans JG, Halsted CH (2006) Interactions among polymorphisms in folate-metabolizing genes and serum total homocysteine concentrations in a healthy elderly population. Am J Clin Nutr 83:708–713

Duthie SJ, Narayanan S, Brand GM, Pirie L, Grant G (2002) Impact of folate deficiency on DNA stability. J Nutr 132:2444S–2449S

Duthie SJ, Narayanan S, Sharp L, Little J, Basten G, Powers H (2004) Folate, DNA stability and colo-rectal neoplasia. Proc Nutr Soc 63:571–578

Elmore CL, Wu X, Leclerc D, Watson ED, Bottiglieri T, Krupenko NI, Krupenko SA, Cross JC, Rozen R, Gravel RA, Matthews RG (2007) Metabolic derangement of methionine and folate metabolism in mice deficient in methionine synthase reductase. Mol Genet Metab 91:85–97

Erbe RW (1979) Genetic aspects of folate metabolism. Adv Hum Genet 9:293–354, 367–369

Esfahani ST, Cogger EA, Caudill MA (2003) Heterogeneity in the prevalence of methylenetetrahydrofolate reductase gene polymorphisms in women of different ethnic groups. J Am Diet Assoc 103:200–207

Esteller M (2008) Epigenetics in cancer. N Engl J Med 358:1148–1159

Fenech M (2001) The role of folic acid and Vitamin B12 in genomic stability of human cells. Mutat Res 475:57–67

Finkelstein JD (2000) Pathways and regulation of homocysteine metabolism in mammals. Semin Thromb Hemost 26:219–225

Finkelstein JD (2007) Metabolic regulatory properties of S-adenosylmethionine and S-adenosylhomocysteine. Clin Chem Lab Med 45:1694–1699

Finnell RH, Shaw GM, Lammer EJ, Rosenquist TH (2008) Gene-nutrient interactions: Importance of folic acid and vitamin B12 during early embryogenesis. Food Nutr Bull 29, S86–S98; discussion S99–S100

Fodinger M, Dierkes J, Skoupy S, Rohrer C, Hagen W, Puttinger H, Hauser AC, Vychytil A, Sunder-Plassmann G (2003) Effect of glutamate carboxypeptidase II and reduced folate carrier polymorphisms on folate and total homocysteine concentrations in dialysis patients. J Am Soc Nephrol 14:1314–1319

Fowler B, Whitehouse C, Wenzel F, Wraith JE (1997) Methionine and serine formation in control and mutant human cultured fibroblasts: Evidence for methyl trapping and characterization of remethylation defects. Pediatr Res 41:145–151

Fredriksen A, Meyer K, Ueland PM, Vollset SE, Grotmol T, Schneede J (2007) Large-scale population-based metabolic phenotyping of thirteen genetic polymorphisms related to one-carbon metabolism. Hum Mutat 28:856–865

Friso S, Choi SW (2002) Gene-nutrient interactions and DNA methylation. J Nutr 132:2382S–2387S

Frosst P, Blom HJ, Milos R, Goyette P, Sheppard CA, Matthews RG, Boers GJ, den Heijer M, Kluijtmans LA, van den Heuvel LP, et al (1995) A candidate genetic risk factor for vascular disease: A common mutation in methylenetetrahydrofolate reductase. Nat Genet 10:111–113

Goyette P, Sumner JS, Milos R, Duncan AM, Rosenblatt DS, Matthews RG, Rozen R (1994) Human methylenetetrahydrofolate reductase: Isolation of cDNA, mapping and mutation identification. Nat Genet 7:195–200

Goyette P, Christensen B, Rosenblatt DS, Rozen R (1996) Severe and mild mutations in cis for the methylenetetrahydrofolate reductase (MTHFR) gene, and description of five novel mutations in MTHFR. Am J Hum Genet 59:1268–1275

Gregory JF (1995) The bioavailability of folate. In: Bailey LB (ed) Folate in health and disease. Marcel Dekker, New York, pp 195–235

Gueant-Rodriguez RM, Gueant JL, Debard R, Thirion S, Hong LX, Bronowicki JP, Namour F, Chabi NW, Sanni A, Anello G, Bosco P, Romano C, Amouzou E, Arrieta HR, Sanchez BE, Romano A, Herbeth B, Guilland JC, Mutchinick OM (2006) Prevalence of methylenetetrahydrofolate reductase 677T and 1298C alleles and folate status: a comparative study in Mexican, West African, and European populations. Am J Clin Nutr 83:701–707

Guenther BD, Sheppard CA, Tran P, Rozen R, Matthews RG, Ludwig ML (1999) The structure and properties of methylenetetrahydrofolate reductase from Escherichia coli suggest how folate ameliorates human hyperhomocysteinemia. Nat Struct Biol 6:359–365

Halsted CH, Ling EH, Luthi-Carter R, Villanueva JA, Gardner JM, Coyle JT (1998) Folylpoly-gamma-glutamate carboxypeptidase from pig jejunum. Molecular characterization and relation to glutamate carboxypeptidase II. J Biol Chem 273:20417–20424

Halsted CH, Wong DH, Peerson JM, Warden CH, Refsum H, Smith AD, Nygard OK, Ueland PM, Vollset SE, Tell GS (2007) Relations of glutamate carboxypeptidase II (GCPII) polymorphisms to folate and homocysteine concentrations and to scores of cognition, anxiety, and depression in a homogeneous Norwegian population: The Hordaland Homocysteine Study. Am J Clin Nutr 86:514–521

Hamid A, Wani NA, Kaur J (2009) New perspectives on folate transport in relation to alcoholism-induced folate malabsorption – association with epigenome stability and cancer development. FEBS J 276:2175–2191

Hannon-Fletcher MP, Armstrong NC, Scott JM, Pentieva K, Bradbury I, Ward M, Strain JJ, Dunn AA, Molloy AM, Kerr MA, McNulty H (2004) Determining bioavailability of food folates in a controlled intervention study. Am J Clin Nutr 80:911–918

Harmon DL, Woodside JV, Yarnell JW, McMaster D, Young IS, McCrum EE, Gey KF, Whitehead AS, Evans AE (1996) The common 'thermolabile' variant of methylene tetrahydrofolate reductase is a major determinant of mild hyperhomocysteinaemia. QJM 89:571–577

Heil SG, Van der Put NM, Waas ET, den Heijer M, Trijbels FJ, Blom HJ (2001) Is mutated serine hydroxymethyltransferase (SHMT) involved in the etiology of neural tube defects? Mol Genet Metab 73:164–172

Herbig K, Chiang EP, Lee LR, Hills J, Shane B, Stover PJ (2002) Cytoplasmic serine hydroxymethyltransferase mediates competition between folate-dependent deoxyribonucleotide and S-adenosylmethionine biosyntheses. J Biol Chem 277:38381–38389

Herrmann W, Knapp JP (2002) Hyperhomocysteinemia: A new risk factor for degenerative diseases. Clin Lab 48:471–481

Holm PI, Hustad S, Ueland PM, Vollset SE, Grotmol T, Schneede J (2007) Modulation of the homocysteine-betaine relationship by methylenetetrahydrofolate reductase 677 C→t genotypes and B-vitamin status in a large-scale epidemiological study. J Clin Endocrinol Metab 92:1535–1541

Hustad S, Midttun O, Schneede J, Vollset SE, Grotmol T, Ueland PM (2007) The methylenetetrahydrofolate reductase 677C→T polymorphism as a modulator of a B vitamin network with major effects on homocysteine metabolism. Am J Hum Genet 80:846–855

Jacob RA, Gretz DM, Taylor PC, James SJ, Pogribny IP, Miller BJ, Henning SM, Swendseid ME (1998) Moderate folate depletion increases plasma homocysteine and decreases lymphocyte DNA methylation in postmenopausal women. J Nutr 128:1204–1212

Jacques PF, Kalmbach R, Bagley PJ, Russo GT, Rogers G, Wilson PW, Rosenberg IH, Selhub J (2002) The relationship between riboflavin and plasma total homocysteine in the Framingham Offspring cohort is influenced by folate status and the C677T transition in the methylenetetrahydrofolate reductase gene. J Nutr 132:283–288

Jang H, Mason JB, Choi SW (2005) Genetic and epigenetic interactions between folate and aging in carcinogenesis. J Nutr 135:2967S–2971S

Kalmbach RD, Choumenkovitch SF, Troen AP, Jacques PF, D'Agostino R, Selhub J (2008) A 19-base pair deletion polymorphism in dihydrofolate reductase is associated with increased unmetabolized folic acid in plasma and decreased red blood cell folate. J Nutr 138:2323–2327

Kang SS, Wong PW, Zhou JM, Sora J, Lessick M, Ruggie N, Grcevich G (1988a) Thermolabile methylenetetrahydrofolate reductase in patients with coronary artery disease. Metabolism 37:611–613

Kang SS, Zhou J, Wong PW, Kowalisyn J, Strokosch G (1988b) Intermediate homocysteinemia: A thermolabile variant of methylenetetrahydrofolate reductase. Am J Hum Genet 43:414–421

Kirke PN, Mills JL, Molloy AM, Brody LC, O'Leary VB, Daly L, Murray S, Conley M, Mayne PD, Smith O, Scott JM (2004) Impact of the MTHFR C677T polymorphism on risk of neural tube defects: Case-control study. BMJ 328:1535–1536

Klerk M, Verhoef P, Clarke R, Blom HJ, Kok FJ, Schouten EG (2002) MTHFR 677C→T polymorphism and risk of coronary heart disease: A meta-analysis. JAMA 288:2023–2031

Klerk M, Lievers KJ, Kluijtmans LA, Blom HJ, den Heijer M, Schouten EG, Kok FJ, Verhoef P (2003) The (2756)A>G variant in the gene encoding methionine synthase: Its relation with plasma homocysteine levels and risk of coronary heart disease in a Dutch case-control study. Thromb Res 110:87–91

Kluijtmans LA, Young IS, Boreham CA, Murray L, McMaster D, McNulty H, Strain JJ, McPartlin J, Scott JM, Whitehead AS (2003) Genetic and nutritional factors contributing to hyperhomocysteinemia in young adults. Blood 101:2483–2488

Kotb M, Kredich NM (1990) Regulation of human lymphocyte S-adenosylmethionine synthetase by product inhibition. Biochim Biophys Acta 1039:253–260

Kraus JP (1998) Biochemistry and molecular genetics of cystathionine beta-synthase deficiency. Eur J Pediatr 157(Suppl 2):S50–S53

Kutzbach C, Stokstad EL (1971) Mammalian methylenetetrahydrofolate reductase. Partial purification, properties, and inhibition by S-adenosylmethionine. Biochim Biophys Acta 250:459–477

Lathrop Stern L, Shane B, Bagley PJ, Nadeau M, Shih V, Selhub J (2003) Combined marginal folate and riboflavin status affect homocysteine methylation in cultured immortalized lymphocytes from persons homozygous for the MTHFR C677T mutation. J Nutr 133:2716–2720

Leclerc D, Campeau E, Goyette P, Adjalla CE, Christensen B, Ross M, Eydoux P, Rosenblatt DS, Rozen R, Gravel RA (1996) Human methionine synthase: cDNA cloning and identification of mutations in patients of the cblG complementation group of folate/cobalamin disorders. Hum Mol Genet 5:1867–1874

Lee BM, Mahadevan LC (2009) Stability of histone modifications across mammalian genomes: Implications for 'epigenetic' marking. J Cell Biochem 108:22–34

LeGros L, Halim AB, Chamberlin ME, Geller A, Kotb M (2001) Regulation of the human MAT2B gene encoding the regulatory beta subunit of methionine adenosyltransferase, MAT II. J Biol Chem 276:24918–24924

Li D, Karp N, Wu Q, Wang XL, Melnyk S, James SJ, Rozen R (2008) Mefolinate (5-methyltetrahydrofolate), but not folic acid, decreases mortality in an animal model of severe methylenetetrahydrofolate reductase deficiency. J Inherit Metab Dis 31:403–411

Lievers KJ, Kluijtmans LA, Boers GH, Verhoef P, den Heijer M, Trijbels FJ, Blom HJ (2002) Influence of a glutamate carboxypeptidase II (GCPII) polymorphism ((1561)C→T) on plasma homocysteine, folate and vitamin B(12) levels and its relationship to cardiovascular disease risk. Atherosclerosis 164:269–273

Lillycrop KA, Phillips ES, Torrens C, Hanson MA, Jackson AA, Burdge GC (2008) Feeding pregnant rats a protein-restricted diet persistently alters the methylation of specific cytosines in the hepatic PPAR alpha promoter of the offspring. Br J Nutr 100:278–282

Lim U, Peng K, Shane B, Stover PJ, Litonjua AA, Weiss ST, Gaziano JM, Strawderman RL, Raiszadeh F, Selhub J, Tucker KL, Cassano PA (2005) Polymorphisms in cytoplasmic serine hydroxymethyltransferase and methylenetetrahydrofolate reductase affect the risk of cardiovascular disease in men. J Nutr 135:1989–1994

Lindenbaum J, Allen RH (1995) Clinical spectrum and diagnosis of folate deficiency. In: Bailey LB (ed) Folate in health and disease. Marcel Dekker, New York, pp 43–73

Linnebank M, Janosik M, Kozich V, Pronicka E, Kubalska J, Sokolova J, Linnebank A, Schmidt E, Leyendecker C, Klockgether T, Kraus JP, Koch HG (2004) The cystathionine beta-synthase (CBS) mutation c.1224–2A>C in Central Europe: Vitamin B6 nonresponsiveness and a common ancestral haplotype. Hum Mutat 24:352–353

Low PS, Antony AC (2004) Folate receptor-targeted drugs for cancer and inflammatory diseases. Adv Drug Deliv Rev 56:1055–1058

Ma J, Stampfer MJ, Giovannucci E, Artigas C, Hunter DJ, Fuchs C, Willett WC, Selhub J, Hennekens CH, Rozen R (1997) Methylenetetrahydrofolate reductase polymorphism, dietary interactions, and risk of colorectal cancer. Cancer Res 57:1098–1102

Ma DW, Finnell RH, Davidson LA, Callaway ES, Spiegelstein O, Piedrahita JA, Salbaum JM, Kappen C, Weeks BR, James J, Bozinov D, Lupton JR, Chapkin RS (2005) Folate transport gene inactivation in mice increases sensitivity to colon carcinogenesis. Cancer Res 65:887–897

MacFarlane AJ, Perry CA, Girnary HH, Gao D, Allen RH, Stabler SP, Shane B, Stover PJ (2009) Mthfd1 is an essential gene in mice and alters biomarkers of impaired one-carbon metabolism. J Biol Chem 284:1533–1539

Mangoni AA, Jackson SH (2002) Homocysteine and cardiovascular disease: Current evidence and future prospects. Am J Med 112:556–565

Mason JB (2003) Biomarkers of nutrient exposure and status in one-carbon (methyl) metabolism. J Nutr 133(Suppl 3):941S–947S

Matherly LH, Goldman DI (2003) Membrane transport of folates. Vitam Horm 66:403–456

Mathers JC, McKay JA (2009) Epigenetics – potential contribution to fetal programming. Adv Exp Med Biol 646:119–123

Matthews RG, Vanoni MA, Hainfeld JF, Wall J (1984) Methylenetetrahydrofolate reductase. Evidence for spatially distinct subunit domains obtained by scanning transmission electron microscopy and limited proteolysis. J Biol Chem 259:11647–11650

McKeever M, Molloy A, Weir DG, Young PB, Kennedy DG, Kennedy S, Scott JM (1995) An abnormal methylation ratio induces hypomethylation in vitro in the brain of pig and man, but not in rat. Clin Sci (Lond) 88:73–79

McKillop DJ, McNulty H, Scott JM, McPartlin JM, Strain JJ, Bradbury I, Girvan J, Hoey L, McCreedy R, Alexander J, Patterson BK, Hannon-Fletcher M, Pentieva K (2006) The rate of intestinal absorption of natural food folates is not related to the extent of folate conjugation. Am J Clin Nutr 84:167–173

McNulty H, Pentieva K (2004) Folate bioavailability. Proc Nutr Soc 63:529–536

McNulty H, McKinley MC, Wilson B, McPartlin J, Strain JJ, Weir DG, Scott JM (2002) Impaired functioning of thermolabile methylenetetrahydrofolate reductase is dependent on riboflavin status: Implications for riboflavin requirements. Am J Clin Nutr 76:436–441

McNulty H, Dowey le RC, Strain JJ, Dunne A, Ward M, Molloy AM, McAnena LB, Hughes JP, Hannon-Fletcher M, Scott JM (2006) Riboflavin lowers homocysteine in individuals homozygous for the MTHFR 677C→T polymorphism. Circulation 113:74–80

Melse-Boonstra A, Holm PI, Ueland PM, Olthof M, Clarke R, Verhoef P (2005) Betaine concentration as a determinant of fasting total homocysteine concentrations and the effect of folic acid supplementation on betaine concentrations. Am J Clin Nutr 81:1378–1382

Meyer K, Fredriksen A, Ueland PM (2004) High-level multiplex genotyping of polymorphisms involved in folate or homocysteine metabolism by matrix-assisted laser desorption/ionization mass spectrometry. Clin Chem 50:391–402

Miles EW, Kraus JP (2004) Cystathionine beta-synthase: Structure, function, regulation, and location of homocystinuria-causing mutations. J Biol Chem 279:29871–29874

Mills JL, Molloy AM, Parle-McDermott A, Troendle JF, Brody LC, Conley MR, Cox C, Pangilinan F, Orr DJ, Earley M, McKiernan E, Lynn EC, Doyle A, Scott JM, Kirke PN (2008) Folate-related gene polymorphisms as risk factors for cleft lip and cleft palate. Birth Defects Res A Clin Mol Teratol 82:636–643

Min SH, Oh SY, Karp GI, Poncz M, Zhao R, Goldman ID (2008) The clinical course and genetic defect in the PCFT gene in a 27-year-old woman with hereditary folate malabsorption. J Pediatr 153:435–437

Mitchell LE, Adzick NS, Melchionne J, Pasquariello PS, Sutton LN, Whitehead AS (2004) Spina bifida. Lancet 364:1885–1895

Molloy AM (2002) Folate bioavailability and health. Int J Vitam Nutr Res 72:46–52

Molloy AM (2004) Folate and homocysteine interrelationships including genetics of the relevant enzymes. Curr Opin Lipidol 15:49–57

Molloy AM, Weir DG, Kennedy G, Kennedy S, Scott JM (1990) A new high performance liquid chromatographic method for the simultaneous measurement of S-adenosylmethionine and S-adenosylhomocysteine. Concentrations in pig tissues after inactivation of methionine synthase by nitrous oxide. Biomed Chromatogr 4:257–260

Molloy AM, Orsi B, Kennedy DG, Kennedy S, Weir DG, Scott JM (1992) The relationship between the activity of methionine synthase and the ratio of S-adenosylmethionine to S-adenosylhomocysteine in the brain and other tissues of the pig. Biochem Pharmacol 44:1349–1355

Molloy AM, Daly S, Mills JL, Kirke PN, Whitehead AS, Ramsbottom D, Conley MR, Weir DG, Scott JM (1997) Thermolabile variant of 5,10-methylenetetrahydrofolate reductase associated with low red-cell folates: Implications for folate intake recommendations. Lancet 349:1591–1593

Molloy AM, Kirke PN, Brody LC, Scott JM, Mills JL (2008) Effects of folate and vitamin B12 deficiencies during pregnancy on fetal, infant, and child development. Food Nutr Bull 29:S101–S111; discussion S112–S115

Molloy AM, Brody LC, Mills JL, Scott JM, Kirke PN (2009) The search for genetic polymorphisms in the homocysteine/folate pathway that contribute to the etiology of human neural tube defects. Birth Defects Res A Clin Mol Teratol 85:285–294

Morin I, Devlin AM, Leclerc D, Sabbaghian N, Halsted CH, Finnell R, Rozen R (2003) Evaluation of genetic variants in the reduced folate carrier and in glutamate carboxypeptidase II for spina bifida risk. Mol Genet Metab 79:197–200

MRC (1991) Prevention of neural tube defects: Results of the Medical Research Council Vitamin Study. MRC Vitamin Study Research Group. Lancet 338:131–137

Mutchinick OM, Lopez MA, Luna L, Waxman J, Babinsky VE (1999) High prevalence of the thermolabile methylenetetrahydrofolate reductase variant in Mexico: A country with a very high prevalence of neural tube defects. Mol Genet Metab 68:461–467

Nurk E, Tell GS, Refsum H, Ueland PM, Vollset SE (2004) Associations between maternal methylenetetrahydrofolate reductase polymorphisms and adverse outcomes of pregnancy: The Hordaland Homocysteine Study. Am J Med 117:26–31

O'Leary VB, Mills JL, Kirke PN, Parle-McDermott A, Swanson DA, Weiler A, Pangilinan F, Conley M, Molloy AM, Lynch M, Cox C, Scott JM, Brody LC (2003) Analysis of the human folate receptor beta gene for an association with neural tube defects. Mol Genet Metab 79:129–133

O'Leary VB, Pangilinan F, Cox C, Parle-McDermott A, Conley M, Molloy AM, Kirke PN, Mills JL, Brody LC, Scott JM (2006) Reduced folate carrier polymorphisms and neural tube defect risk. Mol Genet Metab 87:364–369

Oppenheim EW, Adelman C, Liu X, Stover PJ (2001) Heavy chain ferritin enhances serine hydroxymethyltransferase expression and de novo thymidine biosynthesis. J Biol Chem 276:19855–19861

Parle-McDermott A, Mills JL, Kirke PN, O'Leary VB, Swanson DA, Pangilinan F, Conley M, Molloy AM, Cox C, Scott JM, Brody LC (2003) Analysis of the MTHFR (1298)A→C and 677C→T polymorphisms as risk factors for neural tube defects. J Hum Genet 48:190–193

Parle-McDermott A, Mills JL, Kirke PN, Cox C, Signore CC, Kirke S, Molloy AM, O'Leary VB, Pangilinan FJ, O'Herlihy C, Brody LC, Scott JM (2005a) MTHFD1 R653Q polymorphism is a maternal genetic risk factor for severe abruptio placentae. Am J Med Genet A 132:365–368

Parle-McDermott A, Pangilinan F, Mills JL, Signore CC, Molloy AM, Cotter A, Conley M, Cox C, Kirke PN, Scott JM, Brody LC (2005b) A polymorphism in the MTHFD1 gene increases a mother's risk of having an unexplained second trimester pregnancy loss. Mol Hum Reprod 11:477–480

Parle-McDermott A, Mills JL, Molloy AM, Carroll N, Kirke PN, Cox C, Conley MR, Pangilinan FJ, Brody LC, Scott JM (2006) The MTHFR (1298)CC and 677TT genotypes have opposite associations with red cell folate levels. Mol Genet Metab 88:290–294

Parle-McDermott A, Pangilinan F, Mills JL, Kirke PN, Gibney ER, Troendle J, O'Leary VB, Molloy AM, Conley M, Scott JM, Brody LC (2007) The 19-bp deletion polymorphism in intron-1 of dihydrofolate reductase (DHFR) may decrease rather than increase risk for spina bifida in the Irish population. Am J Med Genet A 143A:1174–1180

Pejchal R, Campbell E, Guenther BD, Lennon BW, Matthews RG, Ludwig ML (2006) Structural perturbations in the Ala→Val polymorphism of methylenetetrahydrofolate reductase: how binding of folates may protect against inactivation. Biochemistry 45:4808–4818

Pepe G, Camacho Vanegas O, Giusti B, Brunelli T, Marcucci R, Attanasio M, Rickards O, De Stefano GF, Prisco D, Gensini GF, Abbate R (1998) Heterogeneity in world distribution of the thermolabile C677T mutation in 5,10-methylenetetrahydrofolate reductase. Am J Hum Genet 63:917–920

Perez Mato I, Sanchez del Pino MM, Chamberlin ME, Mudd SH, Mato JM, Corrales FJ (2001) Biochemical basis for the dominant inheritance of hypermethioninemia associated with the

R264H mutation of the MAT1A gene. A monomeric methionine adenosyltransferase with tripolyphosphatase activity. J Biol Chem 276:13803–13809
Piedrahita JA, Oetama B, Bennett GD, van Waes J, Kamen BA, Richardson J, Lacey SW, Anderson RG, Finnell RH (1999) Mice lacking the folic acid-binding protein Folbp1 are defective in early embryonic development. Nat Genet 23:228–232
Qiu A, Jansen M, Sakaris A, Min SH, Chattopadhyay S, Tsai E, Sandoval C, Zhao R, Akabas MH, Goldman ID (2006) Identification of an intestinal folate transporter and the molecular basis for hereditary folate malabsorption. Cell 127:917–928
Rampersaud GC, Kauwell GP, Bailey LB (2003) Folate: A key to optimizing health and reducing disease risk in the elderly. J Am Coll Nutr 22:1–8
Ravaglia G, Forti P, Maioli F, Martelli M, Servadei L, Brunetti N, Porcellini E, Licastro F (2005) Homocysteine and folate as risk factors for dementia and Alzheimer disease. Am J Clin Nutr 82:636–643
Relton CL, Wilding CS, Pearce MS, Laffling AJ, Jonas PA, Lynch SA, Tawn EJ, Burn J (2004) Gene-gene interaction in folate-related genes and risk of neural tube defects in a UK population. J Med Genet 41:256–260
Robien K, Ulrich CM (2003) 5,10-Methylenetetrahydrofolate reductase polymorphisms and leukemia risk: A HuGE minireview. Am J Epidemiol 157:571–582
Rosenblatt DS (1995) Inherited disorders of folate transport and metabolism. In: Scriver CR, Beaudet AL, Sly WS, Valle D (eds) The metabolic and molecular bases of inherited disease. McGraw-Hill, New York, pp 3111–3128
Ryan BM, Molloy AM, McManus R, Arfin Q, Kelleher D, Scott JM, Weir DG (2001) The methylenetetrahydrofolate reductase (MTHFR) gene in colorectal cancer: role in tumor development and significance of allelic loss in tumor progression. Int J Gastrointest Cancer 30:105–111
Schwab U, Torronen A, Meririnne E, Saarinen M, Alfthan G, Aro A, Uusitupa M (2006) Orally administered betaine has an acute and dose-dependent effect on serum betaine and plasma homocysteine concentrations in healthy humans. J Nutr 136:34–38
Schwahn BC, Wang XL, Mikael LG, Wu Q, Cohn J, Jiang H, Maclean KN, Rozen R (2007) Betaine supplementation improves the atherogenic risk factor profile in a transgenic mouse model of hyperhomocysteinemia. Atherosclerosis 195:e100–e107
Scott JM, Weir DG (1998) Folic acid, homocysteine and one-carbon metabolism: a review of the essential biochemistry. J Cardiovasc Risk 5:223–27
Selhub J (2002) Folate, vitamin B12 and vitamin B6 and one carbon metabolism. J Nutr Health Aging 6:39–42
Shane B (1989) Folylpolyglutamate synthesis and role in the regulation of one-carbon metabolism. Vitam Horm 45:263–335
Sharp L, Little J (2004) Polymorphisms in genes involved in folate metabolism and colorectal neoplasia: a HuGE review. Am J Epidemiol 159:423–443
Shaw GM, Lu W, Zhu H, Yang W, Briggs FB, Carmichael SL, Barcellos LF, Lammer EJ, Finnell RH (2009) 118 SNPs of folate-related genes and risks of spina bifida and conotruncal heart defects. BMC Med Genet 10:49
Shelnutt KP, Kauwell GP, Gregory JF 3rd, Maneval DR, Quinlivan EP, Theriaque DW, Henderson GN, Bailey LB (2004) Methylenetetrahydrofolate reductase 677C→T polymorphism affects DNA methylation in response to controlled folate intake in young women. J Nutr Biochem 15:554–560
Shields DC, Kirke PN, Mills JL, Ramsbottom D, Molloy AM, Burke H, Weir DG, Scott JM, Whitehead AS (1999) The "thermolabile" variant of methylenetetrahydrofolate reductase and neural tube defects: an evaluation of genetic risk and the relative importance of the genotypes of the embryo and the mother. Am J Hum Genet 64:1045–1055
Sibani S, Leclerc D, Weisberg IS, O'Ferrall E, Watkins D, Artigas C, Rosenblatt DS, Rozen R (2003) Characterization of mutations in severe methylenetetrahydrofolate reductase deficiency reveals an FAD-responsive mutation. Hum Mutat 21:509–520

Silaste ML, Rantala M, Sampi M, Alfthan G, Aro A, Kesaniemi YA (2001) Polymorphisms of key enzymes in homocysteine metabolism affect diet responsiveness of plasma homocysteine in healthy women. J Nutr 131:2643–2647

Sinclair KD, Allegrucci C, Singh R, Gardner DS, Sebastian S, Bispham J, Thurston A, Huntley JF, Rees WD, Maloney CA, Lea RG, Craigon J, McEvoy TG, Young LE (2007) DNA methylation, insulin resistance, and blood pressure in offspring determined by maternal periconceptional B vitamin and methionine status. Proc Natl Acad Sci USA 104:19351–19356

Skibola CF, Smith MT, Hubbard A, Shane B, Roberts AC, Law GR, Rollinson S, Roman E, Cartwright RA, Morgan GJ (2002) Polymorphisms in the thymidylate synthase and serine hydroxymethyltransferase genes and risk of adult acute lymphocytic leukemia. Blood 99:3786–3791

Smulders YM, Smith DE, Kok RM, Teerlink T, Gellekink H, Vaes WH, Stehouwer CD, Jakobs C (2007) Red blood cell folate vitamer distribution in healthy subjects is determined by the methylenetetrahydrofolate reductase C677T polymorphism and by the total folate status. J Nutr Biochem 18:693–699

Souto JC, Blanco-Vaca F, Soria JM, Buil A, Almasy L, Ordonez-Llanos J, Martin-Campos JM, Lathrop M, Stone W, Blangero J, Fontcuberta J (2005) A genomewide exploration suggests a new candidate gene at chromosome 11q23 as the major determinant of plasma homocysteine levels: results from the GAIT project. Am J Hum Genet 76:925–933

Spiegelstein O, Mitchell LE, Merriweather MY, Wicker NJ, Zhang Q, Lammer EJ, Finnell RH (2004) Embryonic development of folate binding protein-1 (Folbp1) knockout mice: effects of the chemical form, dose, and timing of maternal folate supplementation. Dev Dyn 231:221–231

Stegmann K, Ziegler A, Ngo ET, Kohlschmidt N, Schroter B, Ermert A, Koch MC (1999) Linkage disequilibrium of MTHFR genotypes 677C/T-(1298)A/C in the German population and association studies in probands with neural tube defects(NTD). Am J Med Genet 87:23–29

Stover PJ (2004) Physiology of folate and vitamin B12 in health and disease. Nutr Rev 62:S3–S12; discussion S13

Summers CM, Hammons AL, Mitchell LE, Woodside JV, Yarnell JW, Young IS, Evans A, Whitehead AS (2008) Influence of the cystathionine beta-synthase 844ins68 and methylenetetrahydrofolate reductase 677C>T polymorphisms on folate and homocysteine concentrations. Eur J Hum Genet 16:1010–1013

Sunden SL, Renduchintala MS, Park EI, Miklasz SD, Garrow TA (1997) Betaine-homocysteine methyltransferase expression in porcine and human tissues and chromosomal localization of the human gene. Arch Biochem Biophys 345:171–174

Surtees R, Leonard J, Austin S (1991) Association of demyelination with deficiency of cerebrospinal-fluid S-adenosylmethionine in inborn errors of methyl-transfer pathway. Lancet 338:1550–1554

Suzuki T, Matsuo K, Hirose K, Hiraki A, Kawase T, Watanabe M, Yamashita T, Iwata H, Tajima K (2008) One-carbon metabolism-related gene polymorphisms and risk of breast cancer. Carcinogenesis 29:356–362

Tanaka T, Scheet P, Giusti B, Bandinelli S, Piras MG, Usala G, Lai S, Mulas A, Corsi AM, Vestrini A, Sofi F, Gori AM, Abbate R, Guralnik J, Singleton A, Abecasis GR, Schlessinger D, Uda M, Ferrucci L (2009) Genome-wide association study of vitamin B6, vitamin B12, folate, and homocysteine blood concentrations. Am J Hum Genet 84:477–482

Tang LS, Finnell RH (2003) Neural and orofacial defects in Folp1 knockout mice [corrected]. Birth Defects Res A Clin Mol Teratol 67:209–218

Taparia S, Gelineau-van Waes J, Rosenquist TH, Finnell RH (2007) Importance of folate-homocysteine homeostasis during early embryonic development. Clin Chem Lab Med 45:1717–1727

Tsai MY, Bignell M, Schwichtenberg K, Hanson NQ (1996) High prevalence of a mutation in the cystathionine beta-synthase gene. Am J Hum Genet 59:1262–1267

Tsai MY, Welge BG, Hanson NQ, Bignell MK, Vessey J, Schwichtenberg K, Yang F, Bullemer FE, Rasmussen R, Graham KJ (1999a) Genetic causes of mild hyperhomocysteinemia in patients with premature occlusive coronary artery diseases. Atherosclerosis 143:163–170

Tsai MY, Yang F, Bignell M, Aras O, Hanson NQ (1999b) Relation between plasma homocysteine concentration, the 844ins68 variant of the cystathionine beta-synthase gene, and pyridoxal-5′-phosphate concentration. Mol Genet Metab 67:352–356

Ubagai T, Lei KJ, Huang S, Mudd SH, Levy HL, Chou JY (1995) Molecular mechanisms of an inborn error of methionine pathway. Methionine adenosyltransferase deficiency. J Clin Invest 96:1943–1947

Ueland PM, Refsum H, Beresford SA, Vollset SE (2000) The controversy over homocysteine and cardiovascular risk. Am J Clin Nutr 72:324–332

Ueland PM, Hustad S, Schneede J, Refsum H, Vollset SE (2001) Biological and clinical implications of the MTHFR C677T polymorphism. Trends Pharmacol Sci 22:195–201

Ueland PM, Holm PI, Hustad S (2005) Betaine: a key modulator of one-carbon metabolism and homocysteine status. Clin Chem Lab Med 43:1069–1075

Ulrich CM, Curtin K, Potter JD, Bigler J, Caan B, Slattery ML (2005) Polymorphisms in the reduced folate carrier, thymidylate synthase, or methionine synthase and risk of colon cancer. Cancer Epidemiol Biomarkers Prev 14:2509–2516

Ulvik A, Vollset SE, Hansen S, Gislefoss R, Jellum E, Ueland PM (2004) Colorectal cancer and the methylenetetrahydrofolate reductase 677C→T and methionine synthase (2756)A→G polymorphisms: A study of 2,168 case-control pairs from the JANUS cohort. Cancer Epidemiol Biomarkers Prev 13:2175–2180

Ulvik A, Ueland PM, Fredriksen A, Meyer K, Vollset SE, Hoff G, Schneede J (2007) Functional inference of the methylenetetrahydrofolate reductase 677C > T and (1298)A > C polymorphisms from a large-scale epidemiological study. Hum Genet 121:57–64

Vargas-Martinez C, Ordovas JM, Wilson PW and Selhub, J. (2002) The glutamate carboxypeptidase gene II (C>T) polymorphism does not affect folate status in the Framingham Offspring cohort. J Nutr 132:1176–1179

Vermeulen SH, van der Vleuten GM, de Graaf J, Hermus AR, Blom HJ, Stalenhoef AF, den Heijer M (2006) A genome-wide linkage scan for homocysteine levels suggests three regions of interest. J Thromb Haemost 4:1303–1307

Vieira AR, Murray JC, Trembath D, Orioli IM, Castilla EE, Cooper ME, Marazita ML, Lennon-Graham F, Speer M (2005) Studies of reduced folate carrier 1 (RFC1) A80G and 5,10-methylenetetrahydrofolate reductase (MTHFR) C677T polymorphisms with neural tube and orofacial cleft defects. Am J Med Genet A 135:220–223

Volcik KA, Shaw GM, Zhu H, Lammer EJ, Laurent C, Finnell RH (2003) Associations between polymorphisms within the thymidylate synthase gene and spina bifida. Birth Defects Res A Clin Mol Teratol 67:924–928

Vollset SE, Refsum H, Irgens LM, Emblem BM, Tverdal A, Gjessing HK, Monsen AL, Ueland PM (2000) Plasma total homocysteine, pregnancy complications, and adverse pregnancy outcomes: The Hordaland Homocysteine study. Am J Clin Nutr 71:962–968

Vyletal P, Sokolova J, Cooper DN, Kraus JP, Krawczak M, Pepe G, Rickards O, Koch HG, Linnebank M, Kluijtmans LA, Blom HJ, Boers GH, Gaustadnes M, Skovby F, Wilcken B, Wilcken DE, Andria G, Sebastio G, Naughten ER, Yap S, Ohura T, Pronicka E, Laszlo A, Kozich V (2007) Diversity of cystathionine beta-synthase haplotypes bearing the most common homocystinuria mutation c.833T>C: a possible role for gene conversion. Hum Mutat 28:255–264

Wald DS, Law M, Morris JK (2002) Homocysteine and cardiovascular disease: Evidence on causality from a meta-analysis. BMJ 325:1202

Walker LS (2007) Regulatory T cells: Folate receptor 4: a new handle on regulation and memory? Immunol Cell Biol 85:506–507

Wasson GR, McGlynn AP, McNulty H, O'Reilly SL, McKelvey-Martin VJ, McKerr G, Strain JJ, Scott J, Downes CS (2006) Global DNA and p53 region-specific hypomethylation in human

colonic cells is induced by folate depletion and reversed by folate supplementation. J Nutr 136:2748–2753
Watkins D, Ru M, Hwang HY, Kim CD, Murray A, Philip NS, Kim W, Legakis H, Wai T, Hilton JF, Ge B, Dore C, Hosack A, Wilson A, Gravel RA, Shane B, Hudson TJ, Rosenblatt DS (2002) Hyperhomocysteinemia due to methionine synthase deficiency, cblG: structure of the MTR gene, genotype diversity, and recognition of a common mutation, P1173L. Am J Hum Genet 71:143–153
Weisberg IS, Park E, Ballman KV, Berger P, Nunn M, Suh DS, Breksa AP 3rd, Garrow TA, Rozen R (2003) Investigations of a common genetic variant in betaine-homocysteine methyltransferase (BHMT) in coronary artery disease. Atherosclerosis 167:205–214
Wilcken B, Bamforth F, Li Z, Zhu H, Ritvanen A, Renlund M, Stoll C, Alembik Y, Dott B, Czeizel AE, Gelman-Kohan Z, Scarano G, Bianca S, Ettore G, Tenconi R, Bellato S, Scala I, Mutchinick OM, Lopez MA, de Walle H, Hofstra R, Joutchenko L, Kavteladze L, Bermejo E, Martinez-Frias ML, Gallagher M, Erickson JD, Vollset SE, Mastroiacovo P, Andria G, Botto LD (2003) Geographical and ethnic variation of the 677C>T allele of 5,10 methylenetetrahydrofolate reductase (MTHFR): findings from over 7000 newborns from 16 areas world wide. J Med Genet 40:619–625
Wilson A, Leclerc D, Rosenblatt DS, Gravel RA (1999a) Molecular basis for methionine synthase reductase deficiency in patients belonging to the cblE complementation group of disorders in folate/cobalamin metabolism. Hum Mol Genet 8:2009–2016
Wilson A, Platt R, Wu Q, Leclerc D, Christensen B, Yang H, Gravel RA, Rozen R (1999b) A common variant in methionine synthase reductase combined with low cobalamin (vitamin B12) increases risk for spina bifida. Mol Genet Metab 67:317–323
Winkelmayer WC, Eberle C, Sunder-Plassmann G, Fodinger M (2003) Effects of the glutamate carboxypeptidase II (GCP2 (1561)C>T) and reduced folate carrier (RFC1 80G>A) allelic variants on folate and total homocysteine levels in kidney transplant patients. Kidney Int 63:2280–2285
Woeller CF, Anderson DD, Szebenyi DM, Stover PJ (2007) Evidence for small ubiquitin-like modifier-dependent nuclear import of the thymidylate biosynthesis pathway. J Biol Chem 282:17623–17631
Xu X, Chen J (2009) One-carbon metabolism and breast cancer: an epidemiological perspective. J Genet Genom 36:203–214
Yamada K, Chen Z, Rozen R, Matthews RG (2001) Effects of common polymorphisms on the properties of recombinant human methylenetetrahydrofolate reductase. Proc Natl Acad Sci USA 98:14853–14858
Yamada K, Gravel RA, Toraya T, Matthews RG (2006) Human methionine synthase reductase is a molecular chaperone for human methionine synthase. Proc Natl Acad Sci USA 103:9476–9481
Yamaguchi T, Hirota K, Nagahama K, Ohkawa K, Takahashi T, Nomura T, Sakaguchi S (2007) Control of immune responses by antigen-specific regulatory T cells expressing the folate receptor. Immunity 27:145–159
Yang QH, Botto LD, Gallagher M, Friedman JM, Sanders CL, Koontz D, Nikolova S, Erickson JD, Steinberg K (2008) Prevalence and effects of gene-gene and gene-nutrient interactions on serum folate and serum total homocysteine concentrations in the United States: findings from the third National Health and Nutrition Examination Survey DNA Bank. Am J Clin Nutr 88:232–246
Yi P, Melnyk S, Pogribna M, Pogribny IP, Hine RJ, James SJ (2000) Increase in plasma homocysteine associated with parallel increases in plasma S-adenosylhomocysteine and lymphocyte DNA hypomethylation. J Biol Chem 275:29318–29323
Zeisel SH (2009) Epigenetic mechanisms for nutrition determinants of later health outcomes. Am J Clin Nutr 89:1488S–1493S
Zhao R, Min SH, Qiu A, Sakaris A, Goldberg GL, Sandoval C, Malatack JJ, Rosenblatt DS, Goldman ID (2007) The spectrum of mutations in the PCFT gene, coding for an intestinal folate transporter, that are the basis for hereditary folate malabsorption. Blood 110:1147–1152

Zhao R, Matherly LH, Goldman ID (2009) Membrane transporters and folate homeostasis: intestinal absorption and transport into systemic compartments and tissues. Expert Rev Mol Med 11:e4

Zhu H, Curry S, Wen S, Wicker NJ, Shaw GM, Lammer EJ, Yang W, Jafarov T, Finnell RH (2005) Are the betaine-homocysteine methyltransferase (BHMT and BHMT2) genes risk factors for spina bifida and orofacial clefts? Am J Med Genet A 135:274–277

Zhu H, Wlodarczyk BJ, Scott M, Yu W, Merriweather M, Gelineau-van Waes J, Schwartz RJ, Finnell RH (2007) Cardiovascular abnormalities in Folr1 knockout mice and folate rescue. Birth Defects Res A Clin Mol Teratol 79:257–268

Chapter 8
Enzymatic and Non-enzymatic Antioxidative Effects of Folic Acid and Its Reduced Derivates

Olaf Stanger and Willibald Wonisch

Abstract A great part of the population appears to have insufficient folate intake, especially subgroups with higher demand, as determined through more sensitive methods and parameters currently in use. As the role of folate deficiency in congenital defects, e.g. in cardiovascular and neurodegenerative diseases, and in carcinogenesis has become better understood, folate has been recognized as having great potential to prevent these many disorders through folate supplementation or fortification for the general population. Folates are essential cofactors in the transfer and utilization of one-carbon groups in the process of DNA-biosynthesis with implications for genomic repair and stability. Folate acts indirectly to lower homocysteine levels and insures optimal functioning of the methylation cycle. Homocysteine was shown to be an independent risk factor for neurodegenerative and cardiovascular disease, which includes peripheral vascular disease, coronary artery disease, cerebrovascular disease and venous thrombosis. In fact, it was long believed that the beneficial effects of folate on vascular function and disease are related directly to the mechanism of homocysteine-diminution. Recent investigations have, however, demonstrated beneficial effects of folates unrelated to homocysteine-diminution, suggesting independent properties. One such mechanism could be free radical scavenging and antioxidant activity, as it is now recognized that free radicals play an important role in the oxidative stress leading to many diseases. It was found that folic acid and, in particular, its reduced derivates act both directly and indirectly to produce antioxidant effects. Folates interact with the endothelial enzyme NO synthase (eNOS) and, exert effects on the cofactor bioavailability of NO and thus, on peroxynitrite formation. Folate metabolism provides an interesting example of gene-environmental interaction.

Keywords Endothelial NO synthase (eNOS) · Folic acid · Glutathione peroxidase · Peroxynitrite · Superoxide anion

O. Stanger (✉)
Royal Brompton and Harefield NHS Foundation, Imperial College of Science, Technology and Medicine, Heart Division, Cardiothoracic Surgery, Sydney Street, London SW3 6NP, United Kingdom
e-mail: o.stanger@rbht.nhs.uk

O. Stanger (ed.), *Water Soluble Vitamins*, Subcellular Biochemistry 56,
DOI 10.1007/978-94-007-2199-9_8, © Springer Science+Business Media B.V. 2012

Abbreviations

Ang	angiotensin
BH4	tetrahydrobiopterin
CAD	coronary artery disease
CβS	cystathionine-β-synthase
DDPH	2,2-diphenyl-1-picrylhydrazyl
DHF	dihydrofolate
DHFR	dihydrofolate reductase
ED	endothelial dysfunction
EDHF	endothelium-derived hyperpolarizing factor
EDRF	endothelial derived relaxing factor
eNOS	Endothelial NO synthase
FRAP	ferric reducing ability of plasma
GPx	glutathione peroxidase
Hcy	homocysteine
HHcy	hyperhomocysteinemia
H_2O_2	hydrogen peroxide
LDL	low density lipoprotein
LPO	lipid peroxidation
MDA	malondialdehyde
MS	methionine synthase
MTHFR	methylenetetrahydrofolate reductase
5-MTHF	5-methyl-tetrahydrofolate
NO	nitric oxide
O_2^-	superoxide anion
$^{\bullet}OH$	hydroxyl radical
$ONOO^-$	peroxynitrite
PG	prostaglandin
RF	relaxing factor
RFC-1	reduced folate carrier-1
ROS	radical oxygen species
SAM	S-adenosylmethionine
SOD	superoxide dismutase
TAC	total antioxidant capacity
TAS	total antioxidant status
TBARS	2-thiobarbituric acid-reactive substances
TEAC	trolox equivalent antioxidant capacity
THF	tetrahydrofolate
TNFα	tumor necrosis factor α
TXB2	thromboxane B2

8.1 Introduction

Folates are essential cofactors in the transfer and utilization of one-carbon-groups and play a key role in the remethylation of methionine, thus providing essential methyl groups for numerous biological reactions. Furthermore, folates donate one-carbon units in the process of DNA -biosynthesis with implications for the regulation of gene expression, transcription, chromatine structure, genomic repair and genomic stability (Stanger 2002).

Tetrahydrofolate (THF) polyglutamates function in cells as a family of coenzymes that activate and carry single carbons on the N5 and/or N10 position. THF-activated carbons are carried at three different oxidation states, e.g. formates as 10-formyl-THF, 5-formyl-THF, or 5,10-methenyl-THF. In addition, THF transports activated formaldehyde as 5,10-methylene-THF and activated methanol as 5-methyl-THF (5-MTHF), the predominant form of folate circulating in the blood-stream and available for cellular uptake.

As folates function as acceptors and donors of one-carbon moieties in reactions involving nucleotide and amino acid metabolism, 5-MTHF is an essential cofactor for the Zn^{2+}- and cobalamin (vitamin B12)-dependent remethylation of homocysteine (Hcy) to methionine catalyzed by methionine synthase (MS). Loss of MS enzyme activity, e.g. through cobalamin deficiency, inhibits nucleotide biosynthesis due to the accumulation of cytoplasmic folate cofactors such as 5-MTHF. This effect is referred to as the "folate or methyl group trap" (Scott and Weir 1981); this elevates Hcy, because remethylation is impaired and Hcy accumulates. Thus folate deficiency consequently elevates homocysteine (hyperhomocysteinemia; HHcy). Although this chapter will not focus on the folate-cobalamin interrelationship in particular, the situation illustrates the difficulty of discriminating between cobalamin and/or folate deficiciency as the cause for HHcy.

Low folate status and mild plasma HHcy (>12[15] μmol/L) are both associated with a significantly increased cardiovascular risk (Stanger et al. 2003, 2009). Various epidemiological studies support a link between folate status and atherosclerotic vascular disease (Pancharuniti et al. 1994; Verhoef et al. 1996; Schwartz et al. 1997; Robinson et al. 1998; Rimm et al. 1998; Bunout et al. 2000). There is evidence for a dose-response relationship between Hcy levels and risk without threshold, even within the normal range (Boushey et al. 1995). Due to the close and inverse relationship between folate bioavailability and Hcy, the individual damaging or protective effects may be difficult to separate, in particular under clinical conditions.

Patients with homocystinuria (see below) represent a "natural" condition with normal folate but extremely elevated Hcy concentrations in plasma that is associated with greatly enhanced cardiovascular risk (Mudd et al. 1985; Wilcken and Wilcken 1997; Yap et al. 2000). Besides the generally accepted toxic effect on vascular structures, HHcy is thought to exert its effects through mechanisms involving radical oxygen species (ROS) and oxidative damage (Loscalzo 1996). Treatment with high-dose vitamins (folate and B-vitamins) dramatically reduces cardiovascular risk but rarely lowers Hcy to normal levels (Wilcken and Wilcken 1997; Yap et al. 2001). In parallel, folates were found to improve endothelial dysfunction (ED) independent of Hc-diminution, thus bringing into question the causative role of Hcy and suggesting that folates have independent antioxidative effects.

8.2 Structure and Metabolism of Folates

8.2.1 Structure

The term "folic acid" usually describes a group of compounds derived from 5,6,7,8-tetrahydropteroyl-γ-glutamate, more commonly referred to as tetrahydrofolate (THF). THF is made up of a pteridine core ring system (2-amino-4-hydroxy-pteridine) that is linked to p-aminobenzoic acid (that is, pteroic acid) by a methylene bridge from the C-6 position of the pyrazine ring, thus forming tetrahydropteroic acid (Fig. 8.1). An additional link attaches glutamate to the p-aminobenzoic acid (with 1–9 residues). The pyrazine ring is fully reduced at the 5,6,7 and 8 positions and forms the biologically active form, referred to as 5,6,7,8-tetrahydrofolate (THF). Dihydrofolate (DHF) is formed through stepwise reduction of the 7 and 8 positions, using two molecules of NADPH. Several other biologically active compounds derive from this basic structure.

Mammals may synthesize the pteridine ring, but are unable to link it with the other compounds. Thus humans depend on dietary intake and synthesis of biologically active folate.

Vital processes in folate disposition include conversion of dietary folylpolyglutamates to monoglutamates, intestinal resorption, receptor and carrier-mediated transport across cell membranes, cellular metabolism and export. Both receptor-mediated and carrier-mediated transport mechanisms are variably expressed in different tissues, thus partially explaining the differential effects of impaired metabolism. Multiple reactions are required to convert dietary folic acid into metabolically active folates. The greater part is attached to enzymes, indicating tight metabolic regulation and limited availability of folates (Donnelly 2001).

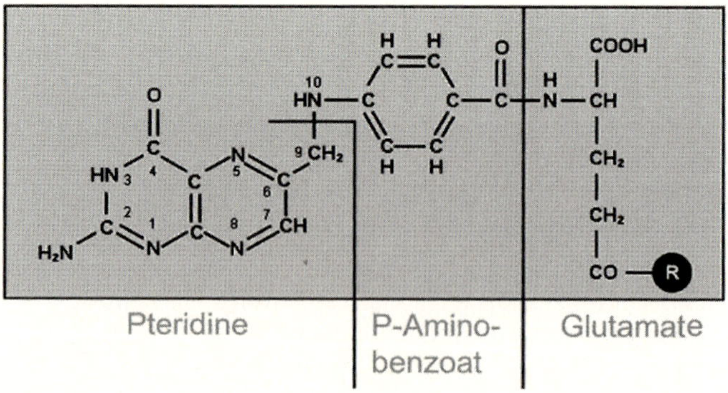

Fig. 8.1 Structure of folic acid

8.2.2 Uptake and Metabolism

Absorbed folate is cleared from the circulation within minutes and is taken up by various tissues and the liver, which is the most important storage organ. Hepatocytes store 10–20% of the folates; the greater part is released after metabolism, mainly into the bile. There, the folates participate in the enterohepatic circulation, with a daily turnover rate of approx. 90 μg (Steinberg et al. 1979). The folate concentration in bile is about the tenfold of serum levels. This means that bile is a readily available endogenous folate source of monoglutamate compounds, which can balance short-term deficiencies, e.g. during fasting. The serum levels of folate and bilirubin are inversely correlated.

THF is enzymatically converted and is mainly available as 5-MTHF, 5,10-methylene-THF and 10-formyl-THF. Alternatively, 5- and 10-formyl-THF derivates can also be converted to 5,10-methenyl-THF, 5,10-methylene-THF and, in an irreversible step, to 5-MTHF (Fig. 8.2).

5-MTHF makes up approx. 40–50% of the total folate polyglutamates in red blood cells. In serum, 5-MTHF is mainly present as monoglutamate, and intracellularly as polyglutamate. Polyglutamates are transported across cellular membranes, ensuring their cellular retention; they are the preferred form in one-carbon metabolism (Shane 1989). 5-MTHF serves as substrate for the cobalamin-dependent

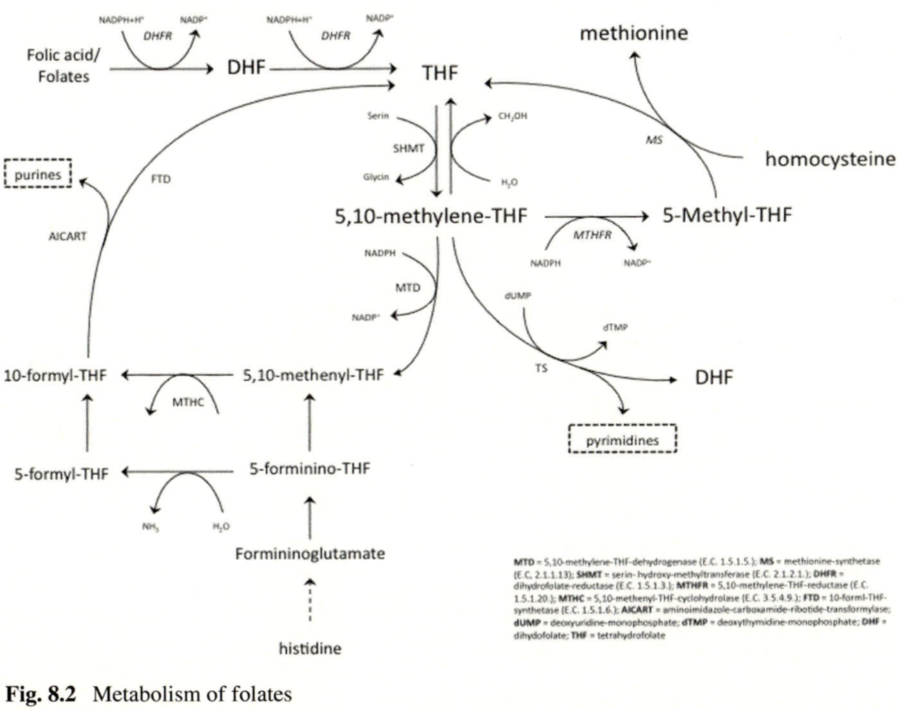

Fig. 8.2 Metabolism of folates

MS (E.C. 2.1.1.13) in the remethylation cycle, passing the methyl group on to Hcy, thus generating methionine and recycling THF.

Receptors and specific carriers are active in transmembranous folate transport. Membrane-located receptors have a high affinity for both folic acid and reduced folates (THF) (Anthony 1996). Three isoforms of such receptors are currently known, one of which is found only in the placenta and most likely plays an important role in folate supply to the embryo through the mother (Prasad et al. 1994). These receptors with high affinity for 5-MTHF are also found in hematopoetic cells and tubular cells of the kidneys (Luhrs and Slomiany 1989). The various carrier mechanisms are characterized by the different affinities for folate compounds and also by specific sensitivity for temperature and pH (Sirotnak and Tolner 1999). There is a carrier system with low affinity but great capacity for reduced folate (RFC-1; reduced folate carrier) in several tissues, but mostly in the brain (Said et al. 1996). The coding gene was demonstrated in luminal epithelial cells of the small intestine (Chiao et al. 1997) and in murine epidermis (Sprecher et al. 1998). Presence and distribution of the RFC-1 gene vary greatly in different tissues (Sirotnak and Tolner 1999). Tissue-specific differences in gene expression could explain local folate deficiency, leading to impairment of local resistance and susceptibility to infections due to dysfunction of immune-competent cells (Sirotnak and Tolner 1999).

8.2.3 Tissue Distribution

Methylenetetrahydrofolate reductase (MTHFR) catalyzes the synthesis of 5-MTHF from 5,10-methylene-THF, which is used for Hcy remethylation to methionine, the precursor of S-adenosylmethionine (SAM). Impaired MTHFR activity will increase Hcy levels and compromise SAM-dependent methylation reactions.

Mild MTHFR deficiency is common in many populations due to a polymorphism at bp 677. The genotype determines the distribution of folates in various tissues (Ghandour et al. 2004) and increased enzyme activity accelerates folate turnover rates, depletes cellular folate concentrations, and may account in part for tissue-specific differences in folate accumulation (Anguera et al. 2003).

To assess how impaired MTHFR activity affects folate metabolism in various tissues in vivo, electrochemical HPLC detection was used to analyze the distribution of folates in plasma, liver, and brain of MTHFR-deficient mice. The most pronounced difference in total folate was observed in plasma (Ghandour et al. 2004). In MTHFR$^{-/-}$ mice, plasma total folate levels were approximately 25% of those in wild-type (MTHFR$^{+/+}$) mice. Only 40% of plasma folate in MTHFR$^{-/-}$ mice was comprised of 5-MTHF, compared to at least 80% in the other 2 genotype groups.

In liver and brain, there were no differences in total folate. However, the proportion of 5-MTHF in both tissues was again markedly reduced in mice with the MTHFR$^{-/-}$ genotype. In this genotype group, 5-MTHF is likely derived from the diet. The study demonstrated reduced total circulatory folate and altered distribution of folate derivatives in liver and brain as a consequence of MTHFR deficiency. Decreased methylfolates and increased nonmethylfolates would affect the flux of one-carbon units between methylation reactions and nucleotide synthesis. This altered flux has implications for several common disorders, including cancer and vascular disease.

Another study sought to determine the effects of depleting cytoplasmic 5-formyl-THF on cellular folate concentrations and folate turnover rates in cell cultures by expressing the human methenyltetrahydrofolate synthetase cDNA in human MCF-7 cells and SH-SY5Y neuroblastoma. Cells with increased methenyltetrahydrofolate synthetase activity exhibited: (1) increased rates of folate turnover, (2) elevated generation of p-aminobenzoylglutamate in culture medium, (3) depressed cellular folate concentrations independent of folic acid concentrations in the medium, and (4) increased average polyglutamate chain lengths of folate cofactors (Anguera et al. 2003). These data indicate that folate catabolism and folate polyglutamylation are competitive reactions that influence cellular folate concentrations, and that increased methenyltetrahydrofolate synthetase activity accelerates folate turnover rates, depletes cellular folate concentrations, and may account in part for tissue-specific differences in folate accumulation.

In conclusion, cellular folate concentrations and tissue distribution are influenced by folate availability, cellular folate transport efficiency, folate polyglutamylation, enzyme activities and folate turnover specifically through degradation.

8.2.4 Oxidative Degradation of Folates

In subjects with HHcy there is usually an inverse relationship to folic acid levels, and supplementation with folic acid will lower Hcy concentrations. The pathogenesis of most – if not all – diseases that are accompanied by moderate HHcy involves cellular immune activation and these patients therefore very often also show a positive correlation between Hcy concentrations and the degree of immune activation, as indicated, e.g. by increased neopterin concentrations. Since neopterin concentrations also serve as an estimate of oxidative stress emerging from immune system activation, this association suggests that cellular immune activation and oxidative stress could be involved in the development of HHcy.

In vitro experiments have shown that superoxide anions (O_2^{-}) may induce folate cleavage (Shaw et al. 1989). Because THF is very susceptible to oxidation, increased oxidative degradation of THFs may become relevant under oxidative stress conditions. Such increased folate catabolism may lead to enhanced folate requirements in patients who are exposed to increased oxidative stress, such as hypercholesterolemic and diabetic patients. Increased oxidative stress may also lead to increased consumption of folate, because of its role as an antioxidant as well as in the regeneration of tetrahydrobiopterin (BH4). In this way, folate deficiency may develop despite adequate dietary intake. HHcy may thus be considered as an indirect consequence of hyperconsumption of antioxidant vitamins (e.g. folates) during prolonged states of immune activation (Widner et al. 2002).

Observations of oxidative degradation of THF and DHF were published very early (Blair 1957; Chippel and Scrimgeour 1970). Indeed, folate cofactors are highly susceptible to oxidative degradation in vitro with the exception of 5-formyl-THF, which may be a storage form of folate. THF, like other reduced pteridines, is readily oxidized (Heales et al. 1988; Odin et al. 1998). Oxidation eliminates the side chain of the pteridine structure, yielding xanthopterin derivates (Murr et al. 1996). The oxidative loss of the side chain is irreversible, and THFs once oxidized in this way can never be recovered by any enzymatic reaction in any organism. Thus, oxidative stress could represent an endogenous reason for THF deficiency even when dietary intake of the vitamin is within the recommended range (Widner et al. 2001; Fuchs et al. 2001).

One of the important examples of the oxidative N-dealkylation of amines is the C9-N10 bond cleavage of folic acid leading to the formation of the amine 2- and 6-substituted pterins (Patro et al. 2005). For this, participation of the superoxide radicals ($O_2^{\bar{}}$), hydroxyl radicals ($^{\cdot}OH$), via a hydrogen atom transfer mechanism or a ferryl-hydroxo complex (Fe^{4+}–OH) have been reported (Patro et al. 2005). Given the relatively high cellular concentrations of free iron and hydrogen peroxide (H_2O_2), the biotoxic hydroxyl radicals may play a major role in oxidative degradation of folic acid in the cellular systems.

The same oxidative degradation was observed with 5-MTHF, the main extracellular folate and the predominant form of the vitamin found in food and blood, yielding a pteridine residue and P-ABG, the latter step resulting in irreversible loss of vitamin activity (Lulock et al. 1995).

Ascorbic acid is efficacious in the facile salvage of 5-MTHF. In serum of folate-saturated healthy men, folate concentrations were measured before and up to 8 h after administration of placebo, 5-MTHF (343 μg + 290 mg ascorbic acid), and 5-MTHF (343 μg + 973 mg ascorbic acid). When 5-MTHF was concurrently administered with either dose of ascorbic acid, the total serum folate response, calculated as area under the curve (AUC), was significantly improved (Verlinde et al. 2008). Thus, administration of a physiological dose of 5-MTHF with ascorbic acid significantly improved the measured folate response in saturated healthy subjects.

Under conditions of increased oxidative stress, compensatory amplification of the redox-sensitive transsulfuration pathway will consume Hcy to generate the endogenous antioxidant glutathione (Tchantchou 2006). Data from clinical conditions in which oxidative stress induces depletion of B vitamins or antioxidants, e.g. prolonged states of inflammation or immune activation (Fuchs et al. 2001) or Alzheimer`s disease (McCaddon et al. 2002), further support the hypothesis that HHcy may be an epiphenomenon (albeit with own tissue-damaging effects) of a more complex pathogenic background. Hcy may be an accurate predictor of folate deficiency under otherwise normal conditions but may not always reflect folate deficiency in the presence of underlying risk factors known to promote increased oxidative stress (Rogers et al. 2007).

As illustrated above, Hcy is elevated in folate deficiency. Both hyperhomocysteinemia and hyperhomocysteinuria have been shown to be independent risk factors for atherosclerotic vascular disease. Considering their close metabolic relationship, clinical studies may not be feasable for distinguishing between HHcy and low folate as a primary etiological factor in cardiovascular disease. Each compound deserves individual attention.

8.3 Homocysteine

8.3.1 Homocystinuria

Homocystinuria may serve as a (natural) model experiment to address the harmful effects of Hcy, and also to question the therapeutic potential of folic acid, betaine and other vitamins associated with Hcy metabolism. Patients with this rare (~1:200.000) recessively inherited disorder of methionine metabolism exhibit extremely elevated Hcy concentrations in plasma (>300 μmol/L) that are associated with greatly enhanced cardiovascular risk (Yap et al. 2000). Patients typically develop occlusive vascular disease in early adulthood or even in childhood. A worldwide review of 627 untreated cases found that a vascular event had occurred in ~50% before the age of 30 (Mudd et al. 1985). This was attributed to both the toxic and vessel damaging effects of homocysteine.

8.3.2 Oxidative Stress in Homocystinuria

The extreme Hcy concentrations observed in homocystinuria are associated with marked oxidative stress indices. In patients homozygous for cystathionine-β-synthase (CβS)-deficiency urinary 8-iso-PGF2α excretion is significantly higher, and urinary 11-dehydro-TCB2 (index of platelet activation) excretion is enhanced as compared to healthy subjects (Davi et al. 2001). It is noteworthy that platelet-active 8-iso-PGF2α formation is a non-enzymatic, free radical-catalyzed peroxidation process of arachidonic acid. In this investigation, urinary 8-iso-PGF2α correlated with plasma homocysteine. These findings suggest that CβS -deficiency stimulates enhanced arachidonic acid peroxidation linking hyperhomocysteinemia with platelet activation, risk for thrombus formation and vascular damage. Folate levels, however, were not measured.

8.3.3 Superoxide Dismutase (SOD)

Superoxide dismutase (SOD) is an important component of the endogenous antioxidant defense opposing the deleterious vascular effects of free radicals. Blood vessels express 3 isoforms of SOD: cytosolic or copper-zinc SOD (CuZn-SOD), manganese Mn-SOD located in mitochondria, and extracellular SOD (EC-SOD) (Faraci and Didion 2004). More than 90% of interstitial SOD is EC-SOD, a copper- and zinc-containing glycoprotein. Furthermore, circulating EC-SOD is in equilibirium with the EC-SOD on the surface of the endothelium and bound to proteoglycan (Stralin et al. 1996). It has been suggested that EC-SOD is the major determinant of nitric oxide (NO) bioavailability in blood vessels (Fukai et al. 2002). NO reacts with superoxide at a rate 3 times faster than dismutation of superoxide by SOD (Darley-Usmar et al. 1995). Due to the high reactivity (superoxide reacts with NO more efficiently than with any other known molecule), the local concentration of SOD is an indirect key determinant of NO -bioactivity (i.e. biological half-life). Thus, a major function of SOD is to protect NO and NO-mediated signaling. Steady-state levels of O_2^- are dependent on both its rate of production as well as activity of the various superoxide dismutases (SODs). Another consequence of SOD activity is formation of H_2O_2. The importance of this ROS within vascular cells is becoming increasingly apparent. H_2O_2 is relatively stable and diffusible (including through cell membranes), compared to many other ROS. H_2O_2 can activate select transcription factors and may also function as an endothelium-derived hyperpolarizing factor (EDHF) in some blood vessels, but has also been suggested to be an ED relaxing factor (RF) without functioning as an EDHF (Miura et al. 2003). In combination with some transition metals like iron or copper, H_2O_2 is metabolized to the most reactive ROS, i.e. hydroxyl radical ($^{\cdot}OH$) through the Fenton reaction. This radical reacts with each adjacent biological structure, leading to cellular damage. H_2O_2 mediated effects and local concentrations are regulated by activity of the various glutathione peroxidases (GPx) or catalase. In homocystinuric patients, there is a highly significant, positive relationship between EC-SOD and total Hcy (Wilcken et al. 2000). In 2 newly diagnosed CβS-deficient patients, Hcy-lowering treatment reduced the associated elevated EC-SOD in each by 50%. This relationship could represent a protective antioxidant response to Hcy-induced oxidative damage in the vasculature. There is evidence for Hcy-related ED in homocystinuric patients that is thought to be due to reduced NO production (Celermajer et al. 1993). O_2^- radicals react avidly with NO to form peroxynitrite ($ONOO^-$), a potent oxidant (see below). Another indication of an unbalanced relationship between NO and ROS comes from the observation that ED in homocystinuric patients is ameliorated with ascorbic acid therapy (Pullin et al. 2002). Besides possible direct free radical scavenging, it was suggested that ascorbic acid may also exert its effects by acting directly upon eNOS, enhancing NO production by stabilizing (Heller et al. 2001) and increasing intracellular BH4 concentrations (Huang et al. 2000).

8.3.4 Treatment Effects of Homocystinuria

Treatment to lower the extremely elevated Hcy levels in homocystinuric patients by using pyridoxine, folic acid, cobalamin and, if necessary, trimethylglycine (betaine) dramatically reduces cardiovascular risk in both pyridoxine-responsive (and the more severe pyridoxine-nonresponsive) patients, although Hcy levels remain well above the normal range. (Wilcken and Wilcken 1997; Yap et al. 2001).

Although ineffective on Hcy levels, vitamin supplementation for 2 weeks produced transiently reduced urinary excretion of urinary 8-iso-PGF2α and 11-dehydro-TXB2 in seven homocystinuric patients (Davi et al. 2001).

There are only few studies on these very rare genetic disorders, but the available data suggest that Hcy -treatment with the vitamins mentioned above counteracts Hcy-induced oxidative stress and they may further provide protection through effects unrelated to Hcy -diminution.

8.3.5 Mild Hyperhomocysteinemia

Several functional pathologies associated with mildly elevated Hcy plasma concentrations (>15 μmol/L) have been linked with oxidative stress. Hcy is readily oxidized when added to plasma, principally as a consequence of auto-oxidation leading to the formation of homocystine, homocysteine-mixed disulfides, and homocysteine thiolactone (Loscalzo 1996). Hcy has also been shown to oxidize low density lipoproteins (LDL) by reactions requiring redox-active transition metal ions (Heinecke et al. 1993). In a study using a rat model, acute methionine load-induced HHcy demonstrated elevation of plasma conjugated dienes, lipid hydroperoxides, and thiobarbituric acid-reactive substances (TBARS) (Durand et al. 1997).

8.3.6 Homocysteine and Lipid Peroxidation

In vivo, Hcy has been found to be associated with increased lipid peroxidation (LPO) as measured by plasma F2-isoprostane concentrations (Voutilainen et al. 1999) and decreased antioxidant activity of SOD on endothelial cell surfaces with impaired endothelial function (Yamamoto et al. 2000). HHcy may exert its effects through increased tumor necrosis factor α (TNFα) expression, which enhances O_2^- induced oxidative stress by upregulating a nox1-based NAD(P)H oxidase and inducible NOS (Ungvari et al. 2003) that is associated with increased protein 3-nitrotyrosine content in smooth muscle cells. The effect of decreased NO-bioavailability under conditions of HHcy can be reversed by administration of superoxide scavengers (Ungvari et al. 2002) and ascorbic acid (Chambers et al. 1999). Incubation with Hcy activated IκB kinases (IKKα and IKKβ), leading to phosphorylation and subsequent degradation of IκBα (Kathy et al. 2004). The consequences were NF-κB nuclear translocation, enhanced NF-κB/DNA binding activity, and increased transcriptional activity. There was marked elevation of O_2^- in Hcy-treated cells.

8.3.7 Glutathione Peroxidase

An elevated level of Hcy was found to decrease the activity of cellular GPx in vivo, an intracellular selenocysteine-containing protein and a key antioxidant enzyme that uses glutathione to reduce H_2O_2 and lipid peroxides to their respective alcohols (Huang et al. 2001). GPx may also act as a peroxynitrite reductase (Sies et al. 1997). In an interesting experiment using the CβS-mouse model, overexpression of GPx could compensate for the adverse effects of Hcy on endothelial function, suggesting that the adverse vascular effects of Hcy are at least partly mediated by oxidative inactivation of NO (Weiss et al. 2001). Purified GPx from human plasma is inhibited by physiological free Hcy (1–5 μmol/L) (Chen et al. 2000). Similarly, in cultured endothelial cells, micromolar concentrations of Hcy are sufficient to decrease the activity of GPx (Weiss et al. 2002; Upchurch et al. 1997). Hcy down-regulates GPx activity by interfering with protein expression (translation) without affecting transcription levels (Handy et al. 2005).

The Hcy-dependent transsulfuration pathway catalyzed by the heme and pyridoxine-dependent CβS leads to cystathionine and subsequently to cysteine and glutathione biosyntheses in addition to providing a catabolic route leading to sulfate. Intracellular glutathione is important for maintenance of the physiological cell redox state and, together with heat shock proteins, plays a key role in mitochondria against adverse effects elicited by Hcy and H_2O_2 (Austin et al. 1998). It is estimated that approximately half of the intracellular glutathione pool in human liver cells is derived from Hcy via the transsulfuration pathway, whose redox sensitivity can be rationalized as an autocorrective response that leads to an increased rate of glutathione synthesis in cells challenged by oxidative stress (Mosharov et al. 2000).

Conditions of chronic oxidative stress may thus impair the synthesis of this major redox buffer. HHcy, induced through methionine loading, is associated with a greater activation of the antioxidant defense as expressed by higher intracellular glutathione content, but was not accompanied by a significant increase of malondialdehyde (MDA) (Campolo et al. 2006). High-dose treatment with 5-MTHF (15 mg/d) improved the antioxidant defense in hyperhomocysteinemic subjects through interaction with glutathione metabolism, i.e. an increased reduction of blood total glutathione that includes oxidized glutathione, protein bound glutathione, glutathione mixed disulfides and reduced free glutathione (Caruso et al. 2006). The reduction of glutathione oxidative forms, however, cannot be explained exclusively by Hcy-diminution or amelioration of glutathione turnover, but may also indicate antioxidative effects of 5-MTHF on the redox state per se, related to glutathione metabolism.

Importantly, it was found that HHcy in healthy young men and elderly men with normal serum folate concentration was not associated with ED or oxidative stress as measured by glutathione, TAS, TBARS, and plasma 8-iso-PGF2α (Hirsch et al. 2004). This finding points toward the hypothesis that folate deficiency is the key factor for ROS and ED, rather than (mild) HHcy. Thus it was postulated that the acute modifications in vascular function induced by folic acid are unrelated to changes in Hcy concentrations. HHcy may, however, help to identify individuals with folate deficiency and target groups that may benefit from supplemental intake.

8.4 Endothelial Dysfunction

ROS play a major role in vascular biology. In general, relatively low concentrations of ROS are thought to act as mediators or modulators of cell signaling and contribute to other key functions, e.g. the regulation of activity of transcription factors and gene expression (Droge 2002). In contrast, higher levels of ROS contribute to vascular dysfunction and abnormal cell growth including hypertrophy of vascular muscle. O_2^- levels are increased in blood vessels in many pathophysiological conditions including hypertension, diabetes, atherosclerosis, hyperhomocysteinemia, heart failure, sepsis, subarachnoid hemorrhage, and Alzheimer's disease, as well as during aging. Multiple lines of evidence have shown that NO signaling plays a major role in vascular biology. Since the initial evidence that superoxide or other ROS inactivate NO or EDRF, many studies have suggested that inactivation of NO by O_2^- contributes to vascular dysfunction under pathophysiological conditions. In addition to inactivating NO and thus preventing NO-mediated signaling, the reaction of NO with O_2^- produces $ONOO^-$, a potent oxidant with cytotoxic potential (Beckman and Koppenol 1996). Emerging evidence suggests that formation of $ONOO^-$ has multiple effects, including (1) selective nitration of tyrosine residues in proteins, such as prostacyclin synthase and Mn-SOD, (2) activation of poly (ADP-ribose) polymerase (PARP) and expression of inducible NO synthase (iNOS), potentially important mediators of vascular dysfunction in disease states, (3) oxidation of BH4, and (4) oxidation of the zinc–thiolate complex in endothelial NOS (eNOS). The latter two effects can produce eNOS "uncoupling", a condition in which the normal flow of electrons within the enzyme is diverted such that eNOS produces O_2^- rather than NO (Faraci and Didion 2004).

ED, assessed as impaired vasodilator response to mechanical or pharmacological stimuli, has been increasingly recognized as a surrogate end point for cardiovascular risk. Several investigations have found an association between ED and myocardial ischemia (Zeiher et al. 1995; Hasdai et al. 1997). Other reports have demonstrated in long-term follow-up studies, that ED is associated with an increased progression of atherosclerotic disease (Schachinger et al. 2000) and a higher incidence of cardiovascular events (Suwaidi et al. 2000). ED refers mainly to reduced bioavailibility of NO and accumulating evidence suggests that ROS are fundamentally involved in HHcy and folate deficiency associated with ED (Stanger et al. 2002). NO production is determined by cofactors such as BH4, which is oxidized and depleted in conditions of oxidant stress, e.g. by $ONOO^-$. THF deficiency contributes to the uncoupling of NO synthase into an O_2^- radical-producing enzyme.

Importantly, several investigators have demonstrated beneficial effects of folates on endothelial function independent of changes in plasma Hcy levels, suggesting that folates may exert direct antioxidative effects and contribute to restoration of impaired NO metabolism. (Usui et al. 1999; Doshi et al. 2002; Stanger and Weger 2003). More specifically, 5-MTHF above all appears to be associated with endothelial function, independent of Hcy levels (Spijkerman et al. 2005). Using the MTHFR gene polymorphism 677C>T, it was recently demonstrated that vascular 5-MTHF, rather than plasma or vascular Hcy, is the key regulator of eNOS coupling and NO bioavailability in human vessels, further suggesting that plasma Hcy is an indirect marker of 5-MTHF rather than a primary regulator of endothelial function (Antoniades et al. 2009).

In a clinical cross-over study, male patients with angiographically documented coronary artery disease (CAD) received 5 mg of folic acid orally for 12 weeks in an on-off fashion and resistance vessel reactivity was repeatedly assessed by venous occlusion plethysmography. Folic acid supplementation significantly increased plasma folate levels 3.5 fold and improved resistance vessel reactivity (peak

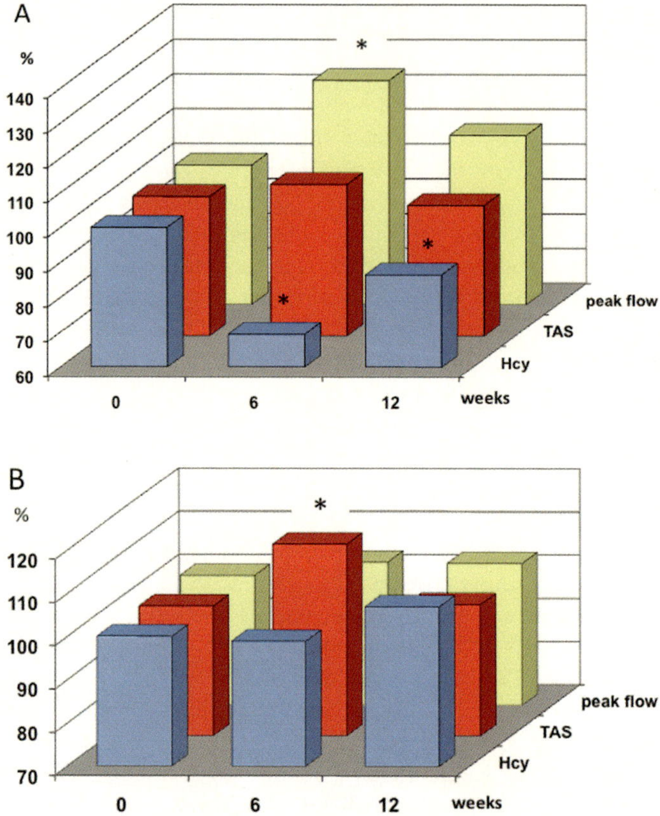

Fig. 8.3 (**a**) Effect of 5 mg/day oral folic acid on TAS, Hcy and resistance vessel reactivity (peak flow) in male patients with CAD after 6 weeks (therapy) and 12 weeks (off-therapy). Changes relative to baseline values (10%). *Indicated significant changes ($p < 0.05$). (**b**) Effect of folate intake on antioxidative capacity as determined by total antioxidant status (TAS) and in resistance vessel reactivity (peak flow) in subjects without significant changes in Hcy levels ($\Delta<2$ μmol/L). Changes relative to baseline values (10%). *Indicated significant changes ($p < 0.05$)

flow), suggesting vasoprotective effects of FA in patients with established ED and atherosclerotic vessel disease (Stanger et al. 2002) (Fig. 8.3a). In subjects without significant changes in Hcy levels ($\Delta<2$ μmol/L), folate administration was associated only with increased antioxidative capacity as determined by total antioxidant status (TAS), suggesting an antioxidative potential for folate (Fig. 8.3b). In other words, unless folate is consumed in the remethylation pathway for Hcy-diminution, folic acid may add antioxidant capacity to plasma.

5-MTHF was suggested to exert direct effects on the enzymatic activity of NOS, which is supported by the identification of a pteridine-binding domain with similarities to the folate binding site of DHF reductase (Hyndman et al. 2002).

Treatment of endothelial cells in vitro with folic acid showed a marked increase of dihydrofolate reductase (DHFR) expression and activity (Gao et al. 2009). Furthermore, folic acid improved the redox status of angiotensin (Ang) II-treated cells by increasing BH4 and NO bioavailability while decreasing O_2^- production. It failed, however, to restore NO levels in DHFR siRNA-transfected or methotrexate pre-treated cells, implicating a specific and intermediate role of DHFR. This has been also demonstrated in mice with oral administration with FA (15 mg/kg/day, 16 days), where endothelial upregulation of DHFR expression and activity occurred in correspondence to improved NO and BH4 bioavailability, and this was highly effective in reducing Ang II infusion-stimulated aortic O_2^- production.

Mice infused with Ang II had markedly increased O_2^- production and reduced NO and BH4 bioavailability. These responses were, however, all significantly reversed by pre-treatment with folic acid. In addition, DHFR expression and activity were increased by FA. This was accompanied by an endothelial downregulation of DHFR. Oral administration of FA significantly improved NO production, while reducing O_2^- levels. Taken together, these data demonstrate that folic acid exerts it beneficial effects in vivo via regulation of DHFR expression and activity. It is important to note that in the endothelium-denuded aortas, DHFR expression was upregulated by Ang II, indicating differential regulation of DHFR in different cell types. This seems consistent as it is known that Ang II stimulates vascular smooth muscle hypertrophy, and that DHFR in general is important for DNA synthesis and cell growth.

In summary, folic acid supplementation was shown to improve endothelial NO bioavailability via upregulation of DHFR expression and activity, and to protect endothelial cells from Ang II-provoked oxidant stress both in vitro and in vivo (Gao et al. 2009).

Growing evidence suggests that potential mechanisms of vascular protection are largely independent of the Hcy-diminution and include a direct superoxide-scavenging capacity of 5-MTHF and effects on eNOS and NO generation. The latter are likely to involve BH4. These effects have been reported for the natural L[6S] and the D[6R] isomer of 5-MTHF, arguing for the possibility that synthetic forms are also biologically active (Stroes et al. 2000; Baggott et al. 2001). Furthermore, beneficial effects of folate administration have also been reported in non-hyperhomocysteinemic subjects, such as the correction of ED in patients with familial hypercholesterolemia (Verhaar et al. 1998) or the prevention of its impairment by postprandial lipaemia (Wilmink et al. 2000).

8.4.1 Interactions

5-MTHF, when given intra-arterially, not only improved flow-mediated dilatation without altering plasma Hcy concentrations but also abolished Hcy-induced increase in intracellular O_2^- generation. Folic acid and BH4 also abolished Hcy-induced intracellular O_2^- production in culture (Doshi et al. 2002). This suggests that folic acid, 5-MTHF and BH4 act in concert to restore endothelial function by suppressing O_2^- production and enhancing NO generation or increasing its half-life. This is supported by the observation that 5-MTHF had no direct effect on in vitro NO production by eNOS but induced a dose-dependent reduction in eNOS- and xanthine oxidase–induced O_2^- generation and reversed impaired endothelium-dependent vasodilation in patients with familial hypercholesterolemia (Verhaar et al. 1998).

It is evident from the preceding discussion that folate/folic acid, H4B, insulin, and O_2^- interact with each other and significantly influence the generation, stability and action of NO. Folate/folic acid, H4B, and insulin suppress O_2^- production and thus prolong the half-life of NO. The importance of ascorbic acid lies in the fact that it enhances eNOS activity by increasing intracellular levels and by chemical stabilization of BH4 (Heller et al. 2001). Ascorbic acid also enhances the release of NO and thus suppresses the formation of increased amounts of total S-nitroso thiols and S-nitroso albumin (Verhaar et al. 1998).

8.5 Folates as Antioxidants

8.5.1 Folate Effects In Vitro

8.5.1.1 Radical Scavenging

The pulse radiolysis technique showed that folic acid can effectively scavenge such free radicals as $CCL_3O_2^{\bullet}$ (a model peroxyl radical), $N_3^{\bullet}$, $SO_4^{\bullet -}$ and $Br_2^{\bullet -}$ (one electron oxidants), physiologically relevant $^{\bullet}OH$ and O_2^- radicals as well (Joshi et al. 2001).

Folic acid and its reduced forms were further used to study their antioxidant capacities in the TEAC, DPPH and FRAP assays (Gliszczynska-Swiglo 2007). In short, the TEAC (trolox equivalent antioxidant capacity) value is based on the ability of the antioxidant to scavenge the blue-green coloured $ABTS^{\bullet}_{+}$ radical cation relative to the $ABTS^{\bullet}_{+}$ scavenging ability of the water-soluble vitamin E analogue, Trolox^{TM} (Rice-Evans and Miller 1994). The antioxidant (radical-scavenging) activity can also be measured using 2,2-diphenyl-1-picrylhydrazyl ($DDPH^{\bullet}$) as a free radical. And finally, the FRAP assay is based on the reduction of a ferric 2,4,6-tripyridyl-s-triazine complex (Fe^{3+}-TPTZ) to the ferrous form (Fe^{2+}-TPTZ) (Benzie and Strain 1996).

The lowest TEAC and FRAP values and the highest IC50 value in the DPPH assay were obtained for folic acid, identifying it as the least potent antioxidant among the folates tested (Gliszczynska-Swiglo 2007). Reduction of folic acid to DHF significantly increased $ABTS^{\bullet}_{+}$ radical cation and $DDPH^{\bullet}$-scavenging activity as well as the ability to reduce Fe^{3+} in the FRAP assay. Further reduction of DHF to THF decreased the TEAC and the FRAP values, indicating a decrease in free-radical scavenging activities of THF as compared to DHF (Gliszczynska-Swiglo 2007). In contrast, the $DDPH^{\bullet}$-scavenging activity of THF compared to that of DHF was almost two-fold higher (IC50 values 59,1 μM and 107 μM, respectively). The

DDPH$^{\cdot}$-scavenging activities of 5-MTHF and THF appear comparable and support data obtained by Rzek et al., using the TEAC test, a lipid peroxidation (LPO) assay (Haenen and Bast 1983) and also measuring the peroxynitrite scavenging capacities (Rzek et al. 2003). Their results indicated that 5-MTHF is the most effective $ONOO^-$ scavenger and THF is the most effective inhibitor of LPO among the folates tested (Rzek et al. 2003).

Lucigenin-enhanced chemiluminescence demonstrated that 5-MTHF could reduce $O_2^{\cdot-}$ generation by 2 $O_2^{\cdot-}$-generating systems: xanthine oxidase/hypoxanthine and eNOS (Verhaar et al. 1998). Folate abolished the Hcy-induced increases in endothelial $O_2^{\cdot-}$ (Doshi et al. 2002), and the $O_2^{\cdot-}$-scavenging capacity of 5-MTHF was confirmed by electron paramagnetic resonance, using 5-(diethoxyphosphoryl)-5-methyl-L-pyrroline N-oxide as a spin trap for $O_2^{\cdot-}$ (Stroes et al. 2000). This method also showed that the scavenging potency of 5-MTHF is about 20-fold lower than that of ascorbic acid. The radical scavenging activity of folic acid and its reduced forms, measured in the TEAC assay, was furthermore found to be strongly pH-dependent. Folic acid is a potent radical scavenger at acidic and basic pH in contrast to neutral pH, whereas reduced forms exhibit their radical scavenger activity predominantly at acidic pH rather than at neutral and basic pH values (Gliszcznska-Swiglo and Muzolf 2007). Thus, folate clearly does exert antioxidant effects.

8.5.1.2 Folates and Lipid Peroxidation

Folic acid has poor antioxidant activity in $ONOO^-$ scavenging and LPO inhibition. However, as folic acid is reduced, its antioxidant capacity increases. When folic acid is reduced to DHF, $ONOO^-$ scavenging becomes more pronounced; further reduction to THF and conversion to 5-MTHF enhance the scavenging activity even more. The fully reduced forms, i.e. THF and 5-MTHF, are the most active compounds in inhibiting LPO. (5-MTHF is relatively less active than THF in protecting against LPO, whereas 5-MTHF is the better $ONOO^-$ scavenger than THF).

The activities of folic acid and derivatives against H_2O_2 generation in linoleic acid (poly unsaturated fatty acid) peroxidation may have an early- "prooxidant" and a strong late- "antioxidant" phase (Higashi-Okai et al. 2007). 5-MTHF, but not folic acid, significantly increased the resistance of VLDL and LDL to oxidation in vitro (McEneny et al. 2007).

In summary, the results suggest that folic acid has a poor protective effect against free radicals and lipid peroxidation. Thus, due to its poor activity against LPO, it is likely that the minor antioxidant capacity of folic acid is physiologically irrelevant, though its physiological metabolites are much more active. Indeed, the fully reduced folates, namely THF and 5-MTHF, are the most active forms in scavenging the DDPH$^{\cdot}$ and $ONOO^-$. THF is the most effective inhibitor of LPO, whereas DHF is most effective in scavenging $ABTS^{\cdot}_{+}$ radical cation and is the best reductant as expressed by the FRAP value.

8.5.2 The Active Sites of Folates

In determining the active part of the molecule responsible for the antioxidant activity of folic acid, it was suggested that the 4-OH group on the pterin ring has an important role (Joshi et al. 2001). It has been found that the side chain, i.e. N-(p-aminobenzoyl)-L-glutaic acid, is more important, since the products formed by nitration of folic acid were identified as 10-nitrofolate and 12-nitrofolate (Nakamura et al. 2002).

To determine the responsible pharmocophore of the reduced forms of folic acid, the antioxidant activities of 4-hydroxy-2,5,6-triaminopyrimidine (and after removing the 5 amino group or the 5 and 6 amino groups) and structurally related compounds (5-6-7-8-tetrahydro-L-biopterin [BH4], pterin, leucopterin) were measured by $ONOO^-$ scavenging and the inhibition of LPO (expressed as IC50) (Rzek et al. 2003). The minor importance of the side chain for the antioxidant activity is suggested by (a) the absence of antioxidant activity of N-(p-aminobenzoyl)-L-glutaic acid, (b) the large difference in activity of the various folates, which all contain the same side chain, and (c) the antioxidant activity of BH4, which lacks the side chain and whose activity is comparable to that of THF. Therefore, and by comparing structurally related compounds, it was found that the antioxidant pharmacophore of THF and 5-MTHF, i.e. 4-hydroxy-2,5,6-triaminopyrimidine, resides in the pterin moiety (Rzek et al. 2003).

It is important to stress that in vitro assays for determination of the antioxidant activity of compounds under investigation do not reflect in vivo conditions because they do not take into account the presence of other compounds, which may interact with other bioactive molecules and therefore influence the antioxidant capacity of a given compound. Moreover, the relevance of potent antioxidants in biological systems may be strongly related to their bioavailability and metabolism.

The actual antioxidant effect of folates in vivo might depend on both reduction by the enzyme MTHFR and consumption, which is related to cobalamin and Hcy levels, tissue distribution and concentration (see Section 8.2.3), interaction with related compounds and susceptibility of the individual redox balance.

8.5.3 Antioxidant Effects of Folic Acid, Trimethylglycine (Betaine), 5-MTHF and Methionine In Vitro and Ex Vivo

Our group recently investigated the antioxidative potential of four different substances that are involved in the Hcy pathway, i.e. folic acid, trimethylglycine (betaine), 5-MTHF and methionine.

We took three different approaches, using established, commercially available biomarkers, to study whether these substances might (a) increase the lag-time of DPHPA, a fluorescent marker, that reflects the water-soluble antioxidant capacity in biological samples (Protein-Ox®), (b) increase peroxidase-activity (EPA®), an endogenous antioxidant, and (c) have a quenching property to reduce total peroxides (TOC®). The tests are described briefly:

(1) Water-soluble antioxidants: The lag time of ex vivo degradation of the fluorophore 1,6-diphenylhexatriene propionic acid (DPHPA) by ROS in serum was determined with a fluorescent diagnostic assay (Protein-Ox®, Tatzber KG, Klosterneuburg, Austria) according to (Mayer et al. 2001), with modifications to the oxidation part. Briefly, 100 μL sera were incubated with 2 nMol DPHPA and kept under argon at 37°C for 1 h. Oxidation was started via a peroxide (0.004%)/peroxidase (10 U/ml) reaction and the time-dependent decrease in fluorescence intensity at 430 nm (excitation at 360 nm) was monitored on a FluoStar Fluorometer (BMG Lab-Technologies; D-77656 Offenburg, Germany). Results were expressed as "lag time" in minutes.
(2) Serum total peroxidase-activity: Peroxidase-activity in serum was determined by the reaction of endogenous peroxidases with hydrogen peroxide, using TMB as the chromogenic substrate (EPA) supplied by Labor Diagnostic Nord (LDN, Nordhorn, Germany) as previously described (Tatzber et al. 2002). Serum peroxidase-activity was expressed as milliunits per millilitre. Serum reference values are in the range of 5 ± 3 mU/mL.
(3) Serum total peroxides: Serum total peroxide concentrations were determined by a rapid enzymatic in vitro diagnostic assay (TOC®) supplied by Labor Diagnostic Nord (LDN, Nordhorn, Germany) as previously described (Tatzber et al. 2002). The test system was based on a peroxide/peroxidase reaction using 3,5,3′,5′-tetramethylbenzidine (TMB) as substrate. Peroxide levels were specified as "μM H_2O_2 equivalents", because of the different contributions of various peroxides to the reaction. Reference values were reported to be less than 200 μM H_2O_2 equivalents (Lindschinger et al. 2004).

Blood was drawn from volunteers, centrifuged and exposed to a range of reference values as well as supra-optimal concentrations of the respective substances. A lyophilisate of each substance was reconstituted in distilled water to achieve the following concentrations: folic acid (7 ng, 18 ng and 18 μg), betaine (35 μmol/L, 146 μmol/L and 200 μmol/L), 5-MTHF (10 ng/mL, 40 ng/mL and 6 μg/mL) and methionine (47 μmol/L and 100 μmol/L).

For in vitro experiments, we applied the reconstituted stock-solutions and a dilution of 1:1 (v/v) with distilled water (Fig. 8.4b and d). For ex vivo experiments, we diluted stock-solutions 1:1 (v/v) with serum in the microtitre-wells for each biomarker (Fig. 8.4a, c, and e). With regard to peroxidase-activity we used the third standard of the EPA-assay (i.e. 10 mu/mL) as a reference.

None of these substances was able to affect significantly the lag-time of water-soluble antioxidants in serum (Fig. 8.4a). With respect to peroxidase-activity, we observed the maximum decrease (30%) at the peak concentration for methionine (100 μmol/L) and for betaine at a concentration of 146 μmol/L (Fig. 8.4b). Furthermore, betaine was able to decrease peroxidase-activity (20%) in a range from 35 to 146 μmol/L, as was also observed for folic acid at the lowest concentration (3.5 ng/ml). Nevertheless, peak and bottom levels for betaine (200 μmol and 17 μmol) did not affect the enzyme-activity in vitro. Just one case of increased enzyme activity was observed at the utmost concentration of 5-MTHF (6 μg/mL).

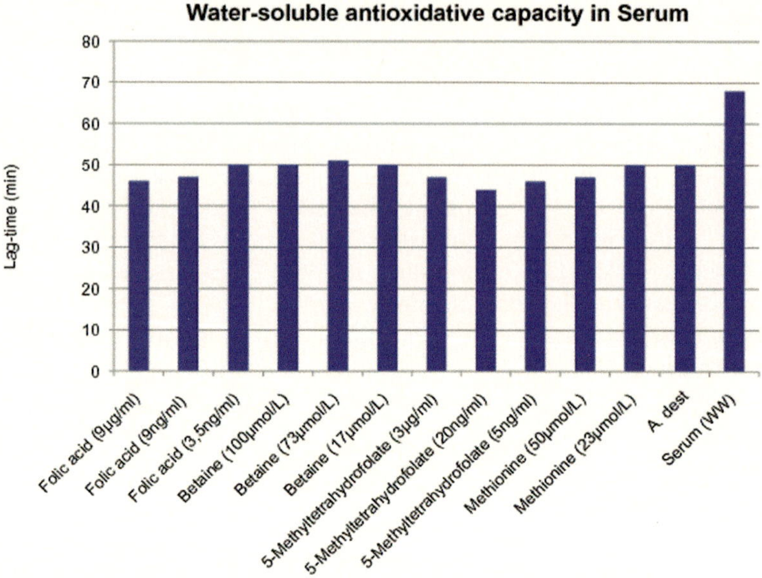

Fig. 8.4 (**a**) Water-soluble antioxidative capacity in serum. Lag-time (min)

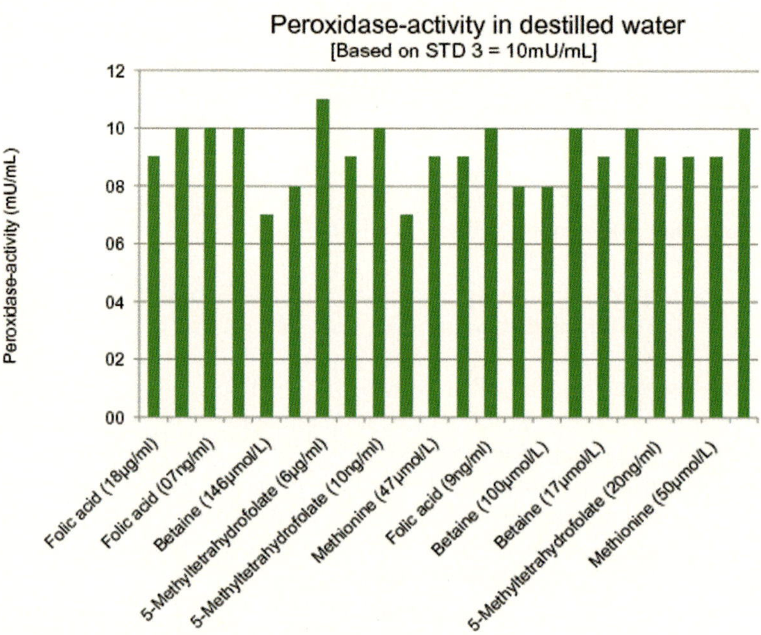

Fig. 8.4 (**b**) Peroxidase-activity in destilled water (mU/mL)

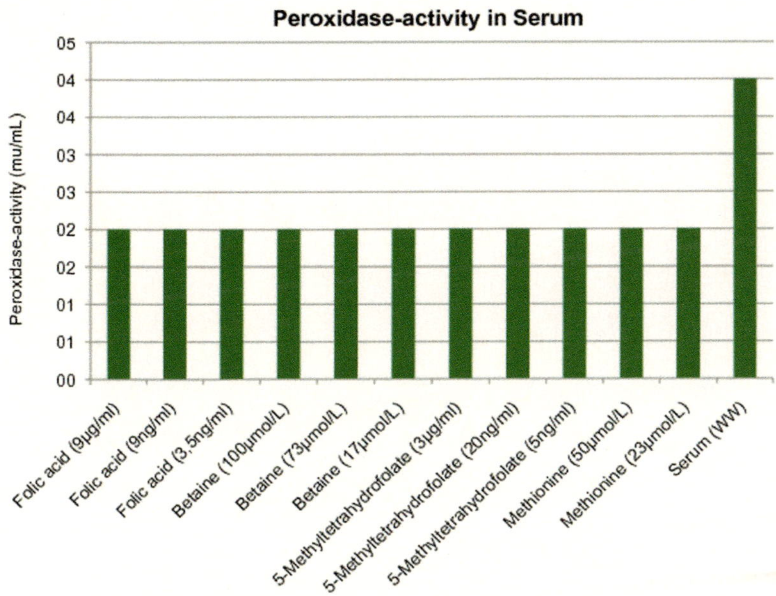

Fig. 8.4 (**c**) Peroxidase-activity in serum (mU/mL)

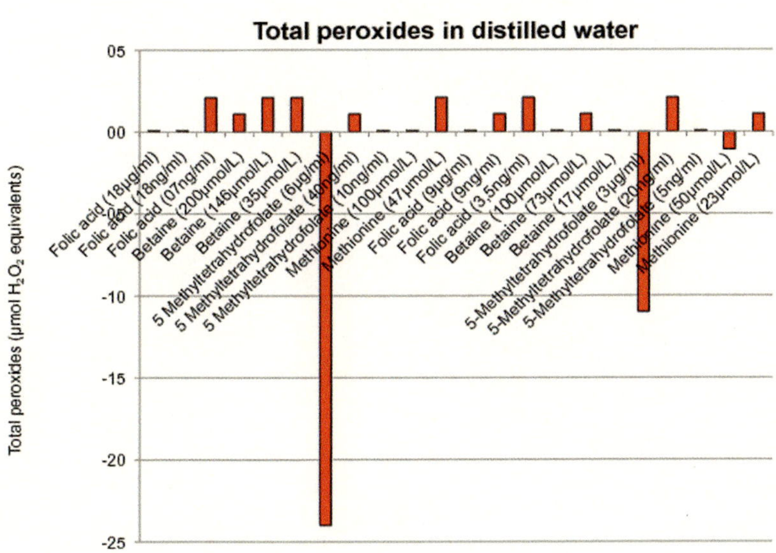

Fig. 8.4 (**d**) Total peroxides in destilled water (µmol H_2O_2 equivalents)

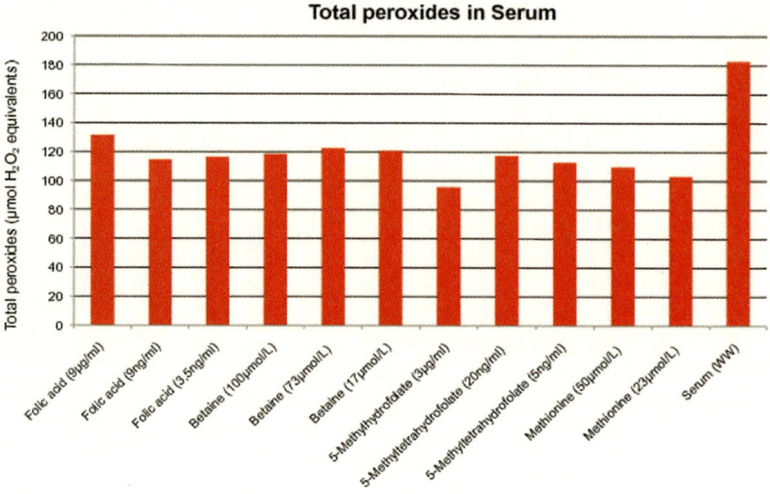

Fig. 8.4 (**e**) Total peroxides in destilled serum (μmol H_2O_2 equivalents)

Nonetheless, peroxidase-activity in serum (ex vivo) was not at all affected by these substances (Fig. 8.4c). The same applied to total peroxides with one exception: 5-MTHF in a micro-molar concentration was able to quench peroxides in a dose-dependent manner (from 3 to 6 μmol/L) as demonstrated in Fig. 8.4d. This effect was confirmed in serum inasmuch as the 3 μmol/L concentration of 5-MTHF was able to quench total peroxides to the same extent as was observed in the in vitro experiment (Fig. 8.4e).

These data suggest that folic acid, betaine, methionine and 5-MTHF have no significant antioxidative effect in vitro and ex vivo, either for the lag-time of water-soluble antioxidants or for the activity of the endogenous antioxidant-enzyme peroxidase.

We did, however, observe a significant quenching effect of 5-MTHF in a micro-molar concentration both in vitro and ex vivo, which supports its capacity to scavenge $ONOO^-$.

8.5.4 Folate Effects In Vivo

After male Wistar rats had been fed with a folate-deficient diet for 4 weeks, a significant elevation of oxidative stress was observed inside the liver mitochondria (mt) with a 77% decrease in mt folate level, a 28% reduction in GPx activity, a 1.2-fold increase in mt protein carbonyls and an accumulated 4834 bp large-scale deletion in mt DNA (Chang et al. 2007). The elicited oxidative injuries in liver mt were associated with a 30% reduction of cytochrome c oxidase activity that coincided with membrane potential dissipation (depolarization) and superoxide overproduction. Interestingly, preincubation with prooxidant copper-sulfate (48 h) in folate-deficient hepatocytes with supplemental folic acid reversed the mt oxidative defects described and diminished O_2^- overproduction. Increased supplemental levels of folic acid strongly correlated with decreased lipid peroxidation and protein oxidative injuries in prooxidant pretreated mitochondria of hepatocytes.

Lipid peroxidation was significantly inhibited by folic acid in Wistar rat liver microsomes (Joshi et al. 2001). Thus, the reduced physiological forms THF and 5-MTHF may act as effective free radical scavengers and LPO inhibitors in vivo. Their free radical scavenging activities expressed as the TEAC and FRAP values are comparable to those of vitamins C and E. The activities of reduced folates in the DPPH assay are merely 2–6 fold lower than those of vitamin C and E.

Oral intake of folic acid (5–10 mg/d/6–8 weeks) increased total antioxidant capacity (TAC) in patients with established CAD (Stanger et al. 2002), in hypercholesterolemic adults under lovastatin treatment (Shidfar et al. 2009), and in hemodialysis patients (Delfino et al. 2007). Activity of folic acid against radical-mediated oxidative damage in human whole blood has been reported previously (Stocker et al. 2003).

In volunteers with manifest atherosclerotic disease and plasma Hcy >20 μmol/L, 2 months of daily oral treatment with 10 mg folic acid were found to significantly increase activities of GPx and SOD, and decrease MDA (Mayer et al. 2002). Wilmink et al. reported that folate treatment prevented the increase in MDA in urine after an acute fat load (Wilmink et al. 2000). Oral treatment with 1 mg folic acid/day in 12 female subjects significantly increased the resistance of very low density lipoprotein (VLDL) and of low density lipoprotein (LDL) against oxidation in (5-MTHF>FA) and in vivo (McEneny et al. 2007).

F2-isoprostanes, a family of bioactive prostaglandin F2-like compounds derived from arachidonic acid through a non-enzymatic process of LPO, are catalyzed by oxygen free radicals on cell membranes and LDL particles. Increased levels of these compounds have been reported in association with myocardial reperfusion injury and with cardiovascular risk conditions (Patrono and Fitzgerald 1997), including severe HHcy because of homozygous CβS-deficiency (Davi et al. 2001).

5-MTHF (15 mg/d/4 weeks) was given to patients with early-onset thrombotic episodes and the 677TT MTHFR genotype. At baseline, patients in the lowest quartile of the plasma folate distribution (<3.9 μg/L) had the highest urinary 8-iso-PGF2α excretion (Coppola et al. 2005). Urinary 8-iso-PGF2α and 11-dehydro-TXB2 excretions were closely correlated. After 5-MTHF supplementation, the urinary 8-iso-PGF2α levels were significantly lowered, their decrease being proportional to baseline values and maximal in patients with lowest pre-supplementation folate levels. On average, 8-iso-PGF2α was 17% lower after 5-MTHF supplementation and no longer significantly different from healthy individuals.

8.5.5 5-MTHF, NO, BH4 and Peroxynitrite

Inflammation of the vascular endothelium is often associated with an "uncoupling" of eNOS, in which eNOS becomes a source of O_2^- concurrent with a reduction in its capacity to generate NO (Stanger and Weger 2003). This phenomenon is associated with oxidation of BH4, an obligate cofactor for coupled activity of eNOS. NOS function is determined by the availability of BH4 versus the abundance of BH2, which is inactive for eNOS cofactor function.

$ONOO^-$ is a powerful oxidizing and nitrating agent. Due to its pKA of 6.8, the anion predominates (80% at pH 7.4) under physiological conditions. The conjugated acid homolyzes at a rate of $0.9\,s^{-1}$ (37°C, pH 7.4), giving free hydroxyl and nitrogen dioxide ($^{\bullet}NO_2$) radicals in 30% yield, which can nitrate aromatic residues such as tyrosine and tryptophan. The preferential biotargets of $ONOO^-$ are metal centers, carbon dioxide, and sulfur and selenium compounds (Ferrer-Sueta and Radi 2009). $ONOO^-$ appears to be the chief mediator of this oxidation and recent evidence indicates that $ONOO^-$ can suppress synthesis of BH4 by promoting the ubiquitination and proteasomal degradation of GTP cyclohydrolase, rate-limiting for BH4- synthesis (Xu et al. 2007). Administration of folic acid or 5-MTHF can "recouple" the activity of eNOS and restore generation of NO, leading to restoration of endothelial function and NO-dependent vasodilation (Stanger et al. 2001, 2003; Stanger 2002; Verhaar et al. 2002; Das 2003) with potential implications for ischemia-reperfusion injury.

In an animal study, folic acid, administered per os before ligation of a coronary artery, caused a dramatic reduction in myocardial infarct area (Moens et al. 2004). Importantly, this effect appeared to be mediated by an NOS-dependent mechanism, as it could be entirely inhibited by administration of a NOS inhibitor.

A recent ex vivo study investigating human arteries and veins presented evidence that exposure to 5-MTHF rapidly boosts the level of BH4 and the BH4/total biopterin ratio in endothelial cells, while reversing the uncoupled behavior of eNOS (Antoniades et al. 2006). The improved NO-mediated endothelium-dependent vasomotor response could not be explained by direct O_2^{-} scavenging by 5-MTHF or by changes in plasma Hcy. Rather, 5-MTHF was shown to be a strong $ONOO^-$ scavenger with increased vascular BH4 availability. This effect may be due to the protective effect of 5-MTHF on $ONOO^-$-induced oxidation of BH4. The findings further suggest that 5-MTHF has primarily specific rather than general antioxidant effects in human atherosclerosis. 5-MTHF could enhance eNOS activity and qualitatively restore eNOS coupling in human vessels. Furthermore, 5-MTHF increased the dimer:monomer ratio in human vessels, suggesting an improvement in eNOS protein dimerization (Antoniades et al. 2006).

Although 5-MTHF may be a rather weak O_2^{-} scavenger, it might increase intracellular BH4 by preventing intracellular oxidation by $ONOO^-$. It may also interact with the pterin-binding site of eNOS in a fashion analogous to BH4 (Hyndman et al. 2002) or directly stimulate BH4-synthesis from BH2. Oxidation of BH4 in the presence of $ONOO^-$ occurs in two steps, the first step yielding a BH3 radical that can be reconverted to BH4 by interaction with ascorbate (Kuzkaya et al. 2003). Further oxidation yields BH2 (7,8-dihydrobiopterin), which promotes eNOS uncoupling due to competition with BH4 for eNOS binding.

Conversely, DHFR (EC 1.5.1.3.) can generate BH4 from BH2. DHFR is a key enzyme in folate metabolism converting folic acid to DHF, and DHF to THF. More recently, the key role for DHFR in eNOS coupling by maintaining the BH4:BH2 ratio, particularly in conditions of low total biopterin availability, was demonstrated (Crabtree et al. 2009).

Folic acid undergoes modification by $ONOO^-$ (Nakamura et al. 2002). The nitrated folate species were identified as 10-nitro-folate and 12-nitro-folate. Furthermore, the radicals derived from $ONOO^-$, nitrogen dioxide and carbonate radical, were demonstrated to inactivate CβS (Celano et al. 2009). CβS catalyzes the condensation of Hcy with serine to form cystathionine and water in the first step of the transsulfuration pathway that yields cysteine. CβS can also catalyze the formation of hydrogen sulfide, a novel gasotransmitter with signalling and cytoprotective effects. Inhibition of the transsulfuration pathway typically leads to HHcy, and to homocystinuria in its extreme form (CβS-deficiency, see above).

Nevertheless, it was recently questioned whether reduced pterins including folates interact directly with $ONOO^-$ (McCarthy et al. 2009). Rather, like uric acid, they may scavenge the nitrogen dioxide and carbonate radicals that evolve from $ONOO^-$ when the latter reacts with carbon dioxide in biological systems. If this is the case, reduced folates may not influence the oxidizing impact of $ONOO^-$ on transition metals or on cysteine and methionine. The reversible oxidation of acidic cysteine groups in proteins is a major route by which $ONOO^-$ (in its protonated form peroxynitrous acid) functions physiologically to modulate signal transmission pathways and enzyme activities. Nontheless, the nitrogen dioxide and carbonate radicals likely are responsible for a high proportion of the cellular damage inflicted by $ONOO^-$, including mutagenesis, nitration, and PARP activation, so the scope of protection afforded by reduced folates may be broad.

References

Anguera MC, Suh JR et al. (2003) Methylenetetrahydrofolate synthetase regulates folate turnover and accumulation. J Biol Chem 278:29856–29862

Anthony AC (1996) Folate receptors. Annu Rev Nutr 16:501–521

Antoniades C, Shirodaria C, Leeson P, Baarholm OA, Van Assche T, Cunnington C, Pillai R, Ratnatunga C, Tousoulis D, Stefanadis C, Refsum H, Channon KM (2009) MTHFR 677C>T polymorphism reveals functional importance for 5-methyltetrahydrofolate, not homocysteine, in regulation of vascular redox state and endothelial function in human atherosclerosis. Circulation 119:2507–2515

Antoniades C, Shirodaria C, Warrick N, Cai S, de Bono J, Lee J, Leeson P, Neubauer S, Ratnatunga C, Pillai R, Refsum H, Channon KM (2006) 5-methyltetrahydrofolate rapidly improves endothelial function and decreases superoxide production in human vessels. Circulation 114:1193–1201

Austin RC, Sood SK, Dorward AM, Singh G, Shaughnessy SG, Pamidi S, Outinen PA, Weitz JI (1998) Homocysteine-dependent alterations in mitochondrial gene expression, function and structure. J Biol Chem 273:30808–30817

Baggott JE, Tamura T, Baker H (2001) Re-evaluation of the metabolism of oral doses of racemic carbon isomers of formyltetrahydrofolate in human subjects. Br J Nutr 85:653–657

Beckman JS, Koppenol WH (1996) Nitric oxide, superoxide, and peroxynitrite: the good, the bad, and the ugly. Am J Physiol 271:C1424–C1437

Benzie IF, Strain JJ (1996) The ferric reducing ability of plasma (FRAP) as a measure of "antioxidant power": the FRAP assay. Analyt Biochem 239:70–76

Blair JA (1957) Some observations on the oxidative degradation of pteroyl-L-glutamic acid. Biochem J 65:209–211

Boushey CJ, Beresford SA, Omenn GS, Motulsky AG (1995) A quantitative assessment of plasma homocysteine as a risk factor for vascular disease. Probable benefits of increasing folic acid intakes. J Am Med Assoc 274:1049–1057

Bunout D, Petermann M, Hirsch S, de la Maza P, Suazo M, Barrera G, Kauffman R (2000) Low serum folate but normal homocysteine levels in patients with atherosclerotic vascular disease and matched healthy controls. Nutrition 16:434–438

Campolo J, Chiara B, Caruso R, De Maria R, Sedda V, Dellanoce C, Parolini M, Cighetti G, Penco S, Baudo F, Parodi O (2006) Methionine challenge paradoxically induces a greater activation of the antioxidant defence in subjects with hyper- vs. normohomocysteinemia. Free Radic Res 40:929–935

Caruso R, Campolo J, Sedda V, De Chiara B, Dellanoce C, Baudo F, Tonini A, Parolini M, Cighetti G, Parodi O (2006) Effect of homocysteine lowering by 5-methyltetrahydrofolate on redox status in hyperhomocysteinemia. J Cardiovasc Pharmacol 47:549–555

Celano L, Gil M, Carballal S, Duran R, Denicola A, Banerjee R, Alvarez B (2009) Inactivation of cystathionine β-synthase with peroxynitrite. Arch Biochem Biophys Arch Biochem Biophys 491:96–105

Celermajer DS, Sorensen K, Ryalls M, Robinson J, Thomas O, Leonard JV, Deanfield JE (1993) Impaired endothelial function occurs in the systemic arteries of children with homozygous homocystinuria but not in their heterozygous parents. J Am Coll Cardiol 22:854–858

Chambers JC, McGregor A, Jean-Marie J, Obeid OA, Kooner JS (1999) Demonstration of rapid onset vascular endothelial dysfunction after hyperhomocysteinemia: an effect reversible with vitamin C therapy. Circulation 99:1156–1160

Chang CM, Yu CC, Lu HT, Chou YF, Huang RF (2007) Folate deprivation promotes mitochondrial oxidative decay: DNA large deletions, cytochrome c oxidase dysfunction, membrane depolarization and superoxide overproduction in rat liver. Br J Nutr 97:855–863

Chen N, Liu Y, Greiner CD, Holtzman JL (2000) Physiologic concentrations of homocysteine inhibit the human plasma GSH peroxidase that reduces organic hydroperoxides. J Lab Clin Med 136:58–65

Chiao JH, Roy K, Tolner B, Yang CH, Sirotnak FM (1997) RFC-1 gene expression regulates folate absorption in mouse small intestine. J Biol Chem 272:11165–11170

Chippel D, Scrimgeour KG (1970) Oxidative degradation of dihydrofolate and tetrahydrofolate. Can J Biochem 48:999–1009

Coppola A, D'Angelo A, Fermo I, Mazzola G, Di Minno M, Cajani A, Sala A, Folco G, Tremoli E, Di Minno G (2005) Reduced in vivo oxidative stress follwoing 5-methyltetrahydrofolate supplementation in patients with early-onset thrombosis and 677TT methylenetetrahydrofolate reductase genotype. Br J Haematol 131:100–108

Crabtree MJ, Tatham AL, Hale AB, Alp NJ, Channon KM (2009) Critical role for tetrahydrobiopterin recycling by dihydrofolate reductase in regulation of endothelial nitric-oxide synthase coupling. J Biol Chem 284:28128–28136

Darley-Usmar V, Wiseman H, Halliwell B (1995) Nitric oxide and oxygen radicals: A question of balance. FEBS Lett 369:131–135

Das U (2003) Folic acid says NO to vascular disease. Nutrition 19:686–692

Davi G, Di Minno G, Coppola A, Andria G, Cerbone AM, Madonna P, Tufano A, Falco A, Marchesani P, Ciabattoni G, Patrono C (2001) Oxidative stress and platelet activation in homozygous homocystinuria. Circulation 104:1124–1128

Delfino VD, Vianna AC, Mocelin AJ, Barbosa DS, Mise RA, Matsuo T (2007) Folic acid therapy reduces plasma homocysteine levels and improves plasma antioxidant capacity in hemodialysis patients. Nutrition 23:242–247

Donnelly JG (2001) Folic acid. Crit Rev Clin Lab Sci 38:183–221

Doshi SN, McDowell IF, Moat SJ, Payne N, Durrant HJ, Lewis MJ, Goodfellow J (2002) Folic acid improves endothelial function in coronary artery disease via mechanisms largely independent of homocysteine lowering. Circulation 105:22–26

Droge W (2002) Free radicals in the physiological control of cell function. Physiol Rev 82:47–95

Durand P, Lussier-Cacan S, Blache D (1997) Acute methionine load-induced hyperhomocysteinemia enhances platelet aggregation, thromboxane biosynthesis, and macrophage-derived tissue factor activity in rats. FASEB J 11:1157–1168

Faraci FM, Didion SP (2004) Vascular protection: superoxide dismutase isoforms in the vessel wall. Arterioscler Thromb Vasc Biol 24:1367–1373

Ferrer-Sueta G, Radi R (2009) Chemical biology of peroxynitrite: kinetics, diffusion, and radicals. ACS Chem Biol 4:161–177

Fuchs D, Jaeger M, Widner B, Wirleitner B, Artner-Dworzak A, Leblhuber F (2001) Is hyperhomocysteinemia due to the oxidative depletion of folate rather than to insufficient dietary intake? Clin Chem Lab Med 39:691–694

Fukai T, Folz RJ, Landmesser U, Harrison DG (2002) Extracellular superoxide dismutase and cardiovascular disease. Cardiovasc Res 55:239–249

Gao L, Chalupsky K, Stefani E, Cai H (2009) Mechanistic insights into folic acid-dependent vascular protection: dihydrofolate reductase (DHFR)-mediated reduction in oxidant stress in endothelial cells and angiotensin II-infused mice: a novel HPLC-based fluorescent assay for DHFR activity. J Mol Cell Cardiol 47:752–760

Ghandour H, Chen Z, Selhub J, Rozen R (2004) Mice deficient in methylenetetrahydrofolate reductase exhibit tissue-specific distribution of folates. J Nutr 134:2975–2978

Gliszczynska-Swiglo A (2007) Folates as antioxidants. Food Chem 101:1480–1483

Gliszcznska-Swiglo A, Muzolf M (2007) PH-dependent radical scavenging activity of folates. J Agric Food Chem 55:8237–8242

Haenen GR, Bast A (1983) Protection against lipid peroxidation by a microsomal glutathione-dependent labile factor. FEBS Lett 159:24–28

Handy DE, Zhang Y, Loscalzo J (2005) Homocysteine down-regulates cellular glutathione peroxidase (GPx1) by decreasing translation. J Biol Chem 280:15518–15525

Hasdai D, Gibbons RJ, Holmes DR, Higano ST, Lerman A (1997) Coronary endothelial dysfunction in humans is associated with myocardial perfusion defects. Circulation 96:3390–3395

Heales SJ, Blair JA, Meinschad C, Ziegler I (1988) Inhibition of monocyte luminol-dependent chemiluminescence by tetrahydrobiopterin, and the free radical oxidation of tetrahydrobiopterin, dihydrobiopterin and dihydroneopterin. Cell Biochem Funct 6:191–195

Heinecke JW, Kawamura M, Suzuki, L, Chait A (1993) Oxidation of low-density lipoprotein by thiols: superoxide-dependent and –independent mechanisms. J Lipid Res 34:2051–2061

Heller R, Unbehaun A, Schellenberg B, Mayer B, Werner-Felmayer G, Werner ER (2001) L-ascorbic acid potentiates endothelial nitric oxide synthesis via a chemical stabilization of tetrahydrobiopterin. J Biol Chem 276:40–47

Higashi-Okai K, Nagino H, Yamada K, Okai Y (2007) Antioxidant and prooxidant activities of B group vitamins in lipid peroxidation. J UOEH 28:359–368

Hirsch S, Ronco AM, Vasquez M, de la Maza MP, Garrido A, Barrera G, Gattas V, Glasinovic A, Leiva L, Bunout D (2004) Hyperhomocysteinemia in healthy young men and elderly men with normal serum folate concentration is not associated with poor vascular reactivity or oxidative stress. Nutrition 134:1832–1835

Huang A, Vita JA, Venema RC, Keaney JF (2000) Ascorbic acid enhances endothelial nitric-oxide synthase activity by increasing intracellular tetrahydrobiopterin. J Biol Chem 275:17399–17406

Huang RF, Hsu YC, Lin HL, Yang FL (2001) Folate depletion and elevated plasma homocysteine promote oxidative stress in rat livers. J Nutr 131:33–38

Hyndman ME, Verma S, Rosenfeld RJ, Anderson TJ, Parsons HG (2002) Interactions of 5-methyltetrahydrofolate and tetrahydrobiopterin on endothelial function. Am J Physiol Heart Circ Physiol 282:2167–2172

Joshi R, Adhikari S, Patro BS, Chattopadhyay S, Mukherjee T (2001) Free radical scavenging behavior of folic acid: evidence for possible antioxidant activity. Free Rad Biol Med 30:1390–1399

Kathy KW, Au-Yeung KW, Woo CW, Sung FL, Yip JC, Siow YL, OK (2004) Hyperhomocysteinemia activates nuclear factor-κB in endothelial cells via oxidative stress. Circ Res 94:28–36

Kuzkaya N, Weissmann N, Harrison DG, Dikalov S (2003) Interaction of peroxynitrite, tetrahydrobiopterin, ascorbic acid, and thiols: implications for uncoupling endothelial nitric-oxide synthase. J Biol Chem 278:22546–22554

Lindschinger M, Nadlinger K, Adelwöhrer N, Holweg K, Wögerbauer M, Birkmayer J, Smolle KH, Wonisch W (2004) Oxidative stress: potential of distinct peroxide determination systems. Clin Chem Lab Med 42:907–914

Loscalzo J (1996) The oxidant stress of hyperhomocyst(e)inemia. J Clin Invest 98:5–7

Luhrs CA, Slomiany BL (1989) A human membrane-associated folate binding protein is anchored by a glycosyl-phosphatidylinositol tail. J Biol Chem 264:21446–21449

Lulock MD, Proestnall M, Daskalakis I, Schorah CJ, Wild J, Levene MI (1995) Nonenzymatic degradation and salvage of dietary folate: physicochemical factors likely to influence bioavailibility. Biochem Mol Med 55:43–53

Mayer B, Schumacher M, Brandstätter H, Wagner FS, Hermetter A (2001) High-throughput fluorescence screening of antioxidative capacity in human serum. Anal Biochem 297:144–153

Mayer O, Simon J, Rosolova H, Hromadka M, Subrt I, Vobrubova I (2002) The effects of folate supplementation on some coagulation parameters and oxidative status surrogates. Eur J Clin Pharmacol 58:1–5

McCaddon A, Regland B, Hudson P, Davies G (2002) Functional vitamin (B)12 deficiency and Alzheimer disease. Neurology 58:1395–1939

McCarthy MF, Barroso-Aranda J, Contreras F (2009) High-dose folate and dietary purines promote scavenging of peroxynitrite-derived radicals. Clinical potential in inflammatory disorders. Med Hypothesis 73:824–834

McEneny J, Couston C, McKibben B, Young IS, Woodside JV (2007) Folate: in vitro and in vivo effects on VLDL and LDL oxidation. Int J Vitam Nutr Res 77:66–72

Miura H, Bosnjak JJ, Ning G, Saito T, Miura M, Gutterman DD (2003) Role of hydrogen peroxide in flow-induced dilation of human coronary arterioles. Circ Res 92: e31–e40

Moens AL, Claeys M, Wuytis FL et al. (2004) Folic acid reduces myocardial ischemia and reperfusion injury by influencing the NO pathway: an in vitro and in vivo study. Circulation 110:III–50

Mosharov E, Cranford MR, Banerjee R (2000) The quantitatively important relationship between homocysteine metabolism and glutathione synthesis by the trassulfuration pathway and its regulation by redox changes. Biochemistry. 39:13005–13011

Mudd SH, Skovby F, Levy HL, Pettigrew KR, Wilcken B, Pyeritz R, Andria G, Boers GH, Bromberg IL, Cerone R et al. (1985) The natural history of homocystinuria due to cystathionine-β-synthase deficiency. Am J Hum Genet 37:1–31

Murr C, Baier-Bitterlich G, Fuchs D, Werner ER, Esterbauer H, Pfleiderer W, Wachter H (1996) Effects of neopterin-derivatives on H_2O_2-induced luminol chemiluminescence: mechanistic aspects. Free Radic Biol Med 21:449–456

Nakamura M, Nagayoshi R, Ijiri K, Nakashima-Matsushita N, Takeuchi T, Matsuyama T (2002) Nitration and chlorination of folic acid by peroxynitrite and hypochlorous acid, and the selective binding of 10-nitro-folate to folate receptor beta. Biochem Biophys Res Comm 297:1238–1244

Odin E, Carlsson G, Frosing R, Gustavsson B, Spears CP, Larsson PA (1998) Chemical stability and human plasma pharmacokinetics of reduced folates. Cancer Invest 447–455

Pancharuniti N, Lewis CA, Sauberlich HE, Perkins LL, Go RC, Alvarez JO, Macaluso M, Acton RT, Copeland RB, Cousins AL et al. (1994) Plasma homocyst(e)ine, folate, and vitamin B-12 concentrations and risk for early-onset coronary artery disease. Am J Clin Nutr 59:940–948

Patro BS, Adhikari S, Mukherjee T, Chattopadhyay S (2005) Possible role of hydroxyl radicals in the oxidative degradation of folic acid. Bioorg Med Chem Lett 15:67–71

Patrono C, Fitzgerald GA (1997) Isoprsotanes: potential markers of oxidant stress in atherothrombotic diseae. Arterioscler Thromb Vasc Biol 17:2309–2315

Prasad PD, Ramamoorthy S, Moe AJ, Smith CH, Leibach FH, Ganapathy V (1994) Selective expression of the high-affinity isoform of the folate receptor (FR-alpha) in the human placental syncytiotrophoblast and choriocarcinoma cells. Biochim Biophys Acta 1223:71–75

Pullin CH, Bonham JR, McDowell IF, Lee PJ, Powers HJ, Wilson JF, Lewis MJ, Moat SJ (2002) Vitamin C therapy ameliorates vascular endothelial dysfunction in treated patients with homocystinuria. J Inherit Metab Dis 25:107–118

Rice-Evans CA, Miller NJ (1994) Total antioxidant status in plasma and body fluids. Methods Enzymol 234:279–293

Rimm EB, Willett WC, Hu FB et al (1998) Folate and vitamin B6 from diet and supplements in relation to risk of coronary heart disease among women. J Am Med Assoc 279:359–364

Rogers EJ, Chen SS, Chan A (2007) Folate deficiency and plasma homocysteine during increased oxidative stress. N Engl J Med 357:421–422

Robinson K, Arheart K, Refsum H, Brattstrom L, Boers G, Ueland P, Rubba P, Palma-Reis R, Meleady R, Daly L et al. (1998) Low circulating folate and vitamin B6 concentrations: risk factors for stroke, peripheral vascular disease, and coronary artery disease. European COMAC Group. Circulation 97:437–443

Rzek BM, Haenen GR, van der Vijgh WJ, Bast A (2003) Tetrahydrofolate and 5-methyltetrahydrofolate are folates with high antioxidant activity. Identification of the antioxidant pharmacophore. FEBS Lett 555:601–605

Said HM, Nguyen TT, Dyer DL, Cowan KH, Rubin SA (1996) Intestinal folate transport: identification of a cDNA involved in folate transport and the functional expression and distribution of its mRNA. Biochim Biophys Acta 1281:164–172

Schachinger V, Britten MB, Zeiher AM (2000) Prognostic impact of coronary vasodilator dysfunction on adverse long-term outcome of coronary heart disease. Circulation 101:1899–1906

Schwartz SM, Siscovick DS, Malinow MR, Rosendaal FR, Beverly RK, Hess DL, Psaty BM, Longstreth WT, Koepsell TD, Raghunathan TE et al. (1997) Myocardial infarction in young women in relation to plasma total homocysteine, folate, and a common variant in the methylenetetrahydrofolate reductase gene. Circulation 96:412–417

Scott JM, Weir DG (1981) The methyl folate trap. Lancet 8242:838–840

Shane B (1989) Folylpolyglutamate synthesis and role in the regulation of one-carbon metabolism. Vitam Horm 45:263–335

Shaw S, Jayatilleke E, Herbert V, Colman N (1989) Cleavage of folates during ethanol metabolism. Biochem J 257:277–280

Shidfar F, Homayounfar R, Fereshtehnejad SM, Kalani A (2009) Effect of folate supplementation on serum homocysteine and plasma antioxidant capacity in hyüercholesterolemic adults under lovastatin treatment: a double-blind randomized controlles clinical trial. Arch Med Red 40:380–386

Sies H, Sharov VS, Klotz LO, Briviba K (1997) Glutathione peroxidase protects against peroxynitrite-mediated oxidations. A new function for selenoproteins as peroxynitrite reductase. J Biol Chem 272:27812–27817

Sirotnak FM, Tolner B (1999) Carrier-mediated membrane transport of folates in mammalian cells. Annu Rev Nutr 19: 91–122

Spijkerman AM, Smulders YM, Kostense PJ, Henry RM, Becker A, Teerlink T, Jakobs C, Dekker JM, Nijpels G, Heine RJ, Bouter LM, Stehouwer CD (2005) S-adenosylmethionine

and 5-methyltetrahydrofolate are associated with endothelial function after controlling for confounding by homocysteine: the Hoorn Study. Arterioscler Thromb Vasc Biol 25:778–784
Sprecher E, Bergman R, Sprecher H, Maor G, Reiter I, Krivoy N, Drori S, Assaraf YG, Friedman-Birnbaum R (1998) Reduced folate carrier (RFC-1) gene expression in normal and psoriatic skin. Arch Dermatol Res 290:656–660
Stanger O (2002) Physiology of folic acid in health and disease. Curr Drug Metab 3:211–223
Stanger O, Fowler B, Pietrzik K, Huemer M, Haschke-Becher E, Semmler A, Lorenzl S, Linnebank M (2009) Homocysteine, folate and vitamin B12 in neuropsychiatric diseases: review and treatment recommendations. Expert Rev Neurother 9:1393–1412
Stanger O, Herrmann W, Pietrzik K, Fowler B, Geisel J, Dierkes J, Weger M (2003) consensus paper on the rational clinical use of homocysteine, folic acid and B-vitamins in cardiovascular and thrombotic diseases: guidelines and recommendations. Clin Chem Lab Med 41:1392–1403
Stanger O, Semmelrock HJ, Wonisch W, Bös U, Pabst E, Wascher TC (2002) Effects of folate treatment and homocysteine lowering on resistance vessel reactivity in atherosclerotic subjects. J Pharmacol Exp Ther 303:158–162
Stanger O, Weger M (2003) Interactions of homocysteine, nitric oxide, folate and radicals in the progressively damaged endothelium. Clin Chem Lab Med 41:1444–1454
Stanger O, Weger M, Renner W, Konetschny R (2001) Vascular dysfunction in hyperhomocyst(e)inemia. Implications for atherothrombotic disease. Clin Chem Lab Med 39:725–733
Steinberg SE, Campbell CL, Hillmann RS (1979) Kinetics of the normal folate enterohepatic cycle. J Clin Invest 64:83–88
Stocker P, Lesgards JF, Vidal N, Chalier F, Prost M (2003) ESR study of a biological assay on whole blood: antioxidant efficiency of various vitamins. Biochim Biophys Acta 1621:1–8
Stralin P, Karlsson K, Johansson BO, Marklund SL (1996) The interstitium of the human arterial wall contains very large amounts of extracellular superoxide dismutase. Arterioscler Thromb Vasc Biol 15:2031–2036
Stroes ES, van Faassen EE, Yo M, Martasek P, Govers R, Rabelink TJ (2000) Folic acid recerts dysfunction of endothelial nitric oxide synthase. Circ Res 86:1129–1134
Suwaidi JA, Hamasaki S, Higano ST, Nishimura RA, Holmes DR, Lerman A (2000) Long-term follow-up of patients with mild coronary artery disease and endothelial dysfunction. Circulation 101:948–954
Tatzber F, Griebenow S, Wonisch W, Winkler R (2002) Dual method for the determination of peroxidase activity and total peroxides – iodide leads to a significant increase of peroxidase activity in human sera. Anal Biochem 316:147–153
Tchantchou F (2006) Homocysteine increase folate oxidative brain homocysteine metabolism and various consequences of folate deficiency. J Alzheimers Dis 421–427
Ungvari Z, Csiszar A, Bagi Z, Koller A (2002) Impaired NO-mediated flow-induced coronary dilation in hyperhomocysteinemia: morphological and functional evidence for increased peroxynitrite formation. Am J Pathol 161:145–153
Ungvari Z, Csiszar A, Edwards JG, Kaminski PM, Wolin MS, Kaley G, Koller A (2003) Increased superoxide production in coronary arteries in hyperhomocysteinemia. Role of tumor necrosis factor-α, NAD(P)H oxidase, and inducible nitric oxide synthase. Arterioscler Thromb Vasc Biol 23:418–424
Upchurch GR, Welch GN, Fabian AJ, Freedman JE, Johnson JL, Keaney JF, Loscalzo J (1997) Homocyst(e)ine decreases bioavailable nitric oxide by a mechanism involving glutathione peroxidase. J Biol Chem 272:17012–17017
Usui M, Matsuoka H, Miyazaki H, Ueda S, Okuda S, Imaizumi T (1999) Endothelial dysfunction by acute hyperhomocyst(e)inemia: restoration by folic acid. Clin Sci 96:235–239
Verhaar MC, Stroes E, Rabelink TJ (2002) Folates and cardiovascular disease. Arterioscler Thromb Vasc Biol 22:6–13
Verhaar MC, Wever RM, Kastelein JJ, van Dam T, Koomans HA, Rabelink TJ (1998) 5-methyltetrahyrofolate, the active form of folic acid, improves endothelial dysfunction in familial hypercholesterolemia. Circulation 97:237–241

Verhoef P, Stampfer MJ, Buring JE, Gaziano JM, Allen RH, Stabler SP, Reynolds RD, Kok FJ, Hennekens CH, Willett WC (1996) Homocysteine metabolism and risk of myocardial infarction: relation with vitamins B6, B12, and folate. Am J Epidemiol 143:845–859

Verlinde PH, Oey I, Hendrickx ME, van Loey AM, Temme EH (2008) L-ascorbic acid improves the serum folate response to an oraal dose of [6S]-5-methyltetrahydrofolic acid in healthy men. Eur J Clin Nutr 62:1224–1230

Voutilainen S, Morrow JD, Roberts LJ, Alfthan G, Alho H, Nyyssönen K, Salonen JT (1999) Enhanced in vivo lipid peroxidation at elevated plasma total homocysteine levels. Arterioscler Thromb Vasc Biol 19:1263–1266

Weiss N, Heydrick S, Zhang YY, Bierl C, Loscalzo J (2002) Cellular redox state and endothelial dysfunction in mildly hyperhomocysteinemic cystathionine beta-synthase-deficient mice. Arterioscler Thromb Vasc Biol 22:34–41

Weiss N, Zhang YY, Heydrick S, Bierl C, Loscalzo J (2001) Overexpression of cellular glutathione peroxidase rescues homocyst(e)ine-induced endothelial dysfunction. Proc Natl Acad Sci USA 98:12503–12508

Widner B, Enzinger C, Laich A, Wirleitner B, Fuchs D (2002) Hyperhomocysteinemia, pteridines and oxidative stress. Curr Drug Metab 3:225–232

Widner B, Fuchs D, Leblhuber F, Sperner-Unterweger B (2001) Does disturbed homocysteine and folate metabolism in depression result from enhanced oxidative stress? J Neurol Neurosurg Psychiatr 70:419

Wilcken DE, Wang XL, Adachi T, Hara H, Duarte N, Gren K, Wilcken B (2000) Relationship between homocysteine and superoxide dismutase in homocystinuria. Possible relevance to cardiovascular risk. Arterioscler Thromb Vasc Biol 20:1199–1202

Wilcken DE, Wilcken B (1997) The natural history of vascular disease in homocystinuria and effects of treatment. J Inherit Metab Dis 20:295–300

Wilmink HW, Stroes ES, Erkelens WD, Gerritsen WB, Wever R, Banga JD, Rabelink JD (2000) Influence of folic acid on postprandial endothelial dysfunction. Arterioscler Thromb Vasc Biol 20:185–188

Xu J, Wu Y, Song P, Zhang M, Wang S, Zou MH (2007) Proteasome-dependent degradation of guanosine 5′- triphosphate cyclohydrolase I causes tetrahydrobiopterin deficiency in diabetes mellitus. Circulation 116:944–953

Yamamoto M, Hara H, Adachi T (2000) Effect of homocysteine on the binding of extracellular-superoxide dismutase on the endothelial cell surface. FEBS Lett 486:159–162

Yap S, Boers GH, Wilcken B, Wilcken DE, Brenton DP, Lee PJ, Walter JH, Howard PM, Naughten ER (2001) Vascular outcome in patients with homocystinuria due to cystathionine beta-synthase deficiency treated chronically: a multicenter observational study. Arterioscler Thromb Vasc Biol 21:2080–2085

Yap S, Naughten ER, Wilcken B, Wilcken DE, Boers GH (2000) Vascular complications of severe hyperhomocysteinemia in patients with homocystinuria due to cystathionine beta-synthase deficiency: effects of homocysteine-lowering therapy. Semin Thromb Hemost 26:335–340

Zeiher AM, Krause T, Schachinger V, Minners J, Moser E (1995) Impaired endothelium-dependent vasodilation of coronary resistance vessels is associated with exercise induced myocardial ischemia. Circulation 91:2345–2352

Chapter 9
Folate-Linked Drugs for the Treatment of Cancer and Inflammatory Diseases

Jun Yang, Erina Vlashi, and Philip Low

Abstract Folic acid, also known as vitamin B9 (Fig. 9.1), is an essential co-enzyme in one-carbon metabolism pathways, including the biosynthesis of nucleotides (i.e. purines, thymidine) and several amino acids. In general, two functionally different systems mediate the cellular uptake of folate: (1) the reduced folate carrier (RFC, Kd ∼ 10–6 M), an anion transporter that delivers folates across the plasma membrane in a bidirectional fashion, and (2) the folate receptor (FR, Kd ∼ 10–10 M), which internalizes folate through active receptor-mediated endocytosis. The RFC, a membrane-spanning anion transporter, is present in virtually all tissues and is responsible for the majority of folate transport in and out of cells. In contrast, FR expression is largely restricted to malignant cells, activated macrophages, and the proximal tubule cells of the kidneys. Because a variety of important diseases are caused by the former two cell types, interest in exploiting FR for drug targeting applications has rapidly increased. And achievement of this targeting objective, primarily through conjugation of drugs to folic acid is believed to enable (1) enhanced net drug uptake by pathologic cells, and more importantly (2) reduction in drug deposition into non-pathologic cells, thereby mitigating collateral toxicity to normal tissues.

Keywords Activated macrophages · Cancer imaging · Folate receptor targeting · Uptake of folic acid · Vitamin-mediated drug delivery

In this chapter, we will first provide an overview of the biology of the folate receptor, followed by a summary of recent advances and developments in FR-targeted cancer therapy. Finally, FR expression in immune cells and its relevance to the treatment of inflammatory and autoimmune diseases will be discussed.

P. Low (✉)
Purdue University, West Lafayette, IN, USA
e-mail: plow@purdue.edu

O. Stanger (ed.), *Water Soluble Vitamins*, Subcellular Biochemistry 56,
DOI 10.1007/978-94-007-2199-9_9, © Springer Science+Business Media B.V. 2012

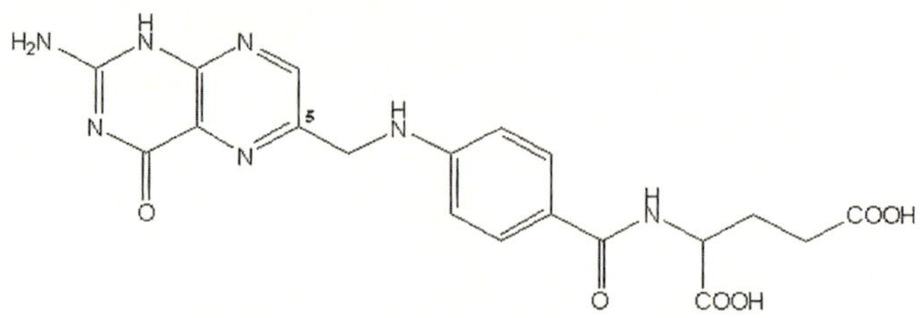

FOLIC ACID
$C_{19}H_{19}N_7O_6$
Mol. Wt.: 441.40

Fig. 9.1 The structure of folic acid

9.1 Biology of the Folate Receptor

The merits for using the folate receptor (FR) as a tumor target are three fold: (i) expression of FR is highly upregulated on cancer cells, (ii) folic acid is a high-affinity, high-specificity ligand for FR and can be chemically modified to enable conjugation with therapeutic and diagnostic agents without impairing its binding to FR, and (iii) FR recycles rapidly between the cell interior and the plasma membrane, allowing delivery of large quantities of folate-linked drugs into FR-expressing cells.

Human folate receptor has at least three well-characterized isoforms (Mr $\sim$ 38 kDa): the α- and β-isoforms are GPI-anchored, whereas the γ isoform is a secreted soluble protein (Shen et al. 1994). The membrane-associated FR-α and -β forms have N-terminal signal sequences of about 20 residues, and characteristic C-terminal hydrophobic segments that serve as the processing signal for GPI-anchor attachment at Ser234 and Asn230, respectively (Yan et al. 1998). Additional acylation may also account for the hydrophobicity of the C-terminal region (Luhrs et al. 1987). All isoforms of FR have 16 conserved cysteine residues, which are believed to form 8 disulfide bridges based on the crystal structure of a closely related protein, namely riboflavin binding protein (Maziarz et al. 1999; Zheng et al. 1988). There are three potential sites for N-linked glycosylation in FR-α (two in FR-β), all of which appear to contribute to cell surface presentation of a functional FR protein. Glycosylation, however appears to be unnecessary for folate binding after FR is already properly folded (Luhrs 1991; Roberts et al. 1998; Shen et al. 1997b). Although capable of mediating uptake of folate (and its conjugates), both α- and β- forms of the FR appear to have somewhat different binding patterns. For example, FR-α binds 6S folinic acid (also known as isovorin, or the 6S diastereoisomer of N5-methyltetrahydrofolate) with an affinity comparable to folic acid, while FR-β displays more than 50-fold lower affinity for the 6S isomer of folinic acid (Wang et al. 1992). Based on studies of a series of chimeric FR variants, such isoform-specific binding selectivity is thought to derive from a sequence divergence involving residues Leu49 in FR-β and Ala49, Val104 and Glu-166 in FR-α (Maziarz et al. 1999; Shen et al. 1997a).

9.2 Expression of the Folate Receptor

Expression of each human FR gene occurs with distinct and narrow tissue specificity, although it has been long believed that elevated FR levels are associated with cells/tissues in need of excess folate (i.e. rapidly dividing cancer cells). According to a comprehensive immunohistochemical analysis of normal tissues by Weitman et al. (1992a) using monoclonal antibodies to FR-α (MOv18 and MOv19), FR-α expression is restricted to certain epithelial cells, including epithelia of the choroid plexus, proximal tubules of the kidney, fallopian tubes, uterus, epididymis, the acinar cells of the breast, submandibular salivary and bronchial glands, type I and II pneumocytes in the lung, and trophoblasts in placenta. Further studies, employing other methods, generally confirm these results (Buist et al. 1995; Garin-Chesa et al. 1993; Stein et al. 1991; Veggian et al. 1989; Wu et al. 1999), and more careful analyses of the anatomic distribution of FR has indicated that the receptor in these tissues is almost exclusively restricted to the apical (lumenally facing) membranes of polarized epithelial cells (Weitman et al. 1992b). Because adherens and tight junctions prevent diffusion of folate conjugates across these epithelia, parenterally administered folate conjugates do not generally gain access to these FR (except in the cases of the kidney and malignant cells). It has, thus been observed that most FRs on normal tissues are inaccessible to folate-targeted agents administered in the blood stream, and therefore, most normal epithelia should remain unaffected by folate therapy.

As noted above, one exception to the rule that normal epithelia are not accessible to folate-linked drugs occurs in normal kidneys, where FR-α is expressed on the apical surface of the proximal tubules. In this case, low molecular weight folate-linked agents, which are filtered into the urine by the glomerulus, are subsequently captured by FR-α on the apical surface of the proximal tubules. These folate conjugates are then transported back to the blood as the FR travels from the apical to the basolateral side of the proximal tubule, a process known as re-absorption transcytosis (Sandoval et al. 2004). The question naturally arises as to whether folate-targeted drugs will be released into the kidney and thus cause renal toxicity. Even though both preclinical and early clinical studies of various folate-targeted therapeutic agents suggest that FR targeting significantly reduces side effects of otherwise toxic compounds without conferring appreciable damage to the kidney (Leamon et al. 2005; Reddy et al. 2006), additional studies are still warranted in order to confirm such conclusions.

Folate receptor expression is also significantly up-regulated in numerous malignancies of epithelial origin, especially ovarian and endometrial cancers. A large number of studies have confirmed that FR-α is consistently over-expressed in ovarian and uterine adenocarcinomas, non-small cell lung cancers, testicular choriocarcinomas, ependymal brain tumors and nonfunctional pituitary adenomas, expressed frequently in malignant pleural mesothelioma and less frequently in breast, colon and renal carcinomas (Buist et al. 1995; Campbell et al. 1991; Evans et al. 2001; Garin-Chesa et al. 1993; Parker et al. 2005; Ross et al. 1994; Toffoli et al. 1997; Veggian et al. 1989; Weitman et al. 1992b; Wu et al. 1999). The ability of FR-α, expressed in malignant tissues, to bind and transport folate into cells has also been demonstrated (Ross et al. 1994; Weitman et al. 1992b), although other speculated functions remain a matter of debate. It also appears from these studies that there is a quantitative increase in FR-α expression associated with the more poorly differentiated tumors (Orr et al. 1995; Wu et al. 1999). In a more recent study, Low and Hartmann (unpublished observations) demonstrated a clear inverse correlation between FR-α levels in primary breast tumors and 4-year survival rates, raising the possibility that FR is somehow involved in tumor progression and may serve as a prognostic marker for aggressive breast cancers. It should also be noted that a considerable variability in the level of FR over-expression can exist among tumor masses within the same patient and among the same tumor types in different patients (Wu et al. 1999).

9.3 Recycling of the Folate Receptor

An important benefit of folate receptor-targeted drug delivery derives from the fact that FR recycles between the cell surface and intracellular organelles constitutively (Kamen and Smith 2004; Sabharanjak and Mayor 2004). Thus, ligand-bound FR will unload folate conjugates upon endocytosis and the resulting vacant receptors will subsequently return to the cell surface, allowing multiple rounds of folate uptake and the consequent amplification of drug delivery (Fig. 9.2). In cancer cells this rate of FR recycling from the cell surface to internal endosomes and back to the cell surface again has been measured at 8–12 h (Paulos et al. 2004b). In contrast, the rate of FR recycling in activated macrophages has been found to require only ~15 min (unpublished observations). Although a considerable number of studies have confirmed the passage of FR through various acidic endosomal compartments, the mechanistic details pertaining to FR trafficking seem to vary significantly depending on the experimental system (e.g. cell lines) employed, and thus the precise mechanisms underlying FR-mediated endocytosis remain a matter of debate.

Studying endocytosis of FR-α in a cultured cell line that endogenously expresses FR, namely KB cells, Rijnboutt and coworkers (Rijnboutt et al. 1996) showed that FR was internalized via non-coated membrane invaginations into fluid-filled endosomes. Endocytic studies in Chinese Hamster Ovary (CHO) cells permanently transfected with FR have demonstrated that FR is segregated from other endocytic membrane molecules (e.g. transferrin) at the onset of endocytosis. In fact, immediately after internalization (within a few minutes) GPI-anchored FR are detected in GPI-anchored protein enriched endocytic compartments (GEEC) that are void of markers of clathrin-mediated endocytosis (i.e. transferrin receptor (TfR) and C6-NBD-sphingomyelin) (Sabharanjak et al. 2002; Turek et al. 1993). To summarize multiple studies, it is now concluded that FR-α is segregated from most transmembrane-anchored receptors at the cell surface and internalized via a clathrin/dynamin-independent, Cdc42-regulated endocytic pathway (Higgs and Pollard 2001; Takenawa and Miki 2001).

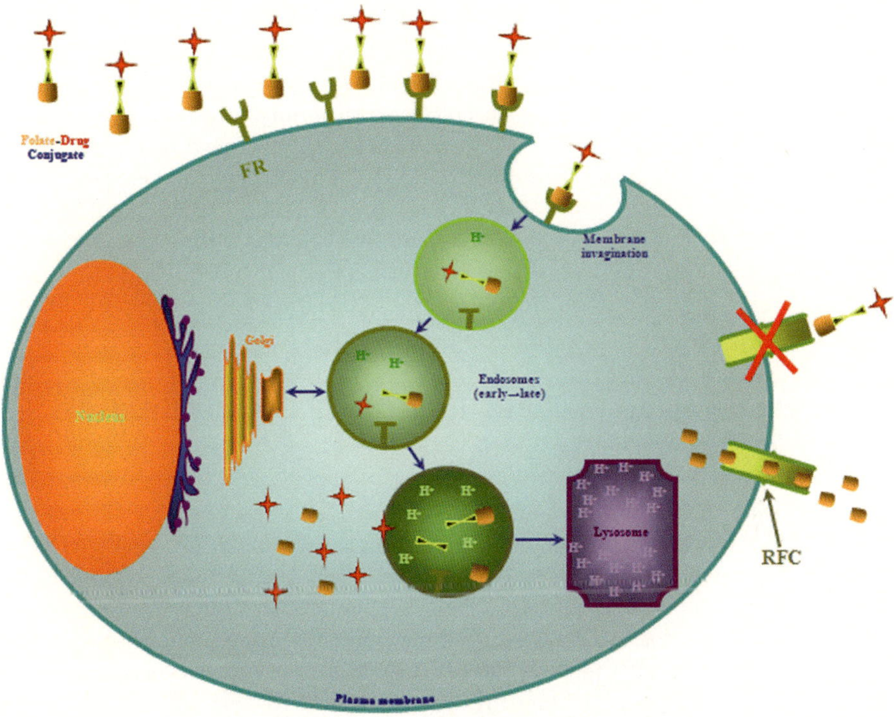

Fig. 9.2 Folate-mediated delivery of therapeutic agents to folate receptor-positive cancer cells. Folates can enter cells via either the reduced folate carrier (RFC), which is present on all cells, or the folate receptor (FR), which is over-expressed on cancer cells. Folate conjugates bind to and enter cells only via the folate receptor. As FR undergoes endocytosis the receptor associated folate–drug conjugates will traffic into the cancer cells. Drugs that require access to intracellular targets can be then delivered in substantial quantities to cytosolic locations via such endocytic pathway

After delivery to the GEECs, FR-α and other GPI-anchored proteins then migrate to a pericentriolar TfR-positive recycling center (REC), from which both eventually return to the plasma membrane (Sabharanjak et al. 2002; Turek et al. 1993). However, comparison of trafficking kinetics of TfR and FR from the REC to the cell surface indicates that GPI-anchored proteins (including FR) are recycled at 3-fold slower rates than their membrane-spanning counterparts (Chatterjee et al. 2001; Mayor et al. 1998), suggesting that the exit of FR from the REC may involve carriers distinct from those for TfR. The question then naturally arises as whether this variability in intracellular trafficking is a consequence of GPI-mediated association with lipid rafts on the cell surface. Consistent with this hypothesis, depletion of lipid raft components (i.e. cholesterol or sphingolipid) has been found to increase the FR recycling rate to a level comparable to TfR by removing the prolonged retention of GPI-anchored proteins in the REC.

Further, Mayor and coworkers (Mayor et al. 1998) have demonstrated that FR (but not a transmembrane isoform) exist in submicron-sized domains (<70 nm)

that are dependent upon cellular cholesterol levels. This highly compact organization of FR on the plasma membrane was also confirmed by Friedrichson and Kurzchalia (Friedrichson and Kurzchalia, 1998) who demonstrated that FR can be chemically crosslinked by an ∼ 11 Å bifunctional crosslinker. In contrast, studies by Kenworthy and colleagues (Kenworthy and Edidin, 1998; Kenworthy et al. 2000) have failed to find evidence for a clustered distribution of FR. It was also shown that the intracellular sorting of GPI-anchored FR is dependent on sphingolipid and cholesterol levels, and the involvement of "raft-like" entities during endocytic trafficking of FR has been implied, although the structural and functional correlates of these rafts remain elusive (Parton and Richards 2003; van Meer and Sprong 2004).

9.4 Folate Receptor-Mediated Targeting of Cytotoxic Agents to FR+ tumors

A wide variety of folate-linked molecules and complexes have been designed to enable selective delivery of therapeutic or diagnostic agents to FR-positive (FR+) cancer cells in vivo, including small molecule chemotherapeutics, radiopharmaceutical agents, immunotherapeutic agents (i.e. antibodies), proteins and protein toxins, liposomes with entrapped drugs, plasmids, aptamers, antisense oligonucleotides and nanoparticles (see Advanced Drug Delivery Reviews, volume 56, issue 8, for a comprehensive collection of reviews on recent advances in folate-targeted drug delivery). Readers interested in folate-targeted cancer imaging agents are also directed to an excellent review by Green et al. (Ke et al. 2004).

The first cytotoxic agents conjugated to folic acid were two protein synthesis inhibitors, namely momordin and Pseudomonas exotoxin (PE), prepared in the early 1990s (Leamon and Low 1992; Leamon et al. 1993). When tested in vitro, folate-momordin (IC50 ∼1 nM) was shown to selectively kill FR+ cancer cells. Furthermore, selective modification of functional residues in the folate-PE conjugate, which resulted in facilitated escape of the toxin from the endosomal compartment following endocytosis, enhanced its cytotoxicity 10-fold (Leamon et al. 1993). Thus, it was demonstrated that if the protein toxin were provided with a release/escape mechanism from the internal vesicular compartment, it would be extremely beneficial to its effective cytotoxicity. This result highlighted a very important conclusion, which has since been confirmed with other folate-drug conjugates: release of the drug, in its functional form, from a folate-drug construct constitutes a critical step in accelerating and enhancing therapeutic efficacy. Due to their extraordinary toxicity (only one or two molecules of a protein toxin are needed to kill a cell), folate-targeted protein toxins have been largely successful in vitro, and moderately successful in treating "liquid" malignancies in vivo (Pan et al. 2002). However, most folate-targeted therapies based on protein toxins and other macromolecular complexes have failed to produce satisfying results when tested in vivo on mice bearing solid tumors. It is now appreciated that large proteins and other macromolecules may penetrate solid tumors too poorly to accumulate in sufficient quantities for a significant therapeutic effect (Dreher et al. 2006; Evans et al. 2001).

Appreciating the fact that a large fraction of the FR+ cancers are in fact solid tumors, conventional low molecular weight chemotherapeutic agents have become an attractive alternative for their treatment and diagnostic imaging (Henne et al. 2006; Ladino et al. 1997; Leamon et al. 2005). Although generally less potent than protein toxins, small chemotherapeutic agents have the benefit that they penetrate solid tumors easily and are rapidly cleared from the body if they are not captured by a cell surface FR. For example, treatment of solid, FR+ tumor-bearing mice with folate-mitomycin C was found to significantly extend their lifespan in the absence of any evidence of toxicity to normal tissues (Leamon et al. 2005). Meanwhile, the administration of the same doses of nontargeted mitomycin-C showed no antitumor effect, and led to profound impairment of renal and hepatic functions. Together with other similar studies, these results constitute evidence that folate receptor-mediated drug delivery selectively targets therapeutic agents to FR+ cancer tissues, thus reducing general toxicity to normal cells.

In an attempt to increase the amount of small chemotherapeutic drugs delivered to an FR+ tumor, folate-targeted liposomes have been employed. Folate can be anchored to the surface of a liposome via the incorporation of a folate-phospholipid conjugate into the liposome bilayer. Folate targeted liposomes have, in fact, been used to deliver a wide variety of compounds to tumor cells, including small molecular weight drugs (Anderson et al. 2001; Ni et al. 2002; Pan and Lee 2005; Pan et al. 2002; Stevens et al. 2004); genes and antisense oligonucleotides (Hofland et al. 2002; Reddy et al. 2002), neutron capture agents (Stephenson et al. 2003; Sudimack et al. 2002), and radionuclides (Henriksen et al. 2004). Often times, folate is conjugated to the distal end of a longer PEGylated phospholipid (Mr $\sim$ 3,350), such that the folate molecule protrudes further from the polyethylene glycol (PEG) coat of the "stealth" liposome (Mr $\sim$ 2,000), thus, allowing for proper presentation of folate to the FR on the target cell. It has also been documented that maximal uptake is achieved at a rather low concentration of the folate-PEG-lipid conjugate (i.e. 0.03% of total lipid) (Reddy et al. 1999a). As mentioned above, it is critical that the folate-cargo is released and escapes from the endosomal compartment after endocytosis occurs. To this end, pH sensitive lipids and peptides have been employed in some cases, to improve endosomal release of the contents of folate-targeted liposomes (Ghaghada et al. 2005; Reddy and Low; 2000; Turk et al. 2002).

9.5 Folate-Receptor Mediated Immunotherapy

Tumor-specific expression and extracellular presentation of the folate receptor renders FR an especially attractive candidate for targeted immunotherapy. Immunotherapeutic strategies involving FR were first explored in studies where two murine monoclonal antibodies were raised against a membrane preparation of ovarian cancer tissue and were found to recognize FR-α (Coney et al. 1991). Subsequent use of these antibodies included: (1) treating with un-conjugated anti-FR antibodies in order to opsonize the FR+ cancer cells in vivo (Buist et al. 1995; van Zanten-Przybysz et al. 2002); (2) concentrating harmful radiation to the site of cancer by administration of anti-FR antibodies conjugated to ionizing radionuclides (131I, 211At, 111I) (Andersson et al. 2001; van Zanten-Przybysz et al. 2000); (3) forcing T cell recognition of FR-expressing tumors via the use of anti-FR/anti-T cell receptor bispecific antibodies (Mezzanzanica et al. 1991); and (4) administration of anti-FR immunocytokine constructs (IL-2/MOv19 scFv) to activate immune cells at the tumor cell surface (Melani et al. 1998). These strategies however have their shortcomings due to the same reason highlighted in the previous section: macromolecules (i.e. antibodies) seem to penetrate solid tumors poorly, thus compromising their therapeutic potential.

To avoid these limitations associated with size, a new strategy involving the targeting of highly immunogenic folate-hapten conjugates to tumor cell surface FR has been recently designed and tested in both animals and humans (Lu and Low 2003). In this multi-step process aimed at increasing the immunogenicity of FR+ tumors, the patient/animal is first vaccinated against a specific hapten (such as fluorescein), and then administered the folate-hapten conjugate which binds to and labels all FR-positive tumor cells with the foreign hapten. The tumor-bound haptens are then seen to be rapidly recognized by the anti-hapten antibodies generated by the vaccination, leading to removal of the antibody-coated tumor cells by Fc receptor-bearing immune cells (Fig. 9.3).

In immune competent mice previously vaccinated against fluorescein, administration of a folate-fluorescein conjugate has been shown to result in complete

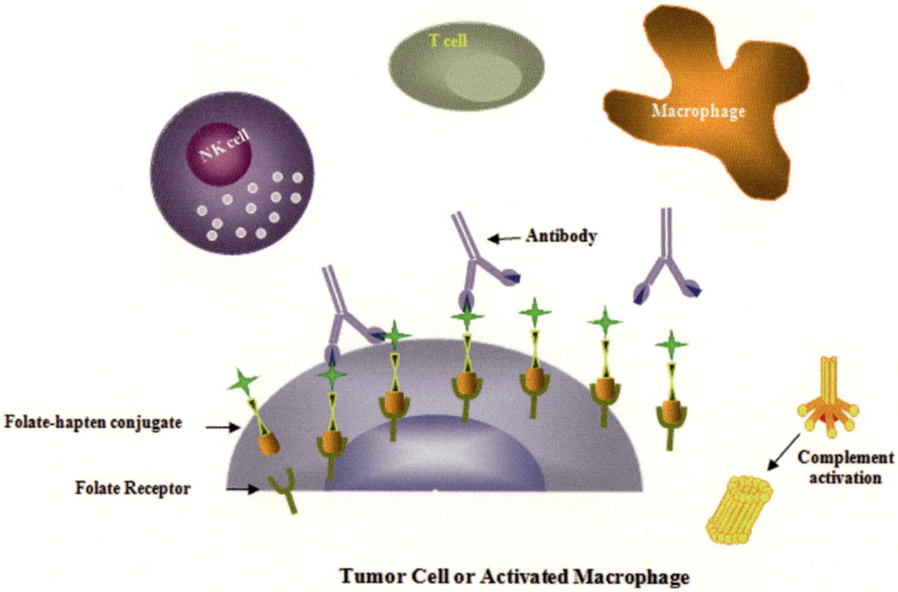

Fig. 9.3 Folate–hapten targeted immunotherapy: retargeting of circulating anti-hapten antibodies to FR-positive tumor cells or activated macrophages via a folate–hapten conjugate. The immune effector cells responsible for the removal of antibody-coated cells are also shown

eradication of established tumors, especially when the immune system was mildly stimulated with cytokines (e.g. IL-2 and IFN-α) (Lu and Low 2002). Perhaps, the most important result to emerge from these studies has been the fact that mice completely cured of their initial tumor burdens have been found to reject subsequent challenges with the same tumor cells, even in the absence of any additional folate-FITC treatment, thus strongly suggesting that a tumor-specific cellular immunity is developed during the initial hapten-mediated folate-targeted therapy, and that this cellular immunity enables subsequent recognition of the same tumor cells (i.e. recurring tumor) without the aid of additional hapten labeling. Recent mechanistic studies have, in fact, confirmed this hypothesis (Lu et al. 2005). They have also shown that depletion of macrophages or CD4+ or CD8+ T cells leads to a substantial compromise of the therapeutic effect, implying involvement of both humoral and cellular immune components (Lu et al. 2005).

9.6 Additional Considerations for the Design of Folate Receptor-Targeted Therapies

Some unusual characteristics of the folate receptor need to be considered when designing a folate conjugate for treatment of cancer. First, most folate conjugates bind FR with sub-nanomolar affinity, often resulting in retention of a significant fraction of the folate-linked agent on FR as it recycles back to the cell surface. Since such a mechanism allows for continuous presentation of the folate-hapten on the surface, it is believed to be beneficial to the immunotherapy. However, in the case of FR-targeted chemotherapy, release of drug inside the cell is crucial for the therapeutic effect. If the folate conjugate returns to the plasma membrane still bound to FR, less drug will not only be delivered to the cell, but fewer FR will be available for the next round of delivery. In order to address this problem, it has been proposed to replace folate with an analogous ligand such as 5-methyltetrahydrofolate, which dissociates more efficiently upon endosomal acidification during the FR recycling (Fig. 9.2). Alternatively, the drug can be attached to folate via a cleavable linker (Fig. 9.4), thus facilitating its release from folate following endocytosis (Leamon and Low 1994; Leamon et al. 1993, 2005).

A variety of releasable linkers, including disulfide bonds, acid sensitive linkers, and enzymatically cleavable bonds, have either been tested or proposed to bridge

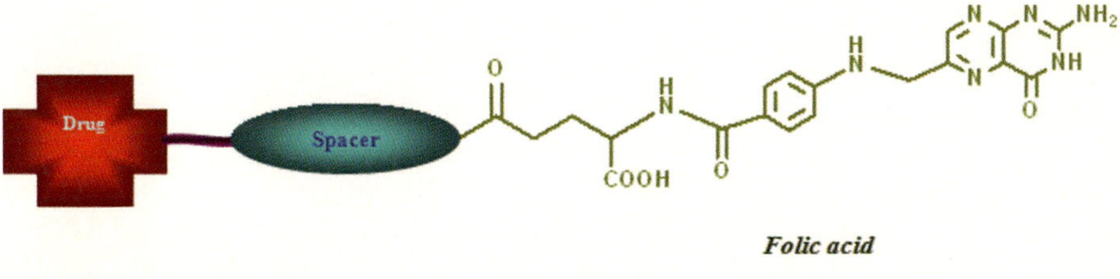

Fig. 9.4 Model fluorescent folate-targeted conjugate for in vitro and in vivo imaging

folate to therapeutic moieties (Hamann et al. 2002; Ladino et al. 1997; Leamon and Reddy 2004; Reddy et al. 2006). Such release mechanisms are especially required in the case when the delivered drug is only active in its unmodified form. Secondly, although many human cancer cells express up to 107 FR/cell, it has also been documented that, in cancer cells, FR recycling is somewhat inefficient. In fact, only ∼1 to 3 × 105 molecules/h are delivered inside many cancer cells via FR-mediated endocytosis (Paulos et al. 2004b). Because weakly active chemotherapeutic agents are ineffective at these concentrations, it is usually necessary to select a highly potent chemotherapeutic drug (i.e. IC50 ∼ 10 nM) to assure measurable therapeutic efficacy. Finally, it is important to point out that once folate is conjugated to the desired drug, the physical/chemical properties of the drug can dictate the pharmacological behavior and distribution of the conjugate, potentially interfering with FR-mediated drug targeting. Thus, depending on the therapeutic moiety, unanticipated problems with serum protein binding, nonspecific uptake by the liver, and undesirable pharmacokinetics can be influenced by the nature of the payload molecule.

9.7 Folate Receptor Expression in Leukocytes

As mentioned previously, expression of human FR-α is highly restricted to certain normal epithelia and tumors of epithelial origin. FR-β, on the other hand, is expressed in the placenta and in cells of hematopoietic origin or associated malignancies, such as leukemias. For example, FR-β is detected in the myelomonocytic lineage and during neutrophil maturation, where it can be co-expressed with CD14, CD33, CD13, and CD11b, but it is not found on cell expressing lymphocyte markers, CD3 and CD19 (Pan et al. 2002). Curiously, the FR-β expressed on most of these myelocytic cells appears to be functionally inactive, since cells expressing FR-β rarely bind radiolabeled folate or its conjugates. It is now clear that while the receptor can be readily detected by antibodies, the receptor is inactive except when the myelocytic cell becomes activated (Reddy et al. 1999b; Ross et al. 1999). In contrast, functional FR-β is consistently expressed on chronic myelogenous leukemia (CML) cells (Ross et al. 1999). In acute myelogenous leukemia (AML), approximately 70% of the leukemias are FR-β positive (Pan et al. 2002), while acute lymphoblastic leukemia (ALL), chronic lymphocytic leukemia (CLL) and hairy cell leukemia cells do not express FR-β (Ross et al. 1999).

A careful examination of FR-expressing cells of the myeloid lineage has revealed that a functional FR-β is often co-expressed with cell surface markers characteristic of a highly activated state (Xia et al. 2009). Consistent with the fact that macrophage/monocyte activation usually occurs only at sites of inflammation, functional FR-β has also been observed on activated (but not quiescent) synovial macrophages in rheumatoid arthritis (RA) in humans (Chen et al. 2005; Nakashima-Matsushita et al. 1999; Reddy et al. 1999b), as well as in a rat arthritic model of the disease (Turk et al. 2002). These studies reveal that FR-β mRNA and FR-β protein are much more abundant in activated monocytes and RA synovial macrophages than in their naïve counterparts from healthy individuals, and that this receptor can bind folate-fluorophore conjugates with high affinity (Turk et al. 2002). Taken together, these data imply that expression of a functional FR-β is associated with activation of macrophages and monocytes during acute and chronic inflammation, although the nature of the activation process and the molecular mechanism underlying acquisition of binding affinity still remain unclear.

9.8 Folate Receptor-Mediated Therapy for Inflammatory Diseases

As summarized above, activated, but not resting macrophages express a functionally active folate receptor, only in this case the receptor isoform is FR-β. Furthermore, these activated cells seem to contribute prominently to many autoimmune and inflammatory diseases such as rheumatoid arthritis, systemic lupus erythematosus, atherosclerosis, multiple sclerosis, Crohn's disease, psoriasis, ulcerative colitis, pulmonary fibrosis, sarcoidosis, scleroderma, and graft versus host disease, etc. The inflammation-associated expression of FR-β has also been demonstrated by imaging studies on both arthritic animals and human patients using EC20, a folate-targeted 99mTc-based radioactive tracer. In these studies, EC20 is seen to concentrate in the arthritic extremities of diseased rats, but with a much reduced intensity in healthy animals. Further, the livers and spleens of arthritic animals also show elevated accumulation of EC20 compared to their normal counterparts (Paulos et al. 2004a). The same results have been obtained in independent experiments on collagen-induced arthritic (CIA) mice. Furthermore, during clinical trials with two different folate-targeted radiopharmaceuticals (folate-DTPA-111In and 99mTc-labeled-folate, EC20) aimed at imaging malignant tissues in cancer patients, arthritic knees, but not healthy knees, of patients also suffering from arthritis can readily be seen in the nuclear images (Fig. 9.5). More recently, analogous results have also been observed in humans with clinically diagnosed rheumatoid arthritis (unpublished observations), suggesting that folate can be exploited for targeting drugs to pathogenic macrophages in a variety of human inflammatory diseases.

As discussed above, the most significant advantage to emerge from FR-mediated cancer therapies has been the surprisingly low level of toxicity to normal tissues. Thus, folate targeting efficiently abrogates uptake of otherwise cytotoxic agents by non-pathologic cells. By the same token, the design of therapeutic strategies to target only the FR+ activated macrophages should result in a reduction of inflammation while minimizing systemic toxicity to nonpathologic cells.

To test this hypothesis, Low et al. have evaluated the aforementioned FR-mediated cancer immunotherapy (Fig. 9.3) on antigen-induced arthritic (AIA) rats. In this application, a folate-hapten conjugate (i.e. folate-FITC) was administered to diseased animals, previously immunized with a KLH-FITC, resulting in significant suppression of both local and systemic inflammation symptoms (e.g. greatly reduced bone and cartilage degradation, reduced paw swelling, smaller spleen size, etc.) in two different advanced-stage models of rheumatoid arthritis. Presumably,

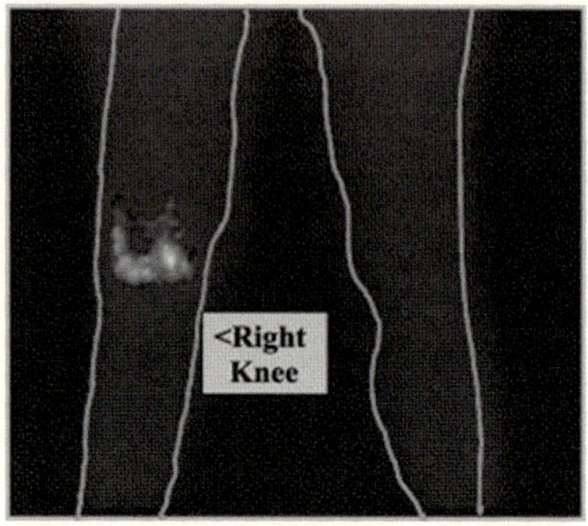

Fig. 9.5 Scintigraphic image of a cancer patient diagnosed with arthritis in the right knee (profile of legs is outlined in *gray*). Four hours following intravenous administration of 2 mg 111In-DTPA-folate, the patient was imaged. This fortuitous observation demonstrated for the first time that folate targeting could be used to deliver drugs to the arthritic extremities of humans with arthritis. Reproduced with permission from Endocyte

specific decoration of FR+ activated macrophages with folate-hapten conjugates promotes selective recognition of the cells by anti-hapten antibodies in immunized animals, leading to elimination of the macrophages via antibody-dependent cell cytotoxicity (ADCC) (Fig. 9.3). As seen in Fig. 9.6 (i) paw swelling, (ii) bone erosion, (iii) systemic inflammation, (iv) splenomegaly, (v) elevation of arthritis score, and (vi) body weight loss associated with untreated arthritis in two different arthritic models (in rats and in mice) can all be essentially prevented/reversed by treatment with folate-hapten conjugates. More importantly, when compared with other standard arthritis treatments in the clinic, such as methotrexate, celecoxib, etanercept and anakinra, folate-hapten therapy yielded better response rates with lower toxicity (Paulos et al. 2006). It should also be pointed out that similar results have been obtained in arthritic dogs, with no observable systemic toxicity even when the folate–hapten treatment lasted for more than 2 years.

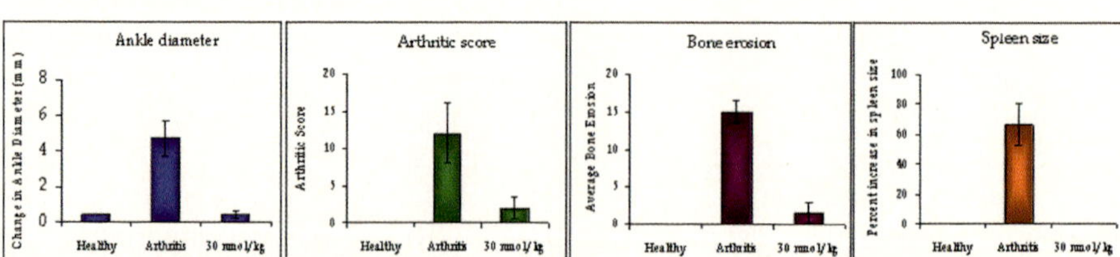

Fig. 9.6 Symptoms associated with rheumatoid arthritis can be prevented/reversed with folate-targeted immunotherapy. The optimum dose for the folate-hapten is as low as 30 nmoles/kg

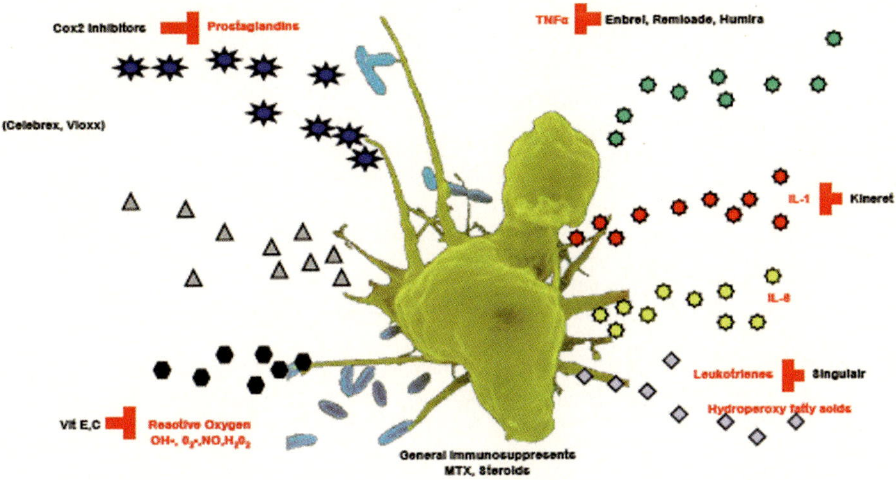

Fig. 9.7 Common anti-inflammatory therapies specifically inhibit a single secretion product of the activated macrophage. In contrast, the folate-targeted therapy enables selective elimination of the activated macrophage responsible for these destructive inflammatory products

9.9 Conclusion

In conclusion, it is now well-accepted that the activated macrophage is the principle contributor to the induction and aggravation of rheumatoid arthritis, as well as many other inflammatory pathologies (listed above). Therefore, this cell constitutes a primary target for therapeutic intervention in numerous inflammatory and autoimmune diseases. The major advantage of the folate-targeting approach lies in the selectivity of the therapy for activated macrophages, a major cause of inflammatory diseases. All other common anti-inflammatory therapies specifically inhibit a single secretion product of the activated macrophage (Fig. 9.7), allowing the macrophage to continue secretion of its other inflammatory mediators. In contrast, the folate-targeted therapy enables selective elimination of a cell population responsible for destructive inflammatory products without damaging other healthy cells required for normal organismal function.

References

Anderson KE, Eliot LA, Stevenson BR, Rogers JA (2001) Formulation and evaluation of a folic acid receptor-targeted oral vancomycin liposomal dosage form. Pharm Res 18:316–322

Andersson H, Palm S, Lindegren S, Back T, Jacobsson L, Leser G, Horvath G (2001) Comparison of the therapeutic efficacy of 211At- and 131I-labelled monoclonal antibody MOv18 in nude mice with intraperitoneal growth of human ovarian cancer. Anticancer Res 21:409–412

Buist MR, Molthoff CF, Kenemans P, Meijer CJ (1995) Distribution of OV-TL 3 and MOv18 in normal and malignant ovarian tissue. J Clin Pathol 48:631–636

Campbell IG, Jones TA, Foulkes WD, Trowsdale J (1991) Folate-binding protein is a marker for ovarian cancer. Cancer Res 51:5329–5338

Chatterjee S, Smith ER, Hanada K, Stevens VL, Mayor S (2001) GPI anchoring leads to sphingolipid-dependent retention of endocytosed proteins in the recycling endosomal compartment. Embo J 20:1583–1592

Chen WT, Mahmood U, Weissleder R, Tung CH (2005) Arthritis imaging using a near-infrared fluorescence folate-targeted probe. Arthritis Res Ther 7:R310–R317

Coney LR, Tomassetti A, Carayannopoulos L, Frasca V, Kamen BA, Colnaghi MI, Zurawski VR Jr (1991) Cloning of a tumor-associated antigen: MOv18 and MOv19 antibodies recognize a folate-binding protein. Cancer Res 51:6125–6132

Dreher MR, Liu W, Michelich CR, Dewhirst MW, Yuan F, Chilkoti A (2006) Tumor vascular permeability, accumulation, and penetration of macromolecular drug carriers. J Natl Cancer Inst 98:335–344

Evans CO, Young AN, Brown MR, Brat DJ, Parks JS, Neish AS, Oyesiku NM (2001) Novel patterns of gene expression in pituitary adenomas identified by complementary deoxyribonucleic acid microarrays and quantitative reverse transcription-polymerase chain reaction. J Clin Endocrinol Metab 86:3097–3107

Friedrichson T, Kurzchalia TV (1998) Microdomains of GPI-anchored proteins in living cells revealed by crosslinking. Nature 394:802–805

Garin-Chesa P, Campbell I, Saigo PE, Lewis JL Jr, Old LJ, Rettig WJ (1993) Trophoblast and ovarian cancer antigen LK26. Sensitivity and specificity in immunopathology and molecular identification as a folate-binding protein. Am J Pathol 142:557–567

Ghaghada KB, Saul J, Natarajan JV, Bellamkonda RV, Annapragada AV (2005) Folate targeting of drug carriers: a mathematical model. J Control Release 104:113–128

Hamann PR, Hinman LM, Hollander I, Beyer CF, Lindh D, Holcomb R, Hallett W, Tsou HR, Upeslacis J, Shochat D, Mountain A, Flowers DA, Bernstein I (2002) Gemtuzumab ozogamicin, a potent and selective anti-CD33 antibody-calicheamicin conjugate for treatment of acute myeloid leukemia. Bioconjug Chem 13:47–58

Henne WA, Doorneweerd DD, Hilgenbrink AR, Kularatne SA, Low PS (2006) Synthesis and activity of a folate peptide camptothecin prodrug. Bioorg Med Chem Lett 16:5350–5355

Henriksen G, Schoultz BW, Michaelsen TE, Bruland OS, Larsen RH (2004) Sterically stabilized liposomes as a carrier for alpha-emitting radium and actinium radionuclides. Nucl Med Biol 31:441–449

Higgs HN, Pollard TD (2001) Regulation of actin filament network formation through ARP2/3 complex: activation by a diverse array of proteins. Annu Rev Biochem 70:649–676

Hofland HE, Masson C, Iginla S, Osetinsky I, Reddy JA, Leamon CP, Scherman D, Bessodes M, Wils P (2002) Folate-targeted gene transfer in vivo. Mol Ther 5:739–744

Kamen BA, Smith AK (2004) A review of folate receptor alpha cycling and 5-methyltetrahydrofolate accumulation with an emphasis on cell models in vitro. Adv Drug Deliv Rev 56:1085–1097

Ke CY, Mathias CJ, Green MA (2004) Folate-receptor-targeted radionuclide imaging agents. Adv Drug Deliv Rev 56:1143–1160

Kenworthy AK, Edidin M (1998) Distribution of a glycosylphosphatidylinositol-anchored protein at the apical surface of MDCK cells examined at a resolution of <100 A using imaging fluorescence resonance energy transfer. J Cell Biol 142:69–84

Kenworthy AK, Petranova N, Edidin M (2000) High-resolution FRET microscopy of cholera toxin B-subunit and GPI-anchored proteins in cell plasma membranes. Mol Biol Cell 11:1645–1655

Ladino CA, Chari RV, Bourret LA, Kedersha NL, Goldmacher VS (1997) Folate-maytansinoids: target-selective drugs of low molecular weight. Int J Cancer 73:859–864

Leamon CP, Low PS (1992) Cytotoxicity of momordin-folate conjugates in cultured human cells. J Biol Chem 267:24966–24971

Leamon CP, Low PS (1994) Selective targeting of malignant cells with cytotoxin-folate conjugates. J Drug Target 2:101–112

Leamon CP, Reddy JA (2004) Folate-targeted chemotherapy. Adv Drug Deliv Rev 56:1127–1141
Leamon CP, Pastan I, Low PS (1993) Cytotoxicity of folate-Pseudomonas exotoxin conjugates toward tumor cells. Contribution of translocation domain. J Biol Chem 268:24847–24854
Leamon CP, Reddy JA, Vlahov IR, Vetzel M, Parker N, Nicoson JS, Xu LC, Westrick E (2005) Synthesis and biological evaluation of EC72: A new folate-targeted chemotherapeutic. Bioconjug Chem 16:803–811
Lu Y, Low PS (2002) Folate targeting of haptens to cancer cell surfaces mediates immunotherapy of syngeneic murine tumors. Cancer Immunol Immunother 51:153–162
Lu Y, Low PS (2003) Immunotherapy of folate receptor-expressing tumors: Review of recent advances and future prospects. J Control Release 91:17–29
Lu Y, Sega E, Low PS (2005) Folate receptor-targeted immunotherapy: Induction of humoral and cellular immunity against hapten-decorated cancer cells. Int J Cancer 116:710–719
Luhrs CA (1991) The role of glycosylation in the biosynthesis and acquisition of ligand-binding activity of the folate-binding protein in cultured KB cells. Blood 77:1171–1180
Luhrs CA, Pitiranggon P, da Costa M, Rothenberg SP, Slomiany BL, Brink L, Tous GI, Stein S (1987) Purified membrane and soluble folate binding proteins from cultured KB cells have similar amino acid compositions and molecular weights but differ in fatty acid acylation. Proc Natl Acad Sci USA 84:6546–6549
Mayor S, Sabharanjak S, Maxfield FR (1998) Cholesterol-dependent retention of GPI-anchored proteins in endosomes. Embo J 17:4626–4638
Maziarz KM, Monaco HL, Shen F, Ratnam M (1999) Complete mapping of divergent amino acids responsible for differential ligand binding of folate receptors alpha and beta. J Biol Chem 274:11086–11091
Melani C, Figini M, Nicosia D, Luison E, Ramakrishna V, Casorati G, Parmiani G, Eshhar Z, Canevari S, Colombo MP (1998) Targeting of interleukin 2 to human ovarian carcinoma by fusion with a single-chain Fv of antifolate receptor antibody. Cancer Res 58:4146–4154
Mezzanzanica D, Canevari S, Colnaghi MI (1991) Retargeting of human lymphocytes against human ovarian carcinoma cells by bispecific antibodies: From laboratory to clinic. Int J Clin Lab Res 21:159–164
Nakashima-Matsushita N, Homma T, Yu S, Matsuda T, Sunahara N, Nakamura T, Tsukano M, Ratnam M, Matsuyama T (1999) Selective expression of folate receptor beta and its possible role in methotrexate transport in synovial macrophages from patients with rheumatoid arthritis. Arthritis Rheum 42:1609–1616
Ni S, Stephenson SM, Lee RJ (2002) Folate receptor targeted delivery of liposomal daunorubicin into tumor cells. Anticancer Res 22:2131–2135
Orr RB, Kreisler AR, Kamen BA (1995) Similarity of folate receptor expression in UMSCC 38 cells to squamous cell carcinoma differentiation markers. J Natl Cancer Inst 87:299–303
Pan XQ, Lee RJ (2005) In vivo antitumor activity of folate receptor-targeted liposomal daunorubicin in a murine leukemia model. Anticancer Res 25:343–346
Pan XQ, Zheng X, Shi G, Wang H, Ratnam M, Lee RJ (2002) Strategy for the treatment of acute myelogenous leukemia based on folate receptor beta-targeted liposomal doxorubicin combined with receptor induction using all-trans retinoic acid. Blood 100:594–602
Parker N, Turk MJ, Westrick E, Lewis JD, Low PS, Leamon CP (2005) Folate receptor expression in carcinomas and normal tissues determined by a quantitative radioligand binding assay. Anal Biochem 338:284–293
Parton RG, Richards AA (2003) Lipid rafts and caveolae as portals for endocytosis: New insights and common mechanisms. Traffic 4:724–738
Paulos CM, Turk MJ, Breur GJ, Low PS (2004a) Folate receptor-mediated targeting of therapeutic and imaging agents to activated macrophages in rheumatoid arthritis. Adv Drug Deliv Rev 56:1205–1217
Paulos CM, Reddy JA, Leamon CP, Turk MJ, Low PS (2004b) Ligand binding and kinetics of folate receptor recycling in vivo: Impact on receptor-mediated drug delivery. Mol Pharmacol 66:1406–1414

Paulos CM, Varghese B, Widmer WR, Breur GJ, Vlashi E, Low PS (2006) Folate-targeted immunotherapy effectively treats established adjuvant and collagen-induced arthritis. Arthritis Res Ther 8:R77

Reddy JA, Dean D, Kennedy MD, Low PS (1999a) Optimization of folate-conjugated liposomal vectors for folate receptor-mediated gene therapy. J Pharm Sci 88:1112–1118

Reddy JA, Haneline LS, Srour EF, Antony AC, Clapp DW, Low PS (1999b) Expression and functional characterization of the beta-isoform of the folate receptor on CD34(+) cells. Blood 93:3940–3948

Reddy JA, Abburi C, Hofland H, Howard SJ, Vlahov I, Wils P, Leamon CP (2002) Folate-targeted, cationic liposome-mediated gene transfer into disseminated peritoneal tumors. Gene Ther 9:1542–1550

Reddy JA, Low PS (2000) Enhanced folate receptor mediated gene therapy using a novel pH-sensitive lipid formulation. J Control Release 64:27–37

Reddy JA, Westrick E, Vlahov I, Howard SJ, Santhapuram HK, Leamon CP (2006) Folate receptor specific anti-tumor activity of folate-mitomycin conjugates. Cancer Chemother Pharmacol 58:229–236

Rijnboutt S, Jansen G, Posthuma G, Hynes JB, Schornagel JH, Strous GJ (1996) Endocytosis of GPI-linked membrane folate receptor-alpha. J Cell Biol 132:35–47

Roberts SJ, Petropavlovskaja M, Chung KN, Knight CB, Elwood PC (1998) Role of individual N-linked glycosylation sites in the function and intracellular transport of the human alpha folate receptor. Arch Biochem Biophys 351:227–235

Ross JF, Chaudhuri PK, Ratnam M (1994) Differential regulation of folate receptor isoforms in normal and malignant tissues in vivo and in established cell lines. Physiologic and clinical implications. Cancer 73:2432–2443

Ross JF, Wang H, Behm FG, Mathew P, Wu M, Booth R, Ratnam M (1999) Folate receptor type beta is a neutrophilic lineage marker and is differentially expressed in myeloid leukemia. Cancer 85:348–357

Sabharanjak S, Mayor S (2004) Folate receptor endocytosis and trafficking. Adv Drug Deliv Rev 56:1099–1109

Sabharanjak S, Sharma P, Parton RG, Mayor S (2002) GPI-anchored proteins are delivered to recycling endosomes via a distinct cdc42-regulated, clathrin-independent pinocytic pathway. Dev Cell 2:411–423

Sandoval RM, Kennedy MD, Low PS, Molitoris BA (2004) Uptake and trafficking of fluorescent conjugates of folic acid in intact kidney determined using intravital two-photon microscopy. Am J Physiol Cell Physiol 287:C517–C526

Shen F, Ross JF, Wang X, Ratnam M (1994) Identification of a novel folate receptor, a truncated receptor, and receptor type beta in hematopoietic cells: cDNA cloning, expression, immunoreactivity, and tissue specificity. Biochemistry 33:1209–1215

Shen F, Wang H, Zheng X, Ratnam M (1997b) Expression levels of functional folate receptors alpha and beta are related to the number of N-glycosylated sites. Biochem J 327(Pt 3):759–764

Shen F, Zheng X, Wang J, Ratnam M (1997a) Identification of amino acid residues that determine the differential ligand specificities of folate receptors alpha and beta. Biochemistry 36:6157–6163

Stein R, Goldenberg DM, Mattes MJ (1991) Normal tissue reactivity of four anti-tumor monoclonal antibodies of clinical interest. Int J Cancer 47:163–169

Stephenson SM, Yang W, Stevens PJ, Tjarks W, Barth RF, Lee RJ (2003) Folate receptor-targeted liposomes as possible delivery vehicles for boron neutron capture therapy. Anticancer Res 23:3341–3345

Stevens PJ, Sekido M, Lee RJ (2004) A folate receptor-targeted lipid nanoparticle formulation for a lipophilic paclitaxel prodrug. Pharm Res 21:2153–2157

Sudimack JJ, Adams D, Rotaru J, Shukla S, Yan J, Sekido M, Barth RF, Tjarks W, Lee RJ (2002) Folate receptor-mediated liposomal delivery of a lipophilic boron agent to tumor cells in vitro for neutron capture therapy. Pharm Res 19:1502–1508

Takenawa T, Miki H (2001) WASP and WAVE family proteins: Key molecules for rapid rearrangement of cortical actin filaments and cell movement. J Cell Sci 114:1801–1809
Toffoli G, Cernigoi C, Russo A, Gallo A, Bagnoli M, Boiocchi M (1997) Overexpression of folate binding protein in ovarian cancers. Int J Cancer 74:193–198
Turek JJ, Leamon CP, Low PS (1993) Endocytosis of folate-protein conjugates: Ultrastructural localization in KB cells. J Cell Sci 106(Pt 1):423–430
Turk MJ, Breur GJ, Widmer WR, Paulos CM, Xu LC, Grote LA, Low PS (2002) Folate-targeted imaging of activated macrophages in rats with adjuvant-induced arthritis. Arthritis Rheum 46:1947–1955
van Meer G, Sprong H (2004) Membrane lipids and vesicular traffic. Curr Opin Cell Biol 16:373–378
van Zanten-Przybysz I, Molthoff CF, Roos JC, Plaizier MA, Visser GW, Pijpers R, Kenemans P, Verheijen RH (2000) Radioimmunotherapy with intravenously administered 131I-labeled chimeric monoclonal antibody MOv18 in patients with ovarian cancer. J Nucl Med 41:1168–1176
van Zanten-Przybysz I, Molthoff C, Gebbinck JK, von Mensdorff-Pouilly S, Verstraeten R, Kenemans P, Verheijen R (2002) Cellular and humoral responses after multiple injections of unconjugated chimeric monoclonal antibody MOv18 in ovarian cancer patients: A pilot study. J Cancer Res Clin Oncol 128:484–492
Veggian R, Fasolato S, Menard S, Minucci D, Pizzetti P, Regazzoni M, Tagliabue E, Colnaghi MI (1989) Immunohistochemical reactivity of a monoclonal antibody prepared against human ovarian carcinoma on normal and pathological female genital tissues. Tumori 75:510–513
Wang X, Shen F, Freisheim JH, Gentry LE, Ratnam M (1992) Differential stereospecificities and affinities of folate receptor isoforms for folate compounds and antifolates. Biochem Pharmacol 44:1898–1901
Weitman SD, Weinberg AG, Coney LR, Zurawski VR, Jennings DS, Kamen BA (1992a) Cellular localization of the folate receptor: Potential role in drug toxicity and folate homeostasis. Cancer Res 52:6708–6711
Weitman SD, Lark RH, Coney LR, Fort DW, Frasca V, Zurawski VR Jr, Kamen BA (1992b) Distribution of the folate receptor GP38 in normal and malignant cell lines and tissues. Cancer Res 52:3396–3401
Wu M, Gunning W, Ratnam M (1999) Expression of folate receptor type alpha in relation to cell type, malignancy, and differentiation in ovary, uterus, and cervix. Cancer Epidemiol Biomarkers Prev 8:775–782
Xia W, Hilgenbrink AR, Matteson EL, Lockwood MB, Cheng JX, Low PS (2009) A functional folate receptor is induced during macrophage activation and can be used to target drugs to activated macrophages. Blood Jan 8;113(2):438–446. Epub 2008 Oct 24
Yan W, Shen F, Dillon B, Ratnam M (1998) The hydrophobic domains in the carboxyl-terminal signal for GPI modification and in the amino-terminal leader peptide have similar structural requirements. J Mol Biol 275:25–33
Zheng DB, Lim HM, Pene JJ, White HB 3rd (1988) Chicken riboflavin-binding protein. cDNA sequence and homology with milk folate-binding protein. J Biol Chem 263:11126–11129

Chapter 10
Folate in Skin Cancer Prevention

J.D. Williams, Elaine L. Jacobson, H. Kim, M. Kim, and M.K. Jacobson

Abstract Skin, the largest, most exposed organ of the body, provides a protective interface between humans and the environment. One of its primary roles is protection against exposure to sunlight, a major source of skin damage where the UV radiation (UVR) component functions as a complete carcinogen. Melanin pigmentation and the evolution of dark skin is an adaptive protective mechanism against high levels of UVR exposure. Recently, the hypothesis that skin pigmentation balances folate preservation and Vitamin D production has emerged. Both micronutrients are essential for reproductive success. Photodegradation of bioactive folates suggests a mechanism for the increased tendency of populations of low melanin pigmentation residing in areas of high UV exposure to develop skin cancers. Folate is proposed as a cancer prevention target for its role in providing precursors for DNA repair and replication, as well as its ability to promote genomic integrity through the generation of methyl groups needed for control of gene expression. The cancer prevention potential of folate has been demonstrated by large-scale epidemiological and nutritional studies indicating that decreased folate status increases the risk of developing certain cancers. While folate deficiency has been extensively documented by analysis of human plasma, folate status within skin has not been widely investigated. Nevertheless, inefficient delivery of micronutrients to skin and photolysis of folate argue that documented folate deficiencies will be present if not exacerbated in skin. Our studies indicate a critical role for folate in skin and the potential to protect sun exposed skin by effective topical delivery as a strategy for cancer prevention.

Keywords Cancer prevention · DNA repair · Folate · Folic acid · Skin · Topical delivery strategy · UV light

---

M.K. Jacobson (✉)
Division of Medicinal Chemistry, Department of Pharmacology and Toxicology, College of Pharmacy and Arizona Cancer Center, University of Arizona, Tucson, AZ, USA

Niadyne Development, Inc., Tucson, AZ, USA
e-mail: mjacobson@pharmacy.arizona.edu

O. Stanger (ed.), *Water Soluble Vitamins*, Subcellular Biochemistry 56,
DOI 10.1007/978-94-007-2199-9_10,

10.1 Human Skin and UV Radiation

The skin provides a direct interface for the human body with the environment and plays many roles in protection against physical, chemical, and microbial insults. The average human skin area exceeds m^2 and is generally less than 2 mm thick (Goldsmith 1991). Skin protects against most solar UV radiation (UVR), regulates body temperature through surface blood flow control and sweating, allows detection of the ambient and physical environments through feel and touch, and participates in social communication through the display of information such as age, health, and ancestry (Jablonski 2004). Human skin is a complex laminar structure comprised of many different cell types organized into the outer epidermal layer and an inner dermal layer. The epidermis provides the barrier properties of the skin and is composed of keratinocytes (structure), melanocytes (pigmentation), Langerhans cells (immune system), and Merkel cells (sense of touch). Keratinocytes are the primary cell type in the epidermis and originate in the stratum basale layer from the continual division of stem cells. They migrate outward through the epidermis while undergoing differentiation until they reach the stratum corneum where they form a layer of dead, flattened, highly keratinized cells called squamous cells. The epidermis is avascular and contributes to the elasticity and toughness of the skin. At the stratum corneum, keratinocytes are continuously shed and replaced (Goldsmith 1991) generating significant metabolic demands for energy and nutrient consumption. The dermis is a dense fibroelastic connective tissue composed of collagen fibers, elastic fibers, and an interfibrillar gel which comprises most of the skin thickness. Fibroblasts are the primary cell type in the dermis and are collagen-rich. Collagen is a major skin component and accounts for the tensile strength of the skin. Interwoven with the collagen is a network of elastic fibers. The dermis appears to be of equal thickness in people with dark or light pigmentation (Taylor 2002).

10.2 UVR, DNA Damage, and Skin Cancer

Skin damage, particularly that derived from sunlight, constitutes a major public health problem. Non-melanoma skin cancers (NMSC) are the most frequent malignancies in the United States with more than 1,000,000 cases diagnosed annually (Karagas et al. 1999). Melanoma skin cancer is the most rapidly increasing type of cancer. Additionally, actinic keratosis (AK), skin lesions that can progress to NMSC are far more prevalent than skin cancers. DNA damage and cellular responses to DNA damage play a central role in skin damage (Ames 2001). Sunlight is the major source of skin damage as it leads to DNA damage directly via formation of pyrimidine dimers and other photoproducts (Ullrich 2002) and indirectly via generation of reactive oxygen species (ROS) and reactive carbonyl species (RCS) by photooxidation and photosensitization reactions (Wondrak et al. 2002a, b) (Wondrak et al. 2003). UVR is a complete carcinogen. UVB (280–320 nm) which penetrates into the epidermis, is responsible for most of the direct DNA damage and is the most effective at initiation of squamous cell carcinoma (SCC). UVA (320–400 nm) which penetrates into the dermis, induces ROS and causes SCC (Pentland et al. 1999; Agar et al. 2004).

Chronic DNA damage results in progressive losses of genomic integrity and end stage skin damage in the form of skin cancer involving altered growth properties of keratinocytes such as unresponsiveness to terminal differentiation signals leading to epidermal hyperplasia and progressively to actinic keratosis (Jeffes and Tang 2000; Lober et al. 2000). Cell populations present in actinic keratosis lesions progress to transformed cell populations that represent epidermal carcinoma in situ (Guenthner et al., 1999; Hurwitz and Monger 1995; Kobayashi et al., 2000). Subsequent cellular changes occur including induction of matrix proteases that facilitate disruption of the integrity of the epidermal barrier leading to invasion of the dermis, the point at which the damage process is diagnosed as SCC. A second major consequence of DNA damage in skin is the suppression of immune responses that would normally detect and remove damaged cells. The consequences of sunlight-induced DNA damage are depicted in Fig. 10.1. While mechanisms of immune suppression extend beyond DNA damage, DNA damage is a major factor. The consequences of genotoxic stress include altered migration, antigen presentation by Langerhans cells, and

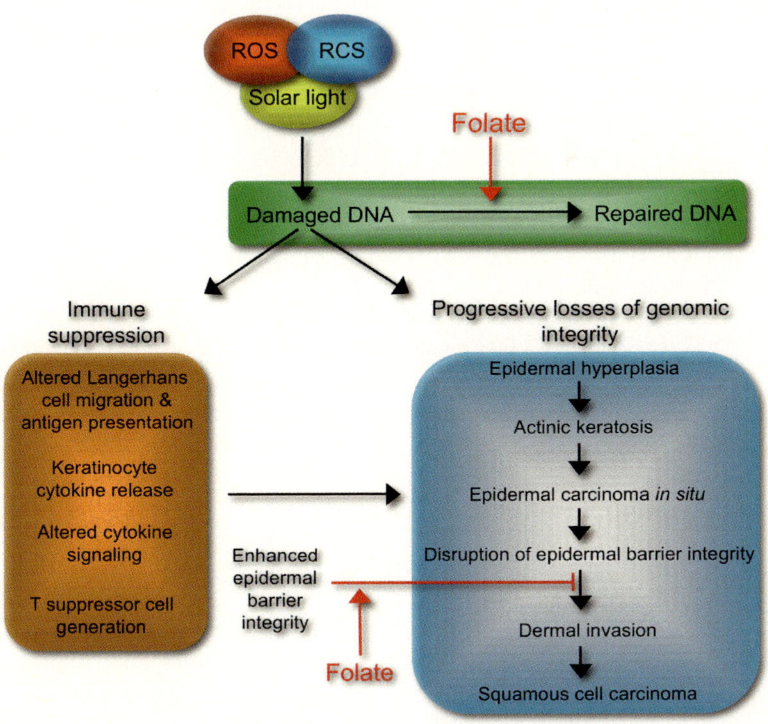

Fig. 10.1 Sunlight-induced DNA damage pathways and opportunities for micronutrient modulation

stimulation of cytokine release by keratinocytes that likely alters cytokine signaling required for normal immune surveillance including generation of T suppressor cells. Given the complexity of damage pathways and the down stream effects, opportunities to modulate the consequences of genotoxic stress include preventing DNA damage, enhancing DNA repair, and strengthening the integrity of the epidermal barrier to prevent migration of transformed cells from the epidermis (depicted in Fig. 10.1). A compelling body of evidence now indicates that several key micronutrients are candidates for skin damage prevention. This chapter will focus on the role of folate in skin health.

10.3 Folate Metabolism

In 1931, it was demonstrated by Lucy Wills that yeast extract was effective in treating tropical macrocytic anemia observed during late pregnancy (Wills 1978). This observation led to the isolation and structural determination of the B vitamin folate, named after the Latin word – folium (leaf). The terms folate and B9 vitamins refer to a large family of chemically similar compounds, the most widely known of which is folic acid. Folic acid or pteroylmonoglutamate (PteGlu), is a synthetic folate analog utilized preferentially due to its enhanced chemical stability and rapid conversion to the bioactive forms upon uptake. Natural folates vary in the one-carbon substituent at the N10 and N5 positions as well as in the number of glutamic acid residues conjugated via gamma glutamyl bonds to form the polyglutamate tail. Natural folates also vary in the oxidation state of the pteridine ring, with the reduced, unsubstituted dihydrofolate (DHF, $H_2PteGlu_n$) and tetrahydrofolate (THF, $H_4PteGlu_n$) forms being particularly prone to biological inactivation due to cleavage of the C-9 and N-10 bond (Blakley 1969). The structure of folic acid and its reduced, native forms are illustrated in Fig. 10.2.

Humans do not synthesize folate, and thus are dependant upon a variety of dietary sources. The overall folate nutritional status of the population is believed to have been positively influenced since the implementation of the United States federal government's requirement to fortify cereal-grain products in 1998 (Dietrich et al. 2005); however, this effect may be waning due to current low carbohydrate diet fads. Natural folates are a mixture of reduced folate polyglutamates which vary in the number of glutamic acid residues. The main folate found in foods is 5methyltetrahydrofolate (5-methyl-$H_4PteGlu_n$, 5MTHF) and the monoglutamyl form is the primary form of folate which enters the circulating plasma (Thien et al. 1977; Herbert et al. 1962).

The transport and metabolism of the synthetic folic acid and it's active folate metabolites have been extensively characterized. Figure 10.3 depicts an overview of folate metabolism which along with the role of folate in disease processes is expertly reviewed in (Lucock 2000).

In the cell, folates serve to carry and transfer one-carbon units of various oxidation levels during biosynthetic reactions that occur within the cell. Of particular

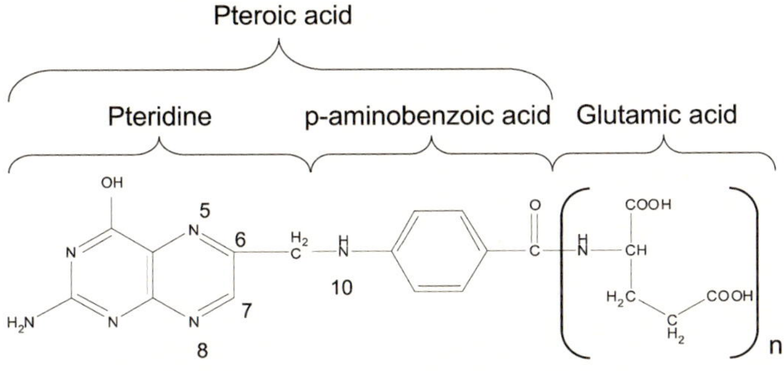

Native folate	Substituent at N-5	N-10
THF	-H at 5,6,7, & 8	-H
DHF	-H at 7 & 8	-H
5FormylTHF	-HCO	-H
5MethylTHF	$-CH_3$	-H
10FormylTHF	-H	-HCO
5,10MethenylTHF	$=CH^+-$	
5,10MethyleneTHF	$-CH_2^-$	

Fig. 10.2 The chemical structure of folic acid and its various derivatives

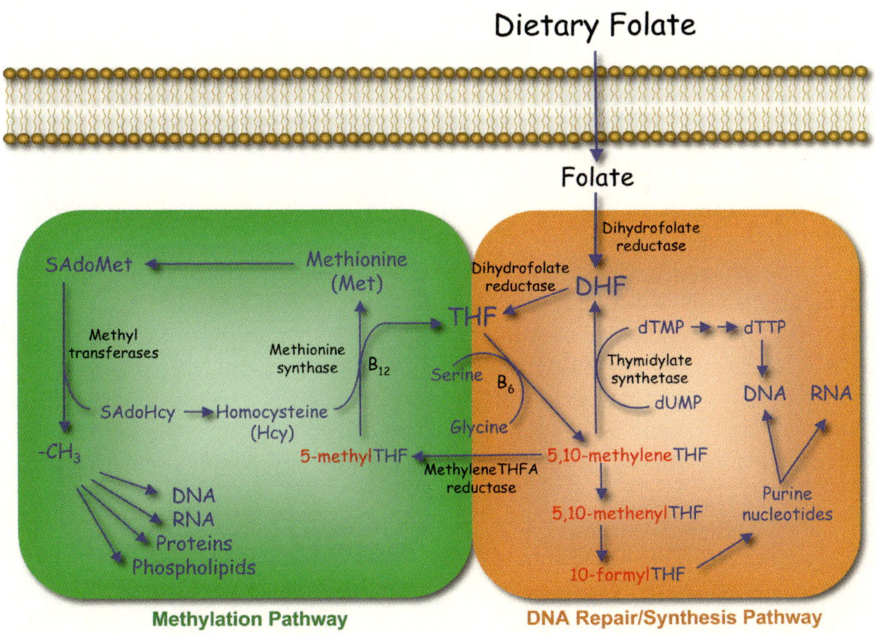

Fig. 10.3 Overview of folate metabolism

importance is the role that folates play in purine and pyrimidine nucleotide biosynthesis to supply precursors for DNA repair and synthesis. Also of importance is the role that 5MTHF plays in the homocysteine (Hcy) remethylation cycle and de novo methionine biosynthesis. De novo synthesized methionine can then be activated by ATP and the enzyme methionine adenosyltransferase to yield the methyl donor Sadenosylmethionine (SAM). SAM serves as the methyl donor for the majority of methyltransferases that modify DNA, RNA, histones and other proteins, contributing to replicational, transcriptional and translational fidelity, mismatch repair, chromatin remodeling, epigenetic modifications and imprinting, which are all topics of great interest and importance in cancer research and aging.

10.4 UVR at the Earth's Surface and Photodegradation of Folates

UVR levels reaching the Earth's surface are affected by numerous factors such as latitude, altitude, season, moisture content, cloud cover, and the depth of the ozone column (Madronich et al. 1998). The shortest UV wavelengths (UV-C, 100–280 nm) are essentially completely blocked or absorbed by atmospheric oxygen (O_2) and ozone (O_3). Wavelengths in the UV-B range (280–320 nm) are partially absorbed by O_3 and thus reach the Earth's surface while UV-A wavelengths (320–400 nm) are only weakly absorbed by O_3 and are most easily transmitted through the atmosphere. At latitudes further away from the Equator, the angle of solar elevation decreases and the thickness of the atmosphere and ozone layer through which sunlight must pass before reaching the Earth's surface increases which results in attenuation of UVR levels. It is now well established that folic acid is photosensitive. Folic acid has absorption peaks at 280 and 350 nm with shoulders at 300 and 370 nm. The absorption peak at 350 nm indicates that folic acid absorbs UVA radiation which is very prevalent at the Earth's surface. Folic acid solutions exposed to biologically relevant levels of UVA radiation demonstrated that UVR exposure results in the cleavage of folic acid to form p-aminiobenzoyl-L-glutamic acid and 6-forml pterin (Off et al. 2005). Bioactive folate derivatives are also UV labile. The most common bioactive folate, 5MTHF, has an absorption maximum at 290 nm. Exposure of 5MTHF to UVB at 312 nm results in oxidation to 5-methyldihydrofolate (5MDHF), which upon continued irradiation is further broken down to a pterin residue and paminobenzoylglutamate (Lucock et al. 2003). While the breakdown of 5MTHF results in irreversible loss of vitamin activity by cleavage of the C9-N10 bond, even oxidation to 5MDHF may decrease vitamin activity. UVB does not penetrate deeply into the skin as depicted in Fig. 10.4, and therefore, does not reach the blood stream. Thus, direct photodegradation of folates may not be the mechanism of biological significance. However, the oxidation of 5MTHF to 5MDHF is stimulated by exposure to light in the presence of photosensitizers (Steindal et al. 2006). This implicates ROS in folate degradation. ROS are produced in human tissues by UVA radiation (Fig. 10.4) and possibly radiation of even longer wavelengths which would penetrate the skin and blood vessels to a much greater extent than UVB. ROS production by longer wavelength radiation is enhanced in the presence of photosensitizers, and numerous endogenous photosensitizers are present in both the skin and blood.

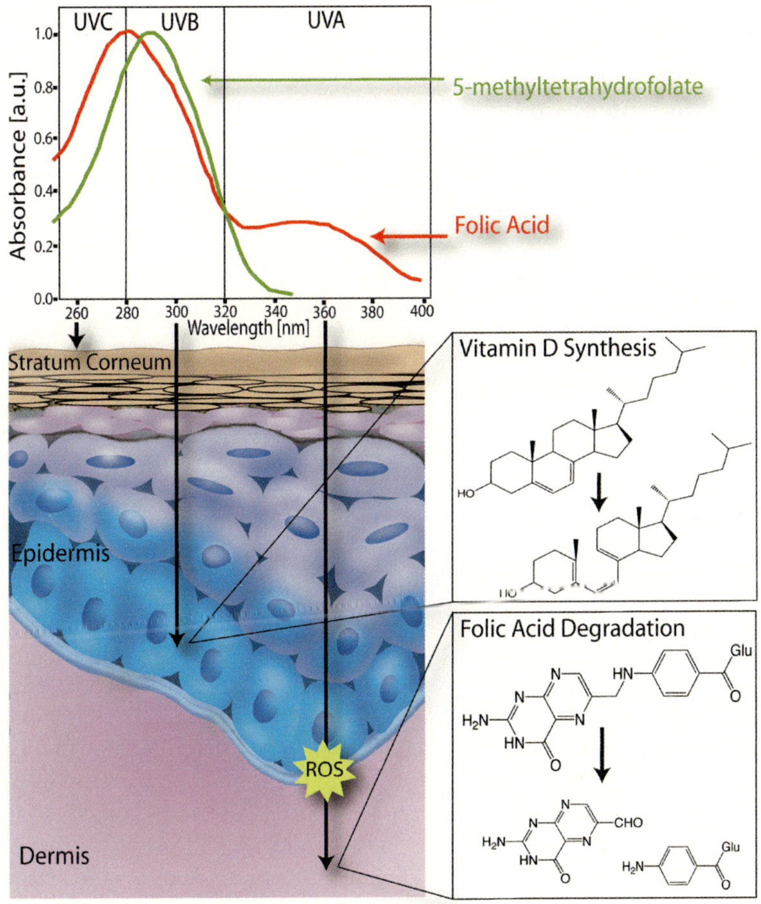

Fig. 10.4 Effects of UV radiation on lightly pigmented skin

10.5 Variations in Human Skin Pigmentation

It has been long recognized that skin colors among indigenous populations show a direct relationship to the intensity of the local UV spectrum. Recently, Jablonski and Chaplin have dramatically advanced the study of the proposed relationships by correlation and regression analyses of quantitative skin color measurements obtained by skin reflectance spectrophotometry with newly available remotely sensed data on levels of UVR, total solar radiation, temperature, humidity, precipitation, and other environmental variables at the Earth's surface (Jablonski 2004). The strongest association to skin color was found to be latitude, which corresponds to an effect of UVR intensity. A strong correlation was found between latitude and annual average minimal erythemal dose of UVR (UVMED), and thus between annual average UVMED and skin reflectance using UVMED data from the Earth's surface (Jablonski and Chaplin 2000). A subsequent study of the influence of minimum, maximum, and seasonal levels of UVR showed that skin reflectance was correlated with autumn levels of UVMED, and that skin reflectance could be almost fully modeled as a linear effect of this variable alone (Jablonski 2004). Dark pigmentation was found to be primarily a function of UVMED (Jablonski and Chaplin 2000), with regression analysis demonstrating that autumn UVMED levels have the strongest effect. Interestingly, their data indicated that skin color is more strongly correlated with UVA, which is consistently higher throughout the year at all latitudes, than with UVB (Jablonski 2004).

10.6 The Evolution of Skin Pigmentation and the Micronutrient Hypothesis

The evolution of human skin color has invited many possible explanations. Enhanced resistance to sunburn is clearly a major hypothesis. Devastating consequences for albinos in topical regions of intense UV light exposure including early fatalities from both SCC and acute injury from solar exposure leading to denudation and resultant infections directly support this reasoning (Cohn 1998). The benefits of lightly pigmented skin have been related to the synthesis of vitamin D. It is estimated that most humans utilize sunlight exposure to obtain nearly 100% of their vitamin D which is optimally stimulated by exposure to UVB photons with wavelengths of 295,300 nm (Holick 2003; Holick et al. 1981). The current hypothesis states that light skin pigmentation is necessary in regions of low UVR in order to permit vitamin D biosynthesis (Loomis 1967) and is an adaptation to resist cold injury (Post et al. 1975). Clearly, melanin pigmentation is an adaptation to some attribute of the physical environment. The genetic characteristic of skin pigmentation is believed to be governed by at least three to five loci which would require continued positive selection for its maintenance. This points to a sustained evolutionary pressure which acts to favor retention of pigmentation characteristics (Cohn 1998). Adaptive explanations for a given phenotypic trait require demonstration that the trait increases the real or probable reproductive success of the organism. Adaptive explanations such as resistance to skin cancer or protection from cold injuries attributed to different levels of melanin pigmentation in human skin have suffered from an inability to demonstrate probable or real differences in survivorship and reproduction of different skin color phenotypes (Jablonski 2004). The original hypothesis that dark skin pigmentation arose to protect against skin cancer induced by UV exposure does not account for selective reproductive pressure as fatal repercussions of skin cancers tend to develop after reproductive age. In this reasoning, the adaptive response of skin pigmentation toward the adequate protection or production of micronutrients which play key roles in the success of the reproductive process has gained much credibility. The evolution of dark skin to protect against folate photo-degradation serves a direct reproductive influence as folate is now known to be essential for fetal development and fertility. The attenuation of skin pigment levels to allow for adequate vitamin D production also serves a direct reproductive influence in fetal bone development and maternal bone health. This model in which skin pigmentation balances availability of essential micronutrients is of much interest as shifting world populations result in many people residing in areas of UV exposure that are significantly different from those to which their skin tone has been adapted.

10.7 Melanin and UV Penetrance

Melanocytes in the epidermis produce the skin's primary pigment, melanin. In mammals there are two types of melanin pigments, eumelanin which is black-brown and pheomelanin which is yellow-reddish (Thody et al. 1991). Melanins are produced in highly organized elliptic membrane bound cytoplasmic organelles of melanocytes called melanosomes. Melanocytes project their dendrites into the neighboring keratinocytes where they then transfer mature melanosomes. Once transferred to the keratinocytes, melanosomes aggregate and are surrounded by a membrane in a melanosome complex (Szabo et al. 1969). The variation in skin color among various races is determined mainly by the number, melanin content, and distribution of melanosomes within the keratinocytes (Jimbow et al. 1976). Natural selection has produced a gradient of skin pigmentations in response to two opposing factors. The first factor results in a cline of photoprotection that grades from darkly pigmented skin at the Equator to lightly pigmented skin near the Poles. The second factor results in a cline of vitamin D photosynthesis that grades from lightly pigmented near the Poles to darkly pigmented at the Equator. In the middle of the two clines we find peoples with enhanced abilities to develop facultative pigmentation according to seasonal UVR levels (Jablonski 2004). The concentration, depth, and distribution of skin melanin is strongly affected by UVR exposure. The tanning response observed in peoples able to develop facultative pigmentation results from distribution of melanosomes throughout the epidermis from the basal layer. Melanin has a broad absorption spectrum stretching from the UV to the near infrared regions. Recently it has been noted that not only the quantity of melanin present but also the depth and distribution of the melanosomes impact the amount of UVR penetrating to the living cells of the epidermis (Nielsen et al. 2006). Within keratinocytes, melanin granules accumulate above the nuclei and act as a sunscreen absorbing harmful UVR before it can reach the nucleus and damage the DNA.

Melanin serves to protect at least in part from mutations that might be caused by UVR exposure (Ohnishi and Mori 1998). In addition to DNA protection, it has also been shown that darkly pigmented skin generates a larger reflectance of wavelengths below 300 nm than lightly pigmented skin due to scattering of light by the spheroid shaped melanosomes (Nielsen et al. 2004). Scattering and absorption of light in this range shows that melanin and melanosomes play an important role as natural sunscreens in moderating UVR related health effects.

10.8 Folate Deficiency

Naturally occurring folates are water soluble, labile compounds that rapidly lose activity in foods over periods of days or weeks, consequently it is estimated that half or even three-quarters of initial folate activity may be lost between harvest and consumption (John Scott 2000). In view of the dietary behavior of many individuals, inadequate folate intake is likely and supplementation of folates is strongly recommended. This usually involves consumption of synthetic folic acid, which is more chemically stable and bioavailable than the natural folates.

10.9 Folate Deficiencies in Skin

Folate deficiency has been extensively documented in human plasma. However, folate status within skin has not been widely investigated. Delivery to the skin via the blood circulation of nutrients taken orally is inherently inefficient since this delivery is distal to other organs, particularly the liver, which removes many agents by first pass metabolism. In addition, the epidermal layer of the skin is non-vascular. The inefficiency of delivery of nutrients to skin argues that documented folate deficiencies will extend to skin. Folate is sensitive to photolysis by exposure to UVR. Exposure to sunlight and UVR in particular is expected to lower folate levels in at least the superficial layers of the skin. Photolysis of folate stores in the skin may be extensive enough to result in systemic depletion of folate. Indeed it has been reported that fair skinned patients undergoing photochemotherapy for dermatological conditions have low serum folate concentrations, suggesting that folate depletion may occur in vivo (Branda and Eaton 1978). However, in another experiment where volunteers were exposed to UVA radiation in a solarium, no connection between UVA exposure and plasma folate status was found (Gambichler et al. 2001). This study has since been criticized for use of non-specific bioassays for folate quantification, which serves to exemplify the point that much research is left to be done (Lucock et al. 2003).

10.10 Folate Deficiencies and Human Disease

Folate deficiencies and genetic folate metabolism alterations have been linked to a wide range of conditions including megaloblastic anemia (Lucock 2000), mood alterations (Godfrey et al. 1990), Alzheimer's disease (Clarke et al. 1998), and numerous athero/thrombogenic phenomena (Brattstrom and Wilcken 2000). Folate deficiencies have also been shown to play a role in Downs syndrome (Hobbs et al. 2000), neural tube defects (MRC vitamin study research group 1991), pregnancy complications (Rajkovic et al. 1997), and male infertility (Wong et al. 2002), which positions folate as a limiting factor in the evolution of the human species. Finally folate deficiencies have been linked to an array of cancers including those of the colon (Slattery et al. 1999), breast (Zhang et al. 1999), pancreas (Stolzenberg-Solomon et al. 2001), stomach (Fang et al. 1997), cervix (Butterworth 1993), bronchus (Kamei et al. 1993), and blood (Skibola et al. 1999).

10.11 Cancer Prevention by Enhancing Genomic Integrity and Repair

The major putative relationship between cancer and folate status relates to the role of folate in providing precursors for DNA repair and synthesis. Folate may also promote genomic integrity through its role in the generation of methyl groups needed for regulation of gene expression via CpG methylation patterns. The cancer protection potential of folates has been demonstrated by large-scale epidemiological and nutritional studies indicating that decreased folate status increases the risk of developing certain cancers. Consistent with a role in DNA repair, chromosome breaks and centrosome abnormalities have been observed in patients deficient in folate (Heath 1966; Chen et al. 1989). In vitro, DNA strand breakage and uracil misincorporation increased in a time and concentration dependent manner after human lymphocytes were cultured with decreasing amounts of folate (Duthie and Hawdon 1998). Moreover, folate deficiency impaired DNA excision repair in specific tissues excised from animal models (Choi et al. 1998). These data indicate that folic acid deficiency affects the stability of cellular DNA at the chromosomal and molecular levels (Choi and Mason 2000). Folate supplementation, particularity localized to areas of elevated demand, is thus proposed as a strategy to enhance genomic integrity and prevent the development of cancer.

10.12 Cancer Prevention by Enhancing the Epidermal Barrier

A terminally differentiated epidermal barrier with high integrity is crucial to cancer prevention. Several points need to be considered with regard to micronutrients in epidermal barrier development. First, there is a growing body of evidence indicating that a significant percentage of the American population is deficient in a number of micronutrients and the constant turnover of the epidermis makes this tissue particularly vulnerable to micronutrient depletion. Micronutrient deficiencies observed in plasma are observed also in skin (Peng et al. 1993). Second, the constant renewal of the epidermal compartment places an important nutrient requirement on the organism. Thus, the nutritional status of micronutrients such as folic acid whose bioactive forms play important roles in the generation of components necessary for optimal cell growth and protein expression regulation are important to the integrity of the epidermal barrier. Third, the non-vascular nature of epidermal compartment makes micronutrient delivery to this compartment inherently inefficient. The above considerations lead to the hypothesis that optimal micronutrient status will strengthen integrity of the epidermal barrier which in turn can lead to a decrease in skin cancer. Studies have shown that cell populations with altered growth properties within actinic keratosis lesions can be recognized by immune surveillance and removed. Alternatively, cell populations within such lesions can progress to cell populations (carcinoma in situ) that secrete proteases and other factors that allow escape from the epidermis. Thus, the degree of integrity of the epidermal barrier can be a deciding factor between the ultimate fates of removal or escape of abnormal cell populations from the epidermal compartment.

10.13 Generation of Folate-Restricted Skin Cell Culture Model

HaCaT keratinocytes have been used to create a cell culture model of folate deficiency. The HaCaT cell line derives from normal human abdominal skin and exhibits a differentiation profile comparable with normal human keratinocytes despite an altered and unlimited growth potential. HaCaT cells exhibit UV-B type-specific mutations on the p53 tumor suppressor gene (Stampfer et al. 1993), representative of an early stage precancerous skin cell. To analyze the role of folate in skin cells during exposure to UV photodamage or photooxidative stress, we have developed a cell culture method in which folate concentrations in the culture medium are controlled by restricting folate and dialyzing the fetal bovine serum. This significantly reduces growth rates of HaCaT keratinocytes, as shown in Fig. 10.5. Because these cells cannot synthesize folates de novo, folate levels are rapidly depleted during cell division.

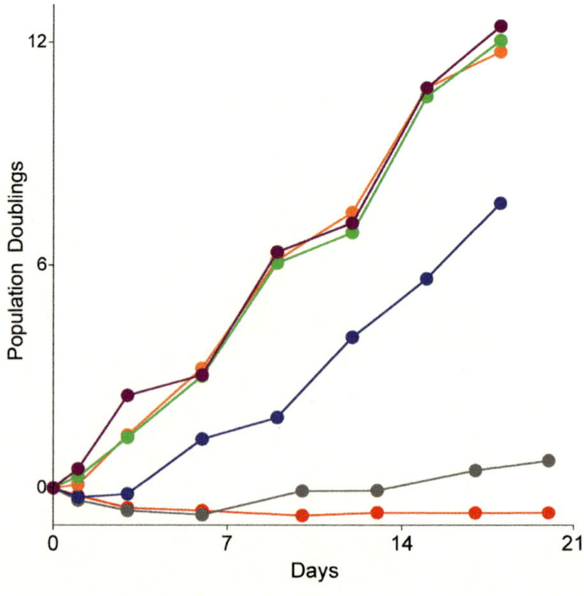

Fig. 10.5 HaCaT keratinocytes were depleted of folate for 14 days. Cell growth was measured after folate repletion at varying concentrations. ● Non-Depleted Control (9100 nM Folate), ● Depleted (9100 nM Folate), ● Depleted (910 nM Folate), ● Depleted (91 nM Folate), ● Depleted (9.1 nM Folate), ● Depleted (0 nM Folate)

We have found that through folate restriction the population doubling times of cells in culture are increased in a direct relationship to folate depletion over a critical range of concentrations (Fig. 10.5). Normal growth rates were reestablished upon folate supplementation. Interestingly, under conditions of complete folate deprivation, HaCaT cells survive without dividing for an extended period of time stalled in the S phase of the cell cycle.

10.14 Topical Micronutrient Delivery for Skin Damage Prevention

Described here is evidence that folate is a candidate for skin damage prevention involving several different mechanisms. However, a major challenge for the development of micronutrient prevention strategies for skin damage is delivery of micronutrients to skin. Folic acid is a good candidate for topical micronutrient delivery based on the roles for the bioactive forms of this nutrient in maintenance of genomic integrity via enhancement of DNA synthesis, DNA repair, and maintenance of epigenetic regulation. Our studies have demonstrated that a topical micronutrient delivery strategy to limit skin damage is feasible. Since folate is hydrophilic, developing a lipophilic form would enhance delivery through the stratum corneum, which is very lipophilic. Pro-folate compounds designed for targeted delivery were synthesized. Once delivered to the epidermis, the abundant esterases present there rapidly cleave the pro-folates to the parent compound. We designed the delivery properties to provide a slow, continuous supply of folate to skin cells to allow increased uptake by the cells. This strategy takes into consideration two distinct barriers that influence the delivery of a micronutrient to skin, lipophilicity of the stratum corneum and skin metabolic activity. The formulation strategy controls the rate of partitioning of the pro-folate into the stratum corneum by optimizing lipophilicity. Figure 10.6 shows a multiple compartment model that served as the framework for the development of this delivery strategy.

The pro-folate must effectively partition from the topical formulation into the stratum corneum. The highly lipophilic nature of the stratum corneum dictates that the profolate be sufficiently lipophilic to effectively partition into the stratum corneum from the delivery vehicle, e.g. a skin cream of lotion (arrow #1 in Fig. 10.6). The pro-folatet must partition from the stratum corneum into the epidermis at an optimal rate to achieve effective delivery to the cellular components of skin. (#2 in Fig. 10.6). Studies of drug structure-penetration relationships have provided useful information concerning this criterion (Tsai et al. 1992; Weber et al. 1994). A correlation between skin permeability and the physicochemical properties of the drug, such as octanol/water partition coefficient (Poct/w) have proven to be of great value in predicting drug transport across skin. A linear correlation between skin permeability of many compounds and their log (Poct/w) has been established (Anderson et al. 1998; Roberts et al. 1978). These findings allow a correlation between prodrug lipophilicity and drug delivery.

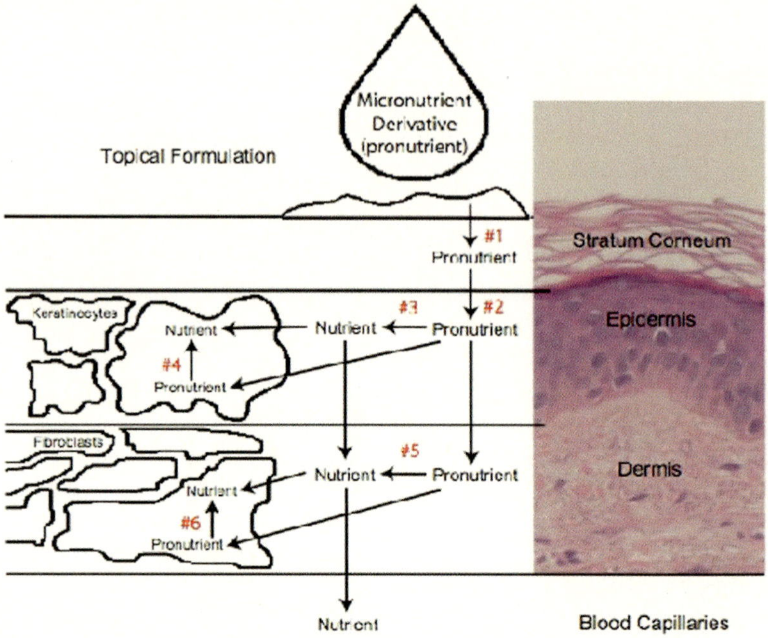

Fig. 10.6 Tropical micronutriment delivery: a multiple compartment model

10.15 Summary

The discussion presented above suggests that a great deal is yet to be learned regarding folate metabolism in skin. Using the in vitro model described along with experiments in human reconstructed skin models and animal models, addressing this dearth of information should be possible. Furthermore, the strategy to optimize folate status by directed pro-folate topical delivery mechanisms in sun exposed skin may provide an opportunity to prevent skin cancer development and progression.

Acknowledgments The research described herein was supported in part by the National Institutes of Health and Niadyne, Inc. ELJ and MKJ are principles in Niadyne, Inc. and conflict of interest management is conducted by the University of Arizona Board of Regents.

References

Agar NS, Halliday GM, Barnetson RS, Ananthaswamy HN, Wheeler M, Jones AM (2004) The basal layer in human squamous tumors harbors more UVA than UVB fingerprint mutations: A role for UVA in human skin carcinogenesis. Proc Natl Acad Sci USA 101:4954–4959

Ames BN (2001) DNA damage from micronutrient deficiencies is likely to be a major cause of cancer. Mutat Res 475:7–20

Anderson BD, Higuchi WI, Raykar PV (1998) Heterogeneity effects on permeability–partition coefficient relationships in human stratum corneum. Pharm Res 5:566–573

Blakley RL (1969) The biochemistry of folic acid and related pteridines. In: Tatum EL (ed) Frontiers of biology, vol 13. North Holland Publishing Company, New York
Branda RF, Eaton JW (1978) Skin color and nutrient photolysis: An evolutionary hypothesis. Science 201:625–626.
Brattstrom L, Wilcken DE (2000) Homocysteine and cardiovascular disease: Cause or effect? Am J Clin Nutr 72:315–323
Butterworth CE Jr (1993) Folate status, women's health, pregnancy outcome, and cancer. J Am Coll Nutr 12:438–441
Chen AT, Reidy JA, Annest JL, Welty TK, Zhou HG (1989) Increased chromosome fragility as a consequence of blood folate levels, smoking status, and coffee consumption. Environ Mol Mutagen 13:319–324
Choi SW, Mason JB (2000) Folate and carcinogenesis: An integrated scheme. J Nutr 130:129–132
Choi SW, Kim YI, Weitzel JN, Mason JB (1998) Folate depletion impairs DNA excision repair in the colon of the rat. Gut 43:93–99
Clarke R, Smith AD, Jobst KA, Refsum H, Sutton L, Ueland PM (1998) Folate, vitamin B12, and serum total homocysteine levels in confirmed Alzheimer Disease. Arch Neurol 55:1449–1455
Cohn BA (1998) The vital role of the skin in human natural history. Int J Dermatol 37:821–824
Dietrich M, Brown CJ, Block G (2005) The effect of folate fortification of cereal-grain products on blood folate status, dietary folate intake, and dietary folate sources among adult non-supplement users in the united states. J Am Coll Nutr 24:266–274
Duthie SJ, Hawdon A (1998) DNA instability (strand breakage, uracil misincorporation, and defective repair) is increased by folic acid depletion in human lymphocytes in vitro. FASEB J 12:1491–1497
Fang JY, Xiao SD, Zhu SS, Yuan JM, Qiu DK, Jiang SJ (1997) Relationship of plasma folic acid and status of DNA methylation in human gastric cancer. J Gastroenterol 32:171–175
Gambichler T, Bader A, Sauermann K, Altmeyer P, Hoffmann K (2001) Serum folate levels after UVA exposure: a two-group parallel randomised controlled trial. BMC Dermatol 1:8
Godfrey PS, Toone BK, Carney MW, Flynn TG, Bottiglieri T, Laundy M, Chanarin I, Reynolds EH (1990) Enhancement of recovery from psychiatric illness by methylfolate. Lancet 336:392–395
Goldsmith LA (1991) Physiology, biochemistry, and molecular biology of the skin, 2nd edn, 1 vols. vol. 1. Oxford University Press, New York
Guenthner ST, Hurwitz RM, Buckel LJ, Gray HR (1999) Cutaneous squamous cell carcinomas consistently show histologic evidence of in situ changes: a clinicopathologic correlation. J Am Acad Dermatol 41:443–448
Heath CW Jr (1966) Cytogenetic observations in vitamin B12 and folate deficiency. Blood 27:800–815
Herbert V, Larrabee AR, Buchanan JM (1962) Studies on the identification of a folate compound of human serum. J Clin Invest 41:1134–1138
Hobbs CA, Sherman SL, Yi P, Hopkins SE, Torfs CP, Hine RJ, Pogribna M, Rozen R, James SJ (2000) Polymorphisms in genes involved in folate metabolism as maternal risk factors for down syndrome. Am J Hum Genet 67:623–630
Holick MF (2003) Vitamin D: a millenium perspective. J Cell Biochem 88:296–307
Holick MF, MacLaughlin JA, Doppelt SH (1981) Regulation of cutaneous previtamin D3 photosynthesis in man: skin pigment is not an essential regulator. Science 211:590–593
Hurwitz RM, Monger LE (1995) Solar keratosis: an evolving squamous cell carcinoma. Benign or malignant? Dermatol Surg 21:184
Jablonski NG (2004) The evolution of human skin and skin color. Annu Rev Anthropol 33:585-C-581
Jablonski NG, Chaplin G (2000) The evolution of human skin coloration. J Hum Evol 39:57–106
Jeffes EW 3rd, Tang EH (2000) Actinic keratosis. Current treatment options. Am J Clin Dermatol 1:167–179
Jimbow K, Quevedo WC Jr, Fitzpatrick TB, Szabo G (1976) Some aspects of melanin biology: 1950–1975. J Invest Dermatol 67:72–89

John Scott FRJF (2000) Folic acid and folates: the feasibility for nutritional enhancement in plant foods. J Sci Food Agric 80:795–824
Kamei T, Kohno T, Ohwada H, Takeuchi Y, Hayashi Y, Fukuma S (1993) Experimental study of the therapeutic effects of folate, vitamin A, and vitamin B12 on squamous metaplasia of the bronchial epithelium. Cancer 71:2477–2483
Karagas MR, Greenberg ER, Spencer SK, Stukel TA, Mott LA (1999) Increase in incidence rates of basal cell and squamous cell skin cancer in New Hampshire, USA. New Hampshire skin cancer study group. Int J Cancer 81:555–559
Kobayashi N, Nakagawa A, Muramatsu T, Yamashina Y, Shirai T, Hashimoto MW, Ishigaki Y, Lehman TA, Modali R, Boukamp P, Stanek J, Bennett WP, Welsh JA, Metcalf RA, Stampfer MJ, Lober BA, Lober CW, Accola J (2000) Actinic keratosis is squamous cell carcinoma. J Am Acad Dermatol 43:881–882
Loomis WF (1967) Skin-pigment regulation of vitamin-D biosynthesis in man. Science 157:501–506
Lucock M (2000) Folic acid: nutritional biochemistry, molecular biology, and role in disease processes. Mol Genet Metab 71:121–138
Lucock M, Yates Z, Glanville T, Leeming R, Simpson N, Daskalakis I (2003) A critical role for B-vitamin nutrition in human developmental and evolutionary biology. Nutr Res 23:1463–1475
Madronich S, McKenzie RL, Bjorn LO, Caldwell MM (1998) Changes in biologically active ultraviolet radiation reaching the Earth's surface. J Photochem Photobiol B 46:5–19
MRC Vitamin Study Research Group (1991) Prevention of neural tube defects: Results of the medical research council vitamin study. Lancet 338:131–137
Nielsen KP, Lu Z, Juzenas P, Stamnes JJ, Stamnes K, Moan J (2004) Reflectance spectra of pigmented and nonpigmented skin in the UV spectral region. Photochem Photobiol 80:450–455
Nielsen KP, Zhao L, Stamnes JJ, Stamnes K, Moan J (2006) The importance of the depth distribution of melanin in skin for DNA protection and other photobiological processes. J Photochem Photobiol B 82:194–198
Off MK, Steindal AE, Porojnicu AC, Juzeniene A, Vorobey A, Johnsson A, Moan J (2005) Ultraviolet photodegradation of folic acid. J Photochem Photobiol B 80:47–55
Ohnishi T, Mori T (1998) Supranuclear melanin caps reduce ultraviolet induced DNA photoproducts in human epidermis. J Invest Dermatol 110:806–810
Peng YM, Peng YS, Lin Y, Moon T, Baier M (1993) Micronutrient concentrations in paired skin and plasma of patients with actinic keratoses: effect of prolonged retinol supplementation. Cancer Epidemiol Biomarkers Prev 2:145–150
Pentland AP, Schoggins JW, Scott GA, Khan KN, Han R (1999) Reduction of UV-induced skin tumors in hairless mice by selective cox-2 inhibition. Carcinogenesis 20:1939–1944
Post PW, Daniels F Jr, Binford RT Jr (1975) Cold injury and the evolution of "White" skin. Hum Biol 47:65–80.
Rajkovic A, Catalano PM, Malinow MR (1997) Elevated homocyst(e)ine levels with preeclampsia. Obstet Gynecol 90:168–171
Roberts MS, Anderson RA, Swarbrick J, Moore DE (1978) The percutaneous absorption of phenolic compounds: the mechanism of diffusion across the stratum corneum. J Pharm Pharmacol 30:486–490
Skibola CF, Smith MT, Kane E, Roman E, Rollinson S, Cartwright RA, Morgan G (1999) Polymorphisms in the methylenetetrahydrofolate reductase gene are associated with susceptibility to acute leukemia in adults. Proc Natl Acad Sci USA 96:12810–12815
Slattery ML, Potter JD, Samowitz W, Schaffer D, Leppert M (1999) Methylenetetrahydrofolate reductase, diet, and risk of colon cancer. Cancer Epidemiol Biomarkers Prev 8:513–518
Stampfer MR, Fusenig N, Rogan EM et al (1993) P53 mutations in human immortalized epithelial cell lines. Carcinogenesis 14:833–839
Steindal AH, Porojnicu AC, Moan J (2006) Is the seasonal variation in cancer prognosis caused by sun-induced folate degradation? Med Hypotheses. Dec 29, EPub ahead of print

Stolzenberg-Solomon RZ, Pietinen P, Barrett MJ, Taylor PR, Virtamo J, Albanes D (2001) Dietary and other methyl-group availability factors and pancreatic cancer risk in a cohort of male smokers. Am J Epidemiol 153:680–687

Szabo G, Gerald AB, Pathak MA, Fitzpatrick TB (1969) Racial differences in the fate of melanosomes in human epidermis. Nature 222:1081–1082

Taylor SC (2002) Skin of color: biology, structure, function, and implications for dermatologic disease. J Am Acad Dermatol 46:S41–S62

Thien KR, Blair JA, Leeming RJ, Cooke WT, Melikian V (1977) Serum folates in man. J Clin Pathol 30:438–448

Thody AJ, Higgins EM, Wakamatsu K, Ito S, Burchill SA, Marks JM (1991) Pheomelanin as well as eumelanin is present in human epidermis. J Invest Dermatol 97:340–344

Tsai R-S, El Tayar N, Carrupt P-A, Testa B (1992) Physicochemical properties and transport behaviour of piribedil: Considerations on its membrane-crossing potential. Int J Pharm 80:39–49

Ullrich SE (2002) Photoimmune suppression and photocarcinogenesis. Front Biosci 7:684–703

Weber H, Meyer-Trumpener K, Lippold BC (1994) Esters of naproxen as potential prodrugs for skin penetration. 2. Penetration behavior in excised mice skin. Arch Pharm 327:681–686

Wills L (1978) Nutrition classics. British Medical Journal 1:1059–64, 1931. Treatment of "Pernicious anaemia of pregnancy" and "Tropical anaemia" With special reference to yeast extract as a curative agent. By lucy wills. Nutr Rev 36:149–151

Wondrak GT, Cervantes-Laurean D, Roberts MJ, Qasem JG, Kim M, Jacobson EL, Jacobson MK (2002a) Identification of alpha-dicarbonyl scavengers for cellular protection against carbonyl stress. Biochem Pharmacol 63:361–373

Wondrak GT, Roberts MJ, Jacobson MK, Jacobson EL (2002b) Photosensitized growth inhibition of cultured human skin cells: Mechanism and suppression of oxidative stress from solar irradiation of glycated proteins. J Invest Dermatol 119:489–498

Wondrak GT, Roberts MJ, Cervantes-Laurean D, Jacobson MK, Jacobson EL (2003) Proteins of the extracellular matrix are sensitizers of photo-oxidative stress in human skin cells. J Invest Dermatol 121:578–586

Wong WY, Merkus HM, Thomas CM, Menkveld R, Zielhuis GA, Steegers-Theunissen RP (2002) Effects of folic acid and zinc sulfate on male factor subfertility: a double-blind, randomized, placebo-controlled trial. Fertil Steril 77:491–498

Zhang S, Hunter DJ, Hankinson SE, Giovannucci EL, Rosner BA, Colditz GA, Speizer FE, Willett WC (1999) A prospective study of folate intake and the risk of breast cancer. JAMA 281:1632–1637

Chapter 11
Thiamin(e): The Spark of Life

Derrick Lonsdale

Abstract One of the earliest vitamins to be discovered and synthesized, thiamin was originally spelled with an "e". The terminal "e" was dropped when it was found that it was not an amine. It is still spelled with and without the "e" depending on the text. This chapter provides a brief historical review of the association of thiamin with the ancient scourge of beriberi. It emphasizes that beriberi is the model for high calorie malnutrition because of its occurrence in predominantly white rice consuming cultures. Some of the symptomatology of this ancient scourge is described, emphasizing the difference from that seen in starvation. High calorie malnutrition, due to excessive ingestion of simple carbohydrates, is widely encountered in the U.S.A. today. Thiamin deficiency is commonly associated with this, largely because of its cofactor status in the metabolism of glucose. The biochemistry of the three phosphorylated esters of thiamin and the transporters are discussed and the pathophysiology of thiamin deficiency reviewed. The role of thiamin, and particularly its synthetic derivatives as therapeutic agents, is not fully appreciated in Western civilization and a clinical section describes some of the unusual cases described in the scientific literature and some experienced by the author. The possible role of high calorie malnutrition and related thiamin deficiency in juvenile crime is hypothesized.

11.1 History

It is now well known that thiamin deficiency is the major cause of beriberi, a disease that had affected humans for centuries. The name "Kakke" was the term used for the disease in Japan and this word can be found in documents as early as 808 (Inouye and Katsura 1965). Until the 17th century the majority of the population in Japan took unpolished rice as the staple food. Polished rice was associated with relative affluence, since it looked better on the table when served. Epidemics of beriberi have been known to occur in association with increased affluence simply because it was expensive to take the rice to the mill. When white rice was served to friends, it became a signature of the newly acquired affluence. As the ingestion of

D. Lonsdale (✉)
Preventive Medicine Group, Westlake, OH 44145, USA
e-mail: dlonsdale@pol.net

O. Stanger (ed.), *Water Soluble Vitamins*, Subcellular Biochemistry 56,
DOI 10.1007/978-94-007-2199-9_11, © Springer Science+Business Media B.V. 2012

well-milled white rice became nation wide, so the incidence of the disease increased. This is because the B group vitamins are in the discarded husks. The first national statistics on mortality appeared in 1899 and showed a death rate of 20 per 100,000. This dropped to 0.5 in 1959 after its nutritional association was discovered. Of considerable interest to us today, for reasons that will appear later in the chapter, the peak incidence of the disease occurred in August and September every year. Factory workers would take their lunch between factory buildings. If the sun came round so that it shone into the corridor, some workers would get the first symptoms of the disease. It was evident that the sun's rays would stress them sufficiently to initiate these symptoms. It was therefore hardly surprising that the etiology in the early 1900s, before its nutritional association became common knowledge, was considered to be from infection.

The discovery of the relationship of thiamin with malnutrition came in the late decades of the 19th century, but it was many years before scientific knowledge caught up. A Japanese naval surgeon by the name of Takaki studied in England from 1875 to 1880. He noted that beriberi was less common in the British Royal Navy than in navy personnel in Japan where the diet on ships was very different. In 1882 a Japanese naval vessel sailed on a 272 day voyage. On its return, 61% of the crew had succumbed to beriberi. Two years later, another ship completed the same voyage, but was provided with an ample supply of dried milk and meat, giving a carbon to nitrogen ration of 16:1. Only 14 crew members had developed beriberi. Takaki concluded that a lack of nitrogenous food was the cause of the disease, a notable contribution before vitamins were known.

In 1890, Eijkman found that polished rice, given to pigeons, caused polyneuritis and the histopathology was similar to that seen in humans in beriberi. Funk and Cooper isolated an "antiberiberi factor" from rice polishing in 1910 and this was crystallized in 1926 and called Vitamine (Jansen and Donath 1926). It was not until 1936 that thiamin was synthesized (Williams 1938) leading to an explosion of basic science and clinical experimentation. The work of Sir Rudolph Peters exposed the vitally important association of thiamin with what was later to become the science of oxidative metabolism (Peters 1936).

11.2 Clinical Facts About Beriberi and Thiamin Deficiency

Although a great many symptoms of beriberi have been described, none of them is considered to be pathognomonic (Inouye and Katsura 1965). Generally it is classified into the "wet" or edematous type and "dry", both being chronic in their course. The peracute and extremely lethal form is known in Japan as "Shoshin". Infantile beriberi is associated with sudden death and the clinical picture is startlingly similar to that of Sudden Infant Death Syndrome that is familiar to us today (Lonsdale 2001). Wernicke encephalopathy, usually associated with alcoholism, is at least one form of the disease that affects brain function in adults. A complete review of the clinical aspects of beriberi was published and interested readers are encouraged to look in this publication or a similar one since the complete information is readily available (Lonsdale 1975). It is not necessary to provide the details again here, but there are some aspects that need to be emphasized because it focuses on some of our modern day clinical problems that are being sadly overlooked.

Many readers might be of the opinion that the classical forms of nutritional deficiency diseases have faded into the background of interesting history. This has caused their diverse symptoms to be neglected by most modern physicians since vitamin enrichment of many foods automatically erases them from their thinking about differential diagnosis. It must be emphasized that beriberi is a disease where the caloric intake is high, particularly in the form of simple carbohydrate foods such as starch. Indeed, it can be superficially compared to a car engine that is being fuelled by a rich mixture of gasoline with insufficient air and/or inefficient spark plugs. It is the lack of vitamin content, particularly thiamin, that results in inefficient oxidative metabolism. Since the brain and heart have a high requirement for oxygen, it is easy to understand why the brain, nervous system and heart are the organs that are affected primarily.

Early investigators found that the arterial oxygen concentration in beriberi patients was relatively low while venous oxygen concentration was relatively high (Inouye and Katsura 1965). This means, of course, that tissue oxygen exchange by cells is defective in the disease. Edema, a common clinical problem today, is one of the important, but not exclusive, signs of beriberi and is always present in the early stages of the disease. A classical sign is tenderness of the calf muscles when gripped with the hand. Anasarca is, however, rare even though there may be profound edema. It should enter into differential diagnosis when cardiovascular and neurological symptoms, especially if there is edema, are present together. Exertion or mental stress results often in cardiac palpitations and disproportionate dyspnea. The typical cardiac enlargement that is easily found by chest roentgenography can easily be mistaken for a viral myocardiopathy and might even lead to consideration of heart transplant if the correct diagnosis is missed. Cardiac beriberi was reported in England in 1971, (McIntyre and Stanley 1971). The authors emphasized the need for it to be considered in the differential diagnosis of heart failure, even when the cardiac output is low. Pang et al. considered that the Shoshin form of beriberi was probably underdiagnosed in the West (Pang et al. 1986).

Perhaps the most important thing to emphasize is the effect of beriberi on the autonomic nervous system (Inouye and Katsura 1965). This is altered early in the disease and produces changes in vasomotor function. The diastolic blood pressure falls and may even reach zero. The systolic pressure may rise or remain normal. In classic beriberi the passage of blood through the femoral artery could be heard, even by an observer standing near the patient. Beriberi victims could be divided into certain categories by their chemistry. Platt, an early investigator, reported that many victims had normoglycemia and were easily treated with thiamin (Platt 1967). More severe cases had hyperglycemia and responded to thiamin with greater difficulty, while others had hypoglycemia and some did not respond at all to the vitamin. Presumably this would reflect the severity and chronicity of the biochemical changes incurred and this should be noted since hyperglycemia would automatically be considered evidence of diabetes in the modern world. Some patients had achlorhydria and others had hyperchlorhydria. Curiously, under treatment with thiamin, those with achlorhydria became hyperchlorhydric before normal acid content was established. The exact opposite effect was seen in the hyperchlorhydric patients. Although the mechanism is unknown, it has suggested that such phenomena might be related to instability and imbalance of autonomic nervous system signaling. At any rate, it certainly introduces the fact that beriberi symptomatology is extremely complex and not a matter of simply giving a few milligrams of thiamin. In fact, it took huge doses of the vitamin for months to abolish the symptoms. We have found many of these facts to be relevant today, bearing them in mind when confronted with symptoms and signs that can be confusing in a differential diagnosis, as will be discussed later in the clinical section.

In assessing the role of thiamin deficiency in the modern world, it was thought that a description of the basic distribution of the autonomic nervous system was a necessary addition to a book devoted to thiamin metabolism (Lonsdale 1987). It would seem that an example of clinical blindness to the importance of vitamin supplementation and therapy is that an extremely detailed and erudite book entitled "Diseases of the Autonomic Nervous System" was edited and published by Sir Roger Bannister (Bannister 1984). Not one word on malnutrition appeared in that book, in our opinion a very important oversight.

After the discovery of thiamin in its application to treatment of a scourge that had existed for thousands of years and still occurs where rice is a staple food, it was natural that a vast amount of research was initiated. In 1962 a symposium was published to commemorate the synthesis of the vitamin. Zbinden reported 696 published papers reporting attempted thiamin therapy in more than 230 different diseases with varying degrees of success (Zbinden 1962). It is probably pertinent to suggest why such information, important to the health of millions, was more or less abandoned until relatively recently when vitamins have reemerged for their true nutrient value. Because so many different diagnosed conditions sometimes responded to the administration of thiamin as a single entity, it offended the emerging model for disease. Each disease is researched to find a cure that is specific for a given condition, sometimes referred to as the "silver bullet". Thiamin provided symptomatic responses in many different disease entities, indicating that it was addressing the underlying biochemical lesion. Perhaps, with the relatively new explosion in nutritional therapy, it is possible to start thinking that our disease model is catastrophically wrong. Since healing is a function of the body itself, all it requires is energy in order to carry out this role and it seems that a new model for disease is required (Lonsdal 2006).

Vitamins fell into relative disrepute and even today they are often spurned by physicians who regard their pharmaceutical use as absurd. This is because even current teaching often provides the concept that vitamins, as cofactors, only work up to their minute physiological concentration. Pharmaceutical doses are conceived as being discarded and excreted without considering the necessary repair to damage that occurs in the enzyme/cofactor bonding as a result of prolonged deficiency.

11.3 Biochemistry

Thiamin consists of a pyrimidine ring (2,5- dimethyl-6-aminopyrimidine) and a thiazolium ring (4-methyl-5-hydroxy ethyl thiazole) joined by a methylene bridge. The naturally occurring vitamin is found in lean pork and other meats, wheat germ, liver and other organ meats, poultry, eggs, fish, beans and peas, nuts and whole grains. Dairy products, fruit and vegetables are not good sources. The RDA is 0.5 mg per 1,000 kcal. Although this may be adequate for a healthy individual consuming a healthy diet, it may only be a marginal consumption. Considerable losses occur during cooking or other heat-processing of food. Polyphenolic compounds in coffee and tea can inactivate thiamin so that heavy use of these beverages could compromise thiamin nutrition.

11.4 Thiamin Monophosphate

Whether thiamin monophophate (TMP) has a specific function in any cells is still unknown. It is formed from free thiamin to synthesize the monophosphate which is then phosphorylated again by a kinase to form thiamin diphosphate, known as the pyrophosphate (TPP). It may be purely analogous to the stepwise phosphorylation of adenosine, since a third phosphorylation of TPP forms thiamin triphosphate (TTP).

11.5 Thiamin Pyrophosphate (TPP)

Also known as cocarboxylase, this is the form of thiamin that is best known. Its biochemical roles in energy metabolism are multiple. It is a cofactor for enzymatic reactions that cleave alpha-keto acids and it has long been known that it activates decarboxylation of pyruvate in the pyruvate dehydrogenase complex. The complex is a group of enzymes and cofactors that form acetyl CoA. This condenses with oxaloacetate to form citrate, the first component of the citric acid cycle. Since pyruvate is derived from glucose via the Embden-Meyerhof pathway, it should be emphasized that the energy drive from oxidation of glucose is dependent upon TPP. It has the same role in the decarboxylating component of alpha-ketoglutarate dehydrogenase, a link in the sequential metabolism of pyruvate through the citric acid cycle. A third enzyme complex, similar in structure to pyruvate dehydrogenase is that which decarboxylates the three branched chain amino acids, leucine, isoleucine and valine From a clinical standpoint, enzymatic failure and thiamin dependency has been described in both of these enzyme complexes and will be discussed in the clinical section.

A fourth enzyme that requires TPP is transketolase that is present twice in the pentose shunt. The functions of this pathway are to provide pentose phosphate for nucleotide synthesis and to supply reduced NADP for various synthetic pathways. Since the pentose shunt is found in erythrocytes, an easily obtained tissue, measurement of the activity of this enzyme has become used as an important laboratory test for thiamin deficiency. It is to be emphasized that it gives information only on TPP deficiency. It does not indicate deficiency of TTP. The reaction is highly dependent upon the presence of magnesium, an important point in the study of patients in the clinical setting.

11.6 Thiamin Triphosphate (TTP)

Although this component is still poorly understood, there is no doubt that it plays a significant part in thiamin metabolism. In 1938 Minz first suggested a relationship between thiamin and nervous excitation when he observed that thiamin was released into the bathing medium when the pneumoagastric nerve from an ox was stimulated (Minz 1938). In 1979 Cooper and Pincus reviewed the evidence that there was a possibility that thiamin has a function in the nervous system distinct from its activity as a cofactor to enzymes (Cooper and Pincus 1979). Like Minz, they found that nerve stimulation in experimental animal systems resulted in decline of the level of TPP and TTP from the stimulated nerve. The released metabolites were in the form of TMP and free thiamin, making it difficult to interpret the function of the vitamin in nerve conduction. They found that thiamin appeared to be uniformly distributed in nervous tissue and was highly localized in membrane structures. After intracerebroventricular injection of radioactive thiamin into rats, the distribution of the esters was found to be: thiamin 8–12%, TMP 12–14%, TPP 72–74% and TTP 2–3% (Iwata et al. 1985).

Bettendorff and associates reported their research on TTP, using the electric organ of the eel Electrophorus electricus, rat brain vesicles and neuroblastoma cells (Bettendorff et al. 1987, 1993, 1994). They found that 87% of the total thiamin content of the electric organ in the eel was in the form of TTP, suggesting the great importance of this ester in nerve physiology since this organ is an adaptation of a synapse to create a condenser. They indicated that the real substrate of TTP phosphatase is probably a 1:1 complex of Mg^{2+} and TTP. Incubation of rat brain homogenates with thiamin and TPP led to synthesis of TTP that appeared to be an activator of chloride channels having a large unit conductance. In mammalian tissues TTP concentrations are regulated by a specific thiamin triphosphatase (Makarchikov et al. 2003). It must be stated, however, that the role of TTP is still incompletely understood.

11.7 Thiamin Transporter

The SLC gene family of solute carriers is a family of three transporter proteins with significant structural similarity, transporting substrates with different structure and ionic charge. SLC19A1 mediates the transport of reduced folate and its analogs and SLC19A2 mediates the transport of thiamin. SLC19A3 is also capable of transporting thiamin (Ganapathy et al. 2004; Nabokina and Said 2004).

11.8 Pathophysiology of Thiamin Deficiency

The clinical results seen in proven thiamin deficiency are now well known. Its etiological association with beriberi and Wernicke Korsakoff encephalopathy is accepted by all. Largely because of our remaining ignorance of the role of TMP and TTP, however, the pathophysiology is still incompletely understood. There is no doubt that thiamin, in all its forms, has a vital role in many different aspects of energy metabolism since it catalyzes the normal use of oxygen. This is very well illustrated by the fact, discussed in relation to the symptoms of beriberi, that arterial oxygen concentration is relatively low while that of venous oxygen is relatively high. It is also pretty clear that TPP is possibly the rate limiting factor in entry of pyruvate to the citric acid cycle. Thus, besides being compared with a spark plug in an engine. it can also be compared to a throttle or accelerator.

A conference was sponsored by the New York Academy of Sciences in 1982 to discuss the pathophysiology of thiamin deficiency and existing knowledge of the vitamin at that time (Sable and Gubler 1982). In pyrithiamin-induced thiamin deficiency in animals the net levels of phosphocreatine in the lateral vestibular nucleus and other nuclei of brain were increased selectively, suggesting to the author that this represented under utilization of this source of ATP synthesis (McCandless 1982). Gibson and associates reported evidence of deterioration in the cholinergic system with thiamin deficiency (Gibson et al. 1982) and Meador and associates found evidence for a central cholinergic effect from administration of large doses of thiamin (Meador et al. 1993a). Lonsdale (Lonsdale 1982) found that intraperitoneal injection of thiamin tetrahydrofurfuryl disulfide, a biologically active derivative of thiamin, prolonged the period of sound stimulated seizures in DBA/J2 mice weanlings. Audiogenic seizures in rodents are known to be cholinergic in origin and this type of seizure undergoes spontaneous remission after only a few days of sensitivity in this strain of mouse. The increase in severity of the seizures and the prolongation beyond the normal period of audiogenic seizure initiation supported the role of thiamin in its stimulation of cholinergic CNS activity.

Two publications have indicated that thiamin supplementation has a mild clinical effect in Alzheimer's disease (AD) where the abnormality of the cholinergic system is part of the pathophysiology (Blass et al. 1988; Meador et al. 1993b). Mastrogiacoma and associates studied thiamin, its phosphate esters and its metabolizing enzymes in autopsied specimens of brain in AD and controls (Mastrogiacoma et al. 1996). In the AD group the mean levels of free thiamin and TMP were normal. Concentrations of TPP were reduced by 18–21% although the TPP metabolizing enzymes were normal. The authors hypothesized ATP deficiency since that is required for increased phosphorylation in thiamin esters. Thiamin deficiency in rats caused encephalopathy and DNA synthesis decreased significantly in cortex, brain stem, cerebellum and subcortical structures. This was reversible by the administration of thiamin (Henderson and Schenker 1975).

Creatinuria has been noted in beriberi as well as in experimentally induced thiamin deficiency. An observation in rats showed that creatinuria occurred as a result of protein calorie deficiency caused by experimental starvation. Where anorexia was caused by thiamin deprivation the creatinuria occurred much sooner and was statistically greater in concentration (Lonsdale 1987). This was mindful of the greater degree of creatinuria observed in calorically starved T3 injected rats than in calorically sufficient animals injected with T3, suggesting that membrane physiology was involved in these two differently stimulated mechanisms. As will be indicated in the clinical section of this chapter, TTP remains as the outstanding mystery of thiamin metabolism, particularly in brain.

11.9 Laboratory Methodology

As in the case of other vitamins, blood and urine measurements of thiamin are unreliable. By far the best test for thiamin deficiency is measurement of transketolase (Boni et al. 1980; Jeyasingham et al. 1987). As already indicated, this enzyme occurs twice in the pentose shunt, a biochemical pathway that exists in erythrocytes. In the first part of the test, the product of the enzymatic reaction, sedoheptulose-7-phosphate is measured per unit of blood per minute by spectrophotometry A baseline activity is then calculated. The reaction is then repeated after the addition of TPP. The test is then reported in two sections, the first being the baseline activity. This is referred to as transketolase activity (TKA). The normal range of this in our laboratory is 42–82 mU/L/min, originally derived from apparently healthy asymptomatic subjects. In the second part of the test, if it is found that TKA increases as a result of the addition of cofactor, it is reported as a percentage increase over that detected at baseline. This is referred to as the thiamin pyrophosphate effect (TPPE). In our laboratory, this is considered to be acceptable up to 18% acceleration. In theory, there should be no acceleration and the TPPE should be zero if the enzyme is saturated with its cofactor. It is possible that the "normal subjects" had not reached a stage of deficiency that caused symptoms. It should be regarded as a continuum, progressing from ideal thiamin cofactor status, through marginal deficiency to increasingly severe deficiency generating symptoms in the affected subject. This test should be readily available in clinical laboratories, but is often neglected because vitamin deficiencies are generally considered to be rare or even non-existent. It has a number of important relationships. Sometimes the TKA is low while the TPPE is normal and this would suggest that there is some abnormality in the enzyme (Blass and Gibson 1977). In most cases of thiamin deficiency in our experience the TPPE is increased, sometimes remarkably so. In a more severe deficiency state the TKA is low and the TPPE is high. Correlating the patient's symptoms with a fall in the TPPE is an excellent way of proving the clinical effect of thiamin supplementation. This can be complicated since the enzyme complex is also dependent on magnesium and other non caloric nutrients (Eisinger et al. 1994). Magnesium depletion aggravates the clinical effects of thiamin deficiency (Dyckner et al. 1985). Magnesium and calcium deficiency affects the distribution of thiamin in rat brain (Kimura and Itokawa 1977). We have recently found several patients whose TPPE could not be corrected until they had received an adequate supplementation with magnesium (Lonsdale

2006c). There is still much to learn about the inter-relationships between vitamins and minerals in the overall management of oxidative metabolism. An increase in TKA was found in B12 deficient patients but not in those where the anemia was due to folate deficiency (Markannen and Kalliomaki 1966; Wells et al. 1968).

In 1969 Cooper and associates (Cooper et al. 1969) published their finding of TTP deficiency in Leigh's disease, also known as subacute necrotizing encephalomyelopathy. The pathophysiology of this disease is similar, but not identical, to that of Wernicke's disease. Diagnosis depended upon their finding of a substance in urine that reportedly inhibited the formation of TTP (Cooper et al. 1970). The substance was never identified and this important research was eventually discontinued for lack of funding. Urine from several patients suspected of thiamin deficiency symptoms were sent to Cooper's laboratory, were reported positive for this test and responded to thiamin supplementation. None of these patients could be considered to be examples of Leigh's disease (Lonsdale Unpublished observations). It was the only laboratory study ever reported to indicate TTP deficiency and there is presently no known clinically available method of depicting this vitally important link in our knowledge of thiamin metabolism in brain.

11.10 Clinical Application of Thiamin in Modern Nutrition

It is important to emphasize once again that vitamin supplementation fell into disrepute after the studies reported in the middle decades of the last century. Very little clinical research has been done since then and major medical journals have consistently rejected what clinical research has been reported. This book indicates that may be changing, so it is apt to refer back to some of the excellent studies that were published 50 or more years ago. One of the most important ones in the study of pure thiamin deficiency (as distinct from the nutritional aspects of beriberi) was performed in 1943 (Williams et al. 1943) and I shall refer to this in some detail since it has great relevance to modern nutritional deficiencies. Severe thiamin deprivation resulted in depressed mental states, generalized weakness, giddiness, backache, soreness of muscles, insomnia, anorexia, nausea, vomiting, weight loss, poor muscular tone, low blood pressure and bradycardia with the subjects at rest. On exertion, heart palpitation and precordial distress (pseudoangina) occurred. Tachycardia and sinus arrhythmia were observed. The investigators reported electrocardiographic changes and impairment of gastro intestinal motility.

Moderate, prolonged restriction of thiamin, but not of calories, resulted in emotional instability, reflected by irritability, moodiness, quarrelsome behavior, lack of cooperation, vague fears and agitation, mental depression, variable restriction of activity and numerous somatic complaints. The effects on one subject, a 48 year old woman, were described in detail after 120 days of this deprivation. Blood pressure was between 90 and 98 and the diastolic between 50 and 60 mm Hg. Heart rate 50–60 bpm and there was marked sinus arrhythmia. Pallor and giddiness were observed when standing from the sitting position and rising from the squatting position could be accomplished only with assistance. The patellar tendon reflexes were hypoactive but could be increased through reinforcement and the Achilles tendon reflex was absent. The comments of the authors included the statement that symptoms suggestive of dysfunction of the central and peripheral nervous pathways preceded by months the gross signs of neurologic dysfunction.

This is an old study and it would not be possible to perform a dangerous experiment of this nature now. But we should look seriously at the symptomatology since a discerning practitioner will find clinical effects of this nature today. A simple diet history often will produce clues that make the diagnosis surprisingly clear. Because of the widespread nature of marginal malnutrition in the U.S. where there is a heavy emphasis on sweet foods containing the simple carbohydrates, particularly sucrose and fructose, findings of this nature are often relevant. Limited data are available on the relation between thiamin requirements and the intake of simple carbohydrate in human physiology, but a study was reported that investigated the influence of stepwise increases of carbohydrate intake on the status of thiamin in healthy volunteers under isocaloric conditions (Elmadfa et al. 2001). An increase of dietary carbohydrate intake caused a decrease of plasma and urine levels of thiamin without affecting enzyme activities. A publication in 1962 (Bhuvaneswaran and Sreenivaran 1962) discussed problems of thiamin deficiency states and their treatment. Quoted in this paper is a reference to Yudkin who "showed that, despite the absence of thiamin, rats can survive for many months if carbohydrate is excluded from the diet. Polyneuritis and death followed the addition to the diet of as little as 5% carbohydrate".

The major point to make is that a patient with complaints of the nature described by Williams and associates is so often regarded today as having psychosomatic disease and is either referred to a psychiatrist or treated empirically with a drug. Some simple clinical observations and laboratory studies remove the stigma that is usually associated with this diagnosis, at least by the patient. As already mentioned above, diet history is frequently revealing to the point of wonder at the excesses involving sugar, soft drinks and alcohol, to name a few of the myriad temptations surrounding the public.

There are several publications that support the contention that high calorie malnutrition of this nature is responsible for a huge amount of so-called psychosomatic disease (Lonsdale 1975, 1990; Lonsdale and Shamberger 1980). It is interesting to note that I received a call from the editor of the Journal for the 1990 paper. She was nervous about its publication and suggested that the word "possible" should be introduced into the title. A subsequent letter to the editor castigated the Journal for publishing it at all. This is just a reflection of the incredible resistance that has accompanied evidence of vitamin deficiency disease for years in contemporary society. Lonsdale and Shamberger (1980) described 20 adolescents with proven evidence of metabolic disturbance associated with the symptoms that are so common in this age group today (Lonsdale and Shamberger 1980). Transketolase abnormalities occurred in all of them and were corrected by the use of supplementary thiamin. Many of them had abnormalities in serum folate and vitamin B12 that were just as important as the transketolase in interpreting the biochemical effects of malnutrition in these individuals. The editor would only publish this article if it was reviewed by two acknowledged experts in the field who criticized it heavily in two editorials that appeared in the issue.

A revealing note should be added to the question of the concentration of thiamin relative to the intake of glucose. A woman was receiving intravenous hyperalimentation in a hospital setting. The injected fluid contained 24 mg of thiamin a day, together with a total intake of 5% dextrose water, yielding a calculated 1,036 calories from this source alone. She died and the autopsy revealed Wernicke encephalopathy (Lonsdale 1978). This letter to the editor included the statement that there were no less than six cases of Wernicke's encephalopathy found in the same hospital some 2 years previously. All of them had been receiving supplementary vitamin.

Autonomic dysfunction, as already discussed, is an important expression of symptomatology in beriberi and has been associated with modern high calorie malnutrition (Lonsdale 1981). This was noted to be asymmetric in some cases (Lonsdale 1990). One of them was a girl whose main complaint came from her mother. She had violent temper tantrums, but each one was accompanied by a set of unusual symptoms and signs. She would complain of a migraine-like headache. One pupil would dilate, the other one did not. She would become white on one half of her body and pink on the other, mindful of a harlequin. Sweat could be obtained by electrophoresis on one side, but not the other. Sleep spindles were observed on one side of the brain by electroencephalogram, but not on the other. When stressed by an intravenous injection of epinephrine the pulse pressures were widely different in the two arms when it was measured at the same time by two people (Lonsdale 1987). By abolishing her appalling diet and providing her with vitamin supplementation, that included thiamin, she was retested a year later and the asymmetric functional changes had disappeared.

Any pediatrician knows only too well that there are huge numbers of children seen in the U.S. today with symptoms that are considered to be psychological in nature. Relatively few of them are seen as examples of limbic system biochemical abnormalities that have ravaged them from their awful diet (Lonsdale 2001). The limbic system is really the primitive brain and is responsible for our emotional reactions. It can certainly be regarded as a computer that presides over our ability to adapt to the environment. Inefficient metabolism in this part of the brain would be expected to cause untoward emotional reactions and autonomic/endocrine signals that might justifiably be defined as "maladaptive". Though efforts are often made by parents now to provide a nutritious diet, the devastation usually comes from the ad lib ingestion of "junk" (empty calories) that the children get between meals. Thiamin deficiency is relative to the increased calorie content from simple carbohydrates and the reason for referring to this disastrous menace as "high calorie malnutrition". The peer pressure is enormous and the temptations provided by a greedy food industry are all too obvious in the shape of advertising and promotion. Many citizens in the U.S. are deeply concerned with some of the mysteries of behavior that beset the generation of children and adolescents. Vandalism is irrational and is extraordinarily common. Destructive juvenile behavior is written off without considering the underlying mechanisms of anger. Much has been published in regard to the connection between ingestion of "junk" food and juvenile crime (Schauss 1981; Lonsdale 1992a, 1994) It is an unacceptable societal factor at present because it offends the concept that crime is always a cold blooded voluntary act.

A case report (Lonsdale 1992b) dealt in detail with a crime in a young African American who had a crime-free background. The Public Defender felt that the incident was brought about by biochemical changes in the brain of this young man. His diet was appalling, involving a huge amount of carbohydrate and alcohol. Biochemical studies revealed profound changes that included an increase in urinary catecholamines. In thiamin deficient rats, catecholamine contents in the cerebral cortex, in the atria and ventricles of the heart and in the spleen were significantly increased compared with those of control animals. There was a significant reduction of noradrenaline metabolism in these tissues. Monoamine oxidase activities were depressed in tissues in which the catecholamine content had been found raised during thiamin deficiency. A close relationship exists between possible changes in enzyme mechanisms for destroying catecholamines and thiamin deficiency (Iwata et al. 1969a, b).

The possibility of thiamin deficiency secondary to diuretic therapy has received only scant attention. A study showing furosemide induced thiamin deficiency in rats (Yui et al. 1978) prompted a clinical study (Seligman et al. 1991) to ascertain whether this commonly used diuretic would induce thiamin deficiency in 23 furosemide treated patients with congestive heart failure. The authors reported a high TPPE indicating deficient thiamin status in 21 of the patients as compared with only two control patients. Thiamin excretion in urine was higher than in controls. Clinical improvement and normalization of TPPE occurred with thiamin supplementation. The metabolic and hemodynamic features of the failing heart can be further aggravated by thiamin deficiency and the authors noted that it was thus possible that in patients with advanced congestive heart failure, particularly those receiving long-term furosemide therapy, thiamin deficiency is one determinant of their poor clinical status

11.11 Thiamin Dependency

When a cofactor such as thiamin is required in huge doses in order to produce enzymatic function it is termed dependency. It is clinically deceptive because, even if the symptoms are regarded as related to vitamin cofactor action, a physiologic dose of the "missing" vitamin might be prescribed on the basic understanding that it would be curative. When no clinical response is observed, the therapist might conclude that the concept was wrong and the vitamin discontinued.

A 6 year old child with intermittent episodes of cerebellar ataxia proved to be the first case of thiamin dependent defective pyruvic dehydrogenase activity (Lonsdale et al. 1969; Blass 1972). Although thiamin responsiveness could not be demonstrated in vitro, the child's episodes were prevented by the ingestion of 600 mg of thiamin a day. If he should succumb to a simple infection the daily dose of thiamin would have to be doubled. At least one intermittent metabolic disorder has been reported where a "stress factor" such as infection could initiate the disease (Dancis et al. 1967) and thiamin plays a role in the decarboxylation of branched chain amino acids (Elsas and Danner 1982).

Some important facts need to be mentioned in regard to this thiamin dependent child who actually imitated a classic example of childhood beriberi. Episodes of ataxia were almost always initiated by a simple infection. Other "stress factors" included head injury and inoculation. Even an acute ambient temperature change could trigger symptoms. On one occasion he entered an air conditioned store from a 90° ambient temperature and succumbed to asthma, something that he had never experienced before. On yet another occasion the air conditioning in the car was turned on and he succumbed again. The inevitable conclusion was that some form of "stress" would initiate a response where metabolic efficiency was lacking. This is a concept that might apply to many different conditions where metabolic efficiency is marginal, as for example, with excess empty calories in the diet.

One of these ataxic episodes was studied in detail. Without any treatment, it resolved spontaneously over about a 10 day period, reaching a clinical climax at about the 6th day. Daytime urinary concentrations of alanine and pyruvate were much higher by day than night throughout the episode and were inversely proportional to urinary concentrations of aspartate and glutamate. This led to two conclusions: first that the mechanism was related to circadian rhythm and secondly that this represented an abnormal redox where pyruvate was being reversely transaminated to the corresponding keto acids, siphoning off glutamate and aspartate from brain function. With thiamin treatment his optic atrophy and learning disability gradually improved. Although his younger brother had the same defect, he did not experience ataxia. As an adolescent, however, he had a relatively mild head injury that rendered him unconscious. He was taken to an emergency room where it was virtually impossible to get the physician to understand the nature of the underlying risk and his requirement for an injection of thiamin.

11.12 Thiamin Transporter Disease

The SLC19 gene family of solute carriers have been described (Ganapathy et al. 2004; Nabokina and Said 2004), opening up a new aspect of thiamin metabolism. It had long been noted that megaloblastic anemia could occur that was treatable by the use of supplementary thiamin (Rogers et al. 1969). The discovery of the transporters made the etiology much clearer. Thiamin responsive megaloblastic anemia (TRMA) syndrome is now known to be an autosomal recessive disorder with features that include this form of anemia, mild thrombocytopenia, leucopenia, sensorineural deafness and diabetes mellitus (Ozdemir et al. 2002). Mutations in the SLC19A2 gene encoding a high-affinity thiamin transporter protein THTR-1 are responsible for the clinical features of this syndrome (Lagarde et al. 2004).

Three TRMA patients have been reported with heart rhythm abnormalities and structural cardiac anomalies (Lorber et al. 2003). This suggested to the authors that knowledge of the TRMA syndrome might shed light on more commonly observed cardiac disease where an etiology is unclear. These authors also agreed that the harmless nature of thiamin administration might even suggest that the vitamin could be used empirically for routine supplementation in cardiac failure where etiology is unknown.

A mouse model has been generated lacking functional SLC19A2. The mouse was unexpectedly found to have a male-specific sterility phenotype. Spermatogenic failure was reversed by injections of large doses of thiamin, giving a new insight into the possible association of thiamin responsive human infertility (Oishi et al. 2004).

11.13 Thiaminase

In 1941 Fujita (Fujita 1941) was engaged in determining the vitamin content of Japanese foodstuffs. The thiamin content of some shellfish and crustacea was zero. When thiamin was added, it could not be recovered because it was being destroyed by an enzyme that occurred in the tissues. He named the enzyme aneurinase. Subsequently he was able to identify two enzymes that destroyed thiamin and might be relevant to human disease. Thiaminase I (EC2.5.1.2) is produced by Clostridium thiaminolyticum, an anerobic organism found in human small intestine and by Bacillus thiaminolyticus, an aerobic organism that occurs in the colon. This enzyme splits the pyrimidine ring of thiamin from the thiazolium ring at the methylene bridge and adds a base compound to the pyridinium to create an analogue inhibitor of thiamin in its metabolic role. Bacillus aneurinolyticus is aerobic and is found in human colon. It produces thiaminase II (EC 3.5.99.2) that functions in the same way as thiaminase I but does not add a base compound to the pyrimidine ring.

A chapter on thiaminase (Murata 1965) can be found in a book that was written by the Vitamin B Research Committee of Japan. After the discovery of the two enzymes it was to be expected that there would be studies performed to ascertain whether they were relevant in human disease. A case of "thiaminase disease" was reported. The patient had beriberi and potent thiaminase activity was found in the feces. In another patient with "thiaminase disease" a bacterium was found subsequently to be responsible. Mentioned by Murata in this chapter, an anaerobic organism was found in human feces and named Clostridium thiaminolyticum. It was also reported that a strain of Candida isolated from the oral cavity of a human subject produced thiaminase II.

Thiaminase I was found in the ruminal contents of animals affected by cerebrocortical necrosis (CCN) (Edwin and Jackman 1970). When this enzyme is purified it does not have thiamin splitting activity without the presence of a cofactor. When these investigators dialysed CCN rumen liquor against phosphate buffer, all thiaminase activity was lost. It could be restored by the addition of various cofactors, the most potent being nicotinic acid. The subsequent metabolite, pyrimidinyl-nicotinic acid, was isolated. They concluded that this might have antithiamin properties since it has elements of similarity to pyrithiamine and amprolium, both known to be thiamin antagonists (McCandless 1982; Brin 1964)). The disease could be produced experimentally in calves by administration of amprolium (Markson et al. 1974). All 60 strains of Cl sporogenes subsequently examined (Shreeve and Edwin 1974) produced thiaminase. This organism has been recognized under various names since it was first described by Metchnikoff in 1908 and is commonly found in soil and in intestinal contents of man and animals. Shreve and Edwin noted in their paper that Clostridium thiaminolyticum, isolated from human feces was later considered to be a member of the Cl sporogenes species.

An epidemic of seasonal ataxia was reported in Western Nigeria (Adamolekum and Ndububa 1994). This disease was later found to be due to activity of thiaminase I extracted from the pupae of an African silkworm that is consumed as a source of protein (Nishimune et al. 2000). The significance of these enzymes in human disease is unclear and little or no research has been performed on the subject since this early work was reported. A letter to the editor (Loew 1974) suggested that CCN might be a model for the experimental study of human disease, in particular Wernicke's encephalopathy and Leigh's subacute necrotizing encephalomyelopathy, both of which are etiologically associated with aspects of thiamin metabolism.

11.14 Thiamin Derivatives

A notable advance in research of thiamin metabolism in Japan was the discovery of allithiamin (Fujiwara 1965). Allicin, the compound that gives garlic its characteristic odor, is produced from alliin by the action of alliinase during the grinding of fresh garlic bulbs and conjugates with thiamin in an alkaline medium to form allithiamin (2′-methyl-4′amino-pyrimidyl-(5′)methylformamino-5-hydroxy-2-pentenyl-(S) allyl disulfide). It had been found that the aqueous solution thus formed lost the characteristic thiochrome reaction given by thiamin but exerted full thiamin activity when administered to animals. The addition of cysteine restored the thiachrome reaction, thus demonstrating that the molecule was reduced to thiamin. These disulfide derivatives were tested extensively for their chemical and biological properties by the Vitamin B Research Committee of Japan. Their then unknown characteristics were found to be quite different from thiamin. Propyl-allithiamin (TPD) was said to have the most satisfactory actions. This compound, however, produced a profound garlic odor from both animals and human subjects. This later led to synthesis of the tetrahydrofurfuryl disulfide (TTFD) (Fig. 11.1) which did not produce the garlic odor. It is now marketed in Japan as a prescription item named Alinamin F (odorless). The biololgical properties of TPD and TTFD are otherwise the same.

When allithiamin is administered by mouth, urinary excretion occurs exclusively in the form of free thiamin, not in the form of allithiamin. This is because it is easily reduced to thiamin by cysteine or glutathione.

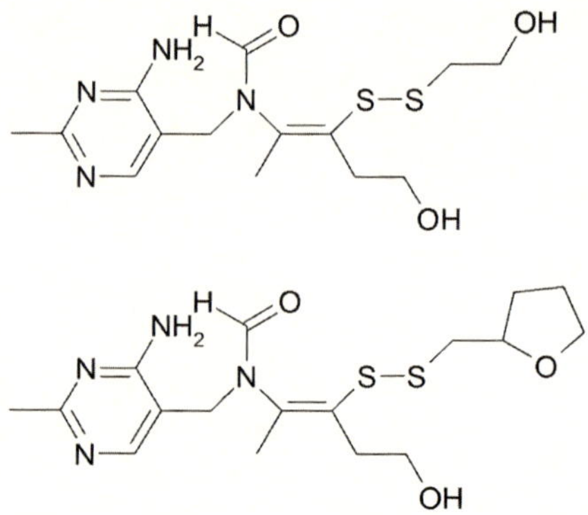

Fig. 11.1 Synthetic allithiamin homologs

Allithiamin + L-cysteine ◊ thiamin + S-allymercapto-L- cysteine

The concentration of excreted urinary thiamin is much greater than occurs with an equivalent dose of a water soluble thiamin salt such as thiamin hydrochloride. This is because its absorption from the intestine is far superior to that of the thiamin salt. After parenteral injection of TPD into healthy male subjects, blood samples showed much greater concentration of thiamin than an equivalent dose of a thiamin salt By the use of 35S-labled TPD, it was shown that the mercaptan fragment remained in the plasma whereas the thiamin segment entered the blood cells. Thus, the open thiazolium ring in allithiamin derivatives closes to form an intact molecule of intracellular thiamin where its vitamin activity is required. When mice were pretreated with TPD in a dose of 1 mg. they found that they were partially protected from potassium cyanide poisoning, given in a dose of 150 μg per 10 g body weight. The death rate in pretreated mice occurred in only 11.7% as compared with 70.6% in those that were not pretreated. It was also found that TPD was effective in prevention of trichloroethylene or lead intoxication in rabbits. It also resulted in striking reduction in liver damage after administration of carbon tetrachloride.

An important experiment was reported in dogs. A segment of jejunum was disconnected from its mesenteric innervation. When TPD was applied topically or given intravenously a marked stimulatory effect on peristalsis was observed. This effect did not occur at all with a thiamin salt. Body weight curves of albino rats were far superior when given TPD than with thiamin hydrochloride. A poorly known therapeutic use of TTFD may well be in its anti-inflammatory effect. Intraperitoneally injected TTFD and TPD showed a strong inhibitory effect in carrageenin induced rat paw edema (Kitzushima 1967). This effect was supported by finding that TTFD reversed the gradual increase in coronary blood flow in the heart-lung preparation of a dog by inhibiting the arachidonic acid cascade activation (Matsui et al. 1985).

All the allithiamin homologues are disulfide derivatives and their easy reduction and consequent absorption of intracellular thiamin is dependent on this. A large number of S-acyl derivatives were synthesized and the Vitamin B Research Committee selected three of them for detailed study (Fig. 11.2).

Figures 11.1 and 11.2 reproduced with permission from Lonsdale D. A review of the biochemistry and clinical benefits of thiamin(e) and its derivatives. eCAM 2006;3(1):49–59.

All of the disulfide (S-S) and S-acyl (S-ac) derivatives were shown to be absorbed from the intestine, at a rate comparable to TPD, far more readily than a thiamin salt. It was found, however, that the S-ac forms were not absorbed into blood cells any better than thiamin hydrochloride. Neither did they have the preventive effect against the various toxins that TPD pretreated animals experienced. They ascribed this major difference to the absence of the S-S bond in S-ac thiamins. These, unlike the disulfide compounds, require enzymatic action in liver or kidney to be reduced to thiamine. An important aspect of the metabolism of these derivatives is what happens to the fragment left outside the cell on hydrolysis. The mercaptan from TTFD has been well studied (Kikuchi et al. 1970; Fujita et al. 1973a, b). Its pharmaceutical effect, if any, is unknown. It was shown to be non toxic and it breaks down through a series of enzymatic actions to sulfates and sulfones when it is excreted in urine. The fragment from the S-ac derivatives has not been studied to my knowledge.

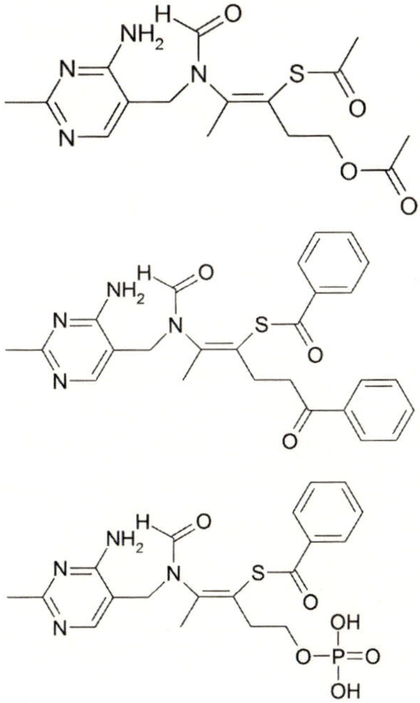

Fig. 11.2 S-acyl-thiamin derivatives

11.15 Clinical Studies

Over the years we have found some unusual aspects of thiamin metabolism in disease. To illustrate the fallacy of relying on a repetitive symptomatology in alerting a physician to the possibility of thiamin deficiency/dependency, studies on two children with recurrent febrile lymphadenopathy were published (Lonsdale 1980). One child had a normal transketolase but was found to have the TTP inhibitor substance in urine (Cooper et al. 1969). His recurrent episodes of high fever and massively enlarged cervical glands had been treated for 2 years with antibiotics on the assumption that they were due to recurrent infection. They ceased with thiamin supplementation and without antibiotic use. To complicate the involved biochemistry, serum folate and B12 were both markedly increased, but fell into the normal range after thiamin was started. The folate and B12 increased and symptoms returned when thiamin was temporarily discontinued. Restoration of thiamin again abolished symptoms: serum folate and B12 concentrations returned again to normal. Approximately a year later his symptoms returned but vanished again when a multivitamin was added as a supplement. The other child had an uncomplicated abnormal transketolase and responded to thiamin supplementation.

Some form of thiamin dependency was depicted in an infant that had the potential clinical hallmarks of a chromosomal abnormality, studied at the age of 6 weeks. Chromosomal testing was normal but the infant had an abnormal transketolase. For this reason, thiamin supplementation was commenced and she began to flourish. Her mother, who has subsequently been found to have symptoms associated with abnormal transketolase, found that her daughter needed more and more thiamin to treat recurrent symptoms as she grew. The dose, given without medical advice by the mother, gradually rose to as high as 7 g a day. Though she was never completely normal in behavior and learning capacity, she was able to graduate from high school where she played an instrument in the marching band. She died at the age of 27 years from what was reported as toxic shock syndrome after an infection.

Another remarkable experience was with a child with recurrent life-threatening croup. His three siblings had had a similar history for the first 4 years of their lives. An abnormal transketolase suggested thiamin treatment and the recurrent croup ceased. One year later, the mother asked whether thiamin supplementation should be continued. It was discontinued on the assumption that he had "outgrown" the problem. Within 3 weeks he required admission to hospital with another attack of croup and thiamin was restored. It could only be assumed that the pathology might have involved the recurrent laryngeal nerve (Lonsdale Unpublished observations)

A family was reported where sleep apnea occurred in three of six siblings between the ages of 18 and 26 months. Twin females first had irregular respiration and episodes of apnea. One twin succumbed to an apneic episode while asleep and the diagnosis of sleep apnea was made in the other twin at a sleep clinic. She also died in an apneic episode 3 months later. A male sibling also died in a similar fashion. The investigators also reported a child with sleep apnea at the age of 7 weeks who died at 31 months. Lesions were confined to the respiratory centers of the lower brain stem. The connection with thiamin was underlined by the discovery of the TTP inhibitor substance (Cooper et al. 1969) in one of these children and other family members (Adickes et al. 1986). Although this cannot be stated as proven, it raises the very important question of the association of thiamin metabolism in Sudden Infant Death Syndrome (SIDS). As long ago as 1944, it was pointed out that sudden death, with a peak incidence at about 4 months of age, the same time frame as is usual in modern SIDS, occurred as a result of thiamin deficiency in breast fed infants of Chinese mothers in Hong Kong (Fehily 1944). Although the etiology in modern SIDS is still argued, it should be possible to see it in terms of the combination of genetics (or epigenetics), poor nutrition and an incidental stress factor (Lonsdale 2001). Certainly thiamin metabolism has been very much underlined but largely ignored as a simple and effective therapeutic intervention in this tragic and preventable event (Jeffrey et al. 1985; Lonsdale and Mercer 1972; Lonsdale 1977, 1990; Lonsdale et al. 1979, 1982). The association of TTP with SIDS was supported by the finding of its deficiency in the phrenic nerve of autopsied SIDS victims (Barker and Jordan 1982). Magnesium deficiency has been reported (Caddell 1972) in SIDS and the strong association with thiamin metabolism already mentioned would make this a logical relationship that should be emphasized strongly.

Four case reports of encephalopathy were published where thiamine metabolism appeared to be involved in the etiology (Lonsdale and Price 1973). All of them had elevation of lactic and pyruvic acids in blood or urine and three of them had TPP inhibitor substance in urine as described by Cooper and Pincus. None of the three had Leigh's disease. One of the children described in this publication was of more than passing interest since it implies that an adequate knowledge of glucose metabolism is often required in perceiving this kind of biochemical lesion. This girl had severe psychomotor retardation and failure to thrive. This followed from a low Apgar score at birth, failure of the suck reflex, cyanosis and vomiting in the neonatal and infancy era. At the age of 9 months she began self mutilation of the lower lip that produced severe damage requiring removal of the upper incisors and transplant of the parotid duct because of excessive drooling. Although she had increased urinary excretion of uric acid and at least one episode of uric acidemia, the genetic defect in Lesch Nyhan syndrome, where compulsive lip biting is characteristic (Lesch and Nyhan 1964), is sex-linked recessive and the defective enzyme, hypoxanthine-guanine phosphoribosyltransferase responsible for this disease was found to be normal in the child's erythrocytes. The fact that she had an increase in pyruvic acid in blood and urine suggested obstruction of pyruvate into the citric acid cycle, perhaps with an overflow of glucose through the pentose pathway, resulting in increased production of purines and uric acid. She was treated empirically with thiamine for this reason. Five months later, the urinary pyruvic acid concentration had decreased significantly and there was definite clinical improvement evidenced by increased activity, interest in her surroundings and alertness. Granted that this did not constitute a solid proof, but it must be regarded as a challenge to others to explore the biochemical etiology of such a disaster as our knowledge increases. Since the body is a biochemical "machine", it is the "biochemical lesion" rather than the symptomatology that must be plumbed for real progress in the treatment of disease.

Branched chain ketoaciduria (Maple syrup urine disease) has been reported due to thiamine dependency, as already mentioned above. The biochemical abnormalities were reversed by administration of pharmacological doses of thiamine (Scriver et al. 1971; Duran et al. 1978). This disease has been reported to occur intermittently (Dancis et al. 1967). Clinical episodes occurred in association with infections in the same way as the intermittent cerebellar ataxia occurred in the case reported by Lonsdale and associates (Lonsdale et al. 1969). The infection is deemed to be an attack that results in mobilization of the metabolic response and is regarded as a form of "stress". This results in an exposure of the underlying defect that is adequate when the subject is not under attack. The concept of "stress" would apply to other forms of attack and is considered to include head injury and even inoculation as described in the episodes of cerebellar ataxia reported by Lonsdale et al. This concept was introduced earlier in the chapter when the first symptoms of beriberi occurred in Chinese workers when they were suddenly exposed to sunlight. As already mentioned, it also strongly suggests that the disaster of SIDS is brought about by the combination of a "stress" event (e.g. a cold), immature brainstem control of vital function and a deficiency of an equally important nutrient such as thiamine or magnesium (Lonsdale 2001). It has been hypothesized that this concept might be studied further by the use of Boolean algebra that involves the interlocking of circles and published as the Three Circles of Health (Lonsdale 1994). The three circles are entitled Stress, Genetics and Nutrition.

11.16 Clinical Use of Thiamine Tetrahydrofurfuryl Disulfide (TTFD)

I have had an Independent Investigator License for the clinical use of TTFD since 1973. Consequently, I have been able to treat hundreds of patients with it and have reported these activities in a number of reviews and case reports, summarized recently (Lonsdale 2006). Its therapeutic benefits are not well known, at least in the U.S.A. where it is not approved by the FDA. Details of an unusual case of cardiac beriberi were published in a book that is now out of print but available on line (Lonsdale 1987, 2006). The boy in question had not been normal since birth and was mildly mentally retarded. The reason for recurrent episodes of gastro enteritis was never solved, but had always been treated as infections. After one of these episodes the parents of the then 13-year old boy took him up into mountainous terrain for a picnic. While sitting in the sun he developed hemiparesis that recovered spontaneously as he was being driven to the nearest hospital. It is to be noted that this onset of symptoms was similar to the onset of first symptoms of beriberi in Chinese factory workers when exposed to sunshine, as discussed previously. The major discovery on arrival at the hospital was a grossly enlarged heart and he was transferred to our institution where his condition was recognized as beriberi heart disease. He responded dramatically to TTFD for a while but gradually became resistant to its effects and died. Unfortunately, the parents were adamant in denying autopsy studies and the basic lesion was never defined, for it was clear that this was not due to simple dietary thiamin deficiency. In the light of more modern knowledge it would be possible to postulate an abnormality in thiamin metabolism such as transporter disease.

Some years ago, there were almost annual epidemics of Reye's syndrome, eventually recognized as severe repercussions from the use of aspirin given to children with one of the many different viral diseases. One child that came in to our institution was an 18-month old girl easily recognized as an example of this syndrome. The treatment given at that time failed and she was comatose and surviving with the application of life sustaining machinery. It was agreed by all that she had little hope of survival. Very large doses of TTFD were given both by mouth and intravenously. The first sign of recovery was reappearance of lip vermilion and pink cheeks. As treatment continued she passed through the stage of the peculiar state of partial consciousness known as coma vigilum. In this state she accepted food and followed the faces of attending personnel with her eyes. Consciousness gradually returned and she was able to walk out of the hospital. There was unfortunately not enough factual evidence for publication of this case (Lonsdale Unpublished observations). Since Reye's syndrome produced its lethal effects by attacking brainstem controls, it can be easily suggested that the recovery in this child was due to restoration of oxidative metabolism in this crucial part of the brain on which depends the survival of each one of us.

It was consideration of the role of thiamin in the brainstem and limbic system that led to its experimental use in the treatment of threatened SIDS (Lonsdale 1977). My colleagues and I were able to show that brainstem auditory evoked potentials (BAEP) were abnormal in many children who presented with symptoms of threatened SIDS (Lonsdale et al. 1979). The concept of "threatened" in this syndrome was, for a long time, unaccepted, since the symptomatology is vague and non specific and the syndrome was regarded as being impossible to forecast. Nevertheless, we encountered a number of infants whose history of life threatening apnea described by the parents had been sometimes regarded in emergency rooms as parental over-reaction when the infant was found to be normal on examination. We described the cases of four infants whose history of near-death episodes ceased and clinical recovery occurred after the administration of TTFD (Lonsdale et al. 1982). In one patient the BAEP was found to be extremely abnormal but partially corrected after administration of intravenous TTFD and went on to a more complete recovery with orally administered TTFD. Although her subsequent life was associated with mild mental problems, she is now a married woman. Of particular interest in this discussion, her mother had consumed a great deal of cola drinks during her pregnancy.

The chromosomal etiology in Down's syndrome is well known as trisomy 21. The biochemical effects are less well known but are brought about by the overexpression of genes occurring on chromosome 21. A clinical study in a group of randomly chosen children with Down's syndrome showed modest improvement in the measured IQ of some of them (Lonsdale and Kissling 1987). A current epidemic of infantile autism is being encountered and our own experience has shown that about 10–20% of these children have an abnormal TPPE, thought to be a significant marker for oxidative stress (Gibson and Zhang 2002). It prompted a pilot study in the treatment of 10 of these children with TTFD (Lonsdale et al. 2002). Because the enteric coated tablets made by Takeda Chemical Industries were not available, the powdered form of TTFD was found to be completely unpalatable and rejected by anyone who tried tasting it, no matter what kind of taste disguise was used. It was therefore administered in the form of rectal suppositories. Using computer-read special symptom forms prepared and tested by the Autism Research Institute in San Diego the symptom scores improved in 8 of the 10 children. This would certainly be a good reason to perform a well designed complete study.

It has now been possible to create a TTFD containing cream that can be applied to the skin, obviating the oral route completely. A by-product of its use is that the patient often will produce a body odor that is constantly likened to the characteristic odor of skunk secretion. The secretion of skunks is known to be due to a mixture of organic mercaptans and the prosthetic group derived from the reduction of TTFD is a mercaptan. By simple trial it has been found that the administration of 10 mg of biotin given each day by mouth will often reduce or remove the odor. Another agent that reduces this odor frequently, discovered by the mother of an autistic child, is the application of fresh lemon juice to the skin. Although scientific study of the mechanism of action has not been performed, it has become clear that this body odor from the patient gradually decreases and may disappear altogether as clinical improvement is observed and the biotin/lemon juice combination is no longer necessary, even though treatment with TTFD may continue. It has been known for some time that thiamin removes lead from experimentally induced lead toxicity in animal studies (Olkowski et al. 1991). Our study of TTFD in autistic children revealed that treatment was associated with increases in urinary lead, arsenic, cadmium and mercury, the SH-reactive heavy metals. This strongly suggests that it is the thiol nature of TTFD that gives it its chelating action of these metals.

An open trial with TTFD was performed on 44 patients with polyneuropathy. Thirty-four patients showed improvement of motor function and some restoration of sensory function. Of 18 patients re-examined electrophysiologically 3 months later, 6 showed remarkable improvement. There were no side effects noted (Djoenaidi and Notermans 1990) The same authors reported three patients with beriberi who presented with different clinical manifestations. The cardiac symptoms responded dramatically to TTFD and there were some improvements in their polyneuropathy (Djoenaidi et al. 1992). In view of the harmless nature of the treatment, these authors also suggested the routine administration of TTFD should be given to all patients in whom heart failure is present without clear evidence of cause. In a 12 week open trial in Alzheimer's disease TTFD at an oral dose of 100 mg/day resulted in a mild beneficial effect on emotional symptoms and intellectual function (Mimori et al. 1996)

A test known as the intravenous olfaction test with TPD is a simple procedure widely used in Japan (Harada et al. 2002). An olfactory stimulus is provided by intravenous injection of TPD. The subject smells n-propyl mercaptan, the prosthetic group derived from hydrolysis. This is discharged from blood into the alveoli and expired. By electroencephalography the authors found alpha-2 and beta-2 waves to be activated over the frontal and temporal regions during the olfactory stimulation. The EEGs returned to pre-stimulus levels after disappearance of the olfactory sensation.

The S acyl derivative known as benzoyl thiamine monophosphate (BTMP), also known as Benfotiamine has recently received attention and is reported to have clinical benefit in the complications of diabetes (Bitsch et al. 1991, Thornalley 2003; Beltramo et al. 2004; Haupt et al. 2005). Alteration of thiamine pharmakoinetics by end stage renal disease improved with administration of BTMP, suggesting that this derivative might be beneficial in end stage renal disease (Frank et al. 1999). It was reported that an open thiazole ring thiol form of thiamine (unspecified) released nitric oxide from S-nitrosoglutathione (Stepuro et al. 2005). This might explain the reported benefit of thiamine derivatives in the microangiopathy of diabetes and certainly deserves further research and clinical trials.

11.17 Conclusion and Hypothesis

The use of thiamin, and its disulfide derivatives in particular, is much neglected in Western medicine. It was one of the earliest vitamins to be discovered and synthesized and it is surprising that it has not been featured more in clinical reports and reviews involving nutritional therapy. The nutritional diseases have been long associated with poverty and starvation. Starvation, however, represents loss of both calorie yielding food components as well as the non-caloric nutrients, usually a slow attrition to death. If calories are maintained without appropriate vitamin and mineral content (empty calories), the outcome is quite different. Beriberi is the classic form of empty calorie disease since it is strongly associated with white rice in Eastern cultures. Modern Western diets, particularly in children and adolescents are loaded with simple carbohydrates, placing a metabolic strain on the mechanisms of oxidative metabolism. The disease outcome is therefore often seen as marginal beriberi.

My experience of more than 30 years of nutritional therapy in the U.S.A. has clearly shown that malnutrition is extremely widespread. Thiamin figures high on the list of nutrient deficiencies, although it can also be said that is more an excess of simple carbohydrates that overwhelms the oxidative system rather than a strictly defined vitamin deficiency. Because of vitamin fortification of processed foods and the relative affluence of our present culture, we are not ready to consider that obscure symptoms, particularly those that are generally termed functional, are of dietary origin. They are frequently the direct result of years of dietary abuse and the subsequent deterioration in enzyme action is not easily repaired. Physiological doses have no effect since the enzyme/cofactor bonding appears to be damaged or partially atrophied. Thus, physicians often become disenchanted when they suspect vitamin deficiency, treat a patient with low-dose supplementation and see no benefit. The RDA of thiamin is 1–1.5 mg per 1,000 kcal but only in a biochemically healthy individual. Like the RDA of vitamin C is the minimal amount to prevent scurvy, so is the RDA of thiamin the minimal to prevent beriberi. As was pointed out in discussion of beriberi, occurring in epidemic form, it took months of high dose thiamin to reverse the symptoms and sometimes it was irreversible.

One important aspect of this kind of malnutrition is the effect that it has on the emotional characteristics of behavior. Anger is easily invoked and temper tantrums may be much more violent and persistent in children. The hypothesis is that the limbic system of the brain becomes much more responsive to incoming stimuli when its oxidative metabolism becomes inefficient. Primitive reflex activity in this system might then occur under stress when the supervisory action of the cognitive brain is overwhelmed. Juvenile crime has been linked with high calorie malnutrition (Schauss 1981; Gray 1987), and the role of thiamin deficiency accentuated (Lonsdale 1992a, 1994) in view of the nature of dietary excesses as considered earlier in this chapter.

Legal circles have addressed this association in only a very limited manner because our concept of criminal behavior is that it is purely volitional and performed in cold blood. It is entirely possible, however, that a perceived anger might give rise to irrational behavior that might explain the presently inexplicable problem of school shootings, for example. To my knowledge no newspaper account has ever included any discussion on diet in any one of these insane incidents since it never enters the consciousness of anyone to make the enquiries. Temporary insanity, at least in the State of Ohio, can be used in court if it can be shown that a criminal "knew what he was doing but was unable to stop himself doing it". The human body is a fuel burning "machine" with the incredible ability to heal itself. Assuming a genetically, or epigenetically, determined cellular "blueprint", all it requires is the appropriate fuel and the catalysts that enable it to go through the complex mechanisms of oxidative metabolism, particularly in the brain. As we move further and further from our biologic origins, so our risks incurred by appalling planet earth stewardship increase accordingly. The aim of the food industry is profit and it appeals to our modern urge to avoid kitchen duty as much as possible by processed convenience foods that provide taste pleasure irrespective of their nutritional quality.

References

Adamolekum B, Ndububa DA (1994) Epidemiology and clinical presentation of a seasonal ataxia in Western Nigeria. J neurol Sci 124(1):95–98

Bannister R (ed) (1984) Autonomic failure: a textbook of clinical disorders of the autonomic nervous system, 2nd edn. Oxford University Press

Barker JN, Jordan F (1982) Phrenic thiamin and neuropathy in sudden infant deaths. In: Sable HZ, Gubler CJ (eds) Thiamin: twenty years of progress. Ann N Y Acad Sci 378: 449–452

Beltramo E, Berrone E, Buttiglieri S, Porta M (2004) Thiamine and benfotiamine prevent increased apoptaosis in endothelial cells and pericytes cultured in high glucose. Diabetes Metab Res Rev 20(4):330–336

Bettendorff L, Michel-Cahay C, Grandfils C, DeRycker C, Schoffeniels E (1987) Thiamine triphosphate and membrane-associated thiamine phosphatases in the electric organ of Electophorus electricus. J Neurochem 49:495–502

Bettendorff L, Kolb HA, Schoffeniels E (1993) Thiamine triphosphate activates anion channels of large unit conduction in neuroblastoma cells. J Membr Biol 136:281–288
Bettendorff L, Hennuy B, De Cherek A, Wins P (1994) Chloride permeability of rat brain vesicles correlates with thiamine trophosphate content. Brain Res 652:157–160
Bhuvaneswaran C, Sreenivaran A (1962) Problems of thiamine deficiency states and their amelioration. Ann N Y Acad Sci 98:576–601
Bitsch R, Wolf M, Moller J et al (1991) Bioavailability assessment of the lipophilic benfotiamine as compared to a water-soluble thiamine derivative. Ann Nutr Metab 35:292–296
Blass J (1972) Abnormalities in pyruvate dehydrogenase and neurologic function. Intern J Neruosci 4:65–69
Blass JP, Gibson GE (1977) Abnormality of a thiamine requiring enzyme in patients with Wernicke-Korsakoff syndrome. N Engl J Med 297:1367–1370
Blass JP, Gleason P, Brush D, DiPonte P, Thaler H (1988) Thiamine and Alzheimer's disease. Arch Neurol 45:833–835
Boni L, Kieckens L, Hendricx A (1980) An evaluation of a modified erythrocyte transketolase assay for assessing thiamine nutritional adequacy. J Nutr Sci Vitaminol 26:507–514
Brin M (1964) The antithiamine effects of amprolium in rats on tissue transketolase activity. Toxicol Appl Pharmacol 6:454–458
Caddell JL (1972) Magnesium deprivation in sudden unexpected infant death. Lancet ii258–262
Cooper JR, Pincus JH (1979) The role of thiamine in nervous tissue. Neurochem Res 4:223–239
Cooper JR, Itokawa Y, Pincus JH (1969) Thiamine triphosphate deficiency in subacute necrotizing encephalomyelopathy. Science 164:72–73
Cooper JR, Pincus JH, Itokawa Y, Piros K (1970) Experiences with phosphoribosyl transferase inhibition in subacute necrotizing encephalomyelopathy. N Engl J Med 283:793–795
Dancis J, Hutzler J, Rokkones T (1967) Intermittent branched chain ketonuria: Variant of maple-syrup-urine disease. New Engl J Med 276:84–89
Djoenaidi W, Notermans SL (1990) Thiamine tetrahydrofurfuryl disulfide in nutritional polyneuropathy. Eur Arch Psychiatry Neurol Sci 239:218–220
Djoenaidi W, Notermans SL, Dunda G (1992) Beriberi caardiomyopathy Eur J Clin Nutr 46:227–234
Duran M, Tielens AGM, Wadman SK et al (1978) Effects of thiamine in a patient with a variant form of branched-chain ketoaciduria. Acta Paediatr Scand 67:367–372
Dyckner T, Elk B, Nyhlin H, Wester PO (1985) Aggravation of thiamine deficiency by magnesium depletion. A case report. Acta Scand 218:129–131
Edwin EE, Jackman R (1970) Thiaminase I in the development of cerebrocortical necrosis in sheep and cattle. Nature 228(5273):772–774
Eisinger J, Bagneres D, Arroyo P, Plantamura A, Ayavou T (1994) Effect of magnesium, high energy phophates, piracetam and thiamin on erythrocyte transketolase. Magnes Res 7:59–61
Elmadfa I, Majchrzak D, Rust P, Genser D (2001) The thiamine status of adult humans depend on carbohydrate intake. Int J Vitam Nutr Res 71:217–221
Elsas LJ, Danner DJ (1982) The role of thiamin in maple syrup urine disease. In: Sable LJ, Gubler CJ (eds) Thiamin: twenty years of progress. Ann N Y Acad Sci 378: 404–420
Fehily L (1944) Human milk intoxication due to B1 avitaminosis. Br Med J 2:590–592
Frank T, Bitsch R, Malwaki J, Stein G (1999) Alteration of thiamine phramacokinetics by end-stage renal disease (ESRD). Int J Clin Pharmacol Ther 37:449–455
Fujita A (1954) Thiaminase. In: Nord FF (ed) Advances in enzymology, vol 15—. Interscience Publishers, pp 389–421
Fujita T, Suzuoki Z (1973a) Enzymatic studies on the metabolism of the tetrahydrofurfuryl mercaptan moiety of thiamine tetrahydrofurfuryl disulfide. I Microsomal S-transmethylase. J Biochem 74:717–722
Fujita T, Suzuoki Z (1973b) Enzymatic studies on the metabolism of the tetrahydrofurfuryl mercaptan moiety of thiamine tetrahydrofurfuryl disulfide. III Oxidative cleavage of the tetrahydrofuran moiety. J Biochem 74:733–738

Fujita T, Suzuoki Z, Kozuka S (1973a) Enzymatic studies on the metabolism of the tetrahydrofurfuryl mercaptan moiety of thiamine tetrahydrofurfuryl disulfide. II Sulfide and sulfoxide oxygenases in microsomes. J Biochem 74:723–732

Fujita T, Teraoka A, Suzuoki Z (1973b) Enzymatic studies on the metabolism of the tetrahydrofurfuryl mercaptan moiety of thiamine tetrahydrofurfuryl disulfide. IV Induction of microsomal S-transmethylase and sulfide and sulfoxide oxygenases in the drug-treated rat. J Biochem 74:739–745

Fujiwara M (1965) Absorption, excretion and fate of thiamine and its derivatives in [the] human body. In: Shimazono N, Katsura E (eds) Thiamine and Beriberi. Igaku Shoin Ltd, Tokyo, pp 179–213

Ganapathy V, Smith SB, Prasad PD (2004) SLC19: the folate/thiamine transporter family. Pflugers Arch 447:641–646

Gibson GE, Barclay L, Blass J (1982) The role of the cholinergic system in thiamin deficiency. In: Sable HZ, Gubler CJ (eds) Thiamin: twenty years of progress. Ann N Y Acad Sci 378: 382–403

Gibson GE, Zhang H (2002) Interactions of oxidative stress with thiamine homeostasis promote neurodegeneration. Neurochem Int 40:493–504

Gray GE (1987) Crime and diet: Is there a relationship? Wld Rev Nutr Diet 49:66–86

Harada H, Rikimaru F, Mori T, Tanaka ME et al (2002) Electroencephalographic changes during intravenous olfactory stimulation in humans. Clin Electroencephalogr 33:189–192

Haupt E, Ledermann H, Kopcke W (2005) Benfotiamine in the treatment of diabetic polyneuropathy—a three–week randomized, controlled pilot study (BEDIP study) Int J Clin Pharmacol Ther 2005;43:71–77. Erratum in: Int J Clin Pharmacol Ther 2005;43:304

Henderson GI, Schenker S (1975) Reversible impairment of cerebral DNA synthesis in thiamine deficiency. J Lab Clin Med 86:77–90

Inouye K, Katsura E (1965) Etiology and pathology of beriberi. In: Thiamine and Beriberi. Igaku Shoin Ltd., Tokyo, pp 1–28

Iwata H, Nishikawa T, Fujimoto S (1969a) Monoamine oxidase activities in tissues of thiamine-deficient rats. J Pharm Pharmac 21:237–240

Iwata H, Watanabe K, Nishikawa T, Ohashi M (1969b) Effects of drugs on behavior, heart rate and catecholamine levels in thiamine-deficient rats. Eur J Pharmacol 6:83–89

Iwata H, Yabushita Y, Doi T, Matsuda T (1985) Synthesis of thiamine triphosphate in rat brain in vivo. Neurochem Res 10:779–787

Jansen BCT, Donath WF (1926) On the isolation of the anti-beriberi vitamin. Proc I Acad Wei Amsterdam 29:1390

Jeffrey HE, McCleary BV, Hensley WJ, Read DJC (1985) Thiamine deficiency- a neglected problem of infants and mothers- possible relationship to sudden infant death syndrome. Aust NZ J Obst Gynaecol 25:198–202

Jeyasingham MD, Pratt O, Burns A, Shaw GK, Thompson A, Marsh A (1987) The activation of red blood cell transketolase in groups of patients especially at risk from thiamin deficiency. Psych Med 117:311–318

Kikuchi S, Nishikawa K, Suzuoki Z (1970) The metabolism of thiamine thetrhydrofurfuryl disulfide in the rat, rabbit and man. Eur J Pharmacol 9:367–373

Kimura M, Itokawa Y (1977) Effects of calcium and magnesium deficiency on thiamine distribution in rat brain and liver. J Neurochem 28:389–393

Lagarde WH, Underwood LE, Moats-Staats BM, Calikoglu AS (2004) Novel mutation in the SLC19A2 gene in an African-American female with thiamine-responsive megaloblastic anemia syndrome. Am J Med Genet 125:299–305

Lesch M, Nyhan WL (1964) A familial disorder of uric acid metabolism and central nervous system function. Am J Med 36:561–570

Loew FM (1974) Possible animal model of subacute necrotizing (Leigh's) encephalomyelopathy J Pediat 85:876–877

Lonsdale D (1975) Thiamine metabolism in disease. Crit Rev Lab Sci 5:289–313

Lonsdale D (1977) Treatment of threatened SIDS with megadose thiamine hydrochloride. Pediat Res 11:379 (Abstr)
Lonsdale D (1978) Wernicke's encephalopathy and hyperalimentation. JAMA 239:1133. (Letter to the editor)
Lonsdale D (1980) Recurrent febrile lymphadenopathy treated with large doses of vitamin B1: Report of two cases. Dev Pharmacol Ther 1:254–264
Lonsdale D (1981) The syndrome of functional dysautonomia. Med Hypoth 7:495–502
Lonsdale D (1982) Effect of thiamine tetrahydrofurfuryl disulfide on audiogenic seizures in DBA/J2 mice. Dev Pharmacol Ther 4:28–36
Lonsdale D (1987a) A nutritionist's guide to the clinical use of vitamin B1. Life Sciences Press, Tacoma, pp. 5–12
Lonsdale D (1987b) Biochemical studies in functional dysautonomia. In: A nutritionist's guide to the clinical use of vitamin B1. Life Sciences Press, Tacoma, WA, pp. 78–115
Lonsdale D (1990a) Hypothesis and case reports: possible thiamin deficiency. J Am Coll Nutr 9:13–17
Lonsdale D (1990b) Asymmetric functional dysautonomia. J Nutr Med 1:59–61
Lonsdale D (1990c) Thiamine deficiency and sudden deaths. Lancet ii:376
Lonsdale D (1992a) Criminal behavior and nutrition. J Adv Med 5(2):494–504
Lonsdale D (1992b) Criminal behavior and nutrition. J Adv Med 5:115–123
Lonsdale D (1994a) Crime and violence: A hypothetical explanation of its relationship with high calorie malnutrition. J Adv Med 7(3):171–180
Lonsdale D (1994b) The three circles of health. In: Why I left orthodox medicine. Hampton Roads Publishing Company, Charlottesville, pp 189–207
Lonsdale D (2001a) Nutritional therapy in children with functional disorders of activity, behavior, attention and learning. Clin Pract Altern Med 2:196–203
Lonsdale D (2001b) Sudden infant death syndrome requires genetic predisposition, some form of stress and marginal malnutrition. Med Hypoth 57:382–386
Lonsdale D. (2006a) http://www.soilandhealth.org/02/0201hyglibcat/0201hyglibcat.html
Lonsdale D (2006b) A review of the biochemistry, metabolism and clinical benefits of thiamin(e) and its derivatives. eCAM 3(1):49–59
Lonsdale D (2006c) Three case reports to illustrate clinical applications in the use of transketolase. eCAM 4(2):247–250
Lonsdale D, Kissling CD (1987) Clinical trials with thiamine tetrahydrofurfuryl disulfide in Down's syndrome. J Orthomolecular Med 1:169–175
Lonsdale D, Mercer RD (1972) Primary hypoventilation syndrome. Lancet ii:487 (Letter to the editor)
Lonsdale D, Price JW (1973) Pyruvic aciduria in the detection of thiamine responsive encephalopathy. Cleve Clin Quart 40:79–88
Lonsdale D, Shamberger RJ (1980) Red cell transketolase as an indicator of nutritional deficiency. Am J Clinc Nutr 33:205–211
Lonsdale D, Faulkner WR, Price JW, Smeby RR (1969) Intermittent cerebellar ataxia associated with hyperpypruvic academia, hyperalaninemia, and hyperalaninuria. Pediatrics 43:1025–1034
Lonsdale D, Nodar RH, Orlowski JP (1979) The effects of thiamine on abnormal brainstem auditory evoked potentials. Cleve Clin Quart 46:83–88
Lonsdale D, Nodar RH, Orlowski JP (1982) Brainstem dysfunction in infants reaponsive to thiamine disulfide; preliminary studies in four patients. Clin EEG 13:82–88
Lonsdale D, Shamberger RJ, Audhya T (2002) Treatment of autistic spectrum children with thiamine tetrahydrofurfuryl disulfide: a pilot study. Neuroendocrinol Lett 23:303–308
Lorber A, Gazit A Z, Khoury A, Schwartz Y, Mandel H (2003) Cardiac manifestations in thiamine-responsive megaloblastic anemia syndrome. Pediatr Cardiol 24:476–481
Makarchikov AE, Lakaye B, Gulyai IE, Czerniecki J, Coumans B, Wins P et al. (2003) Thiamine triphosphatase and thiamine triphophatase activities: from bacteria to mammals. Cell Mol Life Sci 60:1477–1488

Markannen T, Kalliomaki JL (1966) Transketolase activity of blood cells in various clinical conditions. Am J Med Sci 252:564–569
Markson LM, Edwin EE, Lewis G, Richardson C (1974) The production of cerebrocortical necrosis in ruminant calves by the intraruminal administration of amprolium. Br Vet J 130:9–16
Mastrogiacoma P, Bettendorff L, Grisar T, Kish SJ (1996) Brain thiamine, its phosphate esters and its metabolizing enzymes in Alzheimer's disease. Ann Neurol 39:585–591
Matsui K, Hakahara H, Watanabe J, Tamatsu H, Nakayawa M et al (1985) Inhibition by thiamine tetrahydrofurfuryl disulfide (TTFD) of the arachidonic acid cascade-line activation as evidenced in the heart-lung preparation of the dog. Jpn J Pharmacol 39:375–379
McCandless DW (1982) Energy metabolism in the lateral vestibular nucleus in pyrithiamin-induced thiamin deficiency. In: Sable HZ, Guhbler CJ (eds) Thiamin: twenty years of progress. Ann N Y Acad Sci 378: 355–364
McIntyre N, Stanley NN (1971) Cardiac beriberi: two modes of presentation. Br Med J 3:567–569
Meador KJ, Loring D, Nichols M, Zamrini E, Rivner M, Posas H et al (1993a) Preliminary findings of high dose thiamine in dementia of Alzheimer's type. J Geriatr Psychiatry Neurol 6:222–229
Meador KJ, Nichols ME, Franke P, Durkin MW, Oberzin RL, Moore LE et al (1993b) Evidence for a central cholinergic effect of high dose thiamin. Ann Neurol 34:724–726
Mimori Y, Katsuoka H, Nakamura S (1996) Thiamine therapy in Alzheimer's disease. Metab Brain Dis 11(1):89–94
Minz B (1938) Sur la liberation de la vitamin par le tronc isole de nerf pheumogastrique soumis a la excitation electrique. C.R. Soc Biol (Paris) 127:1251–1253
Murata K (1965) Thiaminase. In: Shimazono N, Katsura E (eds) Beriberi and thiamine. Igaku Shoin Ltd., Tokyo, pp 220–254
Nabokina SM, Said HM (2004) Characterization of the 5′-regulating region of the human thiamin transporter SLC19A3: In vitro and in vivo studies. Am J Physiol Gastrointest Liver Physiol 287:G822–G829
Nishimune T, Watanabl Y, Okazaki H, Akai H (2000) Thiamin is decomposed due to Anaphe spp. Entomophagy in seasonal ataxia patients in Nigeria. J Nutr 130:1625–1628
Oishi K, Barchi M, Au AC, Gelb BD, Diaz GA (2004) Male infertility due to germ cell apoptosis in mice lacking the thiamin carrier Tht1. A new insight into the critical role of thiamin in spermatogenesis. Dev Biol 266:299–309
Ozdemir MA, Alcakus M, Kuroglu, Gunes T, Torun YA (2002) TRMA syndrome (thiamine-responsive megaloblastic anemia): a case report and review of the literature. Pediatr Diab 3:205–209
Pang JA, Yardumian A, Davies R, Patterson DL (1986) Shoshin beriberi an underdiagnosed condition? Intensive Care Med 12(5):380–382
Peters RA (1936) The biochemical lesion in vitamin B1 deficiency. Lancet 1:1162–1165
Platt BS (1967) Thiamine deficiency in human beriberi and in Wernicke's encephalopathy. In: Wolstenholme GEW, O'Connor M (eds) Thiamine deficiency. Little Brown and Company, Boston, pp 135–143
Rogers LE, Porter FS, Sidbury JB (1969) Thiamine-responsive megaloblastic anemia. Pediat 74:494–504
Sable HZ, Gubler CJ (eds) (1982) Thiamin: twenty years of progress. Ann NY Acad Sci 378:355–364
Schauss A (1981) Diet, crime and delinquency. Berkeley, CA, pp 1–108
Scriver CR, Mackenzie S, Clow CL, Delvin E (1971) Thiamine-responsive maple-syrup urine disease. Lancet i:310–312
Seligman H, Halkin H, Rauchfleisch S et al (1991) Thiamine deficiency in patients with congestive heart failure receiving long-term furosemide therapy: a pilot study. Am J Med 91:151–155
Shreeve JE, Edwin EE (1974) Thiaminase-producing strains of Cl. Sporogenes associated with outbreaks of cerebrocortical necrosis. Vet Rec 94(15):330
Stepuro AL, Piletskaya TO, Stepura II (2005) Role of thiamine thiol forms in nitric oxide metabolism. Biochemistry (Mosc) 70:339–349

Thornalley PJ (2003) Prevention of incipient diabetic nephropathy by high-dose thiamine and benfotiamine. Diabetes 57:2110–2120

Wells DG, Baylis EM, Holoway L, Marks V (1968) Erythrocyte-transketolase activity in megaloblastic anaemia. Lancet ii:543–545

Williams RR (1938) Chemistry of thiamine (Vitamin B1). JAMA 110:727–731

Williams RD, Mason HL, Marschelle HP, Russell MW (1943) Induced thiamine (Vitamin B1) deficiency in man. Arch Int Med 71:38–53

Yui Y, Fujiwara H, Mitsui H et al (1978) Furosemide-induced thiamine deficiency (abstract). Jpn Circ J 4:744

Zbinden G (1962) Therapeutic use of vitamin B1 in diseases other than beriberi. Ann NY Acad Sci 98:550–561

Chapter 12
Riboflavin in Development and Cell Fate

Hilary J. Powers, B.M. Corfe, and E. Nakano

Abstract Riboflavin (7,8-dimethyl-10-ribitylisoalloxazine; vitamin B2) is a water-soluble vitamin, cofactor derivatives of which (FAD, FMN) act as electron acceptors in the oxidative metabolism of carbohydrate, amino acids and fatty acids and which in the reduced state can donate electrons to complex II of the electron transport chain. This means that riboflavin is essential for energy generation in the aerobic cell, through oxidative phosphorylation. The classic effects of riboflavin deficiency on growth and development have generally been explained in terms of these functions. However, research also suggests that riboflavin may have specific functions associated with cell fate determination, which would have implications for growth and development. In particular, riboflavin depletion interferes with the normal progression of the cell cycle, probably through effects on the expression of regulatory genes, exerted at both the transcriptional and proteomic level.

Keywords Riboflavin · Cell fate · Gene expression · Cell cycle

12.1 Riboflavin Structure, Classical Biochemistry and Conventional Wisdom Regarding Effects of Deficiency

Riboflavin (7,8-dimethyl-10-ribitylisoalloxazine; vitamin B2) is a water-soluble vitamin with a well-established role in redox metabolism. It is obtained from the diet as the 'free' riboflavin and in the forms of flavin mononucleotide (FMN) and flavin adenine dinucleotide (FAD), which participate in numerous redox reactions in intermediary metabolism. This activity is a function of the ability of the isoalloxazine ring of these flavins to lose or gain electrons (Fig. 12.1). Attention has been drawn to the remarkable versatility of flavoproteins, attributed to the wide variation in redox and electronic states in which the isoalloxazine ring of flavins can exist (Miura 2001). FAD and FMN act as electron acceptors in the oxidative metabolism of carbohydrate, amino acids and fatty acids and in the reduced state

H.J. Powers (✉)
Human Nutrition Unit, The School of Medicine, The University of Sheffield, Sheffield S10 2RX, UK
e-mail: h.j.powers@sheff.ac.uk

O. Stanger (ed.), *Water Soluble Vitamins*, Subcellular Biochemistry 56,
DOI 10.1007/978-94-007-2199-9_12,

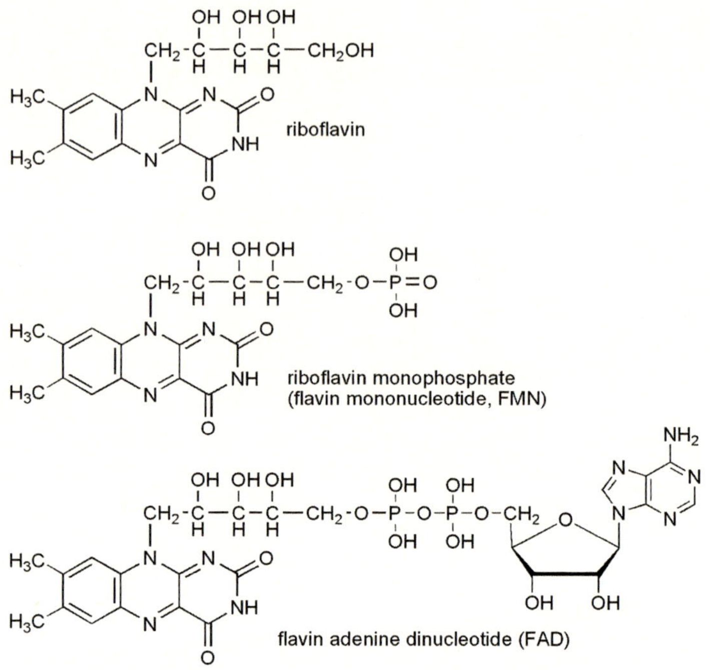

Fig. 12.1 Riboflavin and coenzymes (reprinted from: Human Nutrition, Geissler C and Powers H, eds. P189, 2005, with permission from Elsevier)

can donate electrons to complex II of the electron transport chain. This ensures that dietary riboflavin is essential for energy generation in the aerobic cell, through oxidative phosphorylation, and a lack of riboflavin would therefore be expected to compromise cellular energetics.

12.1.1 What is the Evidence that Riboflavin Deficiency Compromises the Energetics of the Cell?

The importance of riboflavin derivatives to mitochondrial function was demonstrated in the 1970s and 1980s in studies conducted in riboflavin-deficient animals. Mice fed a riboflavin diet from weaning developed morphological abnormalities of hepatic mitochondria, associated with an impaired oxidative metabolism (Hoppel and Tandler 1975). Riboflavin depletion in weanling rats leads to a reduction in the activity of various mitochondrial flavin-dependent enzymes in hepatic tissue, including succinate dehydrogenase and a ketoglutarate dehydrogenase. Riboflavin depletion has the capacity therefore to impact on various different components of the energy-yielding apparatus of the cell and there is evidence of a hierarchic response of the metabolism of an organism to riboflavin depletion. This was first proposed by Prentice and Bates during their work in riboflavin deficient rats

(Prentice and Bates 1981) and has been more fully characterised more recently (Ross and Hansen 1992). The electron transport chain is clearly of high priority to the cell and its activity is preserved at the expense of other flavin-dependent pathways including mitochondrial β-oxidation of fatty acids, although the peroxisomal oxidation of fatty acids seems little affected. Importantly, an early response of animal models to riboflavin depletion is an alteration in the relative activities of the flavin-metabolising enzymes. The fall in flavokinase activity is more pronounced than that of FAD pyrophosphorylase, which ensures that concentrations of FAD are sustained at the expense of FMN (Fass and Rivlin 1969). It is no coincidence that the majority of flavin-dependent enzymes in tissues are FAD-dependent.

Mitochondrial fatty acid oxidation is impaired in riboflavin-deficient rodents through inhibition of FAD – dependent acyl CoA dehydrogenases (Brady and Hoppel 1985; Olpin and Bates 1982). This leads to an accumulation of dicarboxylic acids, which are excreted in the urine (Goodman 1981) and would also explain the altered fatty acid profile in triacylglycerols and phospholipids in riboflavin deficiency (Olpin and Bates 1982). Gene defects leading to flavin-dependent acyl CoA dehydrogenase deficiencies in humans are associated with profound abnormalities of mitochondrial metabolism, characterised by defects in fatty acid ß-oxidation and reduced activity of the electron transport chain (Gianazza et al. 2006). People carrying these gene defects are responsive to high dose riboflavin or of the cofactors FAD or FMN (Antozzi et al. 1994). Genetic acyl CoA dehydrogenase deficiency reported in a human pregnancy, associated with abnormally elevated urinary excretion of riboflavin, led to deaths of the offspring in the neonatal period or early infancy. High riboflavin intake in the third trimester, and in infancy, prevented the syndrome (Harpey et al. 1983).

The mitochondrial oxidation of carboxylic acid intermediates of the tricarboxylic acid cycle and of ketone bodies has also been shown to be impaired in riboflavin-deficient rats (Zaman and Verwilghen 1975). Furthermore, patients with glutaryl CoA dehydrogenase deficiency, an autosomal recessive disorder of mitochondrial amino acid metabolism, have also been shown to be responsive to pharmacological doses of riboflavin (Chalmers et al. 2006).

12.1.2 Implications for Metabolic Rate and Growth

Metabolic rate: As flavoproteins are central to mitochondrial energy-generating pathways, riboflavin status would be expected to be a key determinant of whole body metabolic rate. Oxygen consumption is impaired in hepatic mitochondria of riboflavin-deficient rats (Olpin and Bates 1982) and in mitochondria isolated from brown adipose tissue, which is important for non-shivering thermogenesis (Duerden and Bates 1985). Metabolic rate, measured as basal whole body oxygen consumption, and in response to nor-adrenaline, which is the physiological stimulus to brown adipose tissue thermogenesis, is depressed in riboflavin-deficient suckling rat pups (Patterson and Bates 1989). In contrast, riboflavin depletion in weanling rats elicited an increase in metabolic rate, in association with a marked elevation in

Table 12.1 Effect of riboflavin depletion on weight gain per g food consumed Wistar rats were placed on a riboflavin-deficient diet from weaning and rats receiving a complete diet were weight matched (Williams et al. 1996a)

	Weight gain/g food consumed mean (SEM)	
Days from weaning	Riboflavin-deficient (*n*=6)	Weight-matched (*n*=6)
0	–	–
7	0.26* (0.021)	0.33 (0.018)
14	0.17* (0.021)	0.24 (0.015)
21	0.19* (0.013)	0.25 (0.010)

*Significantly different from weight-matched controls ($P < 0.05$)

mitochondrial GDP binding, which suggests an increase in the thermogenic capacity of brown adipose tissue. The functional effects of riboflavin depletion evidently depend upon the stage of development. Babies born prematurely are at increased risk of riboflavin deficiency but riboflavin supplements given to premature infants did not bring about any change in the rate of 1-13C-octanoic acid metabolism (Patterson et al. 1989).

Growth: Riboflavin depletion in young rats leads to growth impairment, which is partly due to inanition but thought also to be causally related to the role of riboflavin in cellular energetics. When weanling rats are fed a riboflavin deficient diet from weaning and animals receiving a control diet are weight matched to riboflavin deficient animals the latter show a reduced weight gain per gram of diet consumed, indicating an effect on the efficiency of energy production from fuel utilisation (Williams et al. 1996a) (Table 12.1).

Effects of riboflavin deficiency on human growth are less easy to determine, as riboflavin deficiency in isolation is unusual in populations in which riboflavin deficiency is endemic. Riboflavin deficiency in infants and children in developing countries (Bates et al. 1982) is associated with poor growth, which is likely to be a manifestation of numerous chronic nutrient deficiencies including riboflavin (Bates et al. 1993). Nevertheless, a supplement of 0.5 mg riboflavin given to Gambian infants for 12 weeks mitigated the deterioration in growth, measured as length-for-age Z score, seen in infants receiving a placebo (unpublished observations, Fig. 12.2).

12.1.3 Are There Other Reasons Why Riboflavin Deficiency Might Impair Growth?

Effects of riboflavin deficiency on growth are classically related to flavin-dependent redox reactions in the pathways of intermediary metabolism but other mechanisms may also be involved. The oxidative folding of secretory proteins in the endoplasmic reticulum requires flavin-dependent oxidases (Tu et al. 2000). Riboflavin-depletion of Jurkat cells and HepG2 cells in culture is associated with impaired folding and

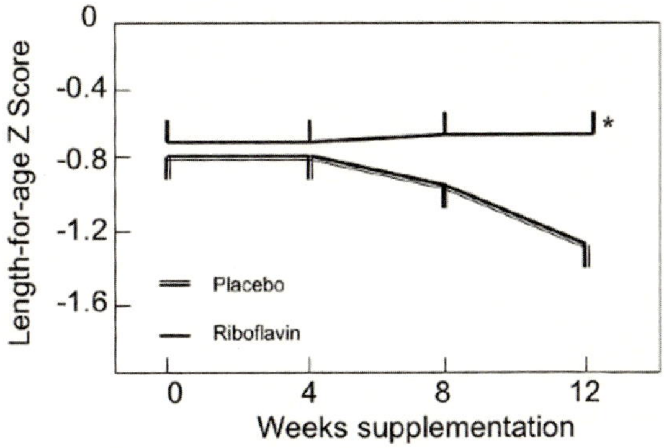

Fig. 12.2 Effect of a daily supplement of 0.5 mg riboflavin on growth in Gambian infants. *Significantly different from placebo ($P < 0.05$)

secretion of proteins and a consequent accumulation of unfolded proteins in the endoplasmic reticulum of these cells. This elicits a stress response (Camporeale and Zempleni 2003; Manthey et al. 2005) known as the Unfolded Protein Response. There are various possible downstream effects of the unfolded protein response, including increased phosphorylation of eukaryotic initiation factor 2α, which leads to an overall suppression of protein synthesis (Pahl 1999).

12.2 Riboflavin in Differentiation and Development

Our understanding of the essentiality of riboflavin for normal fetal development stems from the isolation of a riboflavin carrier protein (RCP) from chicken eggs. RCP is an oestrogen-inducible protein, immunoneutralisation of which leads to fetal resorption in mammalian species (Adiga and Murty 1983). Studies conducted by (Rao et al. 2000) suggested that the fetal resorption observed in immunoneutralised animals was due mainly to abnormal embryo implantation. The requirement for this riboflavin carrier protein during normal fetal development indicated its crucial role in materno-fetal riboflavin transport in avian and mammalian species. Riboflavin carrier proteins have also been measured in amniotic fluid during human pregnancy, (Prasad 1992), as well as in maternal serum (Natraj et al. 1988), and shown to exhibit significant cross-reactivity with chicken riboflavin carrier protein (Visweswariah and Adiga 1987) indicating conservation of the protein during evolution. RCPs are evidently important for embryo growth and development in humans. Riboflavin supply during fetal development seems also to have a rather specific role in skeletal development. There are several reports of a range of skeletal abnormalities in rats and mice born to dams fed a riboflavin deficient diet, including shortening of the digits and defects of the axial skeleton (Noback and Kupperman 1944; Nelson et al. 1956). An interaction has been reported between the occurrence of cleft palate in genetically-susceptible chickens and maternal riboflavin deficiency. Maternal riboflavin deficiency enhanced the effects of the genetic trait and increased the frequency of cleft palate in the offspring (Juriloff and Roberts

1975). Furthermore, the teratogenic effects (including cleft palate) of the phenothiazine derivative T-82 in rats can be significantly mitigated by high dose riboflavin, administered intraperitoneally (Horvath et al. 1976). The human corollary of this is not known although Faron et al. (2001) have suggested that a case of recurrent cleft lip and palate observed in siblings of a patient with malabsorption syndrome may be associated with riboflavin deficiency.

The importance of riboflavin carrier protein(s) to riboflavin transport across cell membranes and intracellular trafficking in humans is not clear. The involvement of a carrier protein in the uptake of riboflavin by cells in culture was first alluded to by Said et al. (1998), who showed a carrier-mediated uptake of riboflavin by HepG2 cells in culture that was upregulated in response to substrate depletion and inhibited by the riboflavin analogues lumichrome and lumiflavin. More recent work reported a receptor-mediated endocytosis pathway for riboflavin uptake into cells (Huang and Swaan 2000; Huang et al. 2003) and this group has further elucidated a role for a riboflavin carrier protein in the uptake and trafficking of riboflavin in mammalian cells, apparently under cAMP regulation (Mason et al. 2006; D'Souza et al. 2006).

12.2.1 Riboflavin Deficiency and Development of the Gastrointestinal Tract

Riboflavin depletion during the weaning process in rats has been shown to have a rapid and reproducible effect on the development of the gastrointestinal tract. Gastrointestinal development is considered to follow a program of structural and functional change from early in fetal development until weaning, and is evidently subject to regulation by nutrient and non-nutrient factors such as hormones and growth factors. Moderate riboflavin depletion from around weaning has a profound impact on the structural and cytokinetic development of the rat small intestine. An early event is crypt hypertrophy, which is associated with increased crypt cellularity (Williams et al. 1996a, b). The rate of crypt bifurcation is also depressed in riboflavin-deficient weanling rats (Yates et al. 2001), which may be a factor in the subsequent failure to increase the size of the villus population. The fact that crypt hypertrophy is not associated with an increase in the size of the proliferative compartment, and neither is it associated with an increased crypt cell proliferation rate, suggests a role for riboflavin in the regulation of apoptosis (Yates et al. 2001). Furthermore, the morphological and cytokinetic effects in the gastrointestinal tract of feeding weanling rats a riboflavin-deficient diet can be reproduced in animals in which the gastrointestinal lumen is deprived of riboflavin whilst systemic riboflavin status is maintained at a normal level through regular intraperitoneal injections of a flavin (Yates et al. 2003). This suggests the presence of an apical receptor/transporter, ligation of which is central to gastrointestinal development. Riboflavin in the lumen may therefore behave as a signalling molecule, and the effects of luminal riboflavin depletion on gastrointestinal development may be partially independent of its role as an essential nutrient (Fig. 12.3).

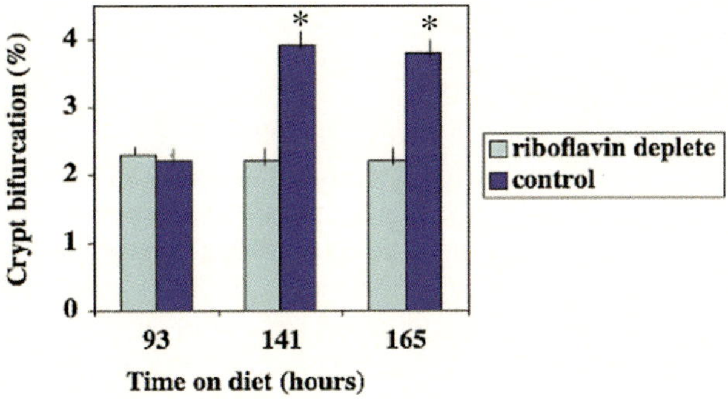

Fig. 12.3 Crypt bifurcation in the duodenum of rats fed a riboflavin deplete diet from weaning or in weight-matched controls fed a complete diet. Values are means with standard errors (Yates et al. 2001). *Significantly different from control (ANOVA and Scheffe test)

Given the evidence that riboflavin deficiency leads to developmental abnormalities during embryogenesis and the specific effects of riboflavin deficiency on cell growth and proliferation during gastrointestinal development, it is hypothesised that riboflavin is a determinant of cell fate, either through direct regulation of proteins of the cell cycle or through effects on biochemical pathways important for cellular proliferation or apoptosis. Such effects would be independent of effects of riboflavin on energy–generating steps of intermediary metabolism.

12.3 Nutrients and Regulation of Proteins of the Cell Cycle

Effects of riboflavin depletion on aspects of growth and development strongly suggest that this vitamin plays an important role in the determination of cell fate. There are plausible mechanisms for such a role. A key apoptosis-inducing protein, (apoptosis-inducing factor-homologous mitochondrion-associated inducer of death, AMID, or p53-responsive gene, PRG3) has 6-hydroxy-FAD as cofactor (Marshall et al. 2005). Accumulation of unfolded proteins and the associated stress response in riboflavin-deficient cells leads to a reduction in the proliferation rate of HepG2 cells, although it is not clear whether the association is causative (Manthey et al. 2005).

The normal cell progresses through a cycle of change comprising four distinct phases. These are, gap 1 (G1), DNA synthesis (S), gap 2 (G2), and mitosis (M). The progression of the cell through these phases is under the regulation of key proteins, the expression of which is influenced by cellular and extracellular factors such as DNA integrity and nutrient availability. Cyclin-dependent kinase complexes regulate the transition of the cell from one phase to another, and these cyclin-CDK complexes are in turn regulated by other gene families including the INK4 and CKI families (Johnson and Walker 1999). The expression of these regulatory proteins can be influenced by nutrients. Retinoic acid is active in regulation of the cell cycle, for example, through effects on the expression of cyclins and CDKs (Alisi et al. 2003). Importantly, the effect of retinoic acid on cell proliferation is specific to the cell type (Chen et al. 2002; Kosaka et al. 2001). During embryogenesis retinoic acid inhibits proliferation of certain cell types, including hepatocytes and smooth muscle cells. In the embryonic mouse retinoic acid upregulates the cyclin-dependent kinase inhibitors, p21 and p27 in primordial endothelial cells, which leads to cell cycle arrest in G1 (Bohnsack and Hirschi 2004). In contrast, retinoic acid stimulates proliferation of other cell types, including cardiomyocytes (Stuckmann et al. 2003). The vitamin D metabolite 1,25(OH)2 has antiproliferative and pro-differentiation effects in different cell types, mediated by increased expression of P21, P27 and other regulatory proteins (Swami et al. 2003). Osteoblasts, haematopoietic cells and keratinocytes all show a depressed proliferation in response to 1,25(OH)2, and an upregulation of a number of cell cycle regulatory proteins, including p21 and p53 (Swami et al. 2003).

12.3.1 Riboflavin and Regulation of the Cell Cycle

Relatively little is known about the role of riboflavin in the regulation of cell proliferation, apoptosis and differentiation. Zempleni's research group have conducted some interesting studies of riboflavin depletion in HepG2 cells that show evidence of an altered cell cycle. They hypothesised that the reported abnormal folding of secretory proteins in riboflavin–deficient cells in culture (Manthey et al. 2005) would lead to a stress response characterised by cell cycle arrest (Werner et al. 2005). HepG2 cells were cultured in RPMI-1640 medium containing either a high concentration of riboflavin (532 nmol/l, control) or 3.1 nmol/l (riboflavin depleted), for 4 days. Media contained 10% dialysed bovine calf serum, which would have contributed little to the total riboflavin content of the medium, although the actual concentration is not reported. Riboflavin depletion was confirmed as a reduction in glutathione reductase activity. Riboflavin-depleted HepG2 cells showed an enhanced binding activity of transcription factors to endoplasmic reticulum stress elements, and an increase in the concentration of the stress protein GADD153, indicative of cell stress (Fig. 12.4). Flow cytometry indicated an alteration in cell cycle progression in riboflavin-depleted cells, with an increase in the proportion of cells in G1 and a reduction in the proportion of cells in S phase and G2 compared with cells incubated in medium containing high concentrations of riboflavin. Decreased cell proliferation rates were inferred from a reduction in 3H-thymidine uptake (Werner et al. 2005). The authors concluded that the accumulation of unfolded proteins in riboflavin-deficient HepG2 cells triggers the unfolded protein response (Sidrauski et al. 2002), with an associated cell cycle arrest. The same group further examined some possible mechanisms for the effects of riboflavin depletion on cell cycle progression in HepG2 cells in culture (Manthey et al. 2006). They tested the hypothesis that cell cycle effects were mediated by oxidative damage to DNA and proteins. They reported an increase in DNA strand breakage, measured using the Comet assay, in response to 4 days growth in riboflavin-deficient medium compared with a high concentration of riboflavin (300 nM). Protein carbonylation also increased in response

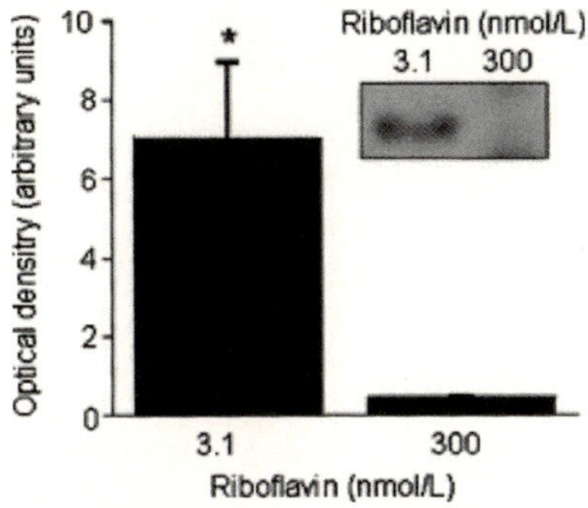

Fig. 12.4 Abundance of GADD153 increases in response to riboflavin deficiency in HepG2 cells. Cells were cultured in media containing 3.1 and 300 nmol/L riboflavin for 4 days. GADD153 was quantified using Western blot analysis and gel densitometry. $^*P<0.05$ vs. 300 nmol/L riboflavin ($n=4$). The *inset* depicts a representative image from Western blot analysis (reprinted from 'Riboflavin deficiency causes protein and DNA damage in HepG2 cells, triggering arrest in G1 phase of the cell cycle' Manthey et al. 2006, with permission from Elsevier)

to riboflavin-depletion, indicating oxidative damage. It is argued that these effects are a direct result of impaired antioxidant protection as glutathione reductase activity falls, with a likely fall in glutathione concentration in the riboflavin-depleted cells.

Thus riboflavin depletion in the HepG2 cell in culture does seem to be associated with an increase in oxidative stress, and downstream effects on cell cycle progression. In these experiments the authors chose to use very high concentrations of riboflavin as a 'control', and these cells may have behaved quite differently from a true physiologic control. The same group conducted gene expression analysis in riboflavin-depleted HepG2 cells, using the Affymetrix Human genome U133A Array. Riboflavin depletion was associated with up or down-regulation of genes, including those known to be important for cell cycle progression and proliferation, and apoptosis. Although genes from both of these functional clusters showed both increased and decreased transcription there was a preponderance of genes showing increased transcription. In contrast, genes important for protein synthesis and degradation were more likely to be down regulated. Taken together, results from this group suggest that riboflavin depletion of HepG2 cells in culture leads to dysregulation of the cell cycle, specifically arrest of cells in G1.

12.3.2 *Effects of Riboflavin Depletion on an In Vitro Model of Enterocyte Proliferation/Differentiation*

In the light of the profound effects of riboflavin deficiency on the development of the gastrointestinal tract in weaning rats we propose that riboflavin has multiple roles in the enterocyte associated with functions in cell fate determination in addition to its accepted role in redox metabolism. In particular, riboflavin regulates the cell cycle so as to influence cell proliferation, differentiation and death of intestinal epithelial cells. In order to test this hypothesis work has been conducted using a human intestinal epithelial cell model (Caco-2 cell, obtained from the European Collection of Cell Cultures, ECACC). Cells were grown in specially prepared DMEM containing no riboflavin. Cells were cultured under three conditions:

(i) Riboflavin-rich medium, consisting of standard DMEM containing 10% foetal calf serum. The riboflavin content of this medium was 1.2 μM, which is higher than the concentration of riboflavin in human plasma following supplementation with moderate dose riboflavin (Powers et al. 2011).
(ii) Physiological medium, consisting of riboflavin-free DMEM containing 10% foetal calf serum. The riboflavin content of this medium was 16 nM. This approximates the riboflavin concentration in plasma from healthy people (Moat et al. 2003).
(iii) Riboflavin-depleted medium, consisting of riboflavin-free DMEM containing 10% dialysed foetal calf serum. The riboflavin content of this medium was 5.5 nM. This approximates the riboflavin concentration in human plasma in the lowest fifth centile of a healthy population (Hustad et al. 2000).

Flavin concentrations were measured in the fetal calf serum before and after dialysis, and in the final medium, by HPLC (Capo-chichi et al. 2000). Following transfer of cells from the riboflavin-rich medium to physiological riboflavin medium the intracellular concentration of FAD fell by about 10%, that of FMN fell by about 30% whilst the riboflavin concentration fell to 50% of that in the control medium, and concentrations were maintained as such over 6 passages. Cell proliferation was unaffected as were basal levels of apoptosis. Transferring cells to riboflavin-depleted medium (5.5 nM) elicited a further progressive decrease in intracellular flavin concentrations to about 50% of that in the physiological cells by 72 h. About 90% of the intracellular flavins were in the form of FAD by this stage. The cell doubling time increased from 0.88 days to 1.62 days during riboflavin depletion.

Riboflavin depletion of Caco-2 cells altered progression of cells through the cell cycle. Following growth in physiological and riboflavin-depleted medium for up to 72 h cells were trypsinised, washed in buffer, fixed in ethanol and incubated with propidium iodide to stain DNA, and analysed by flow cytometry. During growth in riboflavin-depleted medium there was a progressive fall in the proportion of cells in G1 and a progressive increase in the proportion of cells in G2, compared with those cells grown in physiological medium. By 72 h of growth in riboflavin-depleted medium there was a greater proportion of cells in G2 than in G1 with no significant change in S phase, suggesting impaired progression of cells through G2 or M (Fig. 12.5; Table 12.2). Further study has indicated that a more prolonged period of riboflavin depletion leads to accumulation of aneuploid cells, and that the cell cycle alteration is likely to be in M rather than G2 phase (Nakano et al. 2011).

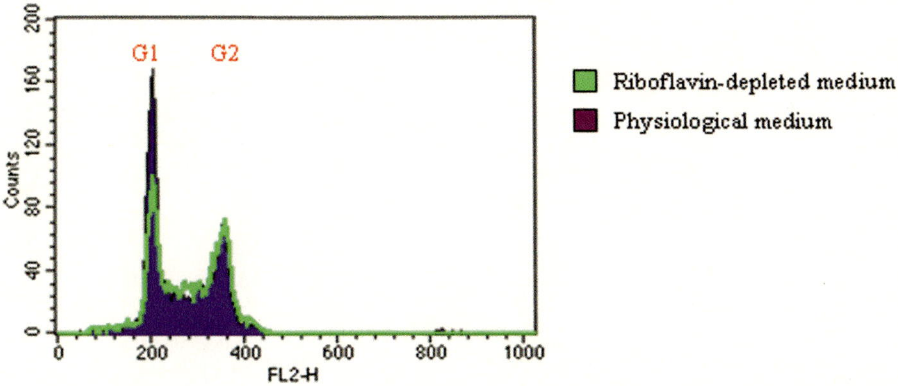

Fig. 12.5 Flow cytometry analysis of Caco-2 cells following growth for 72 h in physiological medium (16 nM riboflavin) and riboflavin-depleted (5.5 nM) medium. Data are shown for a single representative experiment at 72 h

Table 12.2 Effect of riboflavin depletion of CaCo-2 cells on cell cycle phase distribution CaCo-2 cells were grown for 72 h in riboflavin-depleted (5.5 nM riboflavin) or physiological (16 nM riboflavin) medium for 72 h. Values represent the percentage of cells in different stages of the cell cycle, measured using flow cytometry (mean (SEM) of data from 3 experiments). Riboflavin depletion was associated with a significantly greater percentage of cells in G2 ($P = 0.008$) and significantly fewer cells in G1 ($P = 0.011$) compared with cells grown in physiological medium (Wilcoxon non-parametric test)

		Time (hours) in experimental medium		
Growth conditions	Phase of cell cycle	24	48	72
Physiological	G1	44.4 (1.14)	43.4 (0.12)	50.5 (2.41)
Riboflavin-depleted	G1	45.4 (0.12)	37.0 (1.76)	29.6 (0.76)
Physiological	S	26.2 (0.64)	26.9 (1.12)	20.7 (3.36)
Riboflavin-depleted	S	20.2 (0.73)	25.6 (1.27)	29.2 (0.32)
Physiological	G2	25.2 (0.91)	26.2 (1.21)	25.0 (0.81)
Riboflavin-depleted	G2	31.5 (0.86)	31.7 (0.61)	35.4 (0.15)

12.3.3 Apoptosis

Using direct depletion methods we have not been able to detect any change in the basal levels of apoptosis in Caco2 cells. This suggests that riboflavin depletion alone is insufficient to trigger apoptosis. However, when the riboflavin uptake competitor lumiflavin was used to drive down intracellular levels of riboflavin, a marked decrease in attached cell count was identified, at a 4 fold molar excess of lumiflavin (Fig. 12.6). This cell death occurred within 72 h of lumiflavin treatment. This finding seems to correlate with the observation that luminal depletion of riboflavin in systemically riboflavin replete rats drives alterations in crypt morphology attributable

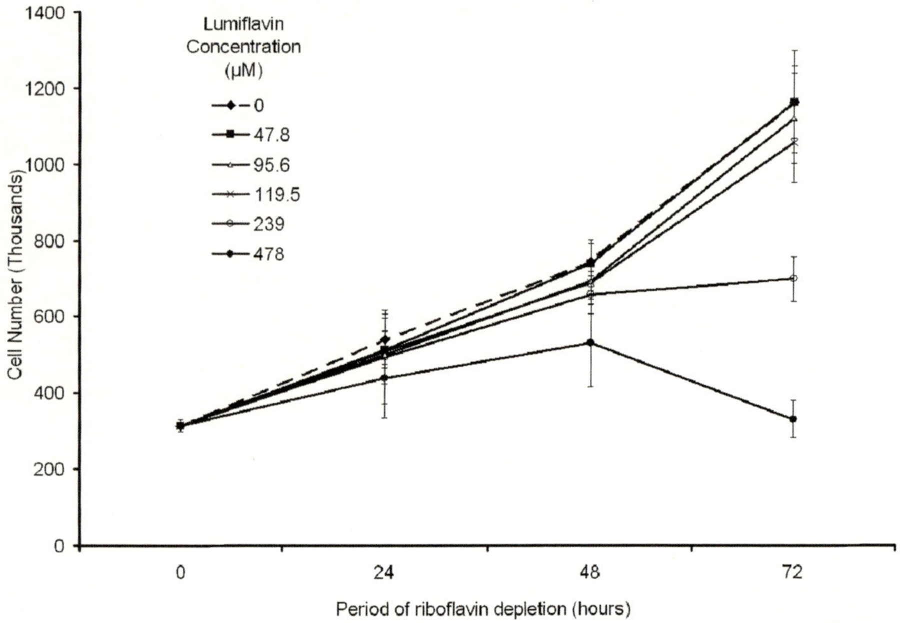

Fig. 12.6 Effects of riboflavin depletion on cell number. Caco-2 cells were grown in the presence of different concentrations of lumiflavin. Data are the mean and SD of cell numbers at 72 h for three independent experiments

to changes in levels of apoptosis. We hypothesize that in this particular context, riboflavin is acting as a specific signal to cells via a hypothetical apical receptor. Ligation of the receptor would be required to maintain cellular viability in vitro and tissue homeostasis in vivo.

The particular effects of riboflavin depletion on progression through the cell cycle appear to be dependent on the cell type. This observation has been made for other nutrients; retinoic acid for example will inhibit proliferation and promote differentiation in embryonic hepatocytes and endothelial cells whilst promoting proliferation in embryonic cardiomyocytes (Stuckmann et al. 2003).

12.3.4 Transcriptional Consequences of Riboflavin Depletion

In order to understand the mechanisms by which riboflavin depletion might inhibit cell cycle progression DNA microarray analysis was undertaken on RNA extracted from cells cultured in riboflavin depleted and physiological medium for up to 72 h. Following 72 h of growth in riboflavin depleted medium 60 genes showed at least a two-fold change and 11 genes showed at least a three-fold change in mRNA concentration (Fig. 12.7). A cluster of these genes is known to

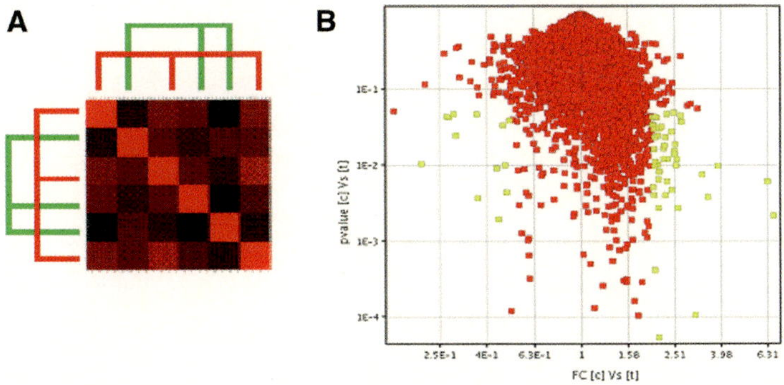

C

	P (All)	P (<0.05)	P (<0.01)	P(<0.005)	P (<0.0001)
FC All	22.283	761	167	83	21
FC>1.1	5.522	665	152	75	19
FC>1.5	808	283	85	47	14
FC>2.0	123	60	23	13	3
FC>3.0	18	11	6	3	1

Fig. 12.7 Alterations in gene expression after riboflavin depletion. Caco2 cells were grown in riboflavin-depleted DMEM (5.5 nM riboflavin) or in medium supplemented with riboflavin to physiological levels (16 nM). RNA was extracted and expression analysed using Affymetrix U133 arrays. **Panel A**: Heat map showing overall similarity between arrays, where *redness* indicates similarity, cluster indicated with a *red* dendogram are the physiological group, the *green* dendogram indicates the depleted group. **Panel B**: Fold change versus probability (volcano plot) or all data, spots in *yellow* are considered as significant. **Panel C**: Numbers of genes reaching increasing levels of probability and increasing levels of fold change, 60 genes reached a *P* value of 0.05 for a FC of >2.0

play a role in the regulation of the cell cycle, including CEACAM1 and cyclin A2. Effects of riboflavin depletion were also seen in genes associated with the stress response (STIP1) and energy generation (AK2, ATP5G1) (Nakano et al. 2011).

In addition to effects of riboflavin depletion on mRNA for particular genes it is possible that some effects occur at the translational or post-translational level. Retinoic acid for example induced the ubiquitination of cyclin D1 in immortalised human bronchial epithelial cells, leading to proteolysis, and subsequent cell cycle arrest, even though mRNA levels for this protein are not influenced (Spinella et al. 1999). Studies are in progress to characterise changes in the expressed proteome in CaCo-2 cells depleted of riboflavin. Preliminary analysis confirms that riboflavin depletion leads to an alteration in the cellular proteome (Fig. 12.8).

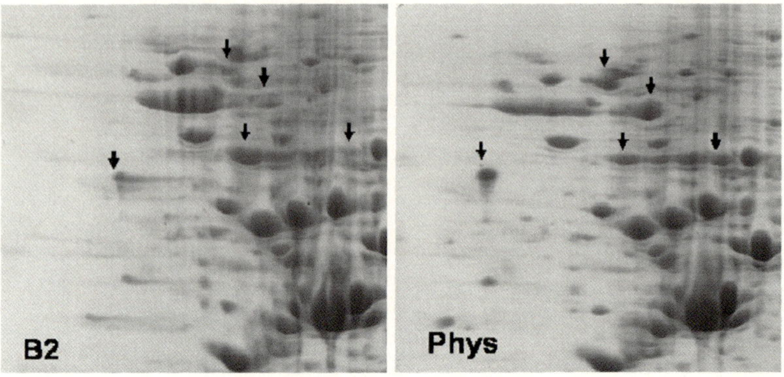

Fig. 12.8 2-D gels of proteins expressed in cells grown in riboflavin–depleted medium (B2) or in a physiological concentration of riboflavin (Phys). The *arrows* indicate proteins that appear to show an altered expression in response to riboflavin depletion. Further analysis will reveal the identity of these proteins

12.4 Conclusions

The requirement for dietary riboflavin to human health has been appreciated for decades. The metabolic basis for this has been hitherto attributed to the well-defined role of flavins in redox metabolism and energy generation. It is becoming apparent that flavins can modulate cell cycle progression, probably through effects on the expression of regulatory genes. There are numerous mechanisms by which flavins may exert such effects, at the transcriptional, translational and post-translational levels, including direct binding to RNA (resulting in a reduction in translational fidelity), and histone modification. It is also possible that influences of flavin concentrations on gene expression are mediated by effects on cellular redox status, bringing us neatly back to classic flavin biochemistry.

The challenge now is to characterise fully the role of dietary riboflavin in the regulation of the cell cycle, to understand the underpinning mechanisms and to evaluate the functional implications in terms of human health.

References

Adiga PR, Murty CV (1983) Vitamin carrier proteins during embryogenic development in birds and mammals. Ciba Found Symp 98:111–136

Alisi A, Leoni S, Lacentani A, Dervirgillis LC (2003) Retinoic acid modulates the cell cycle in rat hepatocytes and HepG2 cells by regulating cyclin-CDK activities. Liver Int 23:179–186

Antozzi C, Garavaglia B, Mora M, Rimoldi M, Morandin L, Ursino E, DiDonato S (1994) Late-onset riboflavin-responsive myopathy with combined multiple acylCoA dehydrogenase and respiratory chain deficiency. Neurology 44:2153–2158

Bates CJ, Prentice AM, Paul AA, Prentice A, Sutcliffe BA, Whitehead RG (1982) Riboflavin status in infants born in rural Gambia and the effect of a weaning food supplement. Trans Roy Soc Trop Med 76:253–258

Bates CJ, Evans PH, Dardenne M, Prentice AM, Lunn PG, Northrop-Clewes CA, Hoare S, Cole TJ, Horan SJ, Longman SC (1993) A trial of zinc supplementation in young rural Gambian children. Brit J Nutr 69 243–255

Bohnsack BL, Hirschi KK (2004) Nutrient regulation of cell cycle progression. Ann Rev Nutr 24:433–453

Brady PS, Hoppel CL (1985) Hepatic peroxisomal and mitochondrial fatty acid oxidation in the riboflavin-deficient rat. Biochem J 299:717–721

Camporeale G, Zempleni J (2003) Oxidative folding of interleukin-2 is impaired in flavin-deficient Jurkat cells, causing intracellular accumulation of interleukin-2 and increased expression of stress response genes. J Nutr 133:668–672

Capo-chichi CD, Gueant JL, Feillet F, Namour F, Vidaillhet M (2000) Analysis of riboflavin and riboflavin cofactor levels in plasma by high performance liquid chromatography. J Chromatogr B Biomed Sci Appl 739(1):219–224

Chalmers RA, Bain MD, Zschocke J (2006) Riboflavin-responsive glutaryl dehydrogenase deficiency. Mol Genet Med 88:29–37

Chen TH, Chang TC, Kang JO, Choudhary, B, Makita, T, Tran C.M., Burch JB, Eid H, Sucor HM (2002) Epicardial induction of fetal cardiomyocyte proliferation via a retinoic acid-inducible trophic factor. Dev Biol 250:198–207

D'Souza VM, Foraker AB, Free RB, Ray. A., Shapiro PS, Swaan PW (2006) cAMP-coupled riboflavin trafficking in placental trophoblasts: A dynamic and ordered process. Biochem 45:6095–6104

Duerden JM, Bates CJ (1985) Effect of riboflavin deficiency on lipid metabolism of liver and brown adipose tissue of suckling rat pups. Brit J Nutr 53:107–115

Faron G, Drouin R, Pedneault L, Poulin LD, Laframboise R, Garrido-Russo M, Fraser WD (2001) Recurrent cleft lip and palate in siblings of a patient with malabsorption syndrome, probably caused by hypovitaminosis A associated with folic acid and vitamin (B2) deficiencies. Teratology 63:161–163

Fass S, Rivlin RS (1969) Regulation of riboflavin-metabolising enzymes in riboflavin deficiency. Amer J Physiol 217:988–991

Gianazza E, Vergani L, Wait R, Brizio C, Brambilla D, Begum S, Giancaspero TA, Conserva F, Eberini I, Bufano D, Angelini C, Pegoraro E, Tramontano A, Barile M (2006) Coordinated and reversible reduction of enzymes involved in terminal oxidative metabolism in skeletal muscle mitochondria from a riboflavin-responsive, multiple acyl CoA dehydrogenase deficiency patient. Electrophoresis 27:1182–1198

Goodman SI (1981) Organic aciduria in the riboflavin-deficient rat. Am J Clin Nutr 34:2434–2437

Harpey JP, Charpentier C, Goodman SI, Darbois Y, Lefebvre G, Sebbah J (1983) Multiple acyl CoA dehydrogenase deficiency occurring in pregnancy and caused by a defect in riboflavin metabolism in the mother. Study of a kindred with seven deaths in infancy: value of riboflavin therapy in preventing this syndrome. J Pediatr 103:394–398

Hoppel CL, Tandler B (1975) Riboflavin and mouse hepatic structure and function. Mitochondrial oxidative metabolism in severe deficiency states. J Nutr 105:562–570

Horvath C, Szony L, Mold K (1976) Preventive effect of riboflavin and ATP on the teratogenic effects of the phenathiozine derivative T-82. Teratology 14:167–170

Huang SN, Swaan PW (2000) Involvement of a receptor-mediated component in cellular translocation of riboflavin J Pharmacol Exp Ther 194:117–125

Huang SN, Phelps MA, Swaan PW (2003) Involvement of endocytic organelles in the subcellular trafficking and localisation of riboflavin. J Pharmacol Exp Ther 306:681–687

Hustad S, Ueland PM, Vollset SE, Zhang Y, Bjorke-Monsen AL, Schneede J (2000) Riboflavin as a determinant of plasma total homocysteine: Effect modification by the methylene tetrahydrofolate reductase C677T polymorphism. Clin Chem 46:1065–1071

Hustad S, McKinley MC, McNulty H, Schneede J, Strain JJ, Scott JM, Ueland PM (2002) Riboflavin, flavin mononucleotide, and flavin adenine dinucleotide in human plasma and erythrocyes at baseline and after low-dose riboflavin supplementation. Clin Chem 48:1571–1577

Johnson DG, Walker, CL (1999) Cyclins and cell cycle checkpoints. Ann Rev Pharmacol Toxicol 39:295–312

Juriloff DM, Roberts CW (1975) Genetics of cleft palate in chickens and the relationship between the occurrence of the trait and maternal riboflavin deficiency. Poult Sci 54:334–346

Kosaka C, Sasaguri T, Komiyama Y, Takahashi H (2001) All-trans retinoic acid inhibits vascular smooth muscle cell proliferation targeting multiple genes for cyclins and cyclin-dependent kinases. Hypertens Res 24:579–588

Manthey KC, Chew YC, Zempleni J (2005) Riboflavin deficiency impairs oxidative folding and secretion of apolipoprotein B-100 in HepG2 cells, triggering stress-response systems. J Nutr 135:978–982

Manthey KC, Rodriguez-Melendez R, Hoi JT, Zempleni J (2006) Riboflavin deficiency causes DNA damage in HepG2 cells, triggering arrest in G1 phase of the cell cycle. J Nutr Biochem. 17:250–256

Marshall KR, Gong M, Wodke L, Lamb JH, Jones DJL., Farmer PB, Scrutton NS, Munro AW (2005) The human apoptosis-inducing protein AMID is an oxidoreductase with a modified flavin cofactor and DNA binding activity. J Biol Chem 280:30735–30740

Mason CW, D'Souza VM, Bareford LM, Phelps MA, Ray A, Swaan PW (2006) Recognition, co-internalisation, and recycling of an avian riboflavin carrier protein in human placental trophoblasts. J Pharmacol Exp Ther 317:465–472

Miura R (2001) Versatility and specificity in flavoproteins: Control mechanisms of flavin reactivity, vol 1. The Chemical Record The Japan Chemical Journal Forum and John Wiley & Sons Inc., pp 83–94

Moat SJ, Ashfield-Watt PAL., Powers HJ, Newcombe RG, McDowell IFW (2003) Effect of riboflavin status on the homocysteine-lowering effect of folate in relation to the MTHFR (C677T) genotype. Clin Chem 49:295–302

Nakano E, Mushtaq S, Heath PR, Lee ES, Bury JP, Riley SA, Powers HJ, Corfe BM (2011) Riboflavin depletion impairs cell proliferation in adult human duodenum: identification of potential effectors. Dig Dis Sci Apr;56(4):1007–1019

Natraj U, George S, Kadam J (1988) Isolation and characterisation of human riboflavin carrier protein and the estimation of its levels during human pregnancy. Reprod Immunol 13:1–16

Nelson MM, Baird CDC, Wright V, Evans HM (1956) Multiple congenital abnormalities in the rat resulting from riboflavin deficiency induced by the antimetabolite galactoflavin. J Nutr 58:125–134

Noback CR, Kupperman HS (1944) Anomalous offspring and growth of Wistar rats maintained on a deficient diet. Proc Exp Biol Med 57:183–185

Olpin SE, Bates CJ (1982) Lipid metabolism in riboflavin-deficient rats 2. Mitochondrial fatty acid oxidation and the microsomal desaturation pathway. Br J Nutr 47:589–596

Pahl HL (1999) Signal transduction from the endoplasmic reticulum to the cell nucleus. Physiol Rev 79:683–700

Patterson BE, Bates CJ (1989) Riboflavin deficiency, metabolic rate and brown adipose tissue function in sucking and weanling rats. Br J Nutr 61:475–483

Patterson BE, Bates CJ, Halliday D, Lucas A (1989) 1-(13)C-octanoate metabolism, energy expenditure and vitamin B2 supplement in premature infants. Acta Paediatr Scand 78:780–781

Powers HJ, Hill MH, Mushtaq S, Dainty JR, Majsak-Newman G, Williams EA (2011) Correcting a marginal riboflavin deficiency improves hematologic status in young women in the United Kingdom (RIBOFEM). Am J Clin Nutr Jun;93(6):1274–1284

Prasad PD (1992) Isolation and characterisation of riboflavin carrier protein from human amniotic fluid. Biochem Int 2:385–395

Prentice AM, Bates CJ (1981) Biochemical evaluation of the erythrocyte glutathione reductase (EC 1.6.4.2) test for riboflavin status. Br J Nutr 45:37–52

Rao J, Seshagiri PB, Shetty G, Ramesh G, Adiga PR (2000) Active immunization against riboflavin carrier protein results in peri-implantation embryonic loss leading to pregnancy termination in rats: use of aternative adjuvants. Indian J Exp Biol 38:863–872

Ross NS, Hansen TP (1992) Riboflavin deficiency is associated with selective preservation of critical flavoenzyme-dependent metabolic pathways. Biofactors 3:185–190

Said HM, Orti A, Ma TY, McCloud E (1998) Riboflavin uptake bythe human-derived liver cells HepG2: mechanism and regulation. J Cell Physiol 176:588–594

Sidrauski C, Brickner JH, Walter P (2002) The unfolded protein response. In: Dalbey RE, von Heijne RE (eds) Protein targeting, transport and translocation. Academic, San Diego, CA, pp 151–179

Spinella MJ, Freemantle SJ, Sekula D, Chang JH, Christie AJ, Dnitrovsky E (1999) Retinoic acid promotes ubiquitination and proteolysis of cycle D1 during induced tumour differentiation. J Biol Chem 274:22013–22018

Stuckmann I, Evans S, Lassar AB (2003) Erythropoietin and retinoic acid, secreted from the epicardium, are required for cardiac myocyte proliferation. Dev Biol 255:334–349

Swami S, Raghavachari N, Muller UR, Bao YP, Fieldman D (2003) Vitamin D growth inhibition of breast cancer cells; gene expression patterns assessed by cDNA microarray. Breast Canc Res Treat 80:49–62

Tu BP, Ho-Schlyer SC, Travers KL, Weissman JS (2000) Biochemical basis of oxidative protein folding in the endoplasmic reticulum. Science 290:1571–1574

Visweswariah SS, Adiga PR (1987) Isolation of riboflavin carrier proteins from pregnant human and umbilical cord serum: Similarities with chicken egg riboflavin carrier protein. Biosci Rep 7:563–571

Werner R, Manthey KC, Griffin JB, Zempleni J (2005) HepG2 cells develop signs of riboflavin deficiency within 4 days of culture in riboflavin-deficient medium. J Nutr Biochem Oct;16(10):617–624

Williams EA, Rumsey RD, Powers HJ (1996a) An investigation into the reversibility of the morphological and cytokinetic changes seen in the small intestine of the riboflavin-deficient rat. Gut 39:220–225

Williams EA, Rumsey RD, Powers HJ (1996b) Cytokinetic and structural responses of the rat small intestine to riboflavin depletion. Brit J Nutr 75:315–324

Yates CA, Evans GS, Pearson T, Powers HJ (2003) Absence of luminal riboflavin disturbs early postnatal development of the gastrointestinal tract. Dig Dis Sci 48:1159–1164

Yates CA, Evans GS, Powers HJ (2001) Riboflavin deficiency: Early effects on post-weaning development of the duodenum in rats. Brit J Nutr 86:593–599

Zaman Z, Verwilghen RL (1975) Effect of riboflavin deficiency on oxidative phosphorylation, flavin enzymes and coenzymes in rat liver. Biochem Biophys Res Commun 67:1192–1198

Chapter 13
Vitamins B6 and Cancer

Sang Woon Choi and Simonetta Friso

Abstract Epidemiologic and laboratory animal studies have suggested that the availability of vitamin B6 modulates cancer risk. The means by which B6 mediates this effect is not known with any surety but it has been reported that high dietary vitamin B6 attenuates and low dietary vitamin B6 increases the risk of cancer. In fact vitamin B6 is widely distributed in foods and overt deficiency of this vitamin is not common. Nevertheless, marginal or secondary vitamin B6 deficiency, which might have an adverse effect on carcinogenesis, is rather common especially among old adults and alcoholics. This chapter addressed currently available information regarding the relationship between vitamin B6 and cancer.

Keywords Cancer · Chemoprevention · One-carbon metabolism · PLP · Supplementation · Vitamin B6

Abbreviations

MTHFR	methylenetetrahydrofolate reductase gene
iNOS	inducible NO synthase
NO	nitric oxide
PLP	pyridoxal-5′-phosphatase
THF	tetrahydrofolate

13.1 Introduction

Vitamin B6, which is the collective term for the family of 2-methyl-3-hydroxy-5-hydroxymethylpyridine compounds, is widely distributed in foods; foods of animal origin such as meat, fish, poultry, eggs, and diary products as well as foods of plant

S.W. Choi (✉)
Tufts University, Medford/Somerville, MA, USA

University of Verona, Verona, Italy

Vitamins and Carcinogenesis Laboratory, Jean Mayer USDA HNRCA, Tufts University, Boston, MA 02111-1524, USA
e-mail: sang.choi@tufts.edu

O. Stanger (ed.), *Water Soluble Vitamins*, Subcellular Biochemistry 56,
DOI 10.1007/978-94-007-2199-9_13, © Springer Science+Business Media B.V. 2012

origin such as whole grain cereal products, white potato and other starchy vegetables, and noncitrus fruits. There are six major derivatives; pyridoxine, pyridoxal and pyridoxamine, and their phosphorylated derivatives, pyridoxine-5′-phosphatase, pyridoxal-5′-phosphatase (PLP) and pyridoxamine 5′-phosphatase. Vitamin B6 in human breast milk is present as pyridoxal and pyridoxal phosphate. In animal-derived foods it is also present largely as pyridoxal phosphate and pyridoxamine phosphate, while in plant derived foods vitamin B6 is present mainly as pyridoxine, pyridoxine phosphate, and pyridoxine glucoside (Clayton 2006). PLP, the major coenzymatic form of vitamin B6, involves in 112 enzyme reactions (3%) among 3,870 enzymes catalogued in the ENZYME data base (Ames et al. 2002), especially for amino acid metabolism, one-carbon metabolism, lipid metabolism, gluconeogenesis, heme biosynthesis, and neurotransmitter biosynthesis. In the U.S. an estimated average requirement of 1.1 mg/day and an RDA of 1.3 mg/day for men and women (19–50 years) were established in 1998. For adults over 50 years, the RDA is 1.5 mg/day for women and 1.7 mg/day for men.

Nowadays the classical symptoms of vitamin B6 deficiencies such as seborrheic dermatitis, microcytic hypochromic anemia, epileptiform convulsions, depression, and confusion are extremely rare, whereas 10–20% of the population and 51% of elderly hospitalized patients have marginal or inadequate vitamin B6 status (Joosten et al. 1993; Selhub et al. 1993; Bender 2003). Secondary deficiencies of vitamin B6 may result from malabsorption, uremia, cancer, heart failure, cirrhosis, and increased metabolic activity associated with alcohol or the use of some medication such as isoniazid, penicillamine, cycloserine, and theophylline. Current interest in vitamin B6 focuses on its role in decreasing circulating homocysteine, a risk factor for vascular disease, or on its role in the risk of cancers.

13.2 Evidence Indicating the Relationship Between Vitamin B6 and Cancer

13.2.1 Epidemiologic Evidence

Old observational studies have found a low circulating vitamin B6 concentration in cancer patients. In 1960 Wachstein et al. reported that PLP concentrations in plasma and leucocytes were decreased in patients with leukemia compared with those in normal subjects (Wachstein et al. 1960). In 1977 Potera et al. reported that plasma PLP was normal in early breast cancer but reduced both in cases of local recurrence ($p < 0.01$) and systemic metastasis ($p < 0.001$) and the difference in plasma PLP between the early breast cancer patients and those with wide-spread disease was also significant ($p < 0.01$), indicating that decrement of plasma PLP concentration is in parallel with cancer progress (Potera et al. 1977). Thereafter, epidemiologic studies have suggested associations of vitamin B6 level in diet or blood with cancer risk, even though results were not always consistent.

13.2.1.1 Colorectal Cancer

In a most recent large Scottish case-control study (2,028 cases and 2,722 controls) an inverse and dose-dependent association was found between the risk of colorectal cancer and the intake of dietary vitamin B6 (OR = 0.77, 95% CI = 0.61–0.98; p for trend = 0.03) (Theodoratou et al. 2008). In two other recent epidemiologic studies conducted in the U.S. (Slattery et al. 1997) and in Europe (Jansen et al. 1999), which investigated the protective effect of high intakes of plant foods against colorectal cancer, high dietary intake of vitamin B6 reduced the risk of colorectal cancer in both men and women (p for trend <0.01) as well as the dietary intake of vitamin B6 was inversely associated with the colorectal cancer mortality. In a population based cohort study Harnack et al. (2002) found that the incidence of proximal colon cancer was lower among those with high dietary intake of vitamin B6 and folate (RR = 0.65, 95% CI = 0.5–0.84) than among those with the lowest intake of these vitamins. In a prospective nested case-control study from the Nurses' Health Study cohort Wei et al. (2005) provided more convincing evidence that both plasma PLP concentration and total vitamin B6 intake were inversely associated with the risk of colon cancer (RR = 0.42, 95% CI = 0.21–0.85 and RR = 0.51, 95% CI = 0.27–0.97, respectively). They also reported a trend of inverse association between plasma PLP concentration and advanced distal colorectal adenoma (RR = 0.65, 95% CI = 0.37–1.11, p for trend = 0.08).

A few studies, which investigated the effect of intake of B vitamins and methylenetetrahydrofolate reductase (MTHFR) genotype on the risk of colorectal cancer, have found an interaction between vitamin B6 intake and MTHFR gene in colonic carcinogenesis. MTHFR gene, which regulates one-carbon pools for DNA synthesis and biological methylation, has a common polymorphism 677 C to T and homozygous variants (677TT) have reduced enzyme activity in vivo and in vitro (Kang et al. 1988; Friso et al. 2002). Le Marchand et al. (Le Marchand et al. 2002) reported the two-way interactions: an apparent direct association between colorectal cancer risk and vitamin B6 intake in subjects with the CC wild-type and a strong inverse association with vitamin B6 intake in subjects with the TT genotype (p for interaction = 0.02) in a population-based case-control study. Slattery et al. (1999) also reported that high levels of intake of folate, vitamin B6, and vitamin B12 were associated with a 30–40% reduction in risk of colon cancer among those with the TT (homozygous variant) relative to those with low levels of intake who were CC wild genotype. Ulrich et al. (Ulrich et al. 1999) also reported that high dietary vitamin B6 intake reduced the risk of colorectal adenoma, a precursor of colorectal cancer, in TT genotype (p for trend = 0.03). In a most recent population based case control study conducted in the UK, significant interactions were found between total vitamin B6 intake and MTHFR 677 C to T polymorphism as well as another common polymorphism, MTHFR 1298 A to C (C677T, p interaction = 0.016; A1298C, p interaction = 0.003) (Sharp et al. 2008). However, the results are not always reproducible. In the other case-control study conducted in Japan neither vitamin B6 nor any genetic polymorphisms in one-carbon metabolism was significantly associated with colorectal cancer (Otani et al. 2005).

Chronic alcohol intake interferes with one-carbon metabolism (Stickel et al. 2000), in which vitamin B6 serves as a coenzyme. Choi et al. (1999) reported that alcohol fed rats had a 33% lower plasma PLP concentrations ($p = 0.02$) and a 16% lower hepatic PLP concentration ($p = 0.03$) than controls along with decreased genomic DNA methylation in the colon ($p < 0.05$). Cravo et al. (Cravo et al. 1996) reported that in chronic alcoholics serum PLP concentration was significantly decreased compared with control ($p < 0.001$). Interestingly, in a population-based cohort study, a long-term intake of dietary vitamin B6 was inversely associated with the risk of colorectal cancer (p for trend = 0.002) and the association was most evident among women with moderate to high alcohol consumption. The multivariate RR for colorectal cancer comparing extreme quintiles of vitamin B6 intake was 0.28 (95% CI = 0.13–0.59) among women who consumed > or = 30 g/week of alcohol (Larsson et al. 2005). These observations suggest that vitamin B6 may play an important role in the prevention of alcohol associated colorectal carcinogenesis.

13.2.1.2 Lung Cancer

In a nested case-control study conducted within the α-tocopherol β-carotene cancer prevention study cohort Hartman et al. (Hartman et al. 2001) reported a significantly lower risk for lung cancer among men who had higher serum vitamin B6 levels. Compared with men with the lowest vitamin B6 concentration, men in the fifth quintile had about one half of the risk of lung cancer (OR = 0.51, 95% CI = 0.23–0.93, p for trend = 0.02). In the most recent case-control study high intake of vitamin B6 reduced the risk of lung cancer among women with MTHFR TT genotype compared with women with MTHFR CC genotype (OR = 0.29, 95% CI = 0.15–0.56) (Shi et al. 2005).

13.2.1.3 Breast Cancer

Compared with colorectal cancer, less convincing is an association between vitamin B6 and breast cancer risk. In a case-control study from the Nurses' Health Study cohort the multivariable RR comparing women in the highest quintile of plasma vitamin B6 levels with those in the lowest quintile was 0.70 (95% CI = 0.48–1.02, p for trend = 0.09) (Zhang et al. 2003). In a population study, which investigated the dietary folate intake and breast cancer risk, a minimal effect of vitamin B6 was found: among women who consumed a low level of vitamin B6, dietary folate was weakly associated with a reduced risk of breast cancer, whereas a striking inverse association was observed for dietary folate intake among those whose intake levels of vitamin B6 were high (Shrubsole et al. 2001), indicating that the protective effect of folate against breast cancer may be modified by dietary intake of vitamin B6, even though dietary intake of vitamin B6 was not independently related to the risk of breast cancer. On the other hand, in a population-based case-control study daily vitamin B6 intake level in breast cancer cases was rather significantly higher than that in controls ($p = 0.03$) (Shrubsole et al. 2004) and two other population-based case-control studies have found no relation of breast caner risk with vitamin B6 intake (Lajous et al. 2006) as well as plasma PLP concentration (Wu et al. 1999). The most recent case control study demonstrated an inverse association between plasma PLP and the risk of breast cancer among postmenopausal women (p for trend = 0.04) (Lin et al. 2008), while in the Nurses' Health Study II vitamin B6 intake was not associated with the risk of breast cancer among premenopausal women ages 26–46 years (Cho et al. 2007).

13.2.1.4 Prostate Cancer

In a case-control study, which investigated the effect of fat intake on the risk of prostate cancer, high dietary intake of vitamin B6 has a tendency to reduce the risk of prostate cancer (OR = 0.77, 95% CI = 0.53–1.12, p for trend = 0.077) and total intake of vitamin B6 including supplements has a reduced risk of this cancer (OR = 0.70, 95% CI = 0.48–1.03, p for trend = 0.029) (Key et al. 1997). In the α-tocopherol β-carotene cancer prevention study conducted in Finland vitamin B6 intake was inversely, but modestly, associated with prostate cancer risk (RR for highest versus lowest quintile = 0.88, 95% CI = 0.72–1.07, p for trend = 0.045) (Weinstein et al. 2006), even though the same group previously reported no association between vitamin B6 and prostate cancer risk in the same cohort (Weinstein et al. 2003). In a recent case-control study conducted in Italy vitamin B6 intake had no relation with this cancer risk (Pelucchi et al. 2005).

13.2.1.5 Pancreas Cancer

In a nested case-control study serum PLP concentrations showed an inverse dose-response relationship with pancreatic cancer risk (OR for highest versus lowest tertile = 0.48, 95% CI = 0.26–0.88, p for trend = 0.02) (Stolzenberg-Solomon et al. 1999). In this study an increased pancreatic cancer risk was also observed with high exposure to cigarettes, which is known to reduce circulating plasma PLP (Vermaak et al. 1990; O'Callaghan et al. 2002), suggesting that maintaining adequate vitamin B6 status may reduce the risk of pancreatic cancer associated with cigarette smoking. However, the same group found no association between vitamin B6 intake and pancreas cancer risk from the α-tocopherol β-carotene cancer prevention study cohort, even though in this cohort cigarette smoking was still associated with an increased risk of this cancer (Stolzenberg-Solomon et al. 2001). Most recently in a prospective nested case-control study using four large prospective cohorts in the US, the plasma concentration of vitamin B6 was not associated with a reduced risk of pancreas cancer (OR for highest versus lowest quartile = 0.91, 95% CI = 0.57–1.46) (Schernhammer et al. 2007).

13.2.1.6 Non-Hodgkin's Lymphoma

In a population-based case-control study of non-African-American adult women and men from four US Surveillance, Epidemiology, and End Results study centers,

Lim et al. (2005) demonstrated that higher intake of vitamin B6 was associated with the reduced risk of non-Hodgkin's lymphoma (OR for highest versus lowest quartile = 0.57, 95% CI = 0.34–0.95), suggesting that vitamin B6 may be protective against this cancer. They further demonstrated that interactions between dietary vitamin B6 and genes involved in one-carbon metabolism were also associated with the reduced risk of non-Hodgkin's lymphoma (Lim et al. 2006). However, from the α-tocopherol β-carotene cancer prevention study cohort same authors found no association between vitamin B6 and the risk of overall non-Hodgkin's lymphoma (Lim et al. 2007).

13.2.2 Evidence from Cultured Cell Studies

High concentrations of vitamin B6 have been reported to suppress growth of animal and human cancer cells; (1) pyridoxine supplemented medium (5 mM) retarded and eventually killed Fu5-5 rat hepatoma cells (DiSorbo and Litwack 1982), (2) pyridoxal supplemented medium (0.5 mM) inhibited the growth of a human malignant melanoma cell line (DiSorbo and Nathanson 1983), (3) either pyridoxine (5 mM) or pyridoxal (0.5 mM) supplemented medium showed an 80% reduction in B16 melanoma cells (DiSorbo and Litwack 1982), and (4) pyridoxine supplemented medium (5 mM) suppressed growth of human hepatoma cell line HepG2 (Molina et al. 1997).

Most recently Shimada et al. (2006) demonstrated that growth of the feline mammary tumor cell line FRM was inhibited by the addition of pyridoxine in a dose dependent manner (1–5 mM) and use of 5 mM pyridoxine caused an almost complete arrest of cell growth. Ren and Melmed (Ren and Melmed 2006) demonstrated that PLP also inhibited cell proliferation in non cancer cell line such as rat pituitary MMQ and GH3 cells in a dose dependent manner (0.01–1 mM). Collectively, vitamin B6 inhibits the growth of cultured cells originated from both cancer and non-cancer tissue.

13.2.3 Evidence from Animal Studies

Even though previous animal studies have suggested a relation between vitamin B6 and carcinogenesis, the most convincing are data from a series of animal studies conducted in a chemical carcinogen model of colon cancer in mice, evaluating the effect of dietary vitamin B6 on colonic carcinogenesis. Komatsu et al. (2001) reported that dietary supplementation with vitamin B6 suppressed colonic tumorigenesis by reducing cell proliferation. Mice were fed diets containing 1, 7, 14 or 35 mg pyridoxine HCl/kg diet. The recommended AIN-93 dietary level is 7 mg pyridoxine HCl/kg diet. After injections of azoxymethane, an alkylating chemical carcinogen, 7, 14, and 35 mg pyridoxine HCl/kg diets significantly suppressed the incidence and number of colon tumor, colonic cell proliferation and expression of c-myc and c-fos proteins in a dose dependent manner. Using the same animal model, Komatsu et al. (2002) then provided observations which suggest that the preventive effect of vitamin B6 against colon tumorigenesis is, in part, mediated by reducing oxidative stress and nitric oxide production. Most recently similar results were reproduced by the same group to determine the high-fat diet effect, indicating that dietary supplementation of vitamin B6 reduces cell proliferation from the early stage of colonic carcinogenesis and this inhibitory effect is markedly enhanced by a high-fat diet (Komatsu et al. 2005).

Shimada et al. (2005) also found that high dietary pyridoxine (35 mg pyridoxine HCl/kg diet) reduced tumor incidence compared with low dietary pyridoxine (7 mg pyridoxine HCl/kg diet) in the chemical carcinogen model of mammary tumor. The number of proliferating cells was also significantly reduced by high dietary pyridoxine.

Animal studies also suggested that tumor tissue might have a different vitamin B6 metabolism compared with normal tissue. In an animal model using highly differentiated transplantable Morris hepatomas and Buffalo strain rats, inoculated tumors in each hind leg were apparently incapable of the complete synthesis of co-enzymatically active PLP from inactive precursor forms, indicating that the metabolism of vitamin B6 was markedly different in the hepatomas than control livers (Thanassi et al. 1981). In the same animal model, compared with pair-fed controls the growth of inoculated Morris hepatomas was significantly decreased when inoculated at 31 days, but not 0, 7 and 15 days, after initiation of vitamin B6 deficient diet. It suggests that impairment in general animal metabolism as a result of vitamin B6 deficiency reduces the tumor growth rather than vitamin B6 deficiency by itself (Tryfiates and Morris 1976).

In an animal study BALB/c mice were fed diets containing pyridoxine at 0.2, 1.2, 7.7 or 74.3 mg/kg. After 4 weeks of diet, mice received a subcutaneous injection of herpes simplex virus type 2-transformed cells to induce tumors. Mice fed 7.7 and 74.3 mg pyridoxine had 14 and 32% less tumor volumes compared with those fed 1.2 mg pyridoxine, respectively and significant inverse associations between tumor volume and tumor PLP concentrations were also found in 1.2, 7.7, and 74.3 mg pyridoxine diet groups (Gridley et al. 1987). Interestingly, mice fed a diet with 0.2 mg pyridoxine had the lowest tumor volume.

13.2.4 Evidence from Clinical Trials

Since animal and human studies have suggested that abnormal tryptophan metabolism can cause bladder cancer, it evoked that supplementation with vitamin B6, which involves in tryptophan metabolism, might reverse the abnormality or prevent recurrences of superficial bladder cancer (Koss and Lavin 1971; Byar and Blackard 1977). In a double blind randomized clinical trial that enrolled 65 patients with biopsy confirmed transitional cell carcinoma of urinary bladder, Lamm et al. demonstrated that a high dose vitamin B6 supplementation (100 mg) with supplementation of vitamin A, C, and E plus Zinc reduced the recurrence rate of bladder tumor (Lamm et al. 1994). However, the other double blind randomized phase III trial demonstrated no effect of supplementation with vitamin B6 alone on the recurrence of transitional cell tumor of the bladder (Newling et al. 1995). To date there is no clinical trial which indicates that vitamin B6 supplementation per se is effective for cancer prevention or reduces the recurrence of cancer.

Zhang et al. (2006) reported an association between intake of vitamin B6 and colorectal cancer risk among women enrolled in a randomized trial on aspirin and vitamin E in disease prevention. Total vitamin B6 intakes were not significantly associated with the risk of colorectal cancer. However, dietary intakes of vitamin B6 were significantly inversely associated with the risk of colorectal cancer among women who were not taking supplements containing vitamin B6. Lim et al. (2005) also reported that total vitamin intake (food plus supplement) is not associated with the risk of non-Hodgkin's lymphoma, while the risk of this cancer is inversely associated with dietary vitamin B6. These observations are consistent with a folate supplementation trial, which demonstrates that the risk of lung cancer is inversely associated with dietary folate but not with intake of fortified folate, supplemented folate, or total folate intake. These observations collectively suggest that supplementation with vitamin B6 in a single synthetic chemical form might not be as effective as high vitamin B6 diet, which contains different forms of vitamin B6 and other nutrients needed for vitamin B6 metabolism.

13.3 Candidate Mechanisms

In 1985 Prior (1985) proposed a theoretical involvement of vitamin B6 in tumor initiation, hypothesizing that vitamin B6 deficiency occurring at the same time as contact with carcinogen could well lead to tumor initiation and subsequent cancer development, because vitamin B6 is a co-enzyme involved in the biosynthesis of thymidine, deficiency of which has been reported to increase replication errors and hence increase mutagenesis. Thereafter, several candidate mechanisms have been suggested but the exact mechanism by which vitamin B6 modulates carcinogenesis is not yet known (Table 13.1).

13.3.1 Antioxidant Defense Mechanism

DNA is continuously exposed to various endogenous and exogenous mutagens. Among them, oxidation is one of the most common vulnerabilities to genetic stability and oxidative DNA damage that have been suggested as an important mechanism in carcinogenesis. By the same token reduced cancer incidence has been found in populations with high dietary antioxidant intake (Poulsen 2005).

In a diabetes cell culture model pyridoxine and pyridoxamine inhibited oxygen radical production (Jain and Lim 2001) and in a rodent study vitamin B6 deficiency caused a decrease in antioxidant defense system and an increase in oxidant stress in liver tissue (Taysi 2005). Moreover, in a chemical carcinogene model of colon cancer Komatsu et al. (2002) demonstrated that vitamin B6 supplementation

Table 13.1 Candidate mechanisms by which vitamin B-6 modulate carcinogenesis

Mechanisms		References
Antioxidant defense mechanism	Inhibition of oxygen radical production	Komatsu et al. (2002), Jain and Lim (2001), and Taysi (2005)
Anti-inflammatory effect	Nitric oxide production	Komatsu et al. (2002), Kojima et al. (1999), Jaiswal et al. (2001), and Yanaka et al. (2005)
Cell proliferation	Expression of *c-myc* and *c-fos*	DiSorbo et al. (1985), Komatsu et al. (2001), Shultz et al. (1988), and Oka (2001)
Angiogenesis	Inhibition of microvessel outgrowth	Matsubara et al. (2001, 2004)
One-carbon metabolism	DNA methylation and synthesis	Le Marchand et al. (2002), Slattery et al. (1999), Ulrich et al. (1999), Shi et al. (2005), and Lim et al. (2007)
Immune function	Humoral and cell-mediated immune responses	Rall and Meydani (1993), Cheng et al. (2006), and Gebhard et al. (1990)
Regulation of steroid hormone	Estrogen-induced gene expression	Allgood and Cidlowski (1992)
Niacin deficiency	DNA repair and genomic stability	Shibata et al. (1995), Kirkland (2003), and Ames (2001)
Inhibition of DNA replication	Replicative DNA polymerases activity	Matsubara et al. (2003) and Mizushina et al. (2003)

suppressed the oxidative stress markers, colonic 8-hydroxyguanosine and 4-hydroxy-2-nonenal, compared with low vitamin B6 diet as well as these two markers were significantly associated with the expression of c-myc and c-fos, indicating that reduced oxidative stress by vitamin B6 supplementation might be in part involved in the mechanism of its preventive effect on tumorigenesis.

13.3.2 Anti-inflammatory Effect

Nitric oxide (NO), which is produced by various cell types such as macrophages, endothelial cells, and certain cancer cells, has been thought to play a role in carcinogenesis (Kojima et al. 1999). NO production is dependent on NO synthase and increased expression of inducible NO synthase (iNOS) has been reported in several types of tumors originated from head and neck, brain, breast, and colon. Upregulation of iNOS activity, which is associated with production of large amounts of NO that can cause tissue damage, is a cardinal feature of inflamed tissues as well as a plausible link between inflammation and cancer initiation, progression, and promotion (Jaiswal et al. 2001). Early on Komatsu et al. (2002) demonstrated that high dietary vitamin B6 significantly reduces the expression of colonic iNOS protein in a chemical carcinogen model, suggesting that the lower NO production by vitamin B6 might have partial responsibility for the reduced tumorigenesis by vitamin B6 supplementation. Most recently in a cultured cell model Yanaka et al. (2005)

demonstrated that pretreatment of RAW cells with vitamin B6 (pyridoxal) inhibits not only lipopolysaccharide-induced expression of iNOS and COX-2 at the mRNA and protein levels but also nuclear translocation of the nuclear factor-kappaB, the proinflammatory transcription factor. It suggests that the anti-inflammatory effect of vitamin B6 is also mediated by suppression of nuclear factor-kappaB activation.

13.3.3 Cell Proliferation

In a human malignant melanoma cell line Shultz et al. (1988) found that pharmacologic concentrations of pyridoxal (0.25–0.5 mM) resulted in significant reductions in cell proliferation, even though physiologic level of pyridoxal (0.005 mM) did not retard cell growth and pyridoxine rather stimulated the growth. On the other hand, the growth of human HepG2 cells was completely inhibited in medium supplemented with pyridoxine in the millimolar range. The growth inhibition of HepG2 cells was accompanied by a marked inhibition of gene expression, protein synthesis, and secretion of serum proteins, particularly albumin. The expression of oncogenes, such as c-fos and c-myc, was considerably reduced in the tumor cells, probably by a similar mechanism to the inhibition of albumin gene expression by vitamin B6 (Oka 2001). Antiproliferative effect of vitamin B6 on cultured cells is also described at the Section 13.2.2.

DiSorbo et al. (1985) reported that mice pretreated with pyridoxal and then injected with B16 melanoma cells had a 62% reduction in tumor weight. Even when tumors were first established in mice and then treated with pyridoxal, a 39% reduction in tumor growth was observed. In a chemical carcinogen model of colon cancer Komatsu et al. (2001) reported that vitamin B6 supplemented diets suppressed tumorigenesis by reducing cell proliferation compared with a low vitamin B6 diet. Interestingly, cell proliferation in all of the colonic epithelium was significantly correlated with the number of tumors and expression of c-myc and c-fos proteins in the colonic crypts were significantly reduced by supplemented vitamin B6. This observation suggests that suppression of cell proliferation by vitamin B6 might be conveyed through reduced expression of cell cycle control genes.

13.3.4 Angiogenesis

In an ex vivo serum-free matrix culture model using rat aortic ring, Matsubara et al. (2001) found that high concentrations of PLP (2.5 and 5 mM) and pyridoxine (5 mM) completely inhibited microvessel outgrowth. Inhibition of PLP on microvessel outgrowth was in a dose-dependent manner within the range of 25–500 μM. These results suggested that the anti-tumor effect of high levels of vitamin B6 might be mediated through suppression of angiogenesis. Most recently same group reported that vitamin B6 suppresses endothelial progenitor cells in circulation, which is associated with tumor induced angiogenesis (Matsubara et al. 2004). Further studies are needed to clarify whether vitamin B6 suppresses carcinogenesis by reducing angiogenesis.

13.3.5 One-Carbon Metabolism

One-carbon metabolism is a network of interrelated biochemical reactions which regulate transfer of the one-carbon moiety into biochemical pathways essential for DNA synthesis (thymidylate and purine synthesis) and biological methylation including DNA methylation (Fig. 13.1) (MacKenzie 1984). Serine hydroxymethyltransferase, a PLP containing enzyme, catalyzes the reversible transfer of methyl group from serine to tetrahydrofolate (THF) to generate glycine and 5,10 methyleneTHF, and this reaction plays a central role in one-carbon metabolism for nucleotide synthesis and remethylation of homocysteine (Shane 1989; Townsend et al. 2004). Homocysteine, a metabolite of one-carbon metabolism, also condenses with serine to form cystathionine in an irreversible reaction catalyzed by a PLP-containing enzyme, cystathionine-β-synthase (transsulfuration) (Craig 2004). Cystathionine is then converted to cysteine, a precursor of glutathione and taurine, catalyzed by another PLP containing enzyme, cystathionine-γ-lyase.

Animal studies demonstrated that vitamin B6 deficiency reduces activities of hepatic serine hydroxymethyltransferase and cystathionine-β-synthase (Martinez et al. 2000), while high dietary vitamin B6 enhances hepatic serine hydroxymethyltransferase enzyme activity (Scheer et al. 2005). Human studies have demonstrated that vitamin B6 deficiency causes a postprandial hyperhomocysteinemia, especially after oral methionine loading, by retarding transsulfuration pathway (Bleie et al. 2004). Taken together, vitamin B6 depletion substantially affects one carbon metabolism that controls both nucleotide synthesis and DNA methylation, alterations of which have been regarded as molecular mechanisms for colonic carcinogenesis.

Evidence suggests that vitamin B6 modulates colorectal carcinogenesis through one-carbon metabolism. As described at the Section 13.2.1.1 recent epidemiologic

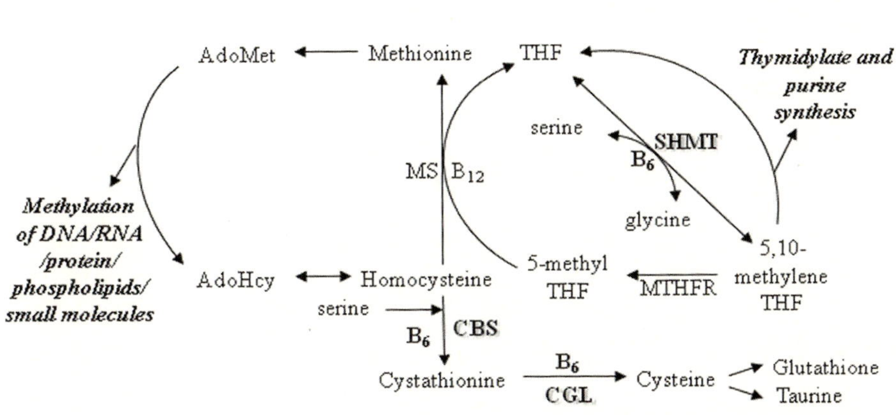

Fig. 13.1 Vitamin B6 in one-carbon metabolism. SHMT: serine hydroxymethyltransferase, CBS: cystathionine-β-synthase, CGL: cystathionine-γ- lyase, MTHFR: methylenetetrahydrofolate reductase, MS: methionine synthase, AdoMet: S-adenosylmethionine, AdoHcy: S-adenosylhomocysteine, THF: tetrahydrofolate

studies have demonstrated that an interaction between vitamin B6 and MTHFR gene can affect the risk of colorectal cancer, revealing the importance of one-carbon metabolism in the vitamin B6 associated carcinogenesis.

13.3.6 Other Mechanisms

13.3.6.1 Immune Function

Even though there is no direct evidence, it emerges that vitamin B6 supplementation might reduce cancer risk by enhancing immune function, because in animal and human studies vitamin B6 deficiency altered both humoral and cell-mediated immune responses (Rall and Meydani 1993) and vitamin B6 supplementation increased total lymphocyte count, T-helper and T-suppressor cell numbers, and the percentage of T-lymphocyte cells and T-suppressors in critically ill patients (Cheng et al. 2006). In an athymic nude mouse model injected with human malignant melanoma xenografts, Gebhard et al. (1990) found that tumor-bearing mice fed a diet containing 61.6 mg pyridoxine/kg had significantly greater lipopolysaccharide-induced spleen cell proliferation compared with mice fed the control diet containing pyridoxine 4.1 mg/kg. However, tumor incidence and progression was not modified by a high level of dietary vitamin B6.

13.3.6.2 Regulation of Steroid Hormone

Vitamin B6 is known to modulate transcriptional activation by multiple members of the steroid hormone receptor super family, which play a critical role in the development of breast cancer. In a cultured cell study estrogen-induced gene expression was reduced by 30% under conditions of elevated intracellular vitamin B6 and enhanced by 85% in vitamin deficiency (Allgood and Cidlowski 1992).

13.3.6.3 Niacin Deficiency

Niacin is largely obtained from the diet but a certain portion is manufactured from tryptophan. Since vitamin B6 involves in this pathway, vitamin B6 deficiency reduces the conversion from tryptophan to niacin (Shibata et al. 1995) and thereby can induce niacin deficiency, which has a potential to influence DNA repair, genomic stability (single- and double-strand breaks, oxidative lesions, or both), and the immune system, eventually having an impact on cancer risk (Kirkland 2003; Ames 2001).

13.3.6.4 Inhibition of DNA Replication

Vitamin B6 has a physiologic role in maintaining the proper structure of the chromosome by controlling the activities of replicative DNA polymerase and DNA topoisomerases I and II (Matsubara et al. 2003). In a cultured cell study PLP inhibited eukaryotic DNA polymerases, DNA polymerase α and ε, and cell proliferation (Mizushina et al. 2003). These effects may relate to the anti-cancer effect of vitamin B6.

13.4 Clinical Implication

In contrast to case-control or prospective cohort studies, most of intervention trials with individual B-vitamins or other nutrients have frequently failed to reach the significant chemopreventive effect, lending that an isolated nutrient in pill form may not have the same biologic effect as a nutrient in the food. The most plausible explanation is that food sources have more physiologic forms as well as other nutrients which can add synergistic or stabilizing effects. It also seems that supplementation with a high-dose single form of vitamin B6 might alter the normal metabolism of other nutrients or target enzyme reactions, because; (1) oral pyridoxine supplementation with 120 mg/day for 5 weeks reduced serum folate concentration as well as total plasma homocysteine concentration in healthy subjects (Mansoor et al. 1999), (2) old studies suggested that supplementation with PLP can suppress thymidylate synthesis which is critical for DNA replication and DNA repair (Huang et al. 1992; Chen et al. 1989). In this regard vitamin B6 supplementation with an isolate form such as pyridoxine might not be sufficient enough to meet the goal of chemoprevention without co-supplementation of substrates for the target enzyme or other nutrients for its metabolism. To clarify whether vitamin B6 supplementation is clinically useful for cancer chemoprevention, it needs well-designed clinical trials considering the optimal dose, timing, and supplement formula as well as subjects' age and gene polymorphisms.

13.5 Conclusion and Future Perspectives

A body of evidence suggests that low dietary vitamin B6 increases and high dietary vitamin B6 reduces the risk of cancer. However, epidemiologic observations are not always consistent and despite decades of studies the exact mechanism(s) by which vitamin B6 conveys the modulation effect on carcinogenesis is not yet known. Since vitamin B6 involves in more than 100 biological enzyme reactions and thereby directly or indirectly affects multiple cellular and molecular pathways implicated in carcinogenesis, it appears that vitamin B6 influences carcinogenesis in a complex fashion. Studies for nutrient and gene interactions in cancer also suggest that common gene polymorphisms modulate the effect of vitamin B6 on carcinogenesis.

Even though current intervention trials do not support the clinical use of vitamin B6 supplementation for cancer chemoprevention, this vitamin still has a high potential to prevent or delay the onset of cancer because many evidences suggest that high dietary vitamin B6 reduces the risk of cancer. Future studies are needed to find the concrete evidence and to define the exact mechanism for vitamin B6 associated carcinogenesis.

References

Allgood VE, Cidlowski JA (1992) Vitamin B6 modulates transcriptional activation by multiple members of the steroid hormone receptor superfamily. J Biol Chem 267:3819–3824

Ames BN (2001) DNA damage from micronutrient deficiencies is likely to be a major cause of cancer. Mutat Res 475:7–20

Ames BN, Elson-Schwab I, Silver EA (2002) High-dose vitamin therapy stimulates variant enzymes with decreased coenzyme binding affinity (increased K(m)): Relevance to genetic disease and polymorphisms. Am J Clin Nutr 75:616–658

Bender D (2003) Vitamin B-6. In: Bender D (ed) Nutritional biochemistry of the vitamins. Cambridge University Press, Cambridge, pp 232–269

Bleie O, Refsum H, Ueland PM, Vollset SE, Guttormsen AB, Nexo E, Schneede J, Nordrehaug JE, Nygard O (2004) Changes in basal and postmethionine load concentrations of total homocysteine and cystathionine after B vitamin intervention. Am J Clin Nutr 80:641–648

Byar D, Blackard C (1977) Comparisons of placebo, pyridoxine, and topical thiotepa in preventing recurrence of stage I bladder cancer. Urology 10:556–561

Chen SC, Daron HH, Aull JL (1989) Inhibition of thymidylate synthase by pyridoxal phosphate. Int J Biochem 21:1217–1221

Cheng CH, Chang SJ, Lee BJ, Lin KL, Huang YC (2006) Vitamin B6 supplementation increases immune responses in critically ill patients. Eur J Clin Nutr 60:1207–1213

Cho E, Holmes M, Hankinson SE, Willett WC (2007) Nutrients involved in one-carbon metabolism and risk of breast cancer among premenopausal women. Cancer Epidemiol Biomarkers Prev 16:2787–2790

Choi SW, Stickel F, Baik HW, Kim YI, Seitz HK, Mason JB (1999) Chronic alcohol consumption induces genomic but not p53-specific DNA hypomethylation in rat colon. J Nutr 129:1945–1950

Clayton PT (2006) B(6)-responsive disorders: a model of vitamin dependency. J Inherit Metab Dis 29:317–326

Craig SA (2004) Betaine in human nutrition. Am J Clin Nutr 80:539–549

Cravo ML, Gloria LM, Selhub J, Nadeau MR, Camilo ME, Resende MP, Cardoso JN, Leitao CN, Mira FC (1996) Hyperhomocysteinemia in chronic alcoholism: correlation with folate, vitamin B-12, and vitamin B-6 status. Am J Clin Nutr 63:220–224

DiSorbo DM, Litwack G (1982) Vitamin B6 kills hepatoma cells in culture. Nutr Cancer 3:216–222

DiSorbo DM, Nathanson L (1983) High-dose pyridoxal supplemented culture medium inhibits the growth of a human malignant melanoma cell line. Nutr Cancer 5:10–15

DiSorbo DM, Wagner R Jr, Nathanson L (1985) In vivo and in vitro inhibition of B16 melanoma growth by vitamin B6. Nutr Cancer 7:43–52

Friso S, Choi SW, Girelli D, Mason JB, Dolnikowski GG, Bagley PJ, Olivieri O, Jacques PF, Rosenberg IH, Corrocher R, Selhub J (2002) A common mutation in the 5,10-methylenetetrahydrofolate reductase gene affects genomic DNA methylation through an interaction with folate status. Proc Natl Acad Sci USA 99:5606–5611

Gebhard KJ, Gridley DS, Stickney DR, Shulz TD (1990) Enhancement of immune status by high levels of dietary vitamin B-6 without growth inhibition of human malignant melanoma in athymic nude mice. Nutr Cancer 14:15–26

Gridley DS, Stickney DR, Nutter RL, Slater JM, Shultz TD (1987) Suppression of tumor growth and enhancement of immune status with high levels of dietary vitamin B6 in BALB/c mice. J Natl Cancer Inst 78:951–959

Harnack L, Jacobs DR Jr, Nicodemus K, Lazovich D, Anderson K, Folsom AR (2002) Relationship of folate, vitamin B-6, vitamin B-12, and methionine intake to incidence of colorectal cancers. Nutr Cancer 43:152–158

Hartman TJ, Woodson K, Stolzenberg-Solomon R, Virtamo J, Selhub J, Barrett MJ, Albanes D (2001) Association of the B-vitamins pyridoxal 5′-phosphate (B(6)), B(12), and folate with lung cancer risk in older men. Am J Epidemiol 153:688–694

Huang S, Parish EJ, Aull JL (1992) Irreversible inhibition of thymidylate synthase by pyridoxine (B6) analogues. J Enzyme Inhib 6:149–156

Jain SK, Lim G (2001) Pyridoxine and pyridoxamine inhibits superoxide radicals and prevents lipid peroxidation, protein glycosylation, and ($Na^+ + K^+$)-ATPase activity reduction in high glucose-treated human erythrocytes. Free Radic Biol Med 30:232–237

Jaiswal M, LaRusso NF, Gores GJ (2001) Nitric oxide in gastrointestinal epithelial cell carcinogenesis: Linking inflammation to oncogenesis. Am J Physiol Gastrointest Liver Physiol 281, G626–634

Jansen MC, Bueno-de-Mesquita HB, Buzina R, Fidanza F, Menotti A, Blackburn H, Nissinen AM, Kok FJ, Kromhout D (1999) Dietary fiber and plant foods in relation to colorectal cancer mortality: the Seven Countries Study. Int J Cancer 81:174–179

Joosten E, van den Berg A, Riezler R, Naurath HJ, Lindenbaum J, Stabler SP, Allen RH (1993) Metabolic evidence that deficiencies of vitamin B-12 (cobalamin), folate, and vitamin B-6 occur commonly in elderly people. Am J Clin Nutr 58:468–476

Kang S-S, Wong PW K., Zhou J, Sora J, Lessick M, Ruggie N, Grcevich G (1988) Thermolabile methylenetetrahydrofolate reductase in patients with coronary artery disease. Metabolism 37:611–613

Key TJ, Silcocks PB, Davey GK, Appleby PN, Bishop DT (1997) A case-control study of diet and prostate cancer. Br J Cancer 76:678–687

Kirkland JB (2003) Niacin and carcinogenesis. Nutr Cancer 46:110–118

Kojima M, Morisaki T, Tsukahara Y, Uchiyama A, Matsunari Y, Mibu R, Tanaka M (1999) Nitric oxide synthase expression and nitric oxide production in human colon carcinoma tissue. J Surg Oncol 70:222–229

Komatsu SI, Watanabe H, Oka T, Tsuge H, Nii H, Kato N (2001) Vitamin B-6-supplemented diets compared with a low vitamin B-6 diet suppress azoxymethane induced colon tumorigenesis in mice by reducing cell proliferation. J Nutr 131:2204–2207

Komatsu S, Watanabe H, Oka T, Tsuge H, Kat N (2002) Dietary vitamin B6 suppresses colon tumorigenesis, 8-hydroxyguanosine, 4-hydroxynonenal, and inducible nitric oxide synthase protein in azoxymethane-treated mice. J Nutr Sci Vitaminol (Tokyo) 48:65–68

Komatsu S, Isobe M, Yanaka N, Kato N (2005) A high-fat diet enhances the inhibitory effect of dietary vitamin B6 on colon cell proliferation in mice. Oncol Rep 14:265–269

Koss LG, Lavin P (1971) Studies of experimental bladder carcinoma in Fischer 344 female rats. I. Induction of tumors with diet low in vitamin B6 containing N-2-fluorenylacetamide after single dose of cyclophosphamide. J Natl Cancer Inst 46:585–595

Lajous M, Lazcano-Ponce E, Hernandez-Avila M, Willett W, Romieu I (2006) Folate, vitamin B(6), and vitamin B(12) intake and the risk of breast cancer among Mexican women. Cancer Epidemiol Biomarkers Prev 15:443–448

Lamm DL, Riggs DR, Shriver JS, vanGilder PF, Rach JF, DeHaven JI (1994) Megadose vitamins in bladder cancer: A double-blind clinical trial. J Urol 151:21–26

Larsson SC, Giovannucci E, Wolk A (2005) Vitamin B6 intake, alcohol consumption, and colorectal cancer: A longitudinal population-based cohort of women. Gastroenterology 128:1830–1837

Le Marchand L, Donlon T, Hankin JH, Kolonel LN, Wilkens LR, Seifried A (2002) B-vitamin intake, metabolic genes, and colorectal cancer risk (United States). Cancer Causes Control 13:239–248

Lim U, Schenk M, Kelemen LE, Davis S, Cozen W, Hartge P, Ward MH, Stolzenberg-Solomon R (2005) Dietary determinants of one-carbon metabolism and the risk of non-Hodgkin's lymphoma: NCI-SEER case-control study, 1998–2000. Am J Epidemiol 162:953–964

Lim U, Wang SS, Hartge P, Cozen W, Kelemen LE, Chanock S, Davis S, Blair A, Schenk M, Rothman N, Lan Q (2007) Gene-nutrient interactions among determinants of folate and one-carbon metabolism on the risk of non-Hodgkin lymphoma: NCI-SEER Case-Control Study. Blood 109:3050–3059

Lim U, Weinstein S, Albanes D, Pietinen P, Teerenhovi L, Taylor PR, Virtamo J, Stolzenberg-Solomon R (2006) Dietary factors of one-carbon metabolism in relation to non-Hodgkin lymphoma and multiple myeloma in a cohort of male smokers. Cancer Epidemiol Biomarkers Prev 15:1109–1114

Lin J, Lee IM, Cook NR, Selhub J, Manson JE, Buring JE, Zhang SM (2008) Plasma folate, vitamin B-6, vitamin B-12, and risk of breast cancer in women. Am J Clin Nutr 87:734–743

MacKenzie R (1984) Biogenesis and interconversion of substituted tetrahydrofolates, vol 1. Wiley, New York

Mansoor MA, Kristensen O, Hervig T, Bates CJ, Pentieva K, Vefring H, Osland A, Berge T, Drablos PA, Hetland O, Rolfsen S (1999) Plasma total homocysteine response to oral doses of folic acid and pyridoxine hydrochloride (vitamin B6) in healthy individuals. Oral doses of vitamin B6 reduce concentrations of serum folate. Scand J Clin Lab Invest 59:139–146

Martinez M, Cuskelly GJ, Williamson J, Toth JP, Gregory JF 3rd (2000) Vitamin B-6 deficiency in rats reduces hepatic serine hydroxymethyltransferase and cystathionine beta-synthase activities and rates of in vivo protein turnover, homocysteine remethylation and transsulfuration. J Nutr 130:1115–1123

Matsubara K, Mori M, Matsuura Y, Kato N (2001) Pyridoxal 5′-phosphate and pyridoxal inhibit angiogenesis in serum-free rat aortic ring assay. Int J Mol Med 8:505–508

Matsubara K, Matsumoto H, Mizushina Y, Lee JS, Kato N (2003) Inhibitory effect of pyridoxal 5′-phosphate on endothelial cell proliferation, replicative DNA polymerase and DNA topoisomerase. Int J Mol Med 12:51–55

Matsubara K, Mori M, Akagi R, Kato N (2004) Anti-angiogenic effect of pyridoxal 5′-phosphate, pyridoxal and pyridoxamine on embryoid bodies derived from mouse embryonic stem cells. Int J Mol Med 14:819–823

Molina A, Oka T, Munoz SM, Chikamori-Aoyama M, Kuwahata M, Natori Y (1997) Vitamin B6 suppresses growth and expression of albumin gene in a human hepatoma cell line HepG2. Nutr Cancer 28:206–211

Mizushina Y, Xu X, Matsubara K, Murakami C, Kuriyama I, Oshige M, Takemura M, Kato N, Yoshida H, Sakaguchi K (2003) Pyridoxal 5′-phosphate is a selective inhibitor in vivo of DNA polymerase alpha and epsilon. Biochem Biophys Res Commun 312:1025–1032

Newling DW, Robinson MR, Smith PH, Byar D, Lockwood R, Stevens I, De Pauw M, Sylvester R (1995) Tryptophan metabolites, pyridoxine (vitamin B6) and their influence on the recurrence rate of superficial bladder cancer. Results of a prospective, randomised phase III study performed by the EORTC GU Group. EORTC Genito-Urinary Tract Cancer Cooperative Group. Eur Urol 27:110–116

O'Callaghan P, Meleady R, Fitzgerald T, Graham I (2002) Smoking and plasma homocysteine. Eur Heart J 23:1580–1586

Oka T (2001) Modulation of gene expression by vitamin B6. Nutr Res Rev 14:257–265

Otani T, Iwasaki M, Hanaoka T, Kobayashi M, Ishihara J, Natsukawa S, Shaura K, Koizumi Y, Kasuga Y, Yoshimura K, Yoshida T, Tsugane S (2005) Folate, vitamin B6, vitamin B12, and vitamin B2 intake, genetic polymorphisms of related enzymes, and risk of colorectal cancer in a hospital-based case-control study in Japan. Nutr Cancer 53:42–50

Pelucchi C, Galeone C, Talamini R, Negri E, Parpinel M, Franceschi S, Montella M, La Vecchia C (2005) Dietary folate and risk of prostate cancer in Italy. Cancer Epidemiol Biomarkers Prev 14:944–948

Potera C, Rose DP, Brown RR (1977) Vitamin B6 deficiency in cancer patients. Am J Clin Nutr 30:1677–1679

Poulsen HE (2005) Oxidative DNA modifications. Exp Toxicol Pathol 57(Suppl 1):161–169

Prior FG (1985) Theoretical involvement of vitamin B6 in tumour initiation. Med Hypotheses 16:421–428

Rall LC, Meydani SN (1993) Vitamin B6 and immune competence. Nutr Rev 51:217–225

Ren SG, Melmed S (2006) Pyridoxal phosphate inhibits pituitary cell proliferation and hormone secretion. Endocrinology 147:3936–3942

Scheer JB, Mackey AD, Gregory JF 3rd (2005) Activities of hepatic cytosolic and mitochondrial forms of serine hydroxymethyltransferase and hepatic glycine concentration are affected by vitamin B-6 intake in rats. J Nutr 135:233–238

Schernhammer E, Wolpin B, Rifai N, Cochrane B, Manson JA, Ma J, Giovannucci E, Thomson C, Stampfer MJ, Fuchs C (2007) Plasma folate, vitamin B6, vitamin B12, and homocysteine and pancreatic cancer risk in four large cohorts. Cancer Res 67:5553–5560

Selhub J, Jacques PF, Wilson PW, Rush D, Rosenberg IH (1993) Vitamin status and intake as primary determinants of homocysteinemia in an elderly population. JAMA 270:2693–2698

Shane B (1989) Folylpolyglutamate synthesis and role in the regulation of one-carbon metabolism. Vitam Horm 45:263–335

Sharp L, Little J, Brockton NT, Cotton SC, Masson LF, Haites NE, Cassidy J (2008) Polymorphisms in the methylenetetrahydrofolate reductase (MTHFR) gene, intakes of folate and related B vitamins and colorectal cancer: A case-control study in a population with relatively low folate intake. Br J Nutr 99:379–389

Shi Q, Zhang Z, Li G, Pillow PC, Hernandez LM, Spitz MR, Wei Q (2005) Sex differences in risk of lung cancer associated with methylene-tetrahydrofolate reductase polymorphisms. Cancer Epidemiol Biomarkers Prev 14:1477–1484

Shibata K, Mushiage M, Kondo T, Hayakawa T, Tsuge H (1995) Effects of vitamin B6 deficiency on the conversion ratio of tryptophan to niacin. Biosci Biotechnol Biochem 59:2060–2063

Shimada D, Fukuda A, Kawaguchi H, Kato N, Yoshida H, Kanouchi H, Oka T (2005) Effect of high dose of pyridoxine on mammary tumorigenesis. Nutr Cancer 53:202–207

Shimada D, Fukuda A, Kanouchi H, Matsumoto M, Oka T (2006) Vitamin B6 suppresses growth of the feline mammary tumor cell line FRM. Biosci Biotechnol Biochem 70:1038–1040

Shrubsole MJ, Jin F, Dai Q, Shu XO, Potter JD, Hebert JR, Gao YT, Zheng W (2001) Dietary folate intake and breast cancer risk: Results from the Shanghai Breast Cancer Study. Cancer Res 61:7136–7141

Shrubsole MJ, Gao YT, Cai Q, Shu XO, Dai Q, Hebert JR, Jin F, Zheng W (2004) MTHFR polymorphisms, dietary folate intake, and breast cancer risk: Results from the Shanghai Breast Cancer Study. Cancer Epidemiol Biomarkers Prev 13:190–196

Shultz TD, Santamaria AG, Gridley DS, Stickney DR, Slater JM (1988) Effect of pyridoxine and pyridoxal on the in vitro growth of human malignant melanoma. Anticancer Res 8:1313–1318

Slattery ML, Potter JD, Coates A, Ma KN, Berry TD, Duncan DM, Caan BJ (1997) Plant foods and colon cancer: An assessment of specific foods and their related nutrients (United States). Cancer Causes Control 8:575–590

Slattery ML, Potter JD, Samowitz W, Schaffer D, Leppert M (1999) Methylenetetrahydrofolate reductase, diet, and risk of colon cancer. Cancer Epidemiol Biomarkers Prev 8:513–518

Stickel F, Choi SW, Kim YI, Bagley PJ, Seitz HK, Russell RM, Selhub J, Mason JB (2000) Effect of chronic alcohol consumption on total plasma homocysteine level in rats. Alcohol Clin Exp Res 24:259–264

Stolzenberg-Solomon RZ, Albanes D, Nieto FJ, Hartman TJ, Tangrea JA, Rautalahti M, Sehlub J, Virtamo J, Taylor PR (1999) Pancreatic cancer risk and nutrition-related methyl-group availability indicators in male smokers. J Natl Cancer Inst 91:535–541

Stolzenberg-Solomon RZ, Pietinen P, Barrett MJ, Taylor PR, Virtamo J, Albanes D (2001) Dietary and other methyl-group availability factors and pancreatic cancer risk in a cohort of male smokers. Am J Epidemiol 153:680–687

Taysi S (2005) Oxidant/antioxidant status in liver tissue of vitamin B6 deficient rats. Clin Nutr 24:385–389

Thanassi JW, Nutter LM, Meisler NT, Commers P, Chiu JF (1981) Vitamin B6 metabolism in Morris hepatomas. J Biol Chem 256:3370–3375

Theodoratou E, Farrington SM, Tenesa A, McNeill G, Cetnarskyj R, Barnetson RA, Porteous ME, Dunlop MG, Campbell H (2008) Dietary vitamin B6 intake and the risk of colorectal cancer. Cancer Epidemiol Biomarkers Prev 17:171–182

Townsend JH, Davis SR, Mackey AD, Gregory JF 3rd (2004) Folate deprivation reduces homocysteine remethylation in a human intestinal epithelial cell culture model: Role of serine in one-carbon donation. Am J Physiol Gastrointest Liver Physiol 286:G588–G595

Tryfiates GP, Morris HP (1976) Growth of Morris hepatoma No. 7794A with and without vitamin B6. Effect of inoculation time. Eur J Cancer 12:9–12

Ulrich CM, Kampman E, Bigler J, Schwartz SM, Chen C, Bostick R, Fosdick L, Beresford SA, Yasui Y, Potter JD (1999) Colorectal adenomas and the C677T MTHFR polymorphism: Evidence for gene-environment interaction? Cancer Epidemiol Biomarkers Prev 8:659–668

Vermaak WJ, Ubbink JB, Barnard HC, Potgieter GM, van Jaarsveld H, Groenewald AJ (1990) Vitamin B-6 nutrition status and cigarette smoking. Am J Clin Nutr 51:1058–1061

Wachstein M, Kellner JD, Ortiz JM (1960) Pyridoxal phosphate in plasma and leukocytes in patients with leukemia and other diseases. Proc Soc Exp Biol Med 105:563–566

Wei EK, Giovannucci E, Selhub J, Fuchs CS, Hankinson SE, Ma J (2005) Plasma vitamin B6 and the risk of colorectal cancer and adenoma in women. J Natl Cancer Inst 97:684–692

Weinstein SJ, Hartman TJ, Stolzenberg-Solomon R, Pietinen P, Barrett MJ, Taylor PR, Virtamo J, Albanes D (2003) Null association between prostate cancer and serum folate, vitamin B(6), vitamin B(12), and homocysteine. Cancer Epidemiol Biomarkers Prev 12:1271–1272

Weinstein SJ, Stolzenberg-Solomon R, Pietinen P, Taylor PR, Virtamo J, Albanes D (2006) Dietary factors of one-carbon metabolism and prostate cancer risk. Am J Clin Nutr 84:929–935

Wu K, Helzlsouer KJ, Comstock GW, Hoffman SC, Nadeau MR, Selhub J (1999) A prospective study on folate, B12, and pyridoxal 5′-phosphate (B6) and breast cancer [In Process Citation]. Cancer Epidemiol Biomarkers Prev 8:209–217

Yanaka N, Koyama TA, Komatsu S, Nakamura E, Kanda M, Kato N (2005) Vitamin B6 suppresses NF-kappaB activation in LPS-stimulated mouse macrophages. Int J Mol Med 16:1071–1075

Zhang SM, Willett WC, Selhub J, Hunter DJ, Giovannucci EL, Holmes MD, Colditz GA, Hankinson SE (2003) Plasma folate, vitamin B6, vitamin B12, homocysteine, and risk of breast cancer. J Natl Cancer Inst 95:373–380

Zhang SM, Moore SC, Lin J, Cook NR, Manson JE, Lee IM, Buring JE (2006) Folate, vitamin B6, multivitamin supplements, and colorectal cancer risk in women. Am J Epidemiol 163:108–115

Chapter 14
Vitamin B6 and Cardiovascular Disease

Simonetta Friso, V. Lotto, R. Corrocher, and Sang Woon Choi

Abstract While overt vitamin B6 deficiency is not a frequent finding nowadays in medical practice, evidence suggests that insufficiency of this vitamin is rather widespread in a quite large portion of the population such as the elderly or in not unusual conditions such as that of alcohol addiction. Moreover, a mild deficiency in B6 vitamin is a state that may be associated with an increased risk of cardiovascular disease. Epidemiologic evidence from case control and prospective studies have suggested that low dietary intake or reduced blood concentrations of vitamin B6 is associated with an increased risk of cardiovascular disease, although most recent trials demonstrated the ineffectiveness of vitamin B6 supplementation on the prevention of cardiovascular events recurrence. Due to limited and somewhat inconsistent data together with the ample variety of critical functions in which vitamin B6 is involved in the human body, it is very challenging to attempt at establishing a cause and effect relationship between vitamin B6 and risk of cardiovascular disease as it is to delineate the exact mechanism(s) by which vitamin B6 may modulate such risk. In the present chapter we review the currently available knowledge deriving from both epidemiological and mechanistic studies designed to define potential candidate mechanisms for the association of vitamin B6 impairment and risk of cardiovascular disease development.

Keywords Vitamin B6 · Coronary artery disease · Stroke peripheral Vascular disease · Atherosclerosis · Supplementation · One-carbon metabolism

Abbreviations

CAD	coronary artery disease
CHD	coronary heart disease
CI	confidence interval
MI	myocardial infarction

S. Friso (✉)
Tufts University, Medford/Somerville, MA, USA

University of Verona School of Medicine, Department of Medicine, 37134 Verona, Italy
e-mail: simonetta.friso@univr.it

O. Stanger (ed.), *Water Soluble Vitamins*, Subcellular Biochemistry 56,
DOI 10.1007/978-94-007-2199-9_14, © Springer Science+Business Media B.V. 2012

OR odds ratio
PLP pyridoxal-5′-phosphate

14.1 Introduction

Following the observation of the presence of atherosclerotic lesions in monkeys fed a pyridoxine-deficient diet (Rinehart and Greenberg 1949; Rinehart and Greenberg 1956), the hypothesis regarding the role of vitamin B6 in atherogenesis has been tested by several subsequent human studies which addressed the significance of this vitamin as an independent risk factor for atherothrombotic disease (Chasan-Taber et al. 1996; Dalery et al. 1995; Folsom et al. 1998; Friso et al. 2004, Kok et al. 1989, Rimm et al. 1998; Robinson et al. 1996, 1998; Verhoef et al. 1996). Recent reports also highlighted the possible relationship of vitamin B6 with cerebrovascular disease and, more specifically, with ischemic stroke and transient ischemic attack (Kelly et al. 2003, 2004).

Vitamin B6 is a coenzyme necessary for a large number of enzymes which are mainly involved in amino acid metabolism but also lipid and carbohydrate metabolic pathways (Shane 2000).

Because vitamin B6 is one of the major determinants of total plasma homocysteine concentrations (Dalery et al. 1995; Miller et al. 1992; Selhub et al. 1993, 1995), a well-known predictor for atherosclerotic disease (Eikelboom et al. 1999; Refsum and Ueland 1998; Stanger et al. 2003), the role of vitamin B6 in coronary artery disease (CAD) risk has been mostly addressed through its coenzymatic function for the degradation of homocystiene via the transsulfuration pathway in one-carbon metabolism (Fig. 14.1) (Selhub and Miller 1992). Measurements of total plasma homocysteine were not available in a number of the early studies which claimed a correlation of vitamin B6 with CAD (Kok et al. 1989; Labadarios et al. 1987; Serfontein et al. 1985; Vermaak et al. 1986, 1987). Therefore, the association of low plasma pyridoxal-5′-phosphate (PLP, the active form of vitamin B6) and coronary artery disease was not adjusted for this factor. However, most recent studies, in which measurement of total plasma homocysteine concentration was available, have clearly demonstrated that the association between vitamin B6 and the risk of CAD is independent from the homocysteine pathway at least for a significant part (Folsom et al. 1998; Friso et al. 2004; Robinson et al. 1995).

The magnitude of the association between low plasma vitamin B6 and CAD is rather mild in most studies (Folsom et al. 1998; Friso et al. 2004; Robinson et al. 1995), although this is not surprising but, rather, it is absolutely compatible with the multifactorial pathogenesis of atherosclerosis, in which there is an interplay of several genetic, nutritional and life style-related factors. Atherosclerosis is a complex disease that represents the final outcome of the interaction of many different contributory agents. Furthermore, the penetrance of different risk factors is variable in each individual. This idea of risk factor is important because it provides the concept of individually-tailored definition of risk and may also help to design specific and more effective intervention strategies.

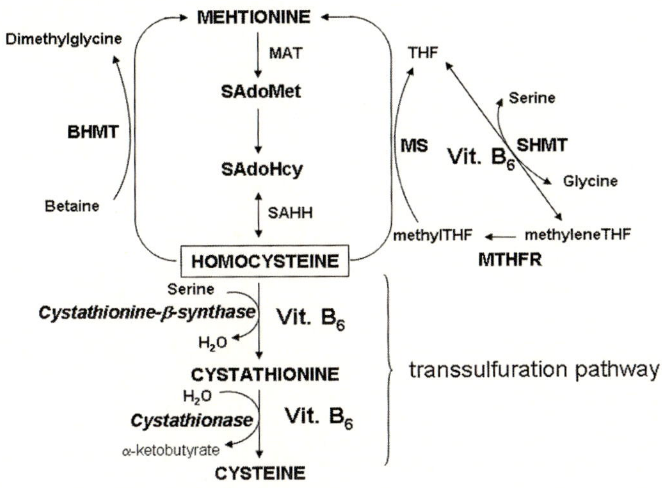

Fig. 14.1 A schematic representation of one-carbon metabolism in which vitamin B6 acts as a major player for its functions as a coenzyme in a number of enzymatic reactions. In particular, in the transsulfuration pathway, vitamin B6 acts as a coenzyme of cystathionine-β-synthase for the conversion of homocysteine to cystathionine and of cystathionase for the synthesis of cysteine from cystathionine. Vitamin B6 functions as a coenzyme for SHMT (serine hydroxymethyltransferase) which catalyzes the conversion of tetrahysdrofolate (THF) to methylenetetrahydrofolate (MTHF). Other abbreviations: MTHFR: methylenetetrahydrofolate reductase; BHMT: betaine hydroxymethyltransferase; MS: methionine synthase; SAdoMet. S-adenosylmethionine; SAdoHcy: S-adenosylhomocysteine; MAT: methionine adenosyl transferase; SAHH: S-adenosylhomocysteine hydrolase

The mechanisms for a vitamin B6-related atherogenesis are still poorly understood. Nevertheless, considering the epidemiologic evidence and understanding some of the potentially implicated mechanisms for the relationship between low vitamin B6 and atherosclerotic disease may facilitate, by correcting the vitamin status, the design of potentially easy-to-perform preventive and therapeutic measures.

In the present chapter, we address the currently available knowledge concerning epidemiologic evidence and propose candidate mechanisms for the association between vitamin B6 and cardiovascular disease risk.

14.2 Epidemiologic Evidence for Vitamin B6 as a Risk Factor for Cardiovascular Disease

14.2.1 Vitamin B6 and Coronary Artery Disease

Several studies report an association between low plasma vitamin B6 and higher risk of coronary artery disease (CAD) (Table 14.1) as well as its major thrombotic complication, myocardial infarction (Dalery et al. 1995; Folsom et al. 1998; Friso

Table 14.1 Epidemiologic studies which evaluated low vitamin B6 as a risk factor for cardiovascular disease

Author and References	Study design	Main findings
Kok FJ et al., Am J Cardiol 1989, Rotterdam, The Netherlands	Case-control	Multivariate adjusted relative odds of MI for subjects in the lowest plasma PLP quartile (< 20 nmol/L) – 5.2 (95% CI 1.4–18.9)
Dalery K et al., Am J Cardiol 1995, Québec, Canada	Case-control	Lower PLP levels in CAD (angiographically documented) than in controls ($P < 0.005$)
Robinson K et al., Circulation 1995, Cleveland, OH, USA	Case-control	In patients with low vitamin B6, OR for CAD adjusted for multiple risk factors including tHcy – 4.3 (95% CI 1.1–16.9, $P < 0.05$)
Chasan-Taber L et al., J Am Coll Nutr 1996, Boston, MA, USA	Case-control	RR of MI for lowest 20th percentile of PLP (< 28.9 nmol/L) in a multivariate model – 1.5 (95% CI 1.0–2.2); if adjusted for folate: RR 1.3 (95% CI 0.9–2.1)
Verhoef P et al., Am J Epidemiol 1996, Boston Area Health Study; Boston, MA, USA	Case-control	Dietary and plasma vitamin B6 lower in cases than controls. OR for the highest quintile of plasma PLP (>88.9 nmol/L) – 0.32 (0.11–0.59) when age and sex adjusted; 0.51 (0.19–1.36) after multiple adjustments
Verhoef P et al., Arterioscler Thromb Vasc Biol 1997, Rotterdam, The Netherlands	Case-control	CAD patients (angiographically documented) had plasma PLP lower than controls but no statistical significance ($P = 0.08$)
Folsom AR et al., Circulation 1998, ARIC Study, Minneapolis, MN, USA	Prospective case-cohort	RR between extreme quintiles of PLP – 0.28 (95% CI 0.1–0.7, $P = 0.001$)
Robinson K et al., Circulation 1998, European COMAC Group	Case-control	PLP< 23.3 nmol/L (20th percentile) associated to increase risk for atherosclerosis, independently of tHcy. (CAD + PAD + CVD: clinic and/or angiography documented)
Siri PW et al., J Am Coll Nutr 1998, Rotterdam, The Netherlands	Case-control	OR for low vitamin B6 – 0.86 (95% CI 0.33–2.22) (patients with severe angiography documented CAD)
Rimm EB et al., JAMA 1998, NHS, Boston, MA, USA	Prospective case-cohort	RR for vitamin B6 intake between extreme quintiles, adjusted for CVD risk factors – 0.67 (95% CI 0.53–0.85, $P = 0.002$). Reduced CAD risk among women who regularly used multiple vitamins (RR = 0.76, 95% CI 0.65–0.90)

Table 14.1 (continued)

Nahlawi M, Am J Cardiol 2002, Cleveland, OH, USA	Follow-up	Vitamin B6 deficiency in 21% of heart transplant recipients with -, and in 9% without CVD complications/death ($P=0.05$). RR in low vitamin B6 ($\leq$20 nmol/L) for CVD events, including death, increased 2.7 times (95% CI 1.2–5.9, $P=0.02$)
Friso S et al., Am J Clin Nutr 2004, Verona, Italy	Case-control	Significant, inverse, graded relation between PLP and both hs-CRP and fibrinogen ($P<0.001$). OR for CAD-risk related to low PLP concentration: 1.89 (95% CI 1.18–3.03; $P=0.008$)
Tavani A, Eur J Nutr 2004, Milan, Italy	Case-control	OR for acute MI risk for vitamin B6 intake – 0.34 (95% CI 0.19–0.60) for the highest versus the lowest tertile
Folsom AR et al., Arch Int Med 2006	Prospective case-cohort	Together with other markers of risk, also vitamin B6 added the most to the AUC (range 0.006–0.011)

MI myocardial infarction; *tHcy* total plasma homocysteine concentrations; *CVD* cardiovascular disease; *PLP* plasma-5′-phosphate; *CAD* coronary artery disease; *PAD* peripheral artery disease; *OR* odds ratio; *RR* relative risk; *AUC* area under the ROC (receiver operating characteristic) curve

et al. 2004; Kok et al. 1989; Labadarios et al. 1987; Nahlawi et al. 2002; Robinson et al. 1995, 1998; Serfontein et al. 1985; Siri et al. 1998; Vermaak et al. 1987). Among the earlier clinical studies conducted on patients affected by CAD, Dalery et al. described in a case-control study conducted in the province of Québec, Canada, that pyridoxal-5′-phosphate (PLP) levels, the active form of vitamin B6, were significantly lower in CAD subjects (150 angiographically documented CAD patients) than controls (584 healthy subjects) (Dalery et al. 1995). In the same year, Robinson et al. reported that plasma concentrations of PLP were lower in CAD patients compared to controls (Robinson et al. 1995). That study was initially designed to evaluate the interrelations between plasma homocysteine, a known risk factor for CAD, and vitamins of the B group which are the major determinants of plasma homocysteine levels. Interestingly, also when plasma levels of homocysteine were included in the multivariate analysis model, an odds ratio of 4.3 persisted for those subjects with low PLP (95% C.I. 1.1–13.7, $P = 0.04$). Robinson et al. concluded that low PLP is a risk factor for CAD even independently of plasma homocysteine levels (Robinson et al. 1995). In this study in which 304 patients with CAD identified by angiographic documentation were compared to 231 controls, the authors reported that vitamin B6 deficiency, defined for PLP levels lower that 20 nmol/L, was seen in about 10% of CAD patients compared to only 2% of controls subjects ($P<0.01$) and yielded an odds ratio of coronary artery disease of 3.8 (C.I. 1.1–13.7, $P=0.04$) even after adjustments for all traditional risk factors for cardiovascular disease.

Additional prospective evidence on the role of vitamin B6 in the increase incidence of coronary heart disease (CHD) was then provided from findings of a nested case-cohort design within the ARIC (The Atherosclerosis Risk in Communities) study. In an analysis primarily intended to determine whether homocysteine-related factors including B6 vitamin are associated with the incidence of CHD over an average of 3.3 years of follow-up, the authors point quite strongly to an independent role of vitamin B6 as a protective factor in case of adequate levels of the vitamin (Folsom et al. 1998). Age, -race, and sex-adjusted geometric mean levels of PLP were significantly lower in incident CHD cases ($N=232$) compared to a reference cohort sample of 537 (of whom 10 were also CHD cases) (19.0 nmol/L vs. 31.5 nmol/L; $P<0.01$). The incidence of CHD was negatively associated with PLP in both genders ($P<0.05$). The relative risk for the highest versus lowest quintile of plasma PLP was 0.28 (95% CI = 0.1–0.7; $P=0.001$). Remarkably, PLP was associated with CHD incidence even after multiple adjustments for other traditional risk-factors including total and HDL cholesterol, hypertension, diabetes and smoking status.

In a multicenter case-control study in Europe, 750 patients with documented vascular disease and 800 controls frequency-matched for age and sex were compared to identify that concentrations of vitamin B6 were significantly lower in case than in control subjects (Robinson et al. 1998). Concentrations of plasma PLP below the lowest 20th percentile for control subjects (corresponding to plasma PLP <23.3 nmol/L) was associated with an increased risk for atherosclerosis and the relationship between vitamin B6 and atherosclerosis was independent of traditional risk factors including plasma total homocysteine levels (Robinson et al. 1998). The risk associated with low vitamin B6 status (odds ratio 1.84; 95% C.I. 1.39–2.42, $P<0.001$) persisted when adjusted for the concentrations of fasting (1.76; 95% C.I. 1.33–2.34, $P<0.001$) as well as post-load total plasma homocysteine concentrations (1.79; 95% C.I. 1.35–2.37, $P<0.001$).

One of the main issues to be considered in the hypothesis of an independent relationship between low PLP and coronary atherosclerotic disease, pertains to the uncertainty on the possible mechanisms involved in such an association. The role of potential candidate mechanisms for a vitamin B6 related atherogenesis have been investigate through cell culture and animal model studies as described below in the present chapter. However, based on the hypothesis of a possible role of vitamin B6 in inflammatory processes some mechanistic hypothesis for the link between low vitamin B6 and coronary heart disease was also done using clinical studies. Indeed, a growing body of evidence supports the theory of atherosclerosis as an inflammatory disease (Libby et al. 2002; Pasceri et al. 2000; Ross 1999). Systemic markers of inflammation, such as fibrinogen and high-sensitivity C-reactive protein have been proposed to be solid predictors of CAD risk (Liuzzo et al. 1994; Ridker et al. 2000). The current understanding is that inflammation plays an essential role at all stages of the atherosclerotic process (Ross 1999) and acute-phase reactants have been proven to be strong and independent risk factors for CAD (Grundy et al. 1999; Harris et al. 2006; Ridker et al. 2004). In patients with rheumatoid arthritis, a disease that share many similarities with coronary atherosclerosis, plasma PLP levels are inversely related to tumor necrosis factor alpha production (Roubenoff et al.

1995) and other major markers of inflammation (Chiang et al. 2003). Most recently, a cross-sectional survey observed also an inverse association between levels of vitamin B6 and plasma fibrinogen concentrations (James et al. 2000; Ji et al. 2003).

Data from 891 participants of the population-based Framingham Heart Study cohort were analyzed precisely to explore a possible relationship between vitamin B6 and a major marker of acute phase. This study highlighted an inverse correlation between plasma PLP and values of C-reactive protein (Friso et al. 2001). This association appeared to be strong and independent of other biomarkers related to vitamin B6 metabolism and suggested a possible function of plasma PLP in inflammatory processes (Friso et al. 2001).

Taking into account that systemic markers of inflammation are solid predictors of CAD risk, a subsequent study was designed to investigate a cohort of angiography-defined severe multivessel CAD patients ($N=475$) compared to 267 subjects free from coronary atherosclerosis with the primary aim of evaluating the relation between plasma PLP and major markers of acute-phase reaction in affecting CAD risk (Friso et al. 2004). That study conducted in an Italian population confirmed the inverse relationship between plasma PLP and both hs-CRP and fibrinogen ($P<0.001$). The prevalence of low PLP (considered as <50th percentile in control population, i.e. 36.3 nmol/L) was significantly higher among CAD patients compared to controls ($P<0.001$). That levels of plasma PLP could be referred to a mild PLP impairment even compared to previous studies which defined a risk for CAD for PLP values as low as 20 nmol/L (Robinson et al. 1995). Yet such moderate impairment is sufficient to confer a higher risk for CAD. Furthermore, the odds ratio for CAD risk related to low PLP was 1.89 (95% CI: 1.18–3.03, $P=0.008$) even in a multivariate model in which hs-CRP, fibrinogen, total plasma homocysteine as well as all major CAD risk factors were included. However, it is to be noted that the magnitude of such association is quite bland but this is not surprising if one considers that atherosclerosis is a very complex disease in which a number of genetic, environmental and life-style factors are involved. Yet, it is interesting to note that low plasma PLP concentrations when considered in combination with hs-CRP or with an increased LDL to HDL-cholesterol ratio, significantly enhanced the risk of disease (Friso et al. 2004).

The association of low PLP and CAD, however, has not been confirmed in all studies. In a study designed to evaluate the role of plasma total homocysteine and B vitamins in the risk of coronary atherosclerotic disease, Verhoef et al. (1997) observed that cases had slightly lower PLP levels compared to controls but this difference was not statistically significant (26.9 vs. 27 nmol/L, $P=0.08$). It is to be noted that population control subjects had lower mean levels of plasma PLP (23.3 nmol/L). The authors considered such finding potentially related to the healthier dietary habits adopted by cases after the cardiology treatment. In such case-control study in which subjects were recruited from men and women undergoing coronary angiography in a Hospital in Rotterdam, the Netherlands, cases ($N=131$) were defined when at least one coronary vessel had an occlusion ≥90% and ≤40% in a second coronary artery and controls ($N=88$) when there was ≤50% vessel occlusion in only one epicardial coronary vessel.

In the same case-control study, conducted in angiography defined patients, Siri et al. (1998) subsequently evaluated whether the association between vitamins of the B group and CAD is mediated by the effect of rising plasma homocysteine levels, observed that there was no difference on plasma PLP levels between cases ($N=131$) and controls ($N=88$). The ORs for low vitamin B6 was 1.23 (95% CI: 0.53–2.83) when age and gender adjustments were performed. However, after multiple adjustments (for age, gender, heavy smoking, diastolic blood pressure, total HDL cholesterol, alcohol intake, body mass index, serum creatinine, vitamin B12 and folate, fasting total homocysteine), the OR was estimated as 0.60 (95% CI: 0.22–1.63) in the lowest quartile (cut-off point for vitamin B6 concentrations was 20.6 nmol/L) and 0.88 (95% CI: 0.34–2.31) for the highest quartile (vitamin B6 concentrations >36.7 nmol/L). The authors, therefore, concluded that vitamin B6 was not clearly related, in their study, to higher risk of atherosclerosis.

As for a more specific association between low PLP and myocardial infarction (MI), the main thrombotic complication of CAD, after a number of early and rather heterogeneous studies that defined an association between low vitamin B6 and MI (Iqbal et al. 2005; Kok et al. 1989; Labadarios et al. 1987; Serfontein et al. 1985; Tavani et al. 2004; Vermaak et al. 1987), further support for such hypothesis was suggested from additional case-control and prospective studies.

In a study conducted among Dutch patients, Kok et al., observed lower vitamin B6 status in patients with acute MI ($N=84$) compared to healthy control subjects ($N=84$) (Kok et al. 1989). The multivariate adjusted relative odds of acute MI (in which homocysteine values were not included) for subjects with plasma PLP levels <20 nmol/L (lowest quartile) was five times higher as compared with those in the highest quartile (relative odds = 5.2, 95% CI: 1.4–18.9).

Data from a nested case-control study using prospectively collected samples from the large male Physician Health Study (14,916 subjects) (Chasan-Taber et al. 1996) showed that for the lowest 20% of vitamin B6 values (PLP levels <28.9 nmol/L), the multivariate relative risk of MI was 1.5 (95% CI: 1.0–2.2). The inclusion of homocysteine in the multivariate model attenuate only slightly the relative risk to 1.3 (95% CI: 0.8–2.1). Adding total homocysteine levels to that model did not add significant predictive value. As stated by the authors, these prospective data from US physicians, although not statistically significant were compatible with the hypothesis that lower dietary intake of vitamin B6 contribute to risk of MI (Chasan-Taber et al. 1996).

In the retrospective setting of the Boston Area Health Study, conducted among men and women at first MI event admitted to the coronary or intensive care units, Verhoef at al., observed that both dietary and plasma levels of vitamin B6 were lower in cases ($N=130$) than population controls ($N=118$) and that there was an inverse association with the risk of MI that was also independent of other known risk factors in a multivariate analysis adjusted for multiple confounders (Verhoef et al. 1996). For plasma PLP, the highest two quintiles (≤88.9 and >88.9 nmol/L, respectively) showed a trend toward an inverse relationship with the risk of MI even when, among other classical risk factors, also strong confounders such as total/high density lipoprotein cholesterol ratio, cigarette smoking, body mass index or family history for diabetes, hypertension or ischemic heart disease were included in the model (OR = 0.32 for PLP levels >88.9 nmol/L). The association between pyridoxal-5′-phosphate and myocardial infarction risk was independent also of plasma folate and homocysteine, suggesting a potential protective effect of vitamin B6 through mechanisms other than those related to homocysteine metabolism.

Some authors reported early on that lower pyridoxal-5′-phosphate levels during the acute phase of myocardial infarction (Kok et al. 1989; Serfontein et al. 1985; Vermaak et al. 1986, 1987), concentrations that returned to baseline after 3–4 days (Vermaak et al. 1988). This observation may need further substantiation from other clinical and mechanistic investigation for a potential part of vitamin B6 in the acute phase of the disease.

In the prospective cohort of the Nurses' Health Study, Rimm et al. in 1998, observed that the relative risk of coronary heart disease between extreme quintiles for vitamin B6 (median intake 4.6 mg/die vs. 1.1 mg/die) was 0.67 (95% CI: 0.53–0.85, $P = 0.002$) after controlling for traditional cardiovascular risk factors including hypertension, smoking, intake of alcohol, fiber, vitamin E and saturated, polyunsaturated and trans fat (Rimm et al. 1998). During 14 years of follow-up of this large (80,082 subjects) prospective cohort study of women, Rimm et al., found graded associations between higher intakes of vitamin B6 and lower risk of CHD where the risk was lowest with higher intake of food and supplement sources combined (relative risk = 0.71; 95% CI: 0.56–0.90). Results from this population of women showed a lower risk for CHD in individuals with higher vitamin B6 levels. Remarkably, in this population of US women average intakes of vitamin B6 (3 mg/day; median = 1.7 mg/d) was well above those from national average (1.51 mg/day) but still a graded reduction in risk was detected with higher intake. The lowest risk was among women with intake of vitamin B6 above 3 mg/d. Therefore, it could be speculated that levels of vitamins that are currently defined sufficient to avoid deficiency of the vitamin among non pregnant women (current RDA 1.6 mg/day), could not be adequate for coronary heart disease prevention.

The prospective ARIC Study, in a recent evaluation, assessed the association of novel markers of risk (Folsom et al. 2006). Of the 19 markers studied, vitamin B6, together with other risk factors such as lipoprotein-associated phospholipase A2, interleukin-6 and soluble thrombomodulin added the most to the area under the receiver operating characteristic curve (Folsom et al. 2006).

In summary, vitamin B6 impairment was reported to be associated to higher risk for CAD (Robinson et al. 1995) in patients with coronary atherosclerosis established by angiographic documentation (Dalery et al. 1995; Friso et al. 2004; Robinson et al. 1995, 1998; Verhoef et al. 1996, 1997). This role for vitamin B6, which emerged as a potentially relevant risk factor for CAD, was also observed in prospective study settings (Chasan-Taber et al. 1996; Folsom et al. 1998, 2006). This association was observed to be independent of plasma homocysteine levels (Folsom et al. 1998, 2006; Friso et al. 2004; Robinson et al. 1995, 1998), a well-recognized CAD risk factor whose plasma concentrations vitamin B6 is a major determinant. The risk for CAD conferred by a mild impairment in plasma pyridoxal-5′-phosphate, the active form of vitamin B6 is rather mild (Folsom et al. 1998, 2006; Friso et al. 2004; Robinson et al. 1995, 1998; Siri et al. 1998, Verhoef et al. 1997) as expected by the multifactorial genesis of this complex disease. Remarkably, in the view of the importance of the concomitant presence of several risk factors in determining an amplification in disease risk, the role of a modest vitamin B6 impairment may have a significant part in atherosclerosis, especially considering that, if not evaluated, it could remain unknown for the absence of the clinically manifested signs of an overt deficiency. The relationship between low PLP and CAD was found to be independent also of inflammatory indexes (Friso et al. 2001; Kelly et al. 2004) which have been reported to be related to both CAD risk and associated to vitamin B6 levels (Friso et al. 2001).

14.2.2 Vitamin B6 and Peripheral Artery Disease

Data on a relationship between vitamin B6 and peripheral artery disease are few (Wilmink et al. 2004). Daily folate and vitamin B6 intakes were found to be lower in cases than in controls and they were independent predictors of peripheral artery occlusive disease. An increase daily vitamin B6 intake by one standard deviation, decreased the risk of peripheral artery disease by 29% (Wilmink et al. 2004).

14.2.3 Vitamin B6 and Stroke

In a study evaluating the association between plasma homocysteine and extracranial carotid artery stenosis, Selhub et al. described a relationship between lower vitamin B6 levels and carotid atherosclerotic disease which diminished when adjusted for homocysteine levels (Selhub et al. 1995). Recent reports from case-control studies highlighted a possible relationship between low plasma levels of vitamin B6 with the onset of cerebrovascular disease and, more specifically, with ischemic stroke and transient ischemic attack (Kelly et al. 2003, 2004; Khatami et al. 1988). Plasma PLP levels were found to be significantly lower in cases ($N = 180$) than controls ($N = 140$) with the evidence of a strong association between stroke/TIA and low PLP (considered as <20 nmol/L), with an OR of 4.6 (95% CI: 1.4–15.1; $P < 0.001$). This association was found to be independent of other vascular risk factors, vitamin supplement use, and also total plasma homocysteine levels. Furthermore, for the highest PLP quintile compared with the lowest, the OR was 0.1 (95% CI: 0.03–0.38; $P < 0.001$), therefore, defining a possible protective influence for higher PLP levels (Kelly et al. 2003; Khatami et al. 1988). The same case-control study was subsequently evaluated for a possible relationship among vitamin B6 status, total plasma homocysteine levels and C-reactive protein (Kelly et al. 2004). The authors observed an adjusted OR of 16.6 (95%CI: 2–139.9; $P = 0.01$) for low PLP in the highest compared with the lowest CRP quartile. In the same year, another group of researcher published a follow-up study (He et al. 2004) which evaluated intakes of vitamin B6, folate and vitamin B12 in relation to risk of ischemic and hemorrhagic stroke. A number of 43,732 men were enrolled, free of cardiovascular disease and diabetes at baseline, and were followed for 4 years and evaluated every 4 years by a semiquantitative food frequency questionnaire. The authors documented 725 incident strokes (455 ischemic, 125 hemorrhagic and 145 of unknown types) during the follow-up of 14 years. Intakes of folate and vitamin B12, adjusted for lifestyle and dietary factors, were associated with a significantly lower risk of ischemic stroke. On the contrary, vitamin B6 was not associated with risk of ischemic stroke (P for trend, not significant) (He et al. 2004).

In another study designed to assess the association of dietary folate, vitamin B6, and vitamin B12 intake with cardiovascular mortality using data from the Spanish National Nutrition Survey, vitamin B6 failed to obtain a definitive result in regard of a possible association of vitamin B6 impairment with cardiovascular mortality, although evinced a limited protective effect that reached statistical significance only with respect to cerebrovascular mortality in men, after multiple adjustments for age, income, total energy and established risk factors (Medrano et al. 2000).

14.3 Candidate Mechanisms for a Vitamin B6 Related Atherogenesis

Since vitamin B6 is an easily available vitamin from both natural and fortified foods, an overt vitamin B6 deficiency is currently a rare condition that is usually secondary to severe malnutrition, alcoholism, or pharmacological therapy such as that with isoniazide (Snider 1980) which may induce a vitamin B6 impairment. Nevertheless, a marginal or frank inadequate vitamin B6 status is nowadays common in the elderly. Recent epidemiologic studies have suggested that low vitamin B6 status is associated with the increased risk of cardiovascular disease but, to date, the exact mechanism is not yet known, even though many candidate mechanisms have been suggested. It appears that a cardioprotective effect of vitamin B6 might be conveyed in a complex fashion because vitamin B6 serves as a coenzyme for a large number of reactions which primarily include enzymes involved in the amino acid but also in the lipid and carbohydrate metabolic pathways.

Indeed, PLP is a coenzyme necessary for a large number of metabolic reactions, and an impaired PLP status have been related likewise to inflammation (Roubenoff et al. 1995), immune function (Meydani et al. 1992; Rall and Meydani 1993), and thrombosis (Packham et al. 1981; Schoene et al. 1986), all pivotal mechanisms throughout all stages of atherosclerotic process. The main candidate mechanisms invoked in the vitamin B6 related atherosclerosis are summarized in Table 14.2. Figure 14.2 shows a scheme for the major mechanisms potentially implicated in a vitamin B6 related atherogenesis.

The two major derivates of vitamin B6 compounds in animal tissues are the coenzyme species pyridoxamine-5′-phosphate (PMP) and pyridoxal-5′-phosphate (PLP) (Shane 2000) which represents the major coenzymatic form in plasma. PLP

Table 14.2 Candidate mechanisms for vitamin B6 related atherogenesis

Mechanism	Description	References
Homocysteine metabolism	Vitamin B6 deficiency induces hyperhomocysteinemia through inhibition of cystathionine-β-synthase	Clarke and Lewington (2002), D'Angelo and Selhub (1997), Eikelboom et al. (1999), Finkelstein (1998), McCully (1970), McCully and Ragsdale (1970), McCully and Wilson (1975), Refsum and Ueland (1998), Refsum et al. (2006), Selhub et al. (1993, 1995), Welch and Loscalzo (1998)
Coagulation pathway	Vitamin B6 affects platelet aggregation by inhibiting ADP receptors and prolong bleeding time, occupancy of glycoprotein IIb/IIIa receptor or down-regulation of glycoprotein IIb gene	Brattstrom et al. (1990), Brownlee et al. (1984), Cattaneo et al. (2001), Chang et al. (1999, 2002), Packham et al. (1981), Khatami et al. (1988), Palareti et al. (1986), Saareks et al. (2002), Schoene et al. (1986), Undas et al. (1999), van Wyk et al. (1992)
Arterial wall degeneration	Impairment of lysyl oxidase activity by vitamin B6 deficiency induces arterial wall degeneration. Vitamin B6 also inhibits endothelial cell proliferation and induce endothelial dysfunction	Bruel et al. (1998), Ji et al. (2003), Levene and Murray (1977), Matsubara et al. (2003, 2004), Murray and Levene (1977), Rodriguez et al. (2002), Woessner (1991)
Cholesterol metabolism	Inhibition of advanced glycation and lipoxidation end-products by Vitamin B6 deficiency induces hypercholesterolemia	Brattstrom et al. (1990), de Gomez Dumm et al. (2003), Degenhardt et al. (2002), Kirsten et al. (1988), Knipscheer et al. (1997), Metz et al. (2003), Onorato et al. (2000), Schuitemaker et al. (2001), Vijayammal and Kurup (1978), Yamada (1961)
Fatty acids composition	Vitamin B6 deficiency impairs conversion of linoleic acid to arachidonic acid and the metabolism of n-3 PUFA from alpha-linoleic acid to eicosapentaenoic and docosahexaenoic acid	Bordoni et al. (1998), Cabrini et al. (2001), Cunnane et al. (1984), Harris et al. (2006), Kirschman and Coniglio (1961), Tsuge et al. (2000)
Inflammation	Low plasma levels of PLP are inversely related to inflammatory biomarkers such as high-sensitivity C-reactive protein and fibrinogen	Folsom et al. (2003), Friso et al. (2001, 2004), James et al. (2000), Roubenoff et al. (1995)
Immunological function	Vitamin B6 deficiency impairs T lymphocyte and macrophage differentiation and interleukins production	Meydani et al. (1992), Rall and Meydani (1993), Roubenoff et al. (1995)

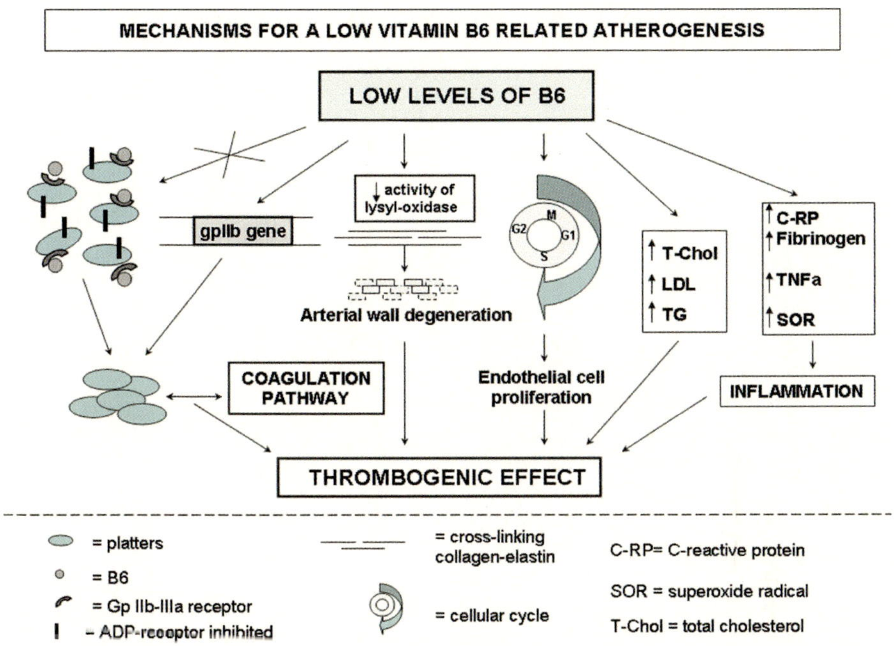

Fig. 14.2 Cartoon showing mechanisms involved in the vitamin B6 related atherogenetic process

is the metabolically active coenzyme produced by the phosphorylation of pyridoxal compound after the oxidation of other vitamin B6 vitamers in the liver. Because it reflects liver PLP levels and stores, plasma PLP has been used, both clinically and experimentally, as a sensitive indicator of tissue vitamin B6 status (Sampson and O'Connor 1989).

14.3.1 Vitamin B6 and Cholesterol Metabolism

Disturbances of cholesterol metabolism is a major risk factor for CAD and it has been suggested that low plasma concentrations of the active vitamin B6 form, PLP, may increase the risk of CAD by affecting cholesterol metabolism. An advanced degree of hypercholesterolemia has been reported in a vitamin B6 deficient animal model, while vitamin B6 supplementation has been shown to decrease plasma cholesterol concentrations (Vijayammal and Kurup 1978) and to protect against the development of atherosclerotic lesions (Yamada 1961). Pyridoxine treatment has also been observed to reduce plasma total cholesterol and low-density lipoprotein cholesterol in atherosclerotic patients (Brattstrom et al. 1990). In patients with chronic renal failure, treatment with pyridoxine induced a significant decrease in total cholesterol and triglycerides levels (de Gomez Dumm et al. 2003; Kirsten et al. 1988) and this effect has been attributed, at least in part, to its inhibitory effect on advanced glycation and lipoxidation end-products (Degenhardt et al. 2002; Metz et al. 2003; Onorato et al. 2000).

However, not all of the evidence supports the lipid lowering effect of vitamin B6. Daily supplementation with 200 mg pyridoxine hydrochloride for 28 days had no significant effect on either total cholesterol or VLDL and LDL cholesterol content among chronic hemodialysis subjects. In a double-blind, randomized, placebo-controlled study, which was conducted in patients with familial hypercholesterolaemia, magnesium-pyridoxal-5′-phosphate-glutamate (MPPG), a derivative of vitamin B6, had no effect on the lipid profile including total cholesterol level (Knipscheer et al. 1997). In a recent placebo-controlled, double-blind, randomized trial, conducted in subjects with serum cholesterol levels between 7.0 and 9.9 mmol/l, 12 months treatment of MMPG was also not effective to reduce total cholesterol level with the exception of a modest effect on reducing serum LDL cholesterol ($P = 0.04$) (Schuitemaker et al. 2001), indicating that vitamin B6 supplementation is not effective to control hypercholesterolemia.

Collectively, evidence suggests that vitamin B6 status is associated with cholesterol metabolism but to date it is not know yet how vitamin B6 affects cholesterol metabolism as well as whether the effect of vitamin B6 on CAD is indeed conveyed through cholesterol metabolism.

14.3.2 Vitamin B6 and Fatty Acids Composition

The metabolic balance among different classes of fatty acids is necessary to support optimal cardiovascular function and tissue levels of these fatty acids have been proposed as risk markers for cardiovascular disease and coronary artery disease (Harris et al. 2006). It has also been described that vitamin B6 affects lipid metabolism by modifying fatty acids composition (Kirschman and Coniglio 1961) particularly in the metabolism of essential fatty acids since vitamin B6 seems particularly involved in the conversion of linoleic acid to arachidonic acid (Cunnane et al. 1984; She et al. 1994) and to impair the metabolism of n-3 PUFA from alpha–linoleic acid to eicosapentaenoic and docosahexaenoic acid with the most significant reduction in the production of docosahexaenoic acid (Tsuge et al. 2000), therefore suggesting an impaired activity of the delta-6-desaturase activity, the enzyme which catalyzes the rate-limiting step in the conversion of linoleic acid to longer and more unsaturated metabolites (Bordoni et al. 1998; Cabrini et al. 2001).

14.3.3 Vitamin B6 and Inflammation

An alternative hypothesis is that the relationship between low PLP and increased CAD risk is due to the common relationships with inflammatory markers. Indeed, a growing body of evidence sustains the theory of inflammation as a major player underlying the pathogenesis of atherosclerosis (Libby et al. 2002; Pasceri et al. 2000; Ross 1999). The current knowledge recognizes that inflammation exerts an essential role at all stages of the atherosclerotic process (Grundy et al. 1999; Libby et al. 2002; Ross 1999) and acute-phase reactants have been proven to be solid and independent risk factors for CAD (Grundy et al. 1999, Ridker 2001; Ridker et al. 2000). Systemic markers of inflammation such as fibrinogen and, particularly, high-sensitivity C-reactive protein (hs-CRP), have been proposed to be strong predictors of CAD risk (Grundy et al. 2004; Kannel et al. 1987; Liuzzo et al. 1994; Ridker 2001; Ridker et al. 2000). In a prospective cohort from the Nurses' Health Study and the Health Professionals Follow-up Study elevated levels of inflammatory markers, particularly CRP, indicates an increased risk of coronary heart disease (Pai et al. 2004).

It has been reported that low plasma levels of PLP were inversely related to C-reactive protein in the population-based Framingham Heart Study cohort (Friso et al. 2001) and this association appeared to be strong and independent of other major biomarkers related to vitamin B6 metabolism (Friso et al. 2001). In healthy middle-aged adults among the Atherosclerosis Risk in Communities (ARIC) Study, an inverse association of plasma PLP with white blood cell count was demonstrated, although in that study there was no association with C-reactive protein (Folsom et al. 2003). In a most recent case-control study with subjects who were characterized by angiography for severe coronary atherosclerosis or CAD-free, a significant, inverse, graded relation was observed between PLP and both hs-CRP and fibrinogen ($P < 0.001$), indicating that vitamin B6 status may affect the risk of CAD by modulating inflammatory process

14.3.4 Vitamin B6 and Immunological Function

There is a bulk of data showing that immune mechanisms are strongly involved in atherogenesis (Binder et al. 2002). Although much remains still to be learned on whether an imbalanced immune system function may be acknowledged as the ultimate trigger behind inflammation in atherogenesis (Libby 2002), several findings have proven the impact of immune signaling system in modulating inflammatory response for the progression of atheroma (Greaves and Channon 2002; Libby and Aikawa 2002). Low concentrations of PLP have been related to alteration of immunological function, including impaired T lymphocyte and macrophage differentiation and interleukins production (Meydani et al. 1992; Rall and Meydani 1993). The inverse relationship between PLP and markers of inflammation together with the finding of an association of low PLP with higher CAD risk may be explained with a role of PLP in affecting the early steps of atherogenesis, in which recruitment of monocytes, T cells, macrophages, and other cells acting for the immune function regulation, promote the cytokine-induced mechanisms for initiating the inflammatory response (Binder et al. 2002; Libby 2002).

14.3.5 Vitamin B6 and Coagulation Pathway

Thrombosis underlies most acute manifestations of coronary atherosclerotic disease (Girelli et al. 2000) and platelet-rich thrombi that form at sites of rupture or fissure of atherosclerotic plaque can cause myocardial infarction (Fuster et al. 1992). PLP has been reported to affect platelet aggregation by inhibiting ADP receptors (Packham et al. 1981; 94) and prolong bleeding time (van Wyk et al. 1992), the occupancy of glycoprotein IIb/IIIa receptor (Chang et al. 2002) or the down-regulation of the expression of glycoprotein IIb gene (Chang et al. 1999). Pyridoxine supplementation has been also described to increase antithrombin III activity (Brattstrom et al. 1990) possibly due to the competitive inhibitory effect of non-enzymatic glycosylation by PLP of lysine residues (Khatami et al. 1988) at the heparin-binding site of antithrombin III (Brattstrom et al. 1990; Brownlee et al. 1984). Another indirect demonstration of a potential thrombogenic effect of low vitamin B6 is the independent association of low PLP with higher risk for deep vein thrombosis (Cattaneo et al. 2001; Lussana et al. 2003).

Other recent studies have shown that treatment with vitamin B6 restores blood coagulation abnormalities in homocystinuric subjects (Palareti et al. 1986) and attenuates the formation of thrombin-antithrombin III complexes and prothrombin fragment 1+2 in subjects treated for mild hyperhomocysteinemia (Undas et al. 1999). The increase in prostacyclin production and the inhibition of thromboxane and leukotriene synthesis have been also described (Saareks et al. 2002).

14.3.6 Vitamin B6 and Arterial Wall Degeneration

Arterial wall degeneration has been proposed as a possible mechanism for cardiovascular disease. Once arterial wall degeneration is initiated, a positive feed back loop may be established whereby degeneration leads to pressure increases, ultimately leading to further vascular degeneration (Zion et al. 2003). The impaired activity of lysyl oxidase, another vitamin B6 – dependent enzyme which is involved in the cross-linking of collagen and elastin, may favor arterial wall degeneration even in conditions of moderate vitamin B6 deficiency (Levene and Murray 1977; Murray and Levene 1977). Reduced activity levels of this enzyme which is also down-regulated by LDL (Rodriguez et al. 2002) result in lower degree of elastin and/or collagen cross-linking, affecting the biochemical properties of extracellular matrixes (Bruel et al. 1998) as well as the susceptibility of collagen and elastin to degradation by proteinases and, thereby, enhances endothelial permeability (Woessner 1991). PLP has been also described to have an inhibitory effect on endothelial cell proliferation (Matsubara et al. 2003, 2004) and reverse endothelial dysfunction induced low-density lipoprotein (Ji et al. 2003).

14.3.7 Vitamin B6 and Anti-oxidant Function

Accumulating evidence suggests that increased oxidant stress, which results from an imbalance between oxidant production and antioxidant defenses, is involved in the pathogenesis of cardiovascular diseases by impairing endothelial function (Heitzer et al. 2001). Several studies have demonstrated that plasma markers of oxidative stress are increased in coronary heart disease or in the presence of its classic risk factors (Stephens et al. 2006) and, furthermore, vitamin B6 has been shown to have antioxidant effect by inhibiting superoxide radical generation, reducing lipid peroxidation (Cabrini et al. 1998; Maranesi et al. 2004; Selvam and Devaraj 1991; Selvam and Ravichandran 1991), and by preventing damage to mitochondrial membrane integrity in a cell culture model (Kannan and Jain 2004). These observations jointly suggest that vitamin B6 may modulate the risk of cardiovascular diseases through its antioxidant effect.

14.3.8 *Vitamin B6 and Homocysteine*

Vitamin B6 is a cofactor for cystathionine-β-synthase and γ-cystathionase, both of which are enzymes involved in the transsulfuration pathway of homocysteine metabolism as well as for serine hydroxymethyl transferase, which catalyzes the methyl transfer for homocysteine remethylation and for nucleotide synthesis pathway (Fig. 14.1, Chapter 7). Cystathionine-β-synthase catalyzes, in an irreversible reaction, the condensation of homocysteine and serine to form cystathionine. The cystathionine is then hydrolyzed by cystathionase to form cysteine and α-ketobutyrate (Fig. 14.1). Thus, by reducing the function of cystathionine-β-synthase vitamin B6 deficiency causes accumulation of S-adenosylhomocysteine as well as export of homocysteine from the cell, leading to hyperhomocysteinemia. This biochemical change induced by vitamin B6 deficiency is quite similar to what is brought about by a well-known genetic disorder which is related to a reduced cystathionine-β-synthase activity and provokes premature atherosclerosis and thrombosis (McCully 1970; McCully and Ragsdale 1970; McCully and Wilson 1975).

Precisely because of the role of PLP as a cofactor in the transsulfuration as well as remethylation pathway of homocysteine (Selhub and Miller 1992), low plasma PLP levels represent a determinant of higher plasma total homocysteine (Selhub et al. 1993), an independent risk factor for occlusive vascular disease (Clarke and Lewington 2002; D'Angelo and Selhub 1997; Eikelboom et al. 1999; Finkelstein 1998; Refsum and Ueland 1998; Refsum et al. 2006; Stanger et al. 2003; Welch and Loscalzo 1998). Therefore, it has been suggested that the role of PLP in atherosclerosis is conveyed through its effect on total plasma homocysteine concentrations (Selhub et al. 1995).

14.4 Vitamin B6 Supplementation and Cardiovascular Disease Prevention

Epidemiologic studies conducted over the past three decades have supported the hypothesis of an association of mild hyperhomocysteinemia with an elevated risk of cardiovascular disease. Subsequently, hyperhomocysteinemia and its association with a deficiency of B-vitamins have offered a unique target for the preventive approach with inexpensive and innocuous naturally occurring B vitamins including vitamin B6.

The most convincing data regarding vitamin B6 supplementation and cardiovascular disease prevention are from the Heart Outcomes Prevention Evaluation (HOPE) 2 Study (Lonn et al. 2006) and the Norwegian Vitamin (NORVIT) trial (Bonaa et al. 2006), both of which are prospective, randomized, double-blind, placebo-controlled multi-center trials. In the (HOPE) 2 study, daily supplementation with 50 mg of vitamin B6 in association with 2.5 mg of folic acid and 1 mg of vitamin B12 for 5 years did not reduce the risk of major cardiovascular events in patients with vascular disease (Lonn et al. 2006). In the NORVIT study daily supplementation with 40 mg of vitamin B6 was not associated with any significant benefit for the recurrent myocardial infarction, stroke, and sudden death attributed to coronary artery disease during 40 months follow up (Bonaa et al. 2006). Combined B vitamin supplementation of 40 mg of vitamin B6 with 0.8 mg folic acid and 0.4 mg of vitamin B12 even tended to be harmful (RR = 1.22, 95%CI = 1.00–1.50, $p = 0.05$). Furthermore, the Vitamins and Thrombosis (VITRO) study, a randomized, double blind, and placebo-controlled, which was designed to investigate the effect of daily supplementation of B vitamins on the secondary prevention of deep vein thrombosis and pulmonary embolism in patients with a first event of venous thrombosis, found that supplementation of 50 mg pyridoxine along with 5 mg folate and 0.4 mg vitamin B12 also did not prevent recurrent venous thrombosis (den Heijer et al. 2007). Collectively, those studies consistently demonstrated that vitamin B6 supplementation is not unequivocally effective for the prevention of recurrence of cardiovascular events including coronary artery disease, stroke, and deep vein thrombosis. Further studies specifically designed to evaluate an effect of vitamin B6 supplementation by considering specific time and dose factors may be needed.

A few small trials, a primary endpoint of which is the progression of atherosclerosis, demonstrated the effectiveness of vitamin B6 supplementation with folate and vitamin B12 on carotid intima-media thickness in hyperhomcysteineic renal transplant or patients at risk to cerebral ischemia (Marcucci et al. 2003; Till et al. 2005). However, results of two double blind, placebo-controlled, randomized trial after coronary angioplasty or stent were inconsistent. Supplementation of vitamin B6 together with folate and vitamin B12 after coronary angioplasty decreased the rate of restenosis and need for revascularization (Schnyder et al. 2001), while supplementation after coronary stenting increased the risk of in-stent restenosis and the need for target-vessel revascularization (Lange et al. 2004).

The negative result of vitamin B6 trials cannot preclude a protective effect in primary prevention as well as of more physiologic doses and forms of B vitamins. However, it could be difficult to demonstrate that vitamin B supplementation that does not work in patients who have had a clinical vascular event would work well in those without a clinical event or with subclinical atherosclerosis. In contrast to the clinical trials that determine the effect of vitamin B6 supplementation on the secondary prevention, clinical trials for the primary prevention needs longer duration as well as larger population to answer whether vitamin B6 supplementation is indeed effective to prevent cardiovascular disease before the first vascular event or during the younger life. Studies are needed to find the optimal dose and combination of vitamin B6 forms that maximize efficacy and minimize adverse effects, to identify subjects who will get the most beneficial effect from interventions with vitamin B6, even on the basis of genetic susceptibility and taking into account other environmental factors to adequately assess the role of B vitamins in the primary prevention of cardiovascular disease.

14.5 Conclusions and Future Perspectives

As illustrated in the present chapter, many observations emphasize the importance of vitamin B6 as a significant, independent risk factor for cardiovascular disease and in some (Robinson et al. 1998), but not all, case-control studies vitamin B6 deficiency is clearly associated with higher risk for cardiovascular disease (Friso et al. 2004; Robinson et al. 1995, 1998), even though clinical trials testified the ineffectiveness of vitamin B6 supplementation on recurrence of vascular events.

Since vitamin B6 is involved in a very large number of physiologic reactions, is likely that vitamin B6 may affect cardiovascular disease in a complex fashion of inter-relationships among several factors and thereby it could be very hard to define the exact mechanism(s) as well as the optimal endpoints for clinical trials.

Current clinical trials indicate that vitamin B6 supplementation is not effective for the prevention of recurrence of cardiovascular disease. However, it is still open to determine whether supplementation with vitamin B6 is effective for the primary prevention of cardiovascular disease. It is also to be evoked that supplementation with vitamin B6 in a single high-dose synthetic form might not be as effective as adequate dietary vitamin B6 intake, since in the latter case, the vitamin is made available through many different physiologic forms and in combination with a number of other nutrients and, particularly, of other B vitamins which are involved in the complex vitamin B6 metabolism.

References

Binder CJ, Chang, M-K, Shaw PX, Miller YI, Hartvigsen K, Dewan A et al. (2002) Innate and acquired immunity in atherogenesis. Nature Med 8:1218–1226

Bonaa KH, Njolstad I, Ueland PM, Schirmer H, Tverdal A, Steigen T, Wang H, Nordrehaug JE, Arnesen E, Rasmussen K, NORVIT Trial Investigators (2006) Homocysteine lowering and cardiovascular events after acute myocardial infarction. N Engl J Med 354(15):1578–1588

Bordoni A, Hrelia S, Lorenzini A, Bergami R, Cabrini L, Biagi PL, Tolomelli B (1998) Dual influence of aging and vitamin B6 deficiency on delta-6-desaturation of essential fatty acids in rat liver microsomes. Prostaglandins Leukot Essent Fatty Acids 58(6):417–420

Brattstrom L, Stavenow L, Galvard H, Nilsson-Ehle P, Berntorp E, Jerntorp P, Elmstahl S, Pessah-Rasmussen H (1990) Pyridoxine reduces cholesterol and low-density lipoprotein and increases

antithrombin III activity in 80-year-old men with low plasma pyridoxal 5-phosphate. Scand J Clin Lab Invest 50(8):873–877
Brownlee M, Vlassara H, Cerami A (1984) Inhibition of heparin-catalyzed human antithrombin III activity by nonenzymatic glycosylation. Possible role in fibrin deposition in diabetes. Diabetes 33(6):532–535
Bruel A, Ortoft G, Oxlund H (1998) Inhibition of cross-links in collagen is associated with reduced stiffness of the aorta in young rats. Atherosclerosis 140(1):135–145
Cabrini L, Bergami R, Fiorentini D, Marchetti M, Landi L, Tolomelli B (1998) Vitamin B6 deficiency affects antioxidant defences in rat liver and heart. Biochem Mol Biol Int 46(4):689–697
Cabrini L, Bergami R, Maranesi M, Carloni A, Marchetti M, Tolomelli B (2001) Effects of short-term dietary administration of marginal levels of vitamin B(6)and fish oil on lipid composition and antioxidant defences in rat tissues. Prostaglandins Leukot Essent Fatty Acids 64(4–5):265–271
Cattaneo M, Lombardi R, Lecchi A, Bucciarelli P, Mannucci PM (2001) Low plasma levels of vitamin B(6) are independently associated with a heightened risk of deep-vein thrombosis. Circulation 104(20):2442–2446
Chang SJ, Chang CN, Chen CW (2002) Occupancy of glycoprotein IIb/IIIa by B-6 vitamers inhibits human platelet aggregation. J Nutr 132(12):3603–3606
Chang SJ, Chuang HJ, Chen HH (1999) Vitamin B6 down-regulates the expression of human GPIIb gene. J Nutr Sci Vitaminol (Tokyo) 45(4):471–479
Chasan-Taber L, Selhub J, Rosemberg IH, Malinow MR, Terry P, Tishler PV, Willet W, Hennekens CH, Stampfer MJ (1996) A prospective study of folate and vitamin B6 and risk of myocardial infarction in US physicians. J Am Coll Nutr 15(2):136–143
Chiang EP, Bagley PJ, Selhub J, Nadeau M, Roubenoff R (2003) Abnormal vitamin B(6) status is associated with severity of symptoms in patients with rheumatoid arthritis. Am J Med 114(4):283–287
Clarke R, Lewington S (2002) Homocysteine and coronary heart disease. Semin Vasc Med 2(4):391–399
Cunnane SC, Manku MS, Horrobin DF (1984) Accumulation of linoleic and gamma-linolenic acids in tissue lipids of pyridoxine-deficient rats. J Nutr 114(10):1754–1761
Dalery K, Lussier-Cacan S, Selhub J, Davignon J, Latour Y, Genest J Jr (1995) Homocysteine and coronary artery disease in French Canadian subjects: relation with vitamins B12, B6, pyridoxal phosphate, and folate. Am J Cardiol 75:1107–1111
D'Angelo A, Selhub J (1997) Homocysteine and thrombotic disease. Blood 90(1):1–11
Degenhardt TP, Alderson NL, Arrington DD, Beattie RJ, Basgen JM, Steffes MW, Thorpe SR, Baynes JW (2002) Pyridoxamine inhibits early renal disease and dyslipidemia in the streptozotocin-diabetic rat. Kidney Int 61(3):939–950
de Gomez Dumm NT, Giammona AM, Touceda LA (2003) Variations in the lipid profile of patients with chronic renal failure treated with pyridoxine. Lipids Health Dis 2:7
den Heijer M, Willems HP, Blom HJ, Gerrits WB, Cattaneo M, Eichinger S, Rosendaal FR, Bos GM (2007) Homocysteine lowering by B vitamins and the secondary prevention of deep vein thrombosis and pulmonary embolism: a randomized, placebo-controlled, double-blind trial. Blood 109(1):139–144
Eikelboom JW, Lonn E, Genest J Jr, Hankey G, Yusuf S (1999) Homocyst(e)ine and cardiovascular disease: a critical review of the epidemiologic evidence. Ann Intern Med 131(5):363–375
Finkelstein JD (1998) The metabolism of homocysteine: pathways and regulation. Eur J Pediatr 1998 157(Suppl 2):S40–S44
Folsom AR, Chambless LE, Ballantyne CM, Coresh J, Heiss G, Wu KK, Boerwinkle E, Mosley TH Jr., Sorlie P, Diao G, Sharrett AR (2006) An assessment of incremental coronary risk prediction using C-reactive protein and other novel risk markers: the atherosclerosis risk in communities study. Arch Intern Med 166(13):1368–1373
Folsom AR, Desvarieux M, Nieto FJ, Boland LL, Ballantyne CM, Chambless LE (2003) B vitamin status and inflammatory markers. Atherosclerosis 169:169–174

Folsom AR, Nieto FJ, McGovern PG, Tsai MY, Malinow MR, Eckfeldt JH et al. (1998) Prospective study of coronary heart disease incidence in relation to fasting total homocysteine, related genetic polymorphisms, and B vitamins: the Atherosclerosis Risk in Communities (ARIC) study. Circulation 98:204–210

Friso S, Girelli D, Martinelli N, Olivieri O, Lotto V, Bozzini C, Pizzolo F, Faccini G, Beltrame F, Corrocher R (2004) Low plasma vitamin B-6 concentrations and modulation of coronary artery disease risk. Am J Clin Nutr 79(6):992–998

Friso S, Jacques PF, Wilson PW, Rosenberg IH, Selhub J (2001) Low circulating vitamin B(6) is associated with elevation of the inflammation marker C-reactive protein independently of plasma homocysteine levels. Circulation 103(23):2788–2791

Fuster V, Badimon L, Badimon JJ, Chesebro JH (1992) The pathogenesis of coronary artery disease and the acute coronary syndromes (2). N Engl J Med 326(5):310–318

Girelli D, Russo C, Ferraresi P, Olivieri O, Pinotti M, Friso S et al. (2000) Polymorphisms in the factor VII gene and the risk of myocardial infarction in patients with coronary artery disease. N Engl J Med 343:774–780

Greaves DR, Channon KM (2002) Inflammation and immune responses in atherosclerosis. Trends Immunol 23:535–541

Grundy SM, Cleeman JI, Merz CN, Brewer HB Jr, Clark LT, Hunninghake DB, Pasternak RC, Smith SC Jr, Stone NJ, Coordinating Committee of the National Cholesterol Education Program (2004) Implications of recent clinical trials for the National Cholesterol Education Program Adult Treatment Panel III Guidelines. J Am Coll Cardiol 44(3):720–732

Grundy SM, Hansen B, Smith SC Jr, Cleeman JI, Kahn RA, American Heart Association, National Heart, Lung, and Blood Institute, American Diabetes Association (2004) Clinical management of metabolic syndrome: report of the American Heart Association/National Heart, Lung, and Blood Institute/American Diabetes Association conference on scientific issues related to management. Arterioscler Thromb Vasc Biol 24(2):e19–e24

Grundy SM, Pasternak R, Greenland P, Smith S Jr, Fuster V (1999) Assessment of cardiovascular risk by use of multiple-risk-factor assessment equations: a statement for healthcare professionals from the American Heart Association and the American College of Cardiology. Circulation 100(13):1481–1492

Harris WS, Assaad B, Poston WC (2006) Tissue omega-6/omega-3 fatty acid ratio and risk for coronary artery disease. Am J Cardiol 98(4A):19i–26i

He K, Merchant A, Rimm EB, Rosner BA, Stampfer MJ, Willett WC, Ascherio A (2004) Folate, vitamin B6, and B12 intakes in relation to risk of stroke among men. Stroke 35(1): 169–174

Heitzer T, Schlinzig T, Krohn K, Meinertz T, Munzel T (2001) Endothelial dysfunction, oxidative stress, and risk of cardiovascular events in patients with coronary artery disease. Circulation 104(22):2673–2678

Iqbal MP, Ishaq M, Kazmi KA, Yousuf FA, Mehboobali N, Ali SA, Khan AH, Waqar MA (2005) Role of vitamins B6, B12 and folic acid on hyperhomocysteinemia in a Pakistani population of patients with acute myocardial infarction. Nutr Metab Cardiovasc Dis 15(2):100–108

James S, Vorster HH, Ventre CS, Kruger HS, Nell TA, Veldman FJ et al. (2000) Nutritional status influences plasma fibrinogen concentration: evidence from the THUSA survey. Thromb Res 98:383–394

Ji Y, Han Y, Diao J, Huang Y, Chen Q, Ferro A (2003) Rabbit aortic endothelial dysfunction by low-density lipoprotein is attenuated by L-arginine, L-ascorbate and pyridoxine. Br J Pharmacol 140(7):1272–1282

Kannan K, Jain SK (2004) Effect of vitamin B6 on oxygen radicals, mitochondrial membrane potential, and lipid peroxidation in H_2O_2-treated U937 monocytes. Free Radic Biol Med 36(4):423–428

Kannel WB, Wolf PA, Castelli WP, D'Agostino RB (1987) Fibrinogen and risk of cardiovascular disease. The Framingham Study. JAMA 258:1183–1186

Kelly PJ, Klister JP, Shih VE, Mandell R, Atassi N, Barron M, Lee H, Silveira S, Furie KL (2004) Inflammation, homocysteine, and vitamin B6 status after ischemic stroke. Stroke 35(1):12–15

Kelly PJ, Shih VE, Kistler JP, Barron M, Lee H, Mandell R, Furie KL (2003) Low vitamin B6 but not homocyst(e)ine is associated with increased risk of stroke and transient ischemic attack in the era of folic acid grain fortification. Stroke 34(6):e51–e54

Khatami M, Suldan Z, David I, Li W, Rockey JH (1988) Inhibitory effects of pyridoxal phosphate, ascorbate and aminoguanidine on nonenzymatic glycosylation. Life Sci 43(21):1725–1731

Kirschman JC, Coniglio JG (1961) The role of pyridoxine in the metabolism of polyunsaturated fatty acids in rats. J Biol Chem 236:2200–2203

Kirsten R, Heintz B, Nelson K, Sieberth HG, Oremek G, Hasford J, Speck U (1988) Magnesium pyridoxal 5-phosphate glutamate reduces hyperlipidaemia in patients with chronic renal insufficiency. Eur J Clin Pharmacol 34(2):133–137

Knipscheer HC, Kindt I, van den Ende A, Nurmohamed MT, Smalbraak H, Mulder WJ, Kastelein JJ (1997) Magnesium pyridoxal-5′-phosphate glutamate, "A vitamin B6 derivative", does not affect lipoprotein levels in patients with familial hypercholesterolaemia. Eur J Clin Pharmacol 51(6):499–503

Kok FJ, Schrijver J, Hofman A, Witteman JC, Kruyssen DA, Remme WJ, Valkenburg HA (1989) Low vitamin B6 status in patients with acute myocardial infarction. Am J Cardiol 63(9):513–516

Labadarios D, Brink PA, Weich HF, Visser L, Louw ME, Shephard GS, van Stuijyenberg ME (1987) Plasma vitamin A, E, C and B6 levels in myocardial infarction. S Afr Med J 71(9):561–563

Lange H, Suryapranata H, De Luca G, Borner C, Dille J, Kallmayer K, Pasalary MN, Scherer E, Dambrink JH (2004) Folate therapy and in-stent restenosis after coronary stenting. N Engl J Med 350(26):2673–2681

Levene CI, Murray JC (1977) The aetiological role of maternal vitamin-B6 deficiency in the development of atherosclerosis. Lancet 1:628–630

Libby P (2002) Inflammation in atherosclerosis. Nature 420:868–874

Libby P, Aikawa M (2002) Stabilization of atherosclerotic plaques: new mechanisms and clinical targets. Nature Med 8:1257–1262

Libby P, Ridker PM, Maseri A (2002) Inflammation and atherosclerosis. Circulation 105:1135–1143

Liuzzo G, Biasucci LM, Gallimore JR, Grillo RL, Rebuzzi AG, Pepys MB et al. (1994) The prognostic value of C-reactive protein and serum amyloid a protein in severe unstable angina. N Engl J Med 331:417–424

Lonn E, Yusuf S, Arnold MJ, Sheridan P, Pogue J, Micks M, McQueen MJ, Probstfield J, Fodor G, Held C, Genest J Jr, Heart Outcomes Prevention Evaluation (HOPE) 2 Investigators (2006) Homocysteine lowering with folic acid and B vitamins in vascular disease. N Engl J Med 354(15):1567–1577

Lussana F, Zighetti ML, Bucciarelli P, Cugno M, Cattaneo M (2003) Blood levels of homocysteine, folate, vitamin B6 and B12 in women using oral contraceptives compared to non-users. Thromb Res 112(1–2):37–41

Maranesi M, Bochicchio D, Zambonin L, Tolomelli B, Cabrini L (2004) Effects of different dietary amounts of LCPUFA n3 and vitamin B6 on lipid composition and antioxidant defences in rat kidney. J Nutr Biochem 15(7):396–401

Marcucci R, Zanazzi M, Bertoni E, Rosati A, Fedi S, Lenti M, Prisco D, Castellani S, Abbate R, Salvadori M (2003) Vitamin supplementation reduces the progression of atherosclerosis in hyperhomocysteinemic renal-transplant recipients. Transplantation 75(9):1551–1555

Matsubara K, Matsumoto H, Mizushina Y, Lee JS, Kato N (2003) Inhibitory effect of pyridoxal 5′-phosphate on endothelial cell proliferation, replicative DNA polymerase and DNA topoisomerase. Int J Mol Med 12(1):51–55

Matsubara K, Mori M, Akagi R, Kato N (2004) Anti-angiogenic effect of pyridoxal 5′-phosphate, pyridoxal and pyridoxamine on embryoid bodies derived from mouse embryonic stem cells. Int J Mol Med 14(5):819–823

McCully KS (1970) Importance of homocysteine-induced abnormalities of proteoglycan structure in arteriosclerosis. Am J Pathol 59(1):181–194

McCully KS, Ragsdale BD (1970) Production of arteriosclerosis by homocysteinemia. Am J Pathol 61(1):1–11

McCully KS, Wilson RB (1975) Homocysteine theory of arteriosclerosis. Atherosclerosis 22(2):215–227

Medrano MJ, Sierram MJ, Almazan J, Olalla MT, Lopez-Abente G (2000) The association of dietary folate, B6, and B12 with cardiovascular mortality in Spain: an ecological analysis. Am J Public Health 90(10):1636–1638

Metz TO, Alderson NL, Chachich ME, Thorpe SR, Baynes JW (2003) Pyridoxamine traps intermediates in lipid peroxidation reactions in vivo: evidence on the role of lipids in chemical modification of protein and development of diabetic complications. J Biol Chem 278(43):42012–42019

Meydani SN, Hayek M, Coleman L (1992) Influence of vitamins E and B6 on immune response. Ann NY Acad Sci 669:125–139

Miller JW, Ribaya-Mercado JD, Russell RM, Shepard DC, Morrow FD, Cochary EF, Sadowski JA, Gershoff SN, Selhub J (1992) Effect of vitamin B-6 deficiency on fasting plasma homocysteine concentrations. Am J Clin Nutr 55(6):1154–1160

Murray JC, Levene CI (1977) Evidence for the role of vitamin B6 as a cofactor of lysyl oxidase. Biochem J 167:463–467

Nahlawi M, Seshadri N, Boparai N, Naso A, Jacobsen DW, McCarthy P, Young J, Robinson K (2002) Usefulness of plasma vitamin B(6), B(12), folate, homocysteine, and creatinine in predicting outcomes in heart transplant recipients. Am J Cardiol 89(7):834–837

Onorato JM, Jenkins AJ, Thorpe SR, Baynes JW (2000) Pyridoxamine, an inhibitor of advanced glycation reactions, also inhibits advanced lipoxidation reactions. Mechanism of action of pyridoxamine. J Biol Chem 275(28):21177–21184

Packham MA, Lam SC, Mustard JF (1981) Vitamin B6 as an antithrombotic agent. Lancet Oct 10; 2(8250):809–810

Pai JK, Pischon T, Ma J, Manson JE, Hankinson SE, Joshipura K, Curhan GC, Rifai N, Cannuscio CC, Stampfer MJ, Rimm EB (2004) Inflammatory markers and the risk of coronary heart disease in men and women. N Engl J Med 351(25):2599–2610

Palareti G, Salardi S, Piazzi S, Legnani C, Poggi M, Grauso F et al. (1986) Blood coagulation changes in homocystinuria: effects of pyridoxine and other specific therapy. J Pediatr 109:1001–1006

Pasceri V, Willerson JT, Yeh ETH (2000) Direct proinflammatory effect of C-reactive protein on human endothelial cells. Circulation 102:2165–2168

Rall LC, Meydani SN (1993) Vitamin B6 and immune competence. Nutr Rev 51:217–225

Refsum H, Nurk E, Smith AD, Ueland PM, Gjesdal CG, Bjelland I, Tverdal A, Tell GS, Nygard O, Vollset SE (2006) The Hordaland Homocysteine Study: a community-based study of homocysteine, its determinants, and associations with disease. J Nutr 136(6 Suppl):1731S–1740S

Refsum H, Ueland PM (1998) Recent data are not in conflict with homocysteine as a cardiovascular risk factor. Curr Opin Lipidol 9(6):533–539

Ridker PM (2001) Role of inflammatory biomarkers in prediction of coronary heart disease. Lancet 358:946–948

Ridker PM, Hennekens CH, Buring JE, Rifai N (2000) C-reactive protein and other markers of inflammation in the prediction of cardiovascular disease in women. N Engl J Med 342:836–843

Ridker PM, Wilson PW, Grundy SM (2004) Should C-reactive protein be added to metabolic syndrome and to assessment of global cardiovascular risk? Circulation 109(23):2818–2825

Rimm EB, Willett WC, Hu FB, Sampson L, Colditz GA, Manson JE, Hennekens C, Stampfer MJ (1998) Folate and vitamin B6 from diet and supplements in relation to risk of coronary heart disease among women. JAMA 279(5):359–364

Rinehart JF, Greenberg LD (1949) Arteriosclerotic lesions in pyridoxine-deficient monkeys. Am J Pathol 25:481–491

Rinehart JF, Greenberg LD (1956) Vitamin B6 deficiency in the rhesus monkey; with particular reference to the occurrence of atherosclerosis, dental caries, and hepatic cirrhosis. Am J Clin Nutr 4(4):318–325; discussion, 325–328

Robinson K, Arheart K, Refsum H, Brattstrom L, Boers G, Ueland P et al. (1998) from the European COMAC Group. Low circulating folate and vitamin B6 concentrations. Risk factors for stroke, peripheral vascular disease, and coronary artery disease. Circulation 97:437–443

Robinson K, Gupta A, Dennis V, Arheart K, Chaudhary D, Green R, Vigo P, Mayer EL, Selhub J, Kutner M, Jacobsen DW (1996) Hyperhomocysteinemia confers an independent increased risk of atherosclerosis in end-stage renal disease and is closely linked to plasma folate and pyridoxine concentrations. Circulation 94(11):2743–2748

Robinson K, Mayer EL, Miller DP, Green R, van Lente F, Gupta A, Kottke-Marchant K, Savon SR, Selhub J, Nissen SE et al. (1995) Hyperhomocysteinemia and low pyridoxal phosphate. Common and independent reversible risk factors for coronary artery disease. Circulation 92(10):2825–2830

Rodriguez C, Raposo B, Martinez-Gonzalez J, Casani L, Badimon L (2002) Low density lipoproteins downregulate lysyl oxidase in vascular endothelial cells and the arterial wall. Arterioscler Thromb Vasc Biol 22(9):1409–1414

Ross R (1999) Atherosclerosis – An inflammatory disease. N Engl J Med 340(2):115–126

Roubenoff R, Roubenoff RA, Selhub J, Nadeau MR, Cannon JG, Freeman LM et al. (1995) Abnormal vitamin B6 status in rheumatoid cachexia. Association with spontaneous tumor necrosis factor alpha production and markers of inflammation. Arthritis Rheum 38:105–109

Saareks V, Ylitalo P, Mucha I, Riutta A (2002) Opposite effects of nicotinic acid and pyridoxine on systemic prostacyclin, thromboxane and leukotriene production in man. Pharmacol Toxicol 90(6):338–342

Sampson D, O'Connor DK (1989) Response of B-6 vitamers in plasma, erythrocytes and tissues to vitamin B6 depletion and repletion in the rat. J Nutr 119:1940–1948

Schnyder G, Roffi M, Pin R, Flammer Y, Lange H, Eberli FR, Meier B, Turi ZG, Hess OM (2001) Decreased rate of coronary restenosis after lowering of plasma homocysteine levels. N Engl J Med 345(22):1593–1600

Schoene NW, Chanmugam P, Reynolds RD (1986) Effect of oral vitamin B6 supplementation on in vitro platelet aggregation. Am J Clin Nutr 43:825–830

Schuitemaker GE, van der Pol GA, Aretz CP, Dinant GJ (2001) A placebo-controlled, double-blind, randomised trial of magnesium-pyridoxal-5′-phosphate-glutamate for hypercholesterolaemia and other clinical-chemical risk factors of cardiovascular disease in a primary care setting. Eur J Clin Pharmacol 56(12):857–863

Selhub J, Jacques PF, Bostom AG, D'Agostino RB, Wilson PW, Belanger AJ et al. (1995) Association between plasma homocysteine concentrations and extracranial carotid-artery stenosis. N Engl J Med 332:286–291

Selhub J, Jacques PF, Wilson PW, Rush D, Rosenberg IH (1993) Vitamin status and intake as primary determinants of homocysteinemia in an elderly population. JAMA 270(22):2693–2698

Selhub J, Miller JW (1992) The pathogenesis of homocysteinemia: interruption of the coordinate regulation by S-adenosylmethionine of the remethylation and transsulfuration of homocysteine. Am J Clin Nutr 55(1):131–138

Selvam R, Devaraj S (1991) Induction of oxalate binding by lipid peroxidation in rat kidney mitochondria. Biochem Int 24(5):857–866

Selvam R, Ravichandran V (1991) Effect of oral methionine and vitamin E on blood lipid peroxidation in vitamin B6 deficient rats. Biochem Int 23(6):1007–1017

Serfontein WJ, Ubbink JB, De Villiers LS, Rapley CH, Becker PJ (1985) Plasma pyridoxal-5-phosphate level as risk index for coronary artery disease. Atherosclerosis 55(3):357–361

Shane B (2000) Folate, Vitamin B12, and Vitamin B6. In: Stipanuk MH (ed) Biochemical and physiological aspects of human nutrition. WB Saunders, Philadelphia, PA, pp. 483–518

She QB, Nagao I, Hayakawa T, Tsuge H (1994) A simple HPLC method for the determination of S-adenosylmethionine and S-adenosylhomocysteine in rat tissues: the effect of vitamin B6 deficiency on these concentrations in rat liver. Biochem Biophys Res Commun 205(3):1748–1754

Siri PW, Verhoef P, Kok FJ (1998) Vitamins B6, B12, and folate: association with plasma total homocysteine and risk of coronary atherosclerosis. J Am Coll Nutr 17(5):435–441

Snider DE Jr (1980) Pyridoxine supplementation during isoniazid therapy. Tubercle 61(4):191–196. Review

Stanger O, Herrmann W, Pietrzik K, Fowler B, Geisel J, Dierkes J, Weger M (2003) DACH-LIGA Homocystein e. V. DACH-LIGA homocystein (German, Austrian and Swiss homocysteine society): consensus paper on the rational clinical use of homocysteine, folic acid and B-vitamins in cardiovascular and thrombotic diseases: guidelines and recommendations. Clin Chem Lab Med 41(11):1392–1403

Stephens JW, Gable DR, Hurel SJ, Miller GJ, Cooper JA, Humphries SE (2006) Increased plasma markers of oxidative stress are associated with coronary heart disease in males with diabetes mellitus and with 10-year risk in a prospective sample of males. Clin Chem 52(3):446–452

Tavani A, Pelucchi C, Parpinel M, Negri E, La Vecchia C (2004) Folate and vitamin B(6) intake and risk of acute myocardial infarction in Italy. Eur J Clin Nutr 58(9):1266–1272

Till U, Rohl P, Jentsch A, Till H, Muller A, Bellstedt K, Plonne D, Fink HS, Vollandt R, Sliwka U, Herrmann FH, Petermann H, Riezler R. (2005) Decrease of carotid intima-media thickness in patients at risk to cerebral ischemia after supplementation with folic acid, Vitamins B6 and B12. Atherosclerosis 181(1):131–135

Tsuge H, Hotta N, Hayakawa T (2000) Effects of vitamin B-6 on (n-3) polyunsaturated fatty acid metabolism. J Nutr 130(2S Suppl):333S–334S

Undas A, Domagala TB, Jankowski M, Szczeklik A (1999) Treatment of hyperhomocysteinemia with folic acid and vitamins B12 and B6 attenuates thrombin generation. Thromb Res 95:281–288

van Wyk V, Luus HG, Heyns AD (1992) The in vivo effect in humans of pyridoxal-5′-phosphate on platelet function and blood coagulation. Thromb Res 66(6):657–668

Verhoef P, Kok FJ, Kruyssen DA, Schouten EG, Witteman JC, Grobbee DE, Ueland PM, Refsum H (1997) Plasma total homocysteine, B vitamins, and risk of coronary atherosclerosis. Arterioscler Thromb Vasc Biol 17(5):989–995

Verhoef P, Stampfer MJ, Buring JE, Gaziano JM, Allen RH, Stabler SP et al. (1996) Homocysteine metabolism and risk of myocardial infarction: relation with vitamins B6, B12, and folate. Am J Epidemiol 143:845–859

Vermaak WJ, Barnard HC, Potgieter GM, Marx JD (1986) Plasma pyridoxal-5′-phosphate levels in myocardial infarction. S Afr Med J 70(4):195–196

Vermaak WJ, Barnard HC, Potgieter GM, Theron HD (1987) Vitamin B6 and coronary artery disease. Epidemiological observations and case studies. Atherosclerosis 63(2–3):235–238

Vermaak WJ, Barnard HC, Van Dalen EM, Potgieter GM, Van Jaarsveld H, Myburgh SJ (1988) Compartmentalization of pyridoxal-5′-phosphate during the acute phase of myocardial infarction. Klin Wochenschr 66(10):428–433

Vijayammal PL, Kurup PA (1978) Pyridoxine and atherosclerosis: role of pyridoxine in the metabolism of lipids and glycosaminoglycans in rats fed normal and high fat, high cholesterol diets containing 16% casein. Aust J Biol Sci 31(1):7–20

Welch GN, Loscalzo J (1998) Homocysteine and atherothrombosis. N Engl J Med 338(15):1042–1050

Wilmink AB, Welch AA, Quick CR, Burns PJ, Hubbard CS, Bradbury AW, Day NE (2004) Dietary folate and vitamin B6 are independent predictors of peripheral arterial occlusive disease. J Vasc Surg 39(3):513–516

Woessner JF Jr (1991) Matrix metalloproteinases and their inhibitors in connective tissue remodeling. FASEB J 5(8):2145–2154

Yamada K (1961) Treatment of arteriosclerosis. Vitamins and sulfuric esters of polysaccharides. Jpn Heart 2:281–296

Zion AS, Bond V, Adams RG, Williams D, Fullilove RE, Sloan RP, Bartels MN, Downey JA, De Meersman RE (2003) Low arterial compliance in young African-American males. Am J Physiol Heart Circ Physiol 285(2):H457–H462

Chapter 15
Vitamin B6: Beyond Coenzyme Functions

Georg T. Wondrak and Elaine L. Jacobson

Abstract Endogenous reactive intermediates such as photoexcited states of tissue chromophores, reactive oxygen species (ROS), reactive carbonyl species (RCS), and transition metal ions are mediators of tissue damage involved in initiation and progression of human pathologies including tumorigenesis, atherosclerosis, diabetes, and neurodegenerative disease. A large body of evidence now suggests that B6 vitamers antagonize the harmful activity of endogenous reactive intermediates fulfilling a very different role than that established as a cofactor for numerous enzymes. In this chapter, the structural basis of vitamin B6 activity as a potent antioxidant, metal chelator, carbonyl scavenger, and photosensitizer is presented and the physiological relevance is discussed.

Keywords Atherosclerosis · Carbonyl scavenger · Diabetes · Glycation · Metal chelator · Ntioxidant · Vitamin B6

15.1 Introduction

The role of endogenous reactive intermediates including photoexcited states of tissue chromophores, reactive oxygen species (ROS), reactive carbonyl species (RCS), transition metal ions, in various human pathologies is now firmly established. These intermediates have been implicated in the initiation and progression of diverse human pathologies including tumorigenesis, atherosclerosis, diabetes, and neurodegenerative disease (Baynes and Thorpe, 2000a; Sasaki et al. 1993; Schumacker 2006; Ulrich and Cerami 2001; Valko et al. 2007; Wondrak et al. 2002a, 2006b). The chemical nature of endogeous reactive species suggests the feasibility of therapeutic intervention by specific molecular antagonists.

In contrast to most other vitamins, vitamin B6 occurs as a family of closely related chemical derivatives comprising six vitamers as shown in Fig. 15.1:

G.T. Wondrak (✉) · E.L. Jacobson (✉)
Department of Pharmacology and Toxicology, College of Pharmacy and Arizona Cancer Center, University of Arizona, Tucson, AZ, USA
e-mail: wondrak@pharmacy.arizona.edu, jacobse@pharmacy.arizona.edu; elaine.Jacobson@pharmacy.arizona.edu

O. Stanger (ed.), *Water Soluble Vitamins*, Subcellular Biochemistry 56,
DOI 10.1007/978-94-007-2199-9_15, © Springer Science+Business Media B.V. 2012

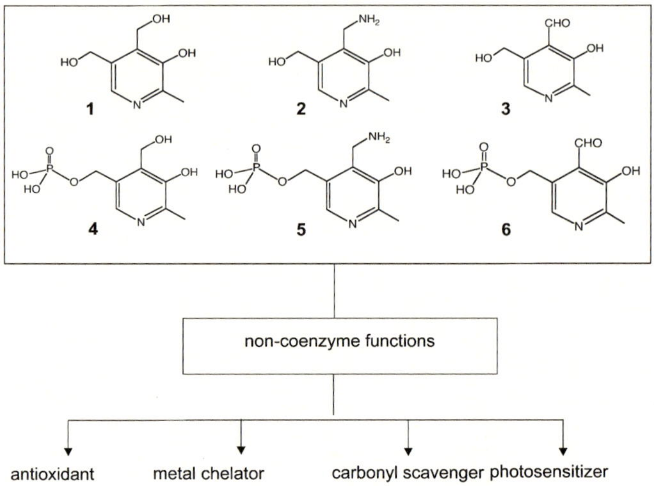

Fig. 15.1 Non-coenzyme functions of vitamin B6. Non-coenzyme functions of B6-vitamers can result from reactivity as antioxidant, metal chelator, carbonyl scavenger, and photosensitizer

pyridoxine (1), pyridoxamine (2), and pyridoxal (3), all of which can be phosphorylated to the corresponding 5′-phosphates (4–6) (McCormick and Chen 1999). The chemical versatility of B6-vitamers is remarkable and has been the subject of many detailed studies. Early findings obtained in the 1940s in the laboratory of Esmond Snell indicated that pyridoxal phosphate dependent enzyme reactions could be catalyzed by pyridoxal in the complete absence of any enzyme, demonstrating the potential of B6-vitamers to participate in spontaneous reactions under physiological conditions (Metzler et al. 1954). More recent studies have confirmed that B6-reactivity is not only essential for coenzyme function during enzymatic transformations including transamination, decarboxylation, racemization, and -elimination, but also allows participation in various non-enzymatic reactions as antioxidants, carbonyl scavengers, metal chelators, and photosensitizers as summarized in Fig. 15.1 (Voziyan and Hudson 2005a).

15.2 B6-Derived Antioxidants

B6 vitamers display spontaneous redox reactivities that are mainly attributable to the presence of a phenolic hydroxyl group in the 3-position of the pyridine ring (Culbertson et al. 2003). The reactions between pyridoxine and several oxygen radicals [hydroxyl (.OH), hydroperoxyl (.OOH), and superoxide (O_2.–)] have been investigated in much detail recently (Matxain et al. 2006). Based on addition reactions to the aromatic ring and H-atom abstraction by radical mechanisms the following reactions with .OH, .OOH, and O_2.– can occur leading to the intermediate formation of B6 free radicals (.B6):

$$B6 + .OH \rightarrow B6 + H_2O \quad (15.1)$$

$$B6 + .OOH \rightarrow B6 + H_2O_2 \quad (15.2)$$

$$B6 + .O_2 \rightarrow B6 + HOO \quad (15.3)$$

Pyridoxine is a highly effective antioxidant in chemical and cellular assays. Using a biological method that allows evaluation of human plasma and red blood cell resistance against free radical-induced hemolysis, potent antioxidant activity of pyridoxine was demonstrated (Stocker et al. 2003). Pyridoxine also exhibited antioxidant activity by suppressing free radical enhanced luminol chemiluminescence. In human erythrocytes, both pyridoxine and pyridoxamine interfere with hyperglycemia-induced superoxide radical formation and prevent lipid peroxidation (Jain and Lim 2001). In another study, pyridoxine and pyridoxamine suppressed hyperglycemia-induced crystalline oxidation in cultured lens cells (Jain et al. 2002). These findings suggest an antioxidant mechanism of action for the beneficial effects of pyridoxine supplementation on clinical symptoms of neuropathy and retinopathy in diabetic patients. In addition to providing protection against ROS including .OH and H_2O_2, B6-vitamers are also effective singlet oxygen ($1O_2$) antagonists.

$1O_2$, a highly reactive form of molecular oxygen that forms by energy transfer upon photoexcitation of organic sensitizer molecules, oxidizes cellular targets with diffusion controlled reaction rates. Importantly, $1O_2$ is a crucial reactive intermediate involved in cellular photodamage that results from photosensitization reactions in skin solar damage and photocarcinogenesis (Wondrak et al. 2006b).

Interestingly, pyridoxine-dependent resistance against self-sensitization and $1O_2$ damage has been demonstrated in the parasitic fungus cercospora keidi (Bilski et al. 2000). Plant cell lysis by this microorganism involves secretion of cercosporin, a perylenequinone photosensitizer that generates $1O_2$ in the presence of ground state $3O_2$ and solar light. Pyridoxine has been identified as an endogenous $1O_2$-antagonist that protects this pathogenic organism against self-sensitization. $1O_2$-antagonism by B6-vitamers seems to involve sacrificial reaction leading to chemical destruction, a mechanism of action that combined with the known activity of B6-vitamers as weak UVA-photosensitizers excludes their use as skin photoprotective agents as discussed below.

15.3 B6-Derived Metal Chelators

Pyridoxamine has long been known to be a moderately potent metal chelator that preferentially forms complexes with Cu^{2+} and Fe^{3+} ions (Williams and Neilands 1954). The bidentate structural feature essential for metal ion binding involves the 4aminomethyl- and the 3-hydroxyl-substituents of the pyridine ring. Chelation involves the phenolate form of the 3-hydroxyl-substituent that displays a high degree of acidity (pKa 3.54) and therefore occurs in the anionic form under physiological conditions, where formation of a 2:1 complex between pyridoxamine and Cu2+ has been observed. Free and inadequately chelated transition metal ions are reactive species that contribute to cellular damage by participation in metal-catalyzed peroxide decay (Fenton reaction), lipid peroxidation, and other redox processes. Accumulating evidence suggests that free iron and copper ion-catalyzed reactions are important in the pathogenesis of various pathologies including ischemia reperfusion damage, neurodegenerative disease, secondary complications of diabetes, and atherosclerosis (Brewer 2007; Tabner et al. 2005). The pathological role of free transition metal ions suggests the feasibility of therapeutic intervention using small molecule transition metal ion chelators including pyridoxamine that bind and neutralize these reactive species. The preventive and therapeutic efficacy of metal chelators against various human pathologies has been documented in many preclinical and clinical studies. Indeed, metal chelation with subsequent suppression of metal ion-catalyzed oxidative and glycoxidative tissue damage seems to contribute to the efficacy of using high doses of pyridoxamine in the prevention and treatment of type II diabetic nephropathy in mouse models of the disease, a therapeutic application currently being evaluated in human phase II clinical studies as discussed below (Voziyan and Hudson 2005b).

15.4 Pyridoxamine as a Carbonyl Scavenger

Among the various B6-vitamers, only pyridoxamine contains the bis-nucleophilic 3hydroxy-4-aminomethyl-pyridine pharmacophore known to impart activity as a potent carbonyl scavenger and anti-glycation agent (Amarnath et al. 2004; Metz et al. 2003b; Onorato et al. 2000; Voziyan and Hudson 2005b). Carbonyl stress mediated by reactive carbonyl species (RCS) is an established source of endogenous tissue damage that occurs in various systemic pathologies, such as diabetes, atherosclerosis, Alzheimer's disease, and chronological aging (Baynes 2001; Baynes and Thorpe 2000b; Ulrich and Cerami 2001; Wondrak et al. 2002b, 2006b). Pyridoxamine is now in advanced clinical studies to test its therapeutic potential as a small molecule inhibitor of tissue carbonyl stress (Giannoukakis 2005; Khalifah et al. 2005). Chemical damage by RCS occurs by spontaneous aminocarbonyl reactions (referred to as 'glycation') between RCS and protein-bound amino groups leading to covalent protein adducts, crosslinking, and formation of protein-epitopes called advanced glycation endproducts (AGEs). The chemistry of AGE-accumulation in human tissue proteins is complex and depends on the crucial involvement of various RCS including mono-, di-, and oligo-carbonyl intermediates (e.g. glyoxal, methylglyoxal, 3-deoxyglucosone, glucosone) that originate from sugar metabolism, sugar autoxidation, and lipid peroxidation (Wondrak et al. 2002a).

AGEs accumulate on extracellular matrix proteins including collagen and elastin as a function of age inducing structural and functional alterations. In addition to protein crosslinking and matrix dysfunction, AGEs can stimulate pathological signaling through a receptor called RAGE involved in chronic inflammation and carcinogenesis (Bierhaus et al. 2000; Wondrak et al. 2005). Apart from the roles of AGE-epitopes as protein crosslinkers and RAGE-ligands, some AGEs with extended heterocyclic chromophores are now established skin photosensitizers that produce ROS upon UVA-irradiation adding to the photooxidative burden of chronologically aged skin and thereby contributing to photocarcinogenesis (Wondrak et al. 2006b). The role of carbonyl stress in the pathology of atherosclerosis, secondary diabetic complications associated with extracellular matrix protein damage including neuropathy, retinopathy, and nephropathy, and skin actinic and chronological damage is now well established (Baynes 2001; Baynes and Thorpe 2000b; Bierhaus et al. 2000, 2005; Ulrich and Cerami 2001; Wondrak et al. 2002b, 2006b). RCS that mediate carbonyl stress are therefore potential molecular targets for therapeutic intervention, and experimental evidence suggests the possibility of therapeutic intervention by rational design of novel RCS-antagonists, called carbonyl scavengers (Reddy and Beyaz 2006; Wondrak et al. 2002a). Carbonyl scavengers are experimental small molecule nucleophilic agents that inactivate RCS by sacrificial covalent adduction. Screening assays for the identification of nucleophilic carbonyl scavengers have been described (Wondrak et al. 2002a), prototype agents have been identified, and their efficacy has been tested in cellular and animal models of carbonyl stress (Giannoukakis 2005; Khalifah et al. 2005; Roberts et al. 2003; Williams 2004; Wondrak et al. 2006a).

Recently, pyridoxamine was identified as a prototype agent to suppress carbonyl stress and glycation reactions (Amarnath et al. 2004; Metz et al. 2003b; Onorato et al. 2000; Voziyan and Hudson 2005b). An abbreviated reaction scheme illustrating the chemistry of carbonyl scavenging by pyridoxamine directed against RCS is shown in Fig. 15.2. Numerous preclinical and some early clinical studies suggest that pyridoxamine is a potent inhibitor of carbonyl stress and tissue glycation damage in vivo (Giannoukakis 2005; Khalifah et al. 2005). Pyridoxamine inhibited the progression of renal disease, decreased hyperlipidemia, and protected against a range of pathological changes in the retinas of streptozotocin-diabetic rats (Degenhardt et al. 2002).

In the Zucker rat model of diabetes and hyperlipidemia, characterized by elevated levels of plasma alpha-dicarbonyl compounds and protein AGEs, pyridoxamine treatment had significant renoprotective effects and effectively decreased plasma levels of RCS including glyoxal and methylglyoxal (Nagaraj et al. 2002). Moreover, RCS-adducts of pyridoxamine were detected in urine of pyridoxamine-treated animals, indicating that the carbonyl-trapping mechanism is operative in vivo (Metz et al. 2003a). In human clinical studies, pyridoxamine has shown a favorable safety profile, and phase 2 clinical studies of this agent (Pyridorin™, Biostratum Inc.) in patients with diabetic nephropathy have been conducted (Giannoukakis 2005).

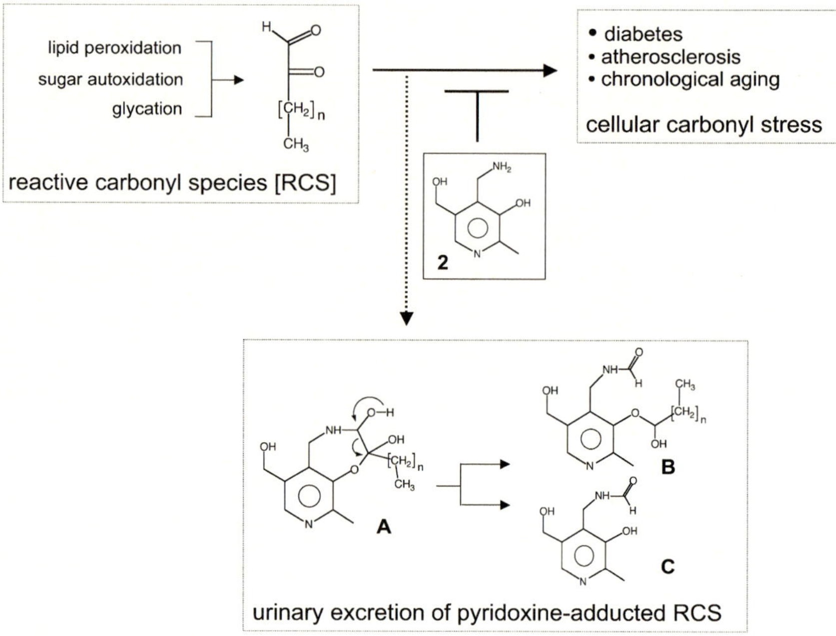

Fig. 15.2 Pyridoxamine as an inhibitor of cellular carbonyl stress. Pyridoxamine traps RCS by covalent adduction thereby interfering with tissue carbonyl stress involved in various human pathologies. RCS-trapping by pyridoxamine involves bis-nucleophilic adduction via the 4-aminomethyl- and 3-hydroxyl-functionalities with formation of products A-C. N-formylpyridoxamine (FAPM, product C) is the major pyridoxamine-adduct detected in urine of diabetic and obese rats treated with pyridoxamine (Metz et al. 2003b)

15.5 B6-Vitamers as Potential Mediators of Skin UVA-Photosensitization

All B6-vitamers share the 3-hydroxypyridine (3-HP) core moiety known to have potent photodynamic activity as recently revealed by a detailed structure activity relationship study of skin cell photosensitization by various endogenous pyridinium derivatives including B6-vitamers (Wondrak et al. 2004). Photosensitization occurs as a consequence of initial formation of excited states of chromophores and their subsequent interaction with substrate molecules (type I photoreaction) or molecular oxygen (type II photoreaction) through energy and/or electron transfer (Foote 1991; Wondrak et al. 2006b). Members of the extended 3-HP class of endogenous skin photosensitizers comprise all B6 vitamers and other 3-HP containing skin chromophores including the collagen crosslink pyridinoline, certain AGEs, and protein adducts of lipid peroxidation products as summarized in Fig. 15.3 (Wondrak et al. 2003, 2006b). Indeed, UVA photosensitization of human skin cells by B6-vitamers (pyridoxine, pyridoxamine, pyridoxal, pyridoxal-5′-phosphate) occurred in the lower micromolar range with dose-dependent inhibition of proliferation, cell

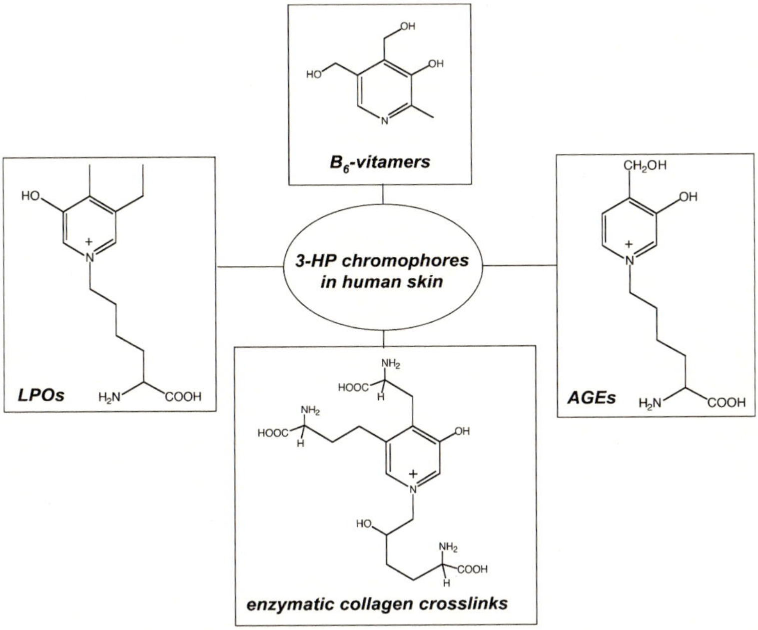

Fig. 15.3 The 3-hydroxypyridine family of potential skin photosensitizers. All B6 vitamers in human skin contain the phototoxic 3-HP moiety activated by solar UVB and UVA radiation. Moreover, 3-HP containing advanced glycation endproducts (AGEs) and lipid peroxidation (LPOs) products formed on skin proteins by reaction of lysine residues with RCS accumulate in skin during chronological and photoaging Paradoxically, 3-HP is also found in the enzymatic skin collagen crosslink pyridinoline, a constitutive component of the human dermal extracellular matrix

cycle arrest in G2/M, induction of apoptosis, intracellular oxidative stress, and p38 MAPkinase activation, all of which were reversible by thiol antioxidant intervention (Wondrak et al. 2004). Obviously, the presence of a phenolic 3-OH substituent is an essential structural requirement for sensitizer activity of B6-vitamers, since 2-HP and 4-HP, which occur predominantly as the tautomeric pyridone structures in aqueous solutions at neutral pH (Wang 1996), display no photosensitizer activity. This is consistent with a free radical mechanism of phototoxicity of biogenic 3-HP-derivatives observed in a recent study using cultured human skin cells (Wondrak et al. 2004). In simple in vitro models of protein photooxidation, B6-vitamers were moderately potent photosensitizers, effecting protein photo-crosslinking and photo-oxidation with incorporation of molecular oxygen. Based on these results, it was concluded that B6-vitamers constitute a novel class of UVA-photosensitizers capable of mediating skin photooxidative damage. Indeed, increased photosensitivity is a known consequence of B6-overdosing in humans (Murata et al. 1998), and phototoxicity of UV-irradiated pyridoxamine was reported as early as 1947 by Shwartzman and Fisher (1947) followed by other reports thereafter (Maeda et al. 2000; Sato et al. 1993).

B6-vitamers may therefore act as endogenous skin photosensitizers with relevance to human skin in vivo, since human skin contains various B6-vitamer forms in significant amounts (approximately 100 nmol/g protein (Coburn et al. 2003)), with pyridoxal-5′-phosphate and pyridoxal being the predominant vitamers in vivo. However, it is important to note that the potential phototoxic activity of B6-vitamers in human skin may be balanced in vivo by their antioxidant and singlet oxygen quenching activity above mentioned, suggesting a complex role in prevention and mediation of skin photodamage to be addressed by future studies.

15.6 Summary

Recent studies suggest that B6-vitamers can illicit therapeutic effects in a variety of human pathologies that depend on ROS, RCS, and transition metal ions including atherosclerosis, diabetes, and other age-related diseases. These non-coenzyme roles of B6-vitamers may therefore represent an opportunity for innovative therapeutic interventions.

Acknowledgements Supported in part by National Institutes of Health Grants (CA 106677-01, SWEHSC pilot research grant [ES06694], and GI Cancer Pilot Grant [SPORE, CA95060]).

References

Amarnath V, Amarnath K, Amarnath K, Davies S, Roberts LJ 2nd (2004) Pyridoxamine: an extremely potent scavenger of 1,4-dicarbonyls. Chem Res Toxicol 17:410–415

Baynes JW (2001) The role of AGEs in aging: causation or correlation. Exp Gerontol 36:1527–1537

Baynes JW, Thorpe SR (2000a, b) Glycoxidation and lipoxidation in atherogenesis. Free Radic Biol Med 28:1708–1716

Bierhaus A, Humpert PM, Morcos M, Wendt T, Chavakis T, Arnold B, Stern DM, Bilski PP, Li MY, Ehrenshaft M, Daub ME, Chignell CF (2000) Vitamin B6 (pyridoxine) and its derivatives are efficient singlet oxygen quenchers and potential fungal antioxidants. Photochem Photobiol 71:129–134

Bierhaus A, Humpert PM, Morcos M, Wendt T, Chavakis T, Arnold B, Stern DM, Nawroth PP (2005) Understanding RAGE, the receptor for advanced glycation end products. J Mol Med 83:876–886

Brewer GJ (2007) Iron and copper toxicity in diseases of aging, particularly atherosclerosis and Alzheimer's disease. Exp Biol Med 232:323–335

Coburn SP, Slominski A, Mahuren JD, Wortsman J, Hessle L, Millan JL (2003) Cutaneous metabolism of vitamin B-6. J Invest Dermatol 120:292–300

Culbertson SM, Enright GD, Ingold KU (2003) Synthesis of a novel radical trapping and carbonyl group trapping anti-AGE agent: a pyridoxamine analogue for inhibiting advanced glycation (AGE) and lipoxidation (ALE) end products. Org Lett 5:2659–2662

Degenhardt TP, Alderson NL, Arrington DD, Beattie RJ, Basgen JM, Steffes MW, Thorpe SR, Baynes JW (2002) Pyridoxamine inhibits early renal disease and dyslipidemia in the streptozotocin-diabetic rat. Kidney Int 61:939–950

Foote CS (1991) Definition of type I and type II photosensitized oxidation. Photochem Photobiol 54:659

Giannoukakis N (2005) Pyridoxamine. Curr Opin Investig Drugs 6:410–418

Jain SK, Lim G (2001) Pyridoxine and pyridoxamine inhibits superoxide radicals and prevents lipid peroxidation, protein glycosylation, and (Na^{+} + K^{+})-ATPase activity reduction in high glucosetreated human erythrocytes. Free Radic Biol Med 30:232–237

Jain AK, Lim G, Langford M, Jain SK (2002) Effect of high-glucose levels on protein oxidation in cultured lens cells, and in crystalline and albumin solution and its inhibition by vitamin B6 and N-acetylcysteine: its possible relevance to cataract formation in diabetes. Free Radic Biol Med 33:1615–1621

Khalifah RG, Chen Y, Wassenberg JJ (2005) Post-Amadori AGE inhibition as a therapeutic target for diabetic complications: a rational approach to second-generation Amadorin design. Ann NY Acad Sci 1043:793–806

Maeda T, Taguchi H, Minami H, Sato K, Shiga T, Kosaka H, Yoshikawa K (2000) Vitamin B6 phototoxicity induced by UVA radiation. Arch Dermatol Res 292:562–567

Matxain JM, Ristila M, Strid A, Eriksson LA (2006) Theoretical study of the antioxidant properties of pyridoxine. J Phys Chem A Mol Spectrosc Kinet Environ Gen Theory 110:13068–13072

McCormick DB, Chen H (1999) Update on interconversions of vitamin B-6 with its coenzyme. J Nutr 129:325–327

Metz TO, Alderson NL, Chachich ME, Thorpe SR, Baynes JW (2003a) Pyridoxamine traps intermediates in lipid peroxidation reactions in vivo: evidence on the role of lipids in chemical modification of protein and development of diabetic complications. J Biol Chem 278:42012–42019

Metz TO, Alderson NL, Thorpe SR, Baynes JW (2003b) Pyridoxamine, an inhibitor of advanced glycation and lipoxidation reactions: a novel therapy for treatment of diabetic complications. Arch Biochem Biophys 419:41–49

Metzler DE, Ikawa M, Snell EE (1954) A general mechanism for vitamin B6-catalyzed reactions. J Am Chem Soc 76:648–652

Murata Y, Kumano K, Ueda T, Araki N, Nakamura T, Tani M (1998) Photosensitive dermatitis caused by pyridoxine hydrochloride. J Am Acad Dermatol 39:314–317

Nagaraj RH, Sarkar P, Mally A, Biemel KM, Lederer MO, Padayatti PS (2002) Effect of pyridoxamine on chemical modification of proteins by carbonyls in diabetic rats: characterization of a major product from the reaction of pyridoxamine and methylglyoxal. Arch Biochem Biophys 402:110–119

Onorato JM, Jenkins AJ, Thorpe SR, Baynes JW (2000) Pyridoxamine, an inhibitor of advanced glycation reactions, also inhibits advanced lipoxidation reactions. J Biol Chem 275:21177–21184

Reddy VP, Beyaz A (2006) Inhibitors of the Maillard reaction and AGE breakers as therapeutics for multiple diseases. Drug Discov Today 11:646–654

Roberts MJ, Wondrak GT, Cervantes-Laurean D, Jacobson MK, Jacobson EL (2003) DNA damage by carbonyl stress in human skin cells. Mutat Res 522:45–56

Sasaki N, Fukatsu R, Tsuzuki K, Hayashi Y, Yoshida T, Fujii N, Koike T, Wakayama I, Sato R, Taguchi H, Maeda T, Yoshikawa K (1993) Pyridoxine toxicity to cultured fibroblasts caused by near-ultraviolet light. J Invest Dermatol 100:266–270

Schumacker PT (2006) Reactive oxygen species in cancer cells: live by the sword, die by the sword. Cancer Cell 10:175–176

Shwartzman G, Fisher A (1947) Studies on antibacterial properties of irradiated pyridoxamine. J Biol Chem 167:345–362

Stocker P, Lesgards JF, Vidal N, Chalier F, Prost M (2003) ESR study of a biological assay on whole blood: antioxidant efficiency of various vitamins. Biochim Biophys Acta 1621:1–8

Tabner BJ, El-Agnaf OM, German MJ, Fullwood NJ, Allsop D (2005) Protein aggregation, metals and oxidative stress in neurodegenerative diseases. Biochem Soc Trans 33:1082–1086

Ulrich P, Cerami A (2001) Protein glycation, diabetes, and aging. Recent Prog Horm Res 56:1–21

Valko M, Leibfritz D, Moncol J, Cronin MT, Mazur M, Telser J (2007) Free radicals and antioxidants in normal physiological functions and human disease. Int J Biochem Cell Biol 39:44–84

Voziyan PA, Hudson BG (2005a) Pyridoxamine as a multifunctional pharmaceutical: targeting pathogenic glycation and oxidative damage. Cell Mol Life Sci 62:1671–1681

Voziyan PA, Hudson BG (2005b) Pyridoxamine: the many virtues of a maillard reaction inhibitor. Ann NY Acad Sci 1043:807–816

Wang J, and Boyd RJ (1996) Tautomeric equilibria of hydroxypyridines in different solvents: ab initio study. J Phys Chem 100:16141–16146

Williams ME (2004) Clinical studies of advanced glycation end product inhibitors and diabetic kidney disease. Curr Diab Rep 4:441–446

Williams VR, Neilands JB (1954) Apparent ionization constants, spectral properties and metal chelation of the cotransaminases and related compounds. Arch Biochem 53:56–70

Wondrak GT, Cervantes-Laurean D, Roberts MJ, Qasem JG, Kim M, Jacobson EL, Jacobson MK (2002a) Identification of alpha-dicarbonyl scavengers for cellular protection against carbonyl stress. Biochem Pharmacol 63:361–373

Wondrak GT, Roberts MJ, Cervantes-Laurean D, Jacobson MK, Jacobson EL (2003) Proteins of the extracellular matrix are sensitizers of photo-oxidative stress in human skin cells. J Invest Dermatol 121:578–586

Wondrak GT, Jacobson MK, Jacobson EL (2005) An emerging molecular targer in melanoma: Cellular carbonyl stress and the inhibition of miochondrial survival pathways by carbonyl scavenger agents. Curr Cancer Ther Rev 1(3):271–276

Wondrak GT, Jacobson MK, Jacobson EL (2006a) Antimelanoma activity of apoptogenic carbonyl scavengers. J Pharmacol Exp Ther 316:805–814

Wondrak GT, Jacobson MK, Jacobson EL (2006b) Endogenous UVA-photosensitizers: mediators of skin photodamage and novel targets for skin photoprotection. Photochem Photobiol Sci 5:215–237

Wondrak GT, Roberts MJ, Jacobson MK, Jacobson EL (2002b) Photosensitized growth inhibition of cultured human skin cells: mechanism and suppression of oxidative stress from solar irradiation of glycated proteins. J Invest Dermatol 119:489–498

Wondrak GT, Roberts MJ, Jacobson MK, Jacobson EL (2004) 3-Hydroxypyridine chromophores are endogenous sensitizers of photooxidative stress in human skin cells. J Biol Chem 279:30009–30020

Chapter 16
Cobalamin Deficiency

Wolfgang Herrmann and Rima Obeid

Abstract Cobalamin (Cbl, vitamin B12) consists of a corrinoid structure with cobalt in the centre of the molecule. Neither humans nor animals are able to synthesize this vitamin. Foods of animal source are the only natural source of cobalamin in human diet. There are only two enzymatic reactions in mammalian cells that require cobalamin as cofactor. Methylcobolamin is a cofactor for methionine synthase. The enzyme methylmalonyl-CoA-mutase requires adenosylcobalamin as a cofactor. Therefore, serum concentrations of homocysteine (tHcy) and methylmalonic acid (MMA) will increase in cobalamin deficiency. The cobalamin absorption from diet is a complex process that involves different proteins: haptocorrin, intrinsic factor and transcobalamin (TC). Cobalamin that is bound to TC is called holotranscobalamin (holoTC) which is the metabolically active vitamin B12 fraction. HoloTC consists 6 and 20% of total cobalamin whereas 80% of total serum cobalamin is bound to another binding protein, haptocorrin. Cobalamin deficiency is common worldwide. Cobalamin malabsorption is common in elderly subjects which might explain low vitamin status. Subjects who ingest low amount of cobalamin like vegetarians develop vitamin deficiency. No single parameter can be used to diagnose cobalamin deficiency. Total serum cobalamin is neither sensitive nor it is specific for cobalamin deficiency. This might explain why many deficient subjects would be overlooked by utilizing total cobalamin as status marker. Concentration of holotranscobalamin (holoTC) in serum is an earlier marker that becomes decreased before total serum cobalamin. Concentrations of MMA and tHcy increase in blood of cobalamin deficient subjects. Despite limitations of these markers in patients with renal dysfunction, concentrations of MMA and tHcy are useful functional markers of cobalamin status. The combined use of holoTC and MMA assays may better indicate cobalamin status than either of them. Because Cbl deficiency is a risk factor for neurodegenerative diseases an early diagnosis of a low B12 status is required which should be followed by an effective treatment in order to prevent irreversible damages.

W. Herrmann (✉)
Department of Clinical Chemistry and Laboratory Medicine, University of Saarland, 66421 Homburg, Germany
e-mail: kchwher@uniklinik-saarland.de

O. Stanger (ed.), *Water Soluble Vitamins*, Subcellular Biochemistry 56,
DOI 10.1007/978-94-007-2199-9_16,

Keywords Cobalamin · Folate trap · Haptocorrin · Holotranscobalamin · Homocysteine · Malabsorption · Methylmalonic acid

16.1 Cobalamin and Its Metabolism

The discovery of cobalamin (Cbl; vitamin B12) was after several years of intensive studies on pernicious anemia, a previously fatal disease. In the early 1920s Minot and Murphy demonstrated that they were able to cure pernicious anaemia by whole liver extract. Later, it was shown that liver is an important source of cobalamin. Cobalamin was then isolated and the crystallized molecule was identified. Cobalamin belongs to a group of compounds of similar chemical structure but completely different biological functions. Cbl consists of a corrinoid molecule with cobalt in the centre of this molecule. The synthetic forms of cobalamin are cyanocobalamin and hydroxycobalamin. There are only two forms of Cbl that have biological activity as cofactors in enzyme reactions (Fig. 16.1). These are adenosylcobalamin (AdoCbl) and methylcobalamin (MeCbl) (Herzlich and Herbert 1988). In mammalian metabolism, Cbl is required for only two key enzymatic reactions. The first reaction occurs in the cytosol and involves synthesis of methionine from homocysteine. Methionine synthase and its cofactor MeCbl catalyze this reaction. The second pathway takes place in the mitochondria and involves isomerisation of methylmalonyl-CoA to succinyl-CoA. This reaction is catalyzed by methylmalonyl-CoA mutase with AdoCbl as a cofactor (Stroinsky and Schneider

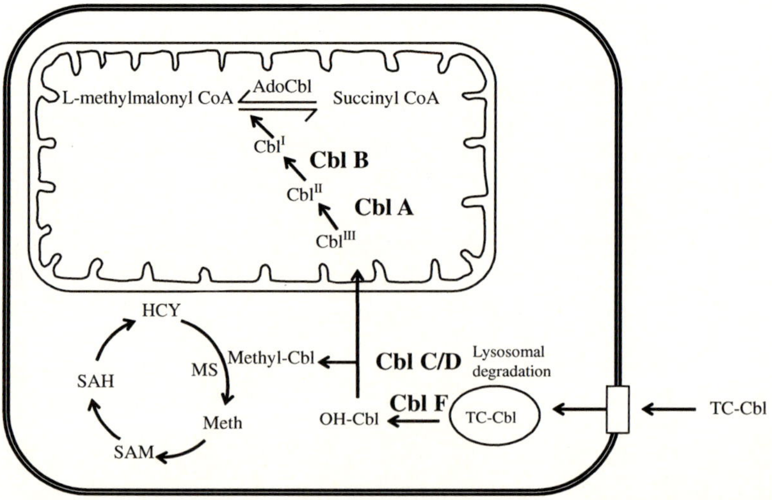

Fig. 16.1 The metabolic pathways enhanced by cobalamin and inherited disorders of cobalamin (see text for details). Hcy; homocysteine, Met; methionine, SAH; S-adenosylhomocysteine, SAM; S-adenosylmethionin, methyl-Cbl; methyl cobalamin, Ado-Cbl; adenosyl cobalamin

1987). The latter reaction is part of the catabolism of odd-chained fatty acids, cholesterol and several amino acids. The excess of methymalonyl-CoA is converted into methylmalonic acid (MMA). Therefore, cobalamin deficiency leads to MMA elevation that is considered a sensitive marker for this deficiency. A reduced flux through the methylmalonyl-CoA mutase reaction is discussed to contribute to subsequent neurological tissue damage.

16.1.1 Sources of Cobalamin

Cobalamin synthesis is very complex and restricted to a certain stain of bacteria. Neither humans, nor animals are able to synthesize this complex molecule. Plants and fungi are thought to neither synthesize, nor to use the vitamin as well. Animals can get cobalamin by consuming foods contaminated with synthesizing bacteria and then incorporate the vitamin into their body organs. Foods of animal source are the only natural source of cobalamin in human diet. The main sources of cobalamin in humans diet are meet (2–5 μg/100 g), fish (2–8 μg/100 g), milk (1.5 μg/100 ml), cheese (1–2 μg/100 g) and eggs (2 μg/100 g) (Scott 1997). Cobalamin is essential for normal maturation and development of all DNA synthesizing cells including blood cells and cells of the central nervous system. The ultimate source of cobalamin in human diet is from foods contaminated with B12-synthesizing bacteria. Not all forms of cobalamin formed by microbes are metabolically active for mammalian cells. Some naturally occurring forms of corrinoids have similar structure but no biological roles in humans. These are produced by some algae (spirulina) and are termed analogous. Cobalamin-analogous may even block the normal metabolism of the vitamin (Herbert and Drivas 1982).

16.1.2 Absorption, Excretion and Haemostasis of Cobalamin

Cobalamin is bound to food proteins and must be released before that the vitamin can be absorbed. This is achieved by the action of gastric acid and proteolytic enzymes in the stomach. In the stomach cobalamin is captured by haptocorrin, an R-binder protein made in the saliva and stomach. In the upper small intestine, pancreatic enzymes and an alkaline pH degrade the haptocorrin-cobalamin complex. The free vitamin is then captured by intrinsic factor, another B12-binding protein. The intrinsic factor-cobalamin complex is transported to the terminal ileum where the complex is recognised and internalised by specific membrane receptors of the enterocytes (intrinsic factor receptor). The receptor-mediated absorption of cobalamin is a saturable process and a maximal amount of 3 μg of the vitamin per meal can be internalized via this pathway. After absorption of cobalamin-intrinsic factor complex into the enterocytes, the complex is degraded and cobalamin is transferred to a third binding protein, transcobalamin (TC).

Transcobalamin is synthesized within the enterocytes and is the only binder that can deliver cobalamin into cells via TC-receptor. The TC-cobalamin complex is released into the portal circulation and is subsequently recognised by TC-receptors that are expressed by all cell types. The part of cobalamin which is bound to TC is named holotranscobalamin (holoTC) (Carmel 1985). Only 6–20% of total plasma cobalamin is present as holoTC (Hall 1977). The remaining part of cobalamin is bound to haptocorrin and is called holohaptocorrin (holoHC) (England et al. 1976). Despite that haptocorrin binds almost 80% of total plasma cobalamin, the functions of this protein are not well investigated.

The bioavailability of cobalamin from vitamin preparations is greater than that from foods. In contrast to the receptor mediated absorption, about 1% of free cobalamin is absorbed by passive diffusion even in the absence of intrinsic factor. Therefore, even patients with pernicious anemia, or those with disturbed gastrointestinal pH or atrophic gastritis may benefit from relatively high doses of oral cobalamin treatment. Furthermore, a considerable amount of cobalamin is secreted into the bile. Two thirds of the secreted cobalamin in the bile is reabsorbed in the ileum. The liver contains most of the body's cobalamin. It is calculated that 2–3 mg of cobalamin are stored in the liver (Markle 1996). The kidney and the brain are also two important organs that accumulate cobalamin. The kidney can release cobalamin in case of short term depletion of the vitamin. Cobalamin is excreted into urine and this can be reabsorbed in the proximal tubules via a specific receptor (megalin) (Moestrup et al. 1996). The major route by which cobalamin is lost from the body is through the feces.

16.2 Cobalamin Deficiency

Frank cobalamin deficiency is common worldwide (Morris et al. 2002; Obeid et al. 2002). In developed countries, cobalamin deficiency is restricted to patients with malabsorption, intestinal resection, or those who do not ingest a sufficient amount of the vitamin (Herrmann et al. 2003; Carmel 1997). Cobalamin deficiency is common in elderly people and in those who ingest a strict vegetarian diet. In contrast, cobalamin deficiency has been reported in children, middle age, and elderly people from developing countries.

Pre-clinical cobalamin deficiency is the state in which metabolic evidence of insufficiency exists without symptoms of anemia or neurological complications. Measurement of serum concentrations of metabolic markers allow detection of a large number of asymptomatic subjects who are depleted of the vitamin.

16.2.1 Cobalamin Deficiency in the Elderly

16.2.1.1 Factors Contributing to Low Vitamin Status with Aging

Cobalamin deficiency in geriatric population has gained a particular importance in recent years (Baik and Russell 1999; Obeid et al. 2004). Serum concentration of cobalamin decreases with age and that of MMA and tHcy increases. On the one hand, the decline in cobalamin status with age could not be explained by a low dietary intake of the vitamin. On the other hand, it is well recognized that low cobalamin status in the elderly is associated with an increased incidence of pernicious anemia (type A atrophic gastritis), and type B atrophic gastritis. In western countries, approximately 2% to 3% of free living elderly people (>60 years) had undiagnosed pernicious anemia (Carmel 1996; Krasinski et al. 1986). Ethnic differences in the incidence of pernicious anemia and the age of onset have been reported (Carmel and Johnson 1978).

16.2.2 Causes of Cobalamin Deficiency

16.2.2.1 Inherited Defects

The most important inherited disorders of cobalamin absorption, transport, metabolism, or utilization are presented in Table 16.1.

Congenital transcobalamin deficiency: This congenital recessive defect causes a lack of the binding protein, TC. Several cases have been described in the literature. Although babies are born asymptomatic, severe haematological and neurological

Table 16.1 Hereditary disorders of cobalamin absorption, transport, metabolism or utilization

Defect	Affected enzyme or step	Metabolic abnormalities	Therapeutic strategies
Congenital pernicious anemia (Intrinsic factor deficiency)	Cobalamin absorption	↑HCY, ↑MMA	Cobalamin + intrinsic factor
Imerslund-Gräsbeck syndrome	Selective cobalamin malabsorption	Proteinurea, ↑HCY, ↑MMA	Cobalamin injection
Transcobalamin II deficiency	Defective transport of cobalamin into the blood stream and into the cells	↑HCY, ↑MMA	High doses of systemic cobalamin
Methylmalonic aciduria	Methylmalonyl CoA mutase deficiency (mut^0, mut^-)	↑MMA, acidosis	Protein restriction (limiting the amino acids that use the propionate pathway)
Cbl A and Cbl B diseases	Failure to synthesize adenosylcobalamin	↑MMA	Cobalamin injections
Cbl E and Cbl G diseases	Failure to synthesize methylcobalamin	↑tHcy, ↓methionine	Cobalamin injections + Betaine
Cbl C and Cbl D diseases	Failure to synthesize adenosylcobalamin and methylcobalamin	↑MMA, ↑tHcy	Cobalamin injections + Betaine
Cbl F disease	Failure to release cobalamin from the lysosome	↑MMA, ↑tHcy	Cobalamin injections

manifestations are expressed few days to weeks after birth. This case is characterized by irritation, failure to thrive, and severe elevation of methylmalonic acid in blood and urine of affected children. Skin fibroblasts or lymphocytes isolated from affected children can not take up cobalamin, in contrast to cells isolated from healthy subjects. In most cases, high doses of cobalamin can improve clinical symptoms (Bibi et al. 1999; Teplitsky et al. 2003; Cooper and Rosenblatt 1987; Hakami et al. 1971). Abnormal transcobalamin has also been reported and is associated with similar symptoms.

Imerslund-Grasbeck-syndrom: This disorder is caused by a defective transport of cobalamin by enterocytes. This disease is characterized by proteinuria and lose of cobalamin in urine (Ben-Ami et al. 1990). Cobalamin malabsorption in this case is not related to intrinsic factor deficiency or to an abnormal cobalain-intrinsic factor complex. The cause behind cobalamin malabsorption and lose is related to a defective cubulin, the receptor for IF-cobalamin that is expressed in the enterocytes and the renal tubulus.

Disorders of cobalamin utilization: There are several inherited defects in cobalamin utilization. These are summarized in Fig. 16.1. Methylmalonic aciduria is related to a functional defect or a deficiency of the mithochondrial enzyme, methylmalonyl CoA mutase. This defect is characterized by severe elevation of blood and urine concentrations of MMA. Cultures of patients fibroblasts have shown that some patients do not response to cobalamin treatment, and others have a residual enzyme activity and response to cobalamin treatment. Cobalamin mutant A and B are related to adenosylcobalamin deficiency and cause methylmalonic aciduria. Cobalamin A defect is related to a failure in a reduction step and the second disorder is caused by a failure to transfer adenosyl to cobalamin.

Cobalamin mutant C, D, and F diseases are associated with combined adenosylcobalamin and methylcobalain deficiencies. Therefore, patients have usually, homocysteinuria, hypomethioninemia, and methylmalonic aciduria. The C and D defects involve a defective reductase step; whereas the F mutant involves inability of releasing cobalamin from the lysosome. The last two disorders in cobalamin utilization are, cobalamin mutant E and G that are related to methylcobalamin deficiency.

Haptocorrin deficiency (deficiency of R binder): This case has been recently reported and is associated with a very low total serum cobalamin. Nevertheless, subjects who express this phenotype remain asymptomatic and concentrations of holotranscobalain remain within the normal range (Lin et al. 2001).

16.2.2.2 Acquired Causes of Cobalamin Deficiency

Food cobalamin malabsorption: Food cobalamin malabsorption is the most common cause of cobalamin deficiency in elderly people. There are several diseases that may cause cobalamin malabsorption. Pernicious anemia (anti IF antibodies) is the most famous disorder of cobalamin absorption. In this case, cobalamin can not be absorbed because of the lack of IF. Antiparietal cell antibodies can cause lack of intrinsic factor and thus cobalamin deficiency.

Type B chronic atrophic gastritis is related to Helicobacter pylori infection and is described to cause cobalamin malabsorption. This disorder results in a low acid-pepsin production and food cobalamin malabsorption. The release of cobalamin from food protein is decreased in case of lowered gastric acidity (Doscherholmen et al. 1977). Importantly, because the production of IF is not affected in Type B atrophic gastritis, those subjects may benefit from crystalline cobalamin. Oral cobalamin therapy (3–5 mg/week) was effective in treatment of food-cobalamin malabsorption (Andres et al. 2001, 2003).

Celiac disease and tropical spree are also associated with cobalamin deficiency. In both cases, recurrent diarrhea causes severe damage to the gastrointestinal tract and interferes with cobalamin absorption by the enterocytes. Importantly, several drugs are known to alter gastrointestinal pH thus causing cobalamin malabsorption. Table 16.2 summarizes the most important acquired causes of cobalamin deficiency.

Cobalamin deficiency related to a low intake of the vitamin: The Recommended Dietary Intake (RDI) of cobalamin for adults is 2.4 μg/day (Institute of Medicine 2000). Recent studies indicated that a total daily intake of 6 μg/day can be necessary for preventing metabolic signs of deficiency in older women (Bor et al. 2006). Chronic low intake of cobalamin can cause severe deficiency. Cobalamin deficiency takes years to develop, even when one stops to ingest the vitamin. This is related to the relatively large body stores of cobalamin, in addition to the effective entero-hepatic circulation that ensures reabsorption of the vitamin. Subjects who adhere to a life long strict vegetarian diet develop cobalamin deficiency (Herrmann et al. 2003; Herbert 1994). Studies have shown that the severity of cobalamin deficiency is related to the degree of animal food restriction and to the duration of the vegetarian lifestyle (Herrmann et al. 2003; Herbert 1994). Because milk and eggs contain a low amount of cobalamin, the intake of cobalamin in a lacto- and lacto-ovo-vegetarian

Table 16.2 Acquired causes of cobalamin deficiency

Disease or condition	Mechanism
Restricted intake	Vegetarians, children of vegetarian mothers, poverty, malnutrition, anorexia nervosa
Increased demands	Bleeding, pregnancy, lactation
Medications	Changing gastrointestinal pH, interaction with vitamin absorption or metabolism
Anti H2 receptor	
Proton pump inhibitors	
Oral contraceptives	
Pernicious anemia (type A atrophic gastritis)	Anti IF antibodies
Antiparital cell antibodies	Lake of IF
Type B atrophic gastritis (H. Pylori)	Changing gastrointestinal pH
Other gastrointestinal morbidities	Interact with the vitamin absorption
Terminal ileal diseases	
Pancreatic insufficiency	
Ileal or gastric resection	
Celiac disease, tropical spree	
Colitis ulcerosa, Morbus Crohn	

diet might be sufficient. Nevertheless, studies have shown that cobalamin status (indicated by MMA, tHcy, holoTC) was lower in lacto- and lacto-ovo-vegetarians compared with omnivorous subjects (Herrmann et al. 2003). In addition, metabolic signs of cobalamin deficiency were more common in vegans compared with lacto- and lacto-ovo-vegetarians (Herrmann et al. 2003). Chronic low intake of cobalamin and its deficiency may be endemic in developing countries and is related to poverty and malnutrition in this case (Refsum et al. 2001).

The RDI for cobalamin is higher for pregnant and lactating women. Cobalamin deficiency may occur in women who do not ingest a sufficient amount of the vitamin. Metabolic and clinical signs of cobalamin deficiency have been reported in newborn babies from strict vegetarian mothers or in breast fed infants from deficient mothers (Bjorke et al. 2001; Schneede et al. 1994; Graham et al. 1992; Dagnelie et al. 1989). The content of cobalamin in mother milk was related to markers of cobalamin deficiency in the infants. These conditions were partly reversible after cobalamin supplementation. In a study that included adolescents who were fed a macrobiotic diet until the age of 6 years, metabolic signs of cobalamin deficiency were observed even several years after ingesting an omnivorous diet (Dhonukshe-Rutten et al. 2005). These data show that metabolic signs of cobalamin deficiency can not be resolved after mild modifications of the diet. Moreover, clinical signs might be irreversible.

16.2.3 Diagnosis of Cobalamin Malabsorption

Schilling test was the first test described for diagnosis of cobalamin malabsorption. This test depends on using an oral dose of radioactively-labeled cobalamin and following the radioactivity in urine or blood of patients. Schilling test can be applied with or without IF. Furthermore, the test has been improved to address the issue of malabsorption of protein-bound cobalamin (Nickoloff 1988; Doscherholmen et al. 1983). The use of radio-labeled cobalamin has limited the use of Schilling test in recent years.

Other approaches have been suggested recently to assess cobalamin absorption. One approach is to measure the response of holoTC (cobalamin saturated part) after a small oral dose of cobalamin (Bor et al. 2005). This test might be also applied with recombinant human IF. Preliminary results indicated that recombinant human IF promotes B12 absorption among patients with evident cobalamin deficiency (Hvas et al. 2001, 2006). Although early results are encouraging, studies that include patients with malabsorption disorders should be conducted before setting this test to the routine use.

16.2.4 Diagnosis of Cobalamin Deficiency

16.2.4.1 Determination of Total Serum Cobalamin

Despite its serious limitations, total serum cobalamin is the first parameter that is measured when cobalamin deficiency is suspected. As mentioned above, 80% of total cobalamin is bound to haptocorrin. This part is metabolically inert and can not be delivered into the cells. Therefore, minor changes in cobalamin status in the blood can not be reflected by concentrations of haptocorrin. Total serum cobalamin has several major limitations. On the one hand, low serum concentrations of cobalamin were not observed in all subjects with clinically evident cobalamin deficiency (false negative). On the other hand, there are many subjects with low serum cobalamin who are not deficient (false positive). Therefore, the low sensitivity and specificity of total cobalamin should be considered, particularly when serum cobalamin concentrations range between 150 and 350 pmol/L (Green 1995; Herrmann et al. 2000, 2001a). Approximately 45% of cobalamin-deficient subjects (MMA > 300 nmol/L and holoTC < 40 pmol/L) would have been overlooked, using cobalamin as a screening parameter. Obviously, the insensitivity of the total cobalamin test can not be resolved by simply shifting the cut-off value towards a higher level, because more subjects with normal B12 status will be falsely diagnosed as deficient (Fig. 16.2).

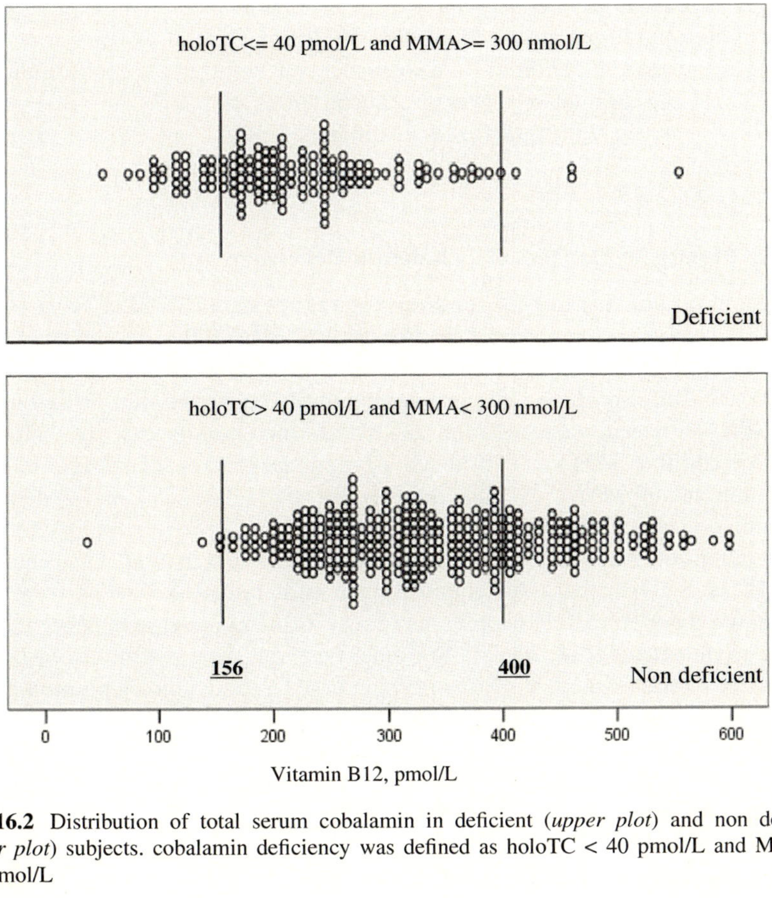

Fig. 16.2 Distribution of total serum cobalamin in deficient (*upper plot*) and non deficient (*lower plot*) subjects. cobalamin deficiency was defined as holoTC < 40 pmol/L and MMA ≥ 300 nmol/L

Because of its low costs compared to other available markers, cobalamin assay can be used as a first line parameter. Nevertheless, the results should be interpreted individually.

16.2.4.2 Determination of Holotranscobalamin

HoloTC represents the biologically active cobalamin fraction. Recent studies indicated that holoTC might be the earliest marker that indicate cobalamin status in blood (Bor et al. 2004).

Currently, the estimation of holoTC has been made possible by several methods (Ulleland et al. 2002; Refsum et al. 2006). Studies on cobalamin-deficient and non-deficient subjects have confirmed that holoTC has a better specificity and sensitivity compared with total B12 (Herrmann et al. 2005).

Limitations of holoTC should be also considered when interpreting the results of the test. Oral contraceptives have been reported to be associated with a 25% lower concentrations of holoTC without affecting serum concentrations of tHcy or MMA (Riedel et al. 2005). Moreover, renal insufficiency is associated with increased concentrations of holoTC in blood (Herrmann et al. 2005; Carmel et al. 2001). This metabolic condition is also associated with higher concentrations of MMA. Therefore, normal or elevated concentrations of holoTC in renal patients or in elderly people with mild renal insufficiency might not exclude cobalamin deficiency.

16.2.4.3 Metabolic Markers of Cobalamin Deficiency

The most important markers for cobalamin status are summarized in Table 16.3.

Plasma concentrations of methylmalonic acid (MMA): Because cobalamin is a cofactor for methylmalonyl-CoA mutase, cobalamin deficiency causes accumulation of methylmalonyl-CoA. The last compound is converted into methylmalonic acid (MMA). Serum concentration of MMA increase in case of cobalamin deficiency and thus MMA is considered a functional marker for cobalamin status. Concentrations of MMA (> 300 nmol/L) are more sensitive and specific for cobalamin deficiency than total serum cobalamin.

Limitations of this metabolic marker should be also kept in mind. Concentrations of MMA increase in renal dysfunction. Even sub-clinical decline in renal function (i.e., common in elderly people) may cause artificial increase in concentrations of MMA (Herrmann et al. 2001a). Intestinal bacterial overgrowth may also cause artificial elevation of MMA because of increased production of propionic acid, a precursor of MMA. Nevertheless, in these cases cobalamin deficiency can not be confirmed or excluded. A significant decrease in concentrations of MMA by cobalamin treatment indicates a pre-clinical deficiency.

Plasma concentrations of tHcy: Concentrations of tHcy increase in cobalamin deficiency because of the role of methylcobalamin as a cofactor for methionine synthase that converts tHcy into methionine. Nevertheless, concentrations of tHcy

Table 16.3 Available markers of B-vitamins deficiency

Parameter	Reference range	Material	Methods	Pre-analytical	Confounding factors
tHcy	< 12 μM	EDTA plasma, serum	GCMS, HPLC, (EIA) enzymatic immuno-assay, LC-MS/MS	Centrifuge and separate within 45 min preferably keep the sample on ice Several stabilizing agents are currently available (DZA)	Renal function (↑) Age (↑) Medications (↑) Hormone therapy (↑→)
MMA	< 0.300 μM	EDTA plasma, serum	GCMS LC_MS/MS	–	Renal function (↑) Age (↑)
Cobalamin	> 156 pmol/L	Serum	Chemiiluminescience, Microbiological assay	–	Renal function (↑)
holoTC	> 40 pmol/L	EDTA plasma, serum	RIA, ELISA	–	Renal function (↑)

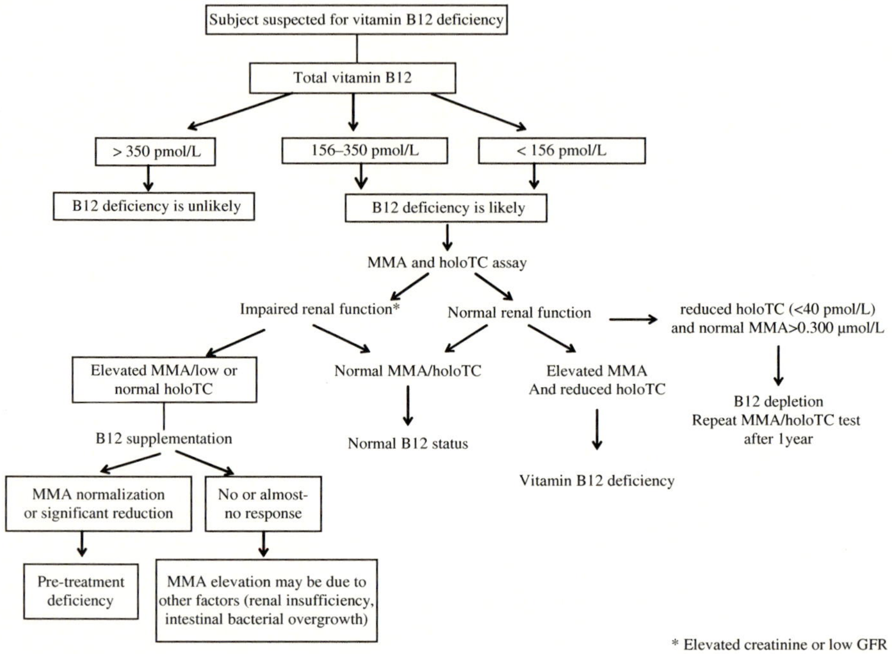

Fig. 16.3 Suggested algorithm for diagnosing cobalamin deficiency

increase also in folate and vitamin B6 deficiency or renal insufficiency (Herrmann et al. 2001b). Folate deficiency has more impact on tHcy compared with cobalamin deficiency. Unlike MMA elevation, tHcy elevation is not common in B12-deficient vegetarians, because of relatively high folate concentrations found in those subjects. High folate status partly compensates for the effect of cobalamin deficiency on tHcy elevation.

Figure 16.3 shows a diagnostic algorithm suggested for diagnosis of cobalamin deficiency.

16.2.4.4 Staging of Cobalamin Deficiency

Cobalamin deficiency is thought to be developed in four stages (Herbert 1994). In the early stages (I and II), the plasma and cell stores become depleted of the vitamin. This is reflected by a low plasma concentration of holoTC (Herbert et al. 1990; Lindgren et al. 1999). If the negative balance continues (stage III), metabolic markers of cobalamin deficiency become elevated (tHcy and MMA). This is explained by impairment of the B12-dependent enzymes (methionine synthase and methylmalonyl-CoA mutase). Clinical signs of cobalamin deficiency become obvious in stage IV (macrocytic anemia, neurological symptoms).

16.2.4.5 Diagnosis of Cobalamin Deficiency Utilizing Available Markers

There is no single parameter that can be reliably used to diagnose cobalamin deficiency. Megaloblastic anemia and neurological symptoms are neither sensitive nor they are specific for cobalamin deficiency. Metabolic signs (elevated MMA and tHcy) of deficiency were observed in the absence of haematological or clinical manifestations. Concentrations of total cobalamin in serum are also too insensitive (Herrmann et al. 2000, 2001b). Measurement of serum concentration of MMA alone or in conjunction with tHcy has partly resolved the demand for a sensitive and a specific test. In general, results of the metabolites should be interpreted with caution, because it is difficult to determine to which extent the impaired kidney function may participate in MMA and tHcy elevation. Subjects with serum cobalamin concentrations up to 350 pmol/L may have increased MMA and low holoTC indicating functional cobalamin deficiency. Studies on vegetarian subjects have shown that the majority of them had low holoTC in addition to metabolic signs indicating cobalamin deficiency (stage III) (elevated tHcy and MMA) (Herrmann et al. 2003).

16.2.4.6 Cobalamin Deficiency Causes Folate Trap

Because of the role of cobalamin in folate metabolism, cobalamin deficiency can cause a secondary folate deficiency. Cobalamin deficiency inhibits the activity of methionine synthase, and causes the retention of 5-MTHF. 5-MTHF becomes trapped because the transfer of the methyl group is inhibited (Fig. 16.1). The level of folate in serum or plasma of cobalamin deficient subjects may be normal to high normal, however, this is mostly as 5-MTHF. This phenomenon is called "folate

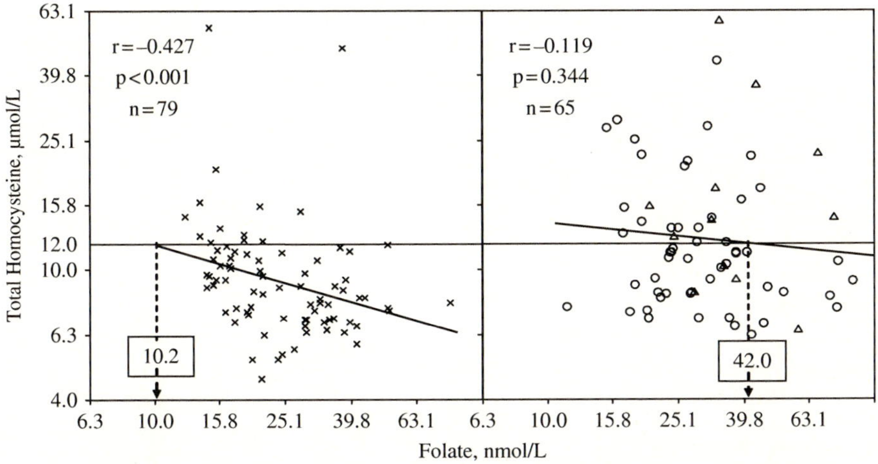

Fig. 16.4 Scatter plot showing concentrations of tHcy in relation to serum folate in omnivorous (*left plot*) and vegetarians (*right plot*). *Numbers* on the axis are anti-log

trap". Folate and cobalamin deficiency have similar clinical (megaloblastic red cells) and metabolic (elevation of tHcy) signs. This explains why cobalamin deficient subjects were frequently treated with folate. This treatment may even relief the hematological symptoms. Nevertheless, neurological signs of cobalamin deficiency may be worsen. Figure 16.4 shows this phenomenon in strict vegetarians compared with omnivorous subjects.

16.3 Cobalamin Deficiency in Renal Patients

Cobalamin deficiency is common in renal patients. Nevertheless, diagnosing cobalamin deficiency in renal patients remains a challenge. Renal insufficiency and hypovolemia cause elevation of MMA and tHcy (Herrmann et al. 2001b; Lindgren et al. 1999; Savage et al. 1994). Unexpectedly low serum concentrations of cobalamin or holoTC are uncommon in patients with renal insufficiency (Herrmann et al. 2005). A very important finding is that renal patients show significant metabolic improvements (reduction of MMA) after treatment with cobalamin (Obeid et al. 2005). The reasons for elevated serum cobalamins (total B12 and holoTC) in renal patients are not known. An abnormal distribution of holoTC (Carmel et al. 2001), a disturbed receptor activity for renal TC uptake and the possibility that TC is functionally altered by renal failure are possible explanations for accumulation of cobalamin in serum of renal patients.

16.3.1 Cobalamin Status and Vegetarian Diet

Cobalamin status is directly correlated with dietary intake and length of time following a vegetarian diet (Chanarin et al. 1985; Miller et al. 1991). According to the strictness of the diet, approximately 30–60% of vegetarian subjects might have metabolic evidence indicating cobalamin deficiency (Donaldson 2000; Rauma et al. 1995). Serum concentrations of MMA, holoTC, tHcy and total cobalamin are related to the type of the diet (Table 16.4). Vegan subjects had the lowest cobalamin status.

Table 16.4 Cobalamin status in vegetarian and non-vegetarian subjects

	Omnivorous $n = 79$	LOV/LV $n = 66$	Vegan $n = 29$
Cobalamin, pmol/L	287 (190/471)	192 (127/450)[a]	148 (99/314)[a,b]
HoloTC, pmol/L	54 (16/122)	23 (3/155)[a]	10 (2/78)[a,b]
MMA, nmol/L	161 (95/357)	355 (138/1948)[a]	708 (193/3470)[a,b]
tHcy, μmol/L	8.8 (5.5/16.1)	10.6 (6.4/27.7)[a]	12.8 (5.9/57.1)[a]
Folate, nmol/L	21.8 (14.5/51.5)	28.8 (16.1/77)[a]	31.8 (19.7/78.1)[a]
Creatinine, μmol/L	71 (53/93)	71 (53/88)	71 (49/88)

Data are median (5th/95th). $p < 0.05$: [a] compared to omnivorous controls; [b] comparing LOV/LV with vegans. LOV/LV; lacto-ovo/lacto-vegetarians

This is expected, because a vegan diet includes no kind animal foods. Although lacto- and lactoovo-vegetarians consume some animal foods (egg, milk, and milk products), metabolic signs of cobalamin deficiency were also common in this group (Herrmann et al. 2003) (Table 16.4). This group of vegetarians show intermediate cobalamin status compared with vegans and omnivorous (Herrmann et al. 2003). Approximately, 45% of subjects with low holoTC and elevated MMA had normal serum B12 (Fig. 16.2). It should be noticed that low doses of cobalamin such as the one usually found in the multivitamin preparations are not likely to prevent the depletion of the vitamin in individuals who ingest little or no animal products.

16.4 Cobalamin Status and Markers in Elderly People

Cobalamin deficiency is common in elderly people. A low dietary intake of cobalamin was not common in cobalamin deficient-elderly people. Nevertheless, several age-related physiological factors may influence negatively the absorption of the vitamin from the intestine. Cobalamin deficiency is associated with a poor cognitive performance and neurological damage. Approximately, 20% of elderly people were found to have a low concentration of holoTC and an elevated MMA indicating a relatively late stage of B12 deficiency. Importantly, a relatively large proportion of elderly subjects were found to have an elevated concentration of MMA and a normal holoTC, in addition to a higher concentration of serum creatinine. This typical profile associated with impaired renal function may delay the diagnosis of cobalamin deficiency. The relation between holoTC and MMA has been examined in relation to age and renal function (Obeid et al. 2004). In general, elderly people display higher MMA concentrations in ranges of holoTC comparable to that in younger subjects. This phenomenon was more pronounced in elderly with renal insufficiency. The last findings suggest that cellular-B12-delivery may be challenged in elderly with renal insufficiency, leading to MMA increment and holoTC retention.

Available evidence suggests that elderly people seem to be resistant to low oral doses of cobalamin (Rajan et al. 2002). Serum concentrations of MMA were not normalized in most B12-deficienct elderly people who received crystalline B12 in doses between 25 and 100 μg. The RDI of B12 for older adults (2.4 μg/day) is far below the dose likely to normalize serum concentrations of the metabolites in elderly people. Taken together, a low cobalamin status in elderly people with renal insufficiency may be overlooked by using holoTC as a sole marker. Moreover, the occurrence of cobalamin deficiency may be overestimated using MMA test. The presence of a low concentration of holoTC and an elevated MMA in elderly people may confirm B12 deficiency and the absence of this combined results may not exclude a deficiency of this vitamin. A significant reduction of MMA after cobalamin treatment might indicate a pre-treatment low status of the vitamin.

Cobalamin deficiency in elderly people was also associated with increased plasma SAH. This marker of methylation status was lowered by high oral dose of cobalamin (1 mg/day for 3 months) (Stabler et al. 1996, 2006).

16.5 Clinical Manifestations of Cobalamin Deficiency

16.5.1 Hematological Symptoms

Megaloblastic anemia is the classical finding of cobalamin deficiency. The underling mechanism of anemia in cobalamin deficiency is a slowed DNA synthesis in rapidly dividing blood cell. This will lead to large red cells (increased mean corpuscular volume, MCV). Nevertheless, megaloblastic anemia is neither sensitive, nor it is specific for cobalamin deficiency. Recent studies have shown that many cobalamin deficient subjects may develop neurological disorders without macrocytosis of red blood cells. Furthermore, because cobalamin and iron deficiencies might coincide in many cases, depending on MCV to diagnose cobalamin deficiency is not reliable. Microcytic anemia of iron deficiency can mask macrocytic anemia of cobalamin deficiency. The result is hypochromic red blood cells of normal size. In addition, the hematological manifestations of cobalamin deficiency can be mistaken by that of folate deficiency. Hypersegmented neutrophils are also common in cobalamin deficient subjects. Nevertheless, in general haematological indices are not reliable in diagnosing early stages of cobalamin deficiency.

16.5.2 Neurological Complications

Concentrations of tHcy > 12.0 μmol/L and that of MMA > 271 nmol/L are common in neuro-psychiatric population even in the absence of haematological manifestations (Lindenbaum et al. 1988). Cobalamin deficiency can cause lesions in spinal cord, peripheral nerves, and cerebrum and improvements have been reported after initiation of vitamin treatment (Masalha et al. 2001; Lorenzl et al. 2003). The most common symptoms are sensory disturbances in the extremities, memory loss, dementia, psychosis.

The association between cobalamin deficiency and depression has been documented in elderly women (Penninx et al. 2000). Metabolically significant cobalamin deficiency was present in approximately 15% of non-depressed women, 17% of mildly depressed women and in 27% of the severely depressed women (Penninx et al. 2000). Women with cobalamin deficiency were 2.05 times as likely to be severely depressed as were non-deficient women. Moreover, peripheral neuropathy occurred in 40% of cobalamin-deficient subjects (Shorvon et al. 1980).

Cobalamin deficiency can cause serious neurological symptoms in infants (Graham et al. 1992; Higginbottom et al. 1978; Kuhne et al. 1991). The most common cause of vitamin deficiency at early age is being born or lactated by a vitamin deficient mother. A serious neurological syndrome and developmental disorders have been described in few exclusively breast-fed infants of strict vegetarian mothers who were cobalamin deficient (Graham et al. 1992; Higginbottom et al. 1978; Kuhne et al. 1991). Vitamin B12 deficiency in infants is associated with a marked developmental regression, a poor brain growth or a poor intellectual outcome. Other signs include impaired communicative reactions and fine and gross motor functions. Low vitamin B12 status had also negative influence on school achievement in schoolchildren.

Prolonged insufficient intake of the vitamin is also a common case of cobalamin deficiency in children. Adolescences previously consumed a macrobiotic diet had lower scores in some measures of cognitive performance as compared to adolescences consumed omnivorous diet from birth onward (Schneede et al. 1994; Dagnelie et al. 1989).

Available studies emphasize the need for early recognition and prevention or treatment of B-vitamins deficiency in children. Although, large scale screening programs are currently not recommended (Refsum et al. 2004), cases suspected for cobalamin deficiency should be individually judged. Cobalamin deficiency should be suspected in children with unexplained neurological symptoms, failure to thrive, or poor intellectual performance. A great attention should be paid to familial factors (maternal vitamin status), and predisposing environmental factors (poverty, vegetarian diet).

16.6 Treatment of Cobalamin Deficiency

A typical western diet provides an average intake of 2–6 μg/day of B12. An additional source of the vitamin is highly recommended for subjects with malabsorption or a low intake of the vitamin. Most elderly people who have already developed metabolic signs indicating cobalamin deficiency (i.e., elevated MMA) require more than 100 μg of oral B12 to normalize serum MMA, which is a larger dose than is available in most standard multivitamins and cobalamin supplements.

Folate treatment for cobalamin deficient subject may delay the diagnosis of B12 deficiency and cause irreversible neurological symptoms. Therefore, addition of cobalamin to folate treatment is strongly recommended. Doses and duration of vitamin supplementation should be individually determined, because malabsorption is a major limiting factor for the bioavailability of the vitamins in elderly people. Table 16.5 shows a few examples on cobalamin doses and responses to treatment in different clinical conditions.

Table 16.5 Examples about vitamin intervention trials

Study	Subjects	Dose, duration of B12	Metabolic effects	Other effects
Stabler et al. (2006)	39 elderly subjects mean age 76.3 y with elevated MMA	1000 μg cyanocobalamin/d, oral for 3 months	MMA (nmol/L); 434 → 240 tHcy (μmol/L); 12.7 → 9.6 SAH (nmol/L); 43 → 37	–
Hvas et al. (2001)	140 subjects with MMA (0.40–2.00 μmol/L)	Placebo-controlled study 1 mg im per week for 4 weeks follow up for 3 months	MMA –61% in treatment group (+4% in placebo) tHcy –23% in treatment group (vs. +2% in placebo)	Subjects with MMA > 0.60 μM showed improved neurological symptoms
Lewerin et al. (2003)	209 elderly mean age 76 y	Oral daily B12 0.5 mg (plus folic acid 0.8 mg and B6 3 mg) for 4 months, placebo controlled	Mean MMA; 0.22 → 0.19 in the treatment group 0.22 → 0.22 in the placebo group	–
Bolann et al. (2000)	51 subjects age 17–87 with baseline MMA > 0.26 μmol/L and a reduction of MMA by >=50% after cobalamin treatment	1 mg cyanocobalamin i.m. twice a week for 2.5 weeks	MMA (μmol/L); 0.88 → 0.14 tHcy (μmol/L); 25.8 → 13.4	–
Andres et al. (2003)	30 patients with food cobalamin malabsorption	250 –1000 μg of oral cyanocobalamin daily for at least 1 month	87% normalized their serum cobalamin 54% corrected their anemia	–
Kuzminski et al. (1998)	38 cobalamin deficient subjects	Either 1 mg i.m. on 9 occasions or 2 mg orally for 120 days Patients were tested after 4 months	In the oral group; B12 = 93 →1005 pg/ml MMA 3850 → 169 nmol/L tHcy 37.2 → 10.6 μmol/L In the i.m. group; B12 = 95 →325 pg/ml MMA 3630 → 265 nmol/L tHcy 40 → 12.2 μmol/L	–
Rajan et al. (2002)	23 elderly (age >64 y) with B12 ≤ 221 pmol/L and MMA > 271 nmol/L	Sequential daily treatment with oral cobalamin 25 μg, 100 μg, and 1000 μg for 6 weeks each	25 μg and 100 μg lowered but did not normalize MMA 1 mg was necessary for normalizing MMA in the majority of the subjects	–

References

Andres E, Kaltenbach G, Noel E, Noblet-Dick M, Perrin AE, Vogel T et al. (2003) Efficacy of short-term oral cobalamin therapy for the treatment of cobalamin deficiencies related to food-cobalamin malabsorption: a study of 30 patients. Clin Lab Haematol 25:161–166

Andres E, Kurtz JE, Perrin AE, Maloisel F, Demangeat C, Goichot B, Schlienger JL (2001) Oral cobalamin therapy for the treatment of patients with food-cobalamin malabsorption. Am J Med 111:126–129

Baik HW, Russell RM (1999) Vitamin B12 deficiency in the elderly. Annu Rev Nutr 19:357–377

Ben-Ami M, Katzuni E, Koren A (1990) Imerslund syndrome with dolichocephaly. Pediatr Hematol Oncol 7:177–181

Bibi H, Gelman-Kohan Z, Baumgartner ER, Rosenblatt DS (1999) Transcobalamin II deficiency with methylmalonic aciduria in three sisters. J Inherit Metab Dis 22:765–772

Bjorke Monsen AL, Ueland PM, Vollset SE, Guttormsen AB, Markestad T, Solheim E, Refsum H (2001) Determinants of cobalamin status in newborns. Pediatrics 108:624–630

Bolann BJ, Solli JD, Schneede J, Grottum KA, Loraas A, Stokkeland M et al. (2000) Evaluation of indicators of cobalamin deficiency defined as cobalamin-induced reduction in increased serum methylmalonic acid. Clin Chem 46:1744–1750

Bor MV, Cetin M, Aytac S, Altay C, Nexo E (2005) Nonradioactive vitamin B12 absorption test evaluated in controls and in patients with inherited malabsorption of vitamin B12. Clin Chem 51:2151–2155

Bor MV, Lydeking-Olsen E, Moller J, Nexo E (2006) A daily intake of approximately 6 microg vitamin B-12 appears to saturate all the vitamin B-12-related variables in Danish postmenopausal women. Am J Clin Nutr 83:52–58

Bor MV, Nexo E, Hvas AM (2004) Holo-transcobalamin concentration and transcobalamin saturation reflect recent vitamin B12 absorption better than does serum vitamin B12. Clin Chem 50:1043–1049

Carmel R (1985) The distribution of endogenous cobalamin among cobalamin-binding proteins in the blood in normal and abnormal states. Am J Clin Nutr 41:713–719

Carmel R (1996) Prevalence of undiagnosed pernicious anemia in the elderly. Arch Intern Med 156:1097–1100

Carmel R (1997) Cobalamin, the stomach, and aging. Am J Clin Nutr 66:750–759

Carmel R, Johnson CS (1978) Racial patterns in pernicious anemia. Early age at onset and increased frequency of intrinsic-factor antibody in black women. N Engl J Med 298:647–650

Carmel R, Vasireddy H, Aurangzeb I, George K (2001) High serum cobalamin levels in the clinical setting – clinical associations and holo-transcobalamin changes. Clin Lab Haematol 23:365–371

Chanarin I, Malkowska V, O'Hea AM, Rinsler MG, Price AB (1985) Megaloblastic anaemia in a vegetarian Hindu community. Lancet 2:1168–1172

Cooper BA, Rosenblatt DS (1987) Inherited defects of vitamin B12 metabolism. Annu Rev Nutr 7:291–320

Dagnelie PC, van Staveren WA, Vergote FJ, Dingjan PG, van den BH, Hautvast JG (1989) Increased risk of vitamin B-12 and iron deficiency in infants on macrobiotic diets. Am J Clin Nutr 50:818–824

Dhonukshe-Rutten RA, Van DM, Schneede J, de Groot LC, van Staveren WA (2005) Low bone mineral density and bone mineral content are associated with low cobalamin status in adolescents. Eur J Nutr 44:341–347

Donaldson MS (2000) Metabolic vitamin B12 status on a mostly raw vegan diet with follow-up using tablets, nutritional yeast, or probiotic supplements. Ann Nutr Metab 44:229–234

Doscherholmen A, Ripley D, Chang S, Silvis SE (1977) Influence of age and stomach function on serum vitamin B12 concentration. Scand J Gastroenterol 12:313–319

Doscherholmen A, Silvis S, McMahon J (1983) Dual isotope Schilling test for measuring absorption of food-bound and free vitamin B12 simultaneously. Am J Clin Pathol 80:490–495

England JM, Down MC, Wise IJ, Linnell JC (1976) The transport of endogenous vitamin B12 in normal human serum. Clin Sci Mol Med 51:47–52

Graham SM, Arvela OM, Wise GA (1992) Long-term neurologic consequences of nutritional vitamin B12 deficiency in infants. J Pediatr 121:710–714

Green R (1995) Metabolite assays in cobalamin and folate deficiency. Baillieres Clin Haematol 8:533–566

Hakami N, Neiman PE, Canellos GP, Lazerson J (1971) Neonatal megaloblastic anemia due to inherited transcobalamin II deficiency in two siblings. N Engl J Med 285:1163–1170

Hall CA (1977) The carriers of native vitamin B12 in normal human serum. Clin Sci Mol Med 53:453–457

Herbert V (1994) Staging vitamin B-12 (cobalamin) status in vegetarians. Am J Clin Nutr 59:1213S–1222S

Herbert V, Drivas G (1982) Spirulina and vitamin B 12. JAMA 248:3096–3097

Herbert V, Fong W, Gulle V, Stopler T (1990) Low holotranscobalamin II is the earliest serum marker for subnormal vitamin B12 (cobalamin) absorption in patients with AIDS. Am J Hematol 34:132–139

Herrmann W, Obeid R, Schorr H, Geisel J (2005) The usefulness of holotranscobalamin in predicting vitamin B12 status in different clinical settings. Curr Drug Metab 6:47–53

Herrmann W, Schorr H, Bodis M, Knapp JP, Muller A, Stein G, Geisel J (2000) Role of homocysteine, cystathionine and methylmalonic acid measurement for diagnosis of vitamin deficiency in high-aged subjects. Eur J Clin Invest 30:1083–1089

Herrmann W, Schorr H, Geisel J, Riegel W (2001a) Homocysteine, cystathionine, methylmalonic acid and B-vitamins in patients with renal disease. Clin Chem Lab Med 39:739–746

Herrmann W, Schorr H, Obeid R, Geisel J (2003) Vitamin B-12 status, particularly holotranscobalamin II and methylmalonic acid concentrations, and hyperhomocysteinemia in vegetarians. Am J Clin Nutr 78:131–136

Herrmann W, Schorr H, Purschwitz K, Rassoul F, Richter V (2001b) Total homocysteine, vitamin B-12, and total antioxidant status in vegetarians. Clin Chem 47:1094–1101

Herzlich B, Herbert V (1988) Depletion of serum holotranscobalamin II. An early sign of negative vitamin B12 balance. Lab Invest 58:332–337

Higginbottom MC, Sweetman L, Nyhan WL (1978) A syndrome of methylmalonic aciduria, homocystinuria, megaloblastic anemia and neurologic abnormalities in a vitamin B12-deficient breast-fed infant of a strict vegetarian. N Engl J Med 299:317–323

Hvas AM, Buhl H, Laursen NB, Hesse B, Berglund L, Nexo E (2006) The effect of recombinant human intrinsic factor on the uptake of vitamin B12 in patients with evident vitamin B12 deficiency. Haematologica 91:805–808

Hvas AM, Ellegaard J, Nexo E (2001) Vitamin B12 treatment normalizes metabolic markers but has limited clinical effect: a randomized placebo-controlled study. Clin Chem 47:1396–1404

Institute of Medicine (2000) Dietary reference intakes for thiamin, riboflavin, niacin, vitamin B6, folate, vitamin B12, pantothenic acid, biotin, and choline. National Academy Press, Washington, DC, pp. 150–195

Krasinski SD, Russell RM, Samloff IM, Jacob RA, Dallal GE, McGandy RB, Hartz SC (1986) Fundic atrophic gastritis in an elderly population. Effect on hemoglobin and several serum nutritional indicators. J Am Geriatr Soc 34:800–806

Kuhne T, Bubl R, Baumgartner R (1991) Maternal vegan diet causing a serious infantile neurological disorder due to vitamin B12 deficiency. Eur J Pediatr 150:205–208

Kuzminski AM, Del Giacco EJ, Allen RH, Stabler SP, Lindenbaum J (1998) Effective treatment of cobalamin deficiency with oral cobalamin. Blood 92:1191–1198

Lewerin C, Nilsson-Ehle H, Matousek M, Lindstedt G, Steen B (2003) Reduction of plasma homocysteine and serum methylmalonate concentrations in apparently healthy elderly subjects after treatment with folic acid, vitamin B12 and vitamin B6: a randomised trial. Eur J Clin Nutr 57:1426–1436

Lin JC, Borregaard N, Liebman HA, Carmel R (2001) Deficiency of the specific granule proteins, R-binder/transcobalamin I and lactoferrin, in plasma and saliva: a new disorder. Am J Med Genet 100:145–151

Lindenbaum J, Healton EB, Savage DG, Brust JC, Garrett TJ, Podell ER et al. (1988) Neuropsychiatric disorders caused by cobalamin deficiency in the absence of anemia or macrocytosis. N Engl J Med 318:1720–1728

Lindgren A, Kilander A, Bagge E, Nexo E (1999) Holotranscobalamin – a sensitive marker of cobalamin malabsorption. Eur J Clin Invest 29:321–329

Lorenzl S, Vogeser M, Muller-Schunk S, Pfister HW (2003) Clinically and MRI documented funicular myelosis in a patient with metabolical vitamin B12 deficiency but normal vitamin B12 serum level. J Neurol 250:1010–1011

Markle HV (1996) Cobalamin. Crit Rev Clin Lab Sci 33:247–356

Masalha R, Chudakov B, Muhamad M, Rudoy I, Volkov I, Wirguin I (2001) Cobalamin-responsive psychosis as the sole manifestation of vitamin B12 deficiency. Isr Med Assoc J 3:701–703

Moestrup SK, Birn H, Fischer PB, Petersen CM, Verroust PJ, Sim RB et al. (1996) Megalin-mediated endocytosis of transcobalamin-vitamin-B12 complexes suggests a role of the receptor in vitamin-B12 homeostasis. Proc Natl Acad Sci USA 93:8612–8617

Miller DR, Specker BL, Ho ML, Norman EJ (1991) Vitamin B-12 status in a macrobiotic community. Am J Clin Nutr 53:524–529

Morris MS, Jacques PF, Rosenberg IH, Selhub J (2002) Elevated serum methylmalonic acid concentrations are common among elderly Americans. J Nutr 132:2799–2803

Nickoloff E (1988) Schilling test: physiologic basis for and use as a diagnostic test. Crit Rev Clin Lab Sci 26:263–276

Obeid R, Jouma M, Herrmann W (2002) Cobalamin status (holo-transcobalamin, methylmalonic acid) and folate as determinants of homocysteine concentration. Clin Chem 48:2064–2065

Obeid R, Kuhlmann MK, Kohler H, Herrmann W (2005) Response of homocysteine, cystathionine, and methylmalonic acid to vitamin treatment in dialysis patients. Clin Chem 51:196–201

Obeid R, Schorr H, Eckert R, Herrmann W (2004) Vitamin B12 status in the elderly as judged by available biochemical markers. Clin Chem 50:238–241

Penninx BW, Guralnik JM, Ferrucci L, Fried LP, Allen RH, Stabler SP (2000) Vitamin B(12) deficiency and depression in physically disabled older women: epidemiologic evidence from the Women's Health and Aging Study. Am J Psychiatry 157:715–721

Rajan S, Wallace JI, Brodkin KI, Beresford SA, Allen RH, Stabler SP (2002) Response of elevated methylmalonic acid to three dose levels of oral cobalamin in older adults. J Am Geriatr Soc 50:1789–1795

Rauma AL, Torronen R, Hanninen O, Mykkanen H (1995) Vitamin B-12 status of long-term adherents of a strict uncooked vegan diet ("living food diet") is compromised. J Nutr 125:2511–2515

Refsum H, Grindflek AW, Ueland PM, Fredriksen A, Meyer K, Ulvik A et al. (2004) Screening for serum total homocysteine in newborn children. Clin Chem 50:1769–1784

Refsum H, Johnston C, Guttormsen AB, Nexo E (2006) Holotranscobalamin and total transcobalamin in human plasma: determination, determinants, and reference values in healthy adults. Clin Chem 52:129–137

Refsum H, Yajnik CS, Gadkari M, Schneede J, Vollset SE, Orning L et al. (2001) Hyperhomocysteinemia and elevated methylmalonic acid indicate a high prevalence of cobalamin deficiency in Asian Indians. Am J Clin Nutr 74:233–241

Riedel B, Bjorke Monsen AL, Ueland PM, Schneede J (2005) Effects of oral contraceptives and hormone replacement therapy on markers of cobalamin status. Clin Chem 51:778–781

Savage DG, Lindenbaum J, Stabler SP, Allen RH (1994) Sensitivity of serum methylmalonic acid and total homocysteine determinations for diagnosing cobalamin and folate deficiencies [see comments]. Am J Med 96:239–246

Schneede J, Dagnelie PC, van Staveren WA, Vollset SE, Refsum H, Ueland PM (1994) Methylmalonic acid and homocysteine in plasma as indicators of functional cobalamin deficiency in infants on macrobiotic diets. Pediatr Res 36:194–201

Scott JM (1997) Bioavailability of vitamin B12. Eur J Clin Nutr 51(Suppl 1):S49–S53

Shorvon SD, Carney MW, Chanarin I, Reynolds EH (1980) The neuropsychiatry of megaloblastic anaemia. Br Med J 281:1036–1038

Stabler SP, Allen RH, Dolce ET, Johnson MA (2006) Elevated serum S-adenosylhomocysteine in cobalamin-deficient elderly and response to treatment. Am J Clin Nutr 84:1422–1429

Stabler SP, Lindenbaum J, Allen RH (1996) The use of homocysteine and other metabolites in the specific diagnosis of vitamin B-12 deficiency. J Nutr 126:1266S–1272S

Stroinsky A, Schneider Z (1987) Cobamide dependant enzymes. In: Schneider Z, Stroinsky A (eds) Comprehensive B-12. Printing house de Gruyter, Berlin, pp. 225–266

Teplitsky V, Huminer D, Zoldan J, Pitlik S, Shohat M, Mittelman M (2003) Hereditary partial transcobalamin II deficiency with neurologic, mental and hematologic abnormalities in children and adults. Isr Med Assoc J 5:868–872

Ulleland M, Eilertsen I, Quadros EV, Rothenberg SP, Fedosov SN, Sundrehagen E, Orning L (2002) Direct assay for cobalamin bound to transcobalamin (holo- transcobalamin) in serum. Clin Chem 48:526–532

Chapter 17
Biochemistry of B12-Cofactors in Human Metabolism

Bernhard Kräutler

Abstract Vitamin B12, the "antipernicious anaemia factor", is a crystallisable cobalt-complex, which belongs to a group of unique "complete" corrinoids, named cobalamins (Cbl). In humans, instead of the "vitamin", two organometallic B12-forms are coenzymes in two metabolically important enzymes: Methyl-cobalamin, the cofactor of methionine synthase, and coenzyme B12 (adenosyl-cobalamin), the cofactor of methylmalonyl-CoA mutase. The cytoplasmatic methionine synthase catalyzes the transfer of a methyl group from N-methyl-tetrahydrofolate to homocysteine to yield methionine and to liberate tetrahydrofolate. In the mitochondrial methylmalonyl-CoA mutase a radical process transforms methylmalonyl-CoA (a remains e.g. from uneven numbered fatty acids) into succinyl-CoA, for further metabolic use. In addition, in the human mitochondria an adenosyl-transferase incorporates the organometallic group of coenzyme B12. In all these enzymes, the bound B12-derivatives engage (or are formed) in exceptional organometallic enzymatic reactions. This chapter recapitulates the physiological chemistry of vitamin B12, relevant in the context of the metabolic transformation of B12-derivatives into the relevant coenzyme forms and their use in B12-dependent enzymes.

Keywords 5′-adenosyl radical · Coenzyme B12 · Methionine synthase · Methylcobalamin · Methylgroup transfer · Methylmalonyl-CoA mutase · Radical reaction · Vitamin B12

Abbreviations and Symbols

Ado	5′-(deoxy)-adenosyl
AdoCbl	5′-deoxy-5′-adenosyl-cobalamin (coenzyme B12, also named adenosyl-cobalamin)
Ado-ImCba	5′-deoxy-5′-adenosyl-imidazolyl-cobamide
B12r	cob(II)alamin
B12s	cob(I)alamin

B. Kräutler (✉)
Institute of Organic Chemistry and Centre of Molecular Biosciences (CMBI), University of Innsbruck, Innsbruck, Austria
e-mail: Bernhard.kraeutler@uibk.ac.at

O. Stanger (ed.), *Water Soluble Vitamins*, Subcellular Biochemistry 56,
DOI 10.1007/978-94-007-2199-9_17,

BDE	(homolytic) bond dissociation energy
Cba	cobamide
Cbl	cobalamin (a DMB-cobamide)
CNCbl	cyano-cobalamin (vitamin B12)
CN-ImCba	cyano-imidazolyl-cobamide
DMB	5,6-dimethylbenzimidazole
H_2OCbl^+	aquo-cobalamin (B12a)
HOCbl	hydroxo-cobalamin
ImCba	imidazolyl-cobamide
MeCbl	methyl-cobalamin
Me-ImCba	methyl-imidazolyl-cobamide
MetH	methionine synthase (from *E. coli*)
MMCM	methylmalonyl-CoA mutase (from *Propionibacterium shermanii*)
NHE	normal hydrogen electrode
NMR	nuclear magnetic resonance
SAM	S-adenosyl-methionine (AdoMet)
UV/Vis	ultraviolet/visible absorbance spectrum

17.1 Introduction

B12-coenzymes are conceivably Nature's most broadly relevant and most complex organometallic cofactors (Eschenmoser 1988; Kräutler et al. 1998; Banerjee 1999). They are required (and are vitamins) in human metabolism, which depends on the binding and uptake of B12, on the controlled transport and metabolic transformation into the relevant B12-cofactors, and the catalysis by B12-dependent enzymes (Yamanishi et al. 2005). B12-derivatives (co)catalyze unique enzymatic reactions that directly depend upon the reactivity of the cobalt coordinated organic ligands (Kräutler and Ostermann 2003; Brown 2005).

About 60 years ago, the red cyanide-containing cobalt-complex vitamin B12 (cyano-cobalamin, CNCbl) was isolated as the (extrinsic) "anti-pernicious anaemia factor" (Fig. 17.1) (Rickes et al. 1948; Smith and Parker 1948). CNCbl is a relatively inert Co(III)-corrin, which crystallizes readily. It is the most important commercially available form of the naturally occurring B12-derivatives, although a direct physiological function of CNCbl appears not to exist (Ellenbogen and Cooper 1984). The physiologically relevant B12-derivatives are the light-sensitive and chemically more labile organometallic cofactors, coenzyme B12 (5′-deoxy-5′-adenosyl-cobalamin, AdoCbl) and methyl-cobalamin (MeCbl), as well as (in a formal sense) the "inorganic" and easily reducible B12-derivative aquo-cobalamin (H_2OCbl^+).

Important historic contributions to the biologically relevant chemistry of B12 have been reviewed earlier (Dolphin 1982; Friedrich 1988). Remarkable scientific advances contributing to the solution of the major "B12-mysteries" were achieved during the last decades (Kräutler et al. 1998; Banerjee 1999; Brown 2005), including the elucidation of the microbial B12-biosynthesis (Battersby 1998; Scott

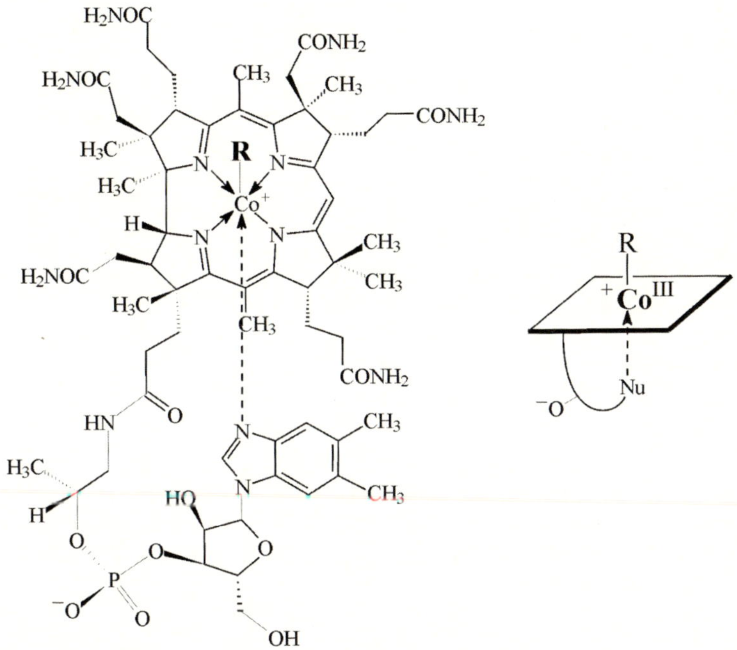

Fig. 17.1 *Left*: structural formulae of selected cobalamins (Cbl); *Right*: symbol used. Vitamin B12 (CNCbl), R = CN; coenzyme B12 (AdoCbl), R = 5′-Ado; methyl-cobalamin (MeCbl), R = CH3; aquo-cobalamin (H_2OCbl^+), R = H_2O^+; cob(II)alamin (B12r), R = e–

et al. 2003), important structural and mechanistic studies of B12-binding proteins (Drennan et al. 1994; Ludwig and Evans 1999; Gruber and Kratky 2002), as well as genetic studies of inborn errors of Cbl metabolism (Rosenblatt and Fenton 1999).

17.2 B12: Structure and Reactivity

17.2.1 B12-Structural Studies

Vitamin B12 (CNCbl), and other B12-derivatives (where the cyanide ligand of CNCbl is replaced by a different "upper" or β-ligand) are all cobalamins (Cbl, 5′,6′-dimethylbenzimidazolyl-cobamides), the B12-derivatives relevant in human metabolism. Pioneering X-ray crystallographic studies of Hodgkin et al. (Hodgkin et al. 1955; Lenhert and Hodgkin 1961) established the structure of CNCbl (and of its unique corrin core), as well as the amazing nature of the organometallic cofactor coenzyme B12 (AdoCbl). Since the time of these landmark analyses, accurate structural studies of a variety of crystalline B12-derivatives have become available, as presented in recent reviews (Kratky and Kräutler 1999; Randaccio et al. 2006).

The replacement of the cobalt coordinated 5,6-dimethylimidazole (DMB) base by a protein-derived imidazole in B12-dependent enzymes, such as methionine synthase (Drennan et al. 1994) and methylmalonyl-CoA mutase (Mancia et al. 1996) (see below), has been a puzzling discovery. To study the effect of an imidazole in B12-derivatives, unnatural imidazolyl-cobamides (ImCba) were made available via "guided" biosynthesis, such as 5′-deoxy-5′-adenosyl-imidazolyl-cobamide (Ado-ImCba) (Brown et al. 2004), cyano-imidazolyl-cobamide (CN-ImCba) and methyl-imidazolyl-cobamide (Me-ImCba), both analyzed by x-ray crystallography (Kräutler et al. 1994; Fasching et al. 2000).

Two structural characteristics of the corrin ligand are the saturated and direct trans-junction between two of its four five-membered rings and the non-planar nature of the corrin core in B12-derivatives. The bulky DMB-base was suggested to be a relevant enhancer to this latter "ligand-folding", which is a main factor also to the variability in the conformation of the corrin ligand (Kratky and Kräutler 1999). Indeed, the folding of the corrin ligand in CN-ImCba (11.3°) is less than half of that of vitamin B12 (CNCbl). This finding could be rationalized, as the less bulky and more nucleophilic imidazole showed less steric interaction with corrin ligand in CN-ImCba, than the DMB base in CNCbl. However, the crystallographic studies could not provide a pertinent structure-based rationalization of the observed displacement of the DMB-base by histidine in the enzymes. Crystallographic information on the

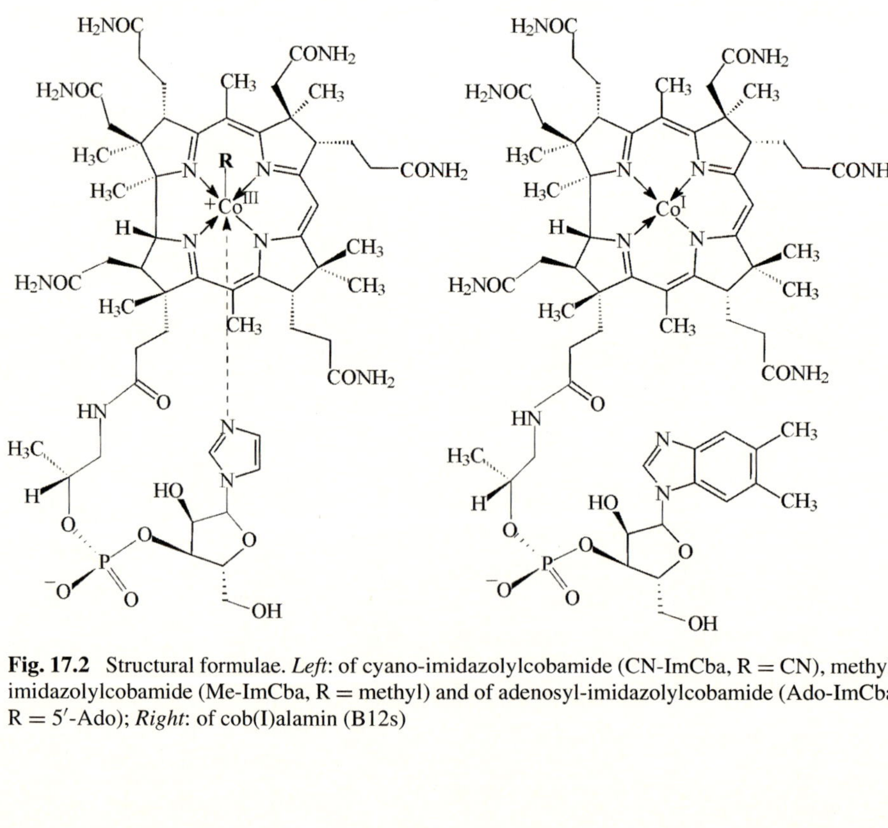

Fig. 17.2 Structural formulae. *Left*: of cyano-imidazolylcobamide (CN-ImCba, R = CN), methyl-imidazolylcobamide (Me-ImCba, R = methyl) and of adenosyl-imidazolylcobamide (Ado-ImCba, R = 5′-Ado); *Right*: of cob(I)alamin (B12s)

structure of cob(II)alamin (B12r) has been of particular interest (Kräutler et al. 1989), as it is the product of Co–C bond homolysis of coenzyme B12 (AdoCbl), relevant for the catalytic cycle of coenzyme B12-dependent enzymes (see below).

Nuclear Magnetic Resonance (NMR) spectroscopy has become an important methodology for the study of (diamagnetic) B12 derivatives, in (aqueous) solution. Frequently, it is complementary to crystal structure analysis and is particularly valuable in such cases, where crystals of the B12-derivative are not obtained (Konrat et al. 1998; Brown 1999). Natural corrinoids from bacteria were often characterized by NMR-spectroscopy (Hoffmann et al. 2000). By applying so called homo- and hetero-nuclear NMR-experiments and high field NMR-spectrometers, the structure of B12-derivatives in aqueous solution has been studied, and their behaviour as dynamic molecules could be analyzed (Summers et al. 1986; Konrat et al. 1998).

17.2.2 B12-Redoxchemistry

Under physiological conditions vitamin B12-derivatives have been observed in the oxidation states Co(III), Co(II), and Co(I), each possessing differing coordination properties and reactivities (Pratt 1972; Kräutler 1998).

Oxidation-reduction processes are of key importance in the metabolism of B12. Electrochemical methods have been crucial in analytical studies of B12-derivatives (Lexa and Savéant 1983) (Fig. 17.3). Axial coordination to the corrin-bound cobalt centre depends on the formal oxidation state of the cobalt ion (Kratky and Kräutler 1999): As a rule, the number of axial ligands decreases in parallel with the cobalt oxidation state. The diamagnetic Co(III) (coordination number 6) typically has two axial ligands bound, the paramagnetic (low spin) Co(II) (coordination number 5) has one axial ligand and the diamagnetic Co(I) (coordination number 4) has no axial ligands bound (Pratt 1972; Lexa and Savéant 1983). Redox reactions of B12-derivatives are, therefore, accompanied by a change in the number of axial ligands. In turn, the presence and the nature of axial ligands strongly influences the (thermodynamic and kinetic features of the) electron transfer of cobalt corrins (Lexa and Savéant 1983; Kräutler 1999).

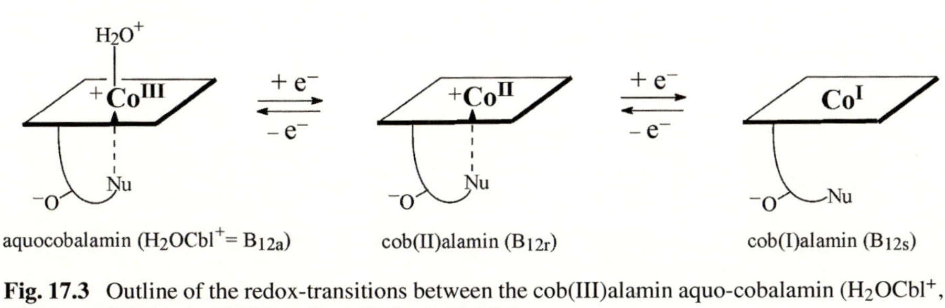

Fig. 17.3 Outline of the redox-transitions between the cob(III)alamin aquo-cobalamin (H_2OCbl^+, B12a), cob(II)alamin (B12r) and cob(I)alamin (B12s)

In Co(III)-corrins, such as vitamin B12 (CNCbl), coenzyme B12 (AdoCbl) and methyl-cobalamin (MeCbl), the corrin-bound cobalt centre is coordinatively saturated, when binding two axial ligands. In contrast, the metal centre in Co(I)-corrins, such as cob(I)alamin (B12s) is highly nucleophilic (with very low basicity) (Lexa and Savéant 1983). The intermediate oxidation level of the Co(II)-corrins, such as cob(II)alamin (B12r), provides a reactive metal-centred radicaloid species (Endicott and Netzel 1979; Kräutler et al. 1989). In neutral aqueous solution, cob(II)alamin (B12r) is present in its "base-on" form (i.e. the DMB-base is coordinating the Co(II)centre). B12r is converted into the "base-off" form, B12r-H^+, by protonation of the DMB-base, with pKa (B12r-H^+) = 2.9, and de-coordination from the Co(II)centre.

A standard potential vs. pH diagram has been established, which correlates the thermodynamics of the H_2OCbl^+-system (B12a /B12r /B12s) (Fig. 17.4) (Lexa and Savéant 1983). In the pH range 2.9–7.8, H2OCbl and (base-on) B12r represent the predominant Co(III)-/Co(II)-redox couple, with a standard potential of +0.20 V. For the Co(II)-/Co(I)-redox system there are two pH-independent standard potentials: at a pH less than 5.6 the Co(II)-/Co(I)-couple (base-off) B12r-H+/B12s-H predominates (standard potential of –0.50 V, at higher pH the redox couple (base-on) B12r / B12s exists with a more negative standard potential of –0.61 V.

The standard potential of the Co(III)-/Co(II)-redox pair of organometallic B12-derivatives, such as coenzyme B12 (AdoCbl) and methyl-cobalamin (MeCbl), is significantly more negative than that of B12r/B12s and out of the reach of biological reductants (Kräutler 1999). One-electron reduction of methyl-cobalamin to methyl-cob(II)alamin occurs at –1.36 V vs. NHE and demethylates MeCbl (at –30°C).

The thermodynamic trends of B12-redox systems can be summarized as:

(i) Intra-molecular coordination of the nucleotide base and/or strongly coordinating or nucleophilic ligands (such as cyanide ions) stabilize the cobalt centre against reduction and shift the Co(III)-/Co(II)-redox couples to more negative potentials.

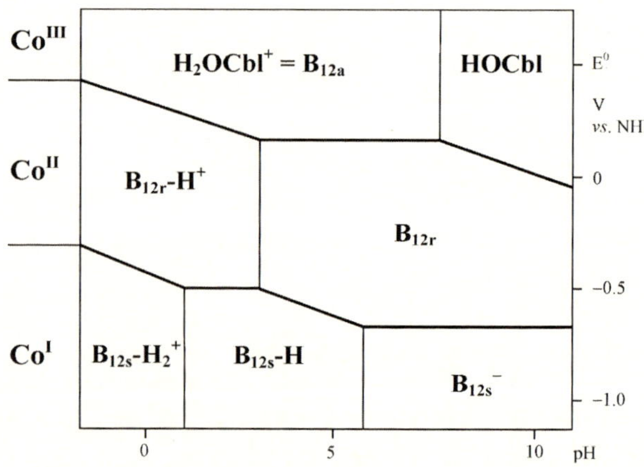

Fig. 17.4 Dependence of standard potentials of the redox system Co(III)-/Co(II)-/Co(I)-corrin (B12a/B12r/B12s) upon pH in aqueous solution (at 22°C), potentials are referenced to the normal hydrogen electrode (NHE) adapted from (Lexa and Savéant 1983)

(ii) The one-electron reduction of organometallic Co(III)-corrins typically occurs at more negative potentials than the Co(II)-/Co(I)-redox couple B12r/B12s and leads to loss of the organometallic ligand.

Electrochemistry is an excellent method for the selective production of reduced B12 forms. Since alkyl halides or alkyl tosylates react quickly and efficiently with Co(I)-corrins (Schrauzer and Deutsch 1969; Pratt 1972), which are cleanly generated at the electrodes, electrochemistry also is a method for the clean synthesis of most organometallic B12 derivatives (Kräutler 1999).

17.2.3 B12-Organometallic Reactivity

The reactivity of B12-derivatives in organometallic (redox-) reactions is the basis for their biological role as cofactors. Formation and cleavage of the Co–C bond in organometallic B12-cofactors are crucial steps in the reactions catalyzed by the B12-dependent enzymes (Kräutler et al. 1998; Banerjee 1999). In solution, cleavage and formation of the Co–C bond have been observed to occur in all of the basic oxidation levels for the cobalt centre of the corrin core (Kräutler and Ostermann 2003; Brown 2005).

Two mechanisms for these organometallic reactions have been found to be the basis for catalysis by the human cobalamin-dependent enzymes (Fig. 17.5):

(i) the homolytic mode, which involves the cleavage or formation of a single axial bond, as is typical of the reactivity of coenzyme B12 (AdoCbl): 5′-Ado-Co(III)-corrin $\rightleftarrows$ Co(II)-corrin + 5′-deoxy-5′-adenosyl radical
(ii) the nucleophile induced, heterolytic mode, which involves the cleavage or formation of two (trans-) axial bonds, as is typical of the reactivity of methylcobalamin (MeCbl): methyl-Co(III)-corrin + nucleophile $\rightleftarrows$ Co(I)-corrin + methylating agent

The homolytic mode of the cleavage of the Co–C bond of coenzyme B12 (AdoCbl) is of particular importance in its role as a cofactor. Indeed, thermal decomposition of AdoCbl in aqueous solution occurs readily at higher temperatures and leads predominantly to an adenosyl radical and cob(II)alamin (B12r) (Fig. 17.6). AdoCbl is thus considered to be a "reversible carrier of an alkyl radical" (or a reversibly functioning "radical source") (Halpern 1985). At room temperature (and in the dark), a neutral aqueous solution of AdoCbl is rather stable, with a calculated half life of 10^{10} s. The strength of the Co–C bond of AdoCbl (the homolytic Co–C bond dissociation energy = BDE) is about 30 kcal/mol (Halpern 1985; Hay and Finke 1986). The strength of the Co–C bond of AdoCbl is hardly influenced by the coordination of the nucleotide (Kräutler 1987, 1998).

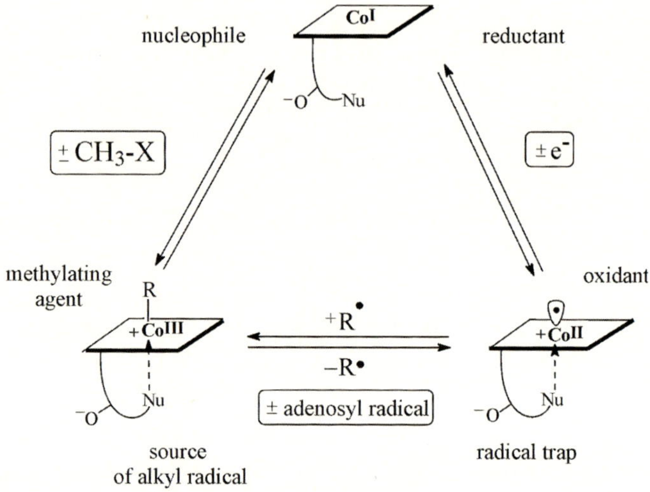

Fig. 17.5 Patterns of reactivity and formal analysis of elementary reaction steps of "complete" corrinoids in organometallic (redox-)transformations, relevant for their cofactor functions in B12-dependent enzymes

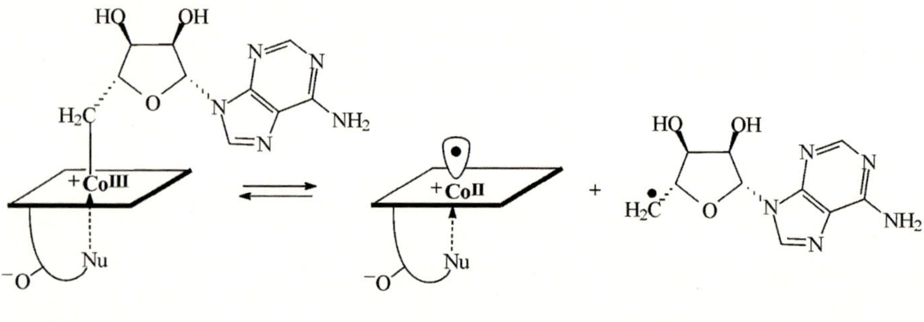

coenzyme B_{12} (AdoCbl) cob(II)alamin (B_{12r}) 5'-deoxy-5'-adenosyl radical

Fig. 17.6 Coenzyme B12 (AdoCbl), a reversibly functioning "radical source" (it provides a 5′-deoxy-5′-adenosly radical); in reverse, the "radical trap" B12r rapidly reacts with (5′-deoxy-5′-adenosyl) radicals to give organo-cob(III)alamins directly (such as AdoCbl)

The (crystal) structure of B12r is surprisingly similar to that of the cobalamin moiety of AdoCbl (Kräutler et al. 1989). The radicaloid B12r has a penta-coordinated Co(II)-centre, so that reactions with alkyl radicals (at its "upper" β-face) give hexa-coordinate organo-cob(III)alamins (such as AdoCbl) and occur with little restructuring of the corrin moiety: B12r thus fulfils the structural criteria of a very efficient "radical trap" and it reacts very fast (near diffusion control) with alkyl radicals (such as the 5′-deoxy-5′-adenosyl radical) (Endicott and Netzel 1979).

The second important type of organometallic reactivity of B12-derivatives represents a heterolytic mode of formation/cleavage of the Co–C bond and concerns the inter-conversion between Co(I)-corrins and methyl-Co(III)-corrins, important in enzyme-catalysed methyl-transfer reactions (Kräutler et al. 1998; Banerjee 1999; Matthews 2001): Co(I)-corrins react as strong nucleophiles with methylating agents (the Co–CH3 bond is formed) and methyl Co(III)-corrins may be demethylated by nucleophiles (the Co–CH3 bond is cleaved).

Alkylation at the corrin-bound Co(I) centre of B12s normally proceeds via the "classical" bimolecular nucleophilic substitution (SN2) mechanism (Fig. 17.7), where the Co(I)-corrin acts as a "supernucleophile" (Schrauzer and Deutsch 1969). However, in certain cases alkylation of Co(I)-corrins occurs via an alternative two-step one-electron transfer path, where Co(I)-corrins act as strong one-electron reducing agents and the process proceeds via Co(II)-corrin intermediates (Kräutler and Caderas 1984; Kräutler 1998).

In Co(I)-corrins, like B12s, the nucleophilicity in alkylation reactions is virtually independent of the presence of the DMB-nucleotide. Co(I)-corrins, both "complete" and "incomplete", preferentially react at their β-face (Kräutler 1998). The immediate product of the β-alkylation is likely to be a penta-coordinate Coβ-alkyl-Co(III)-corrin (Fig. 17.7). Since most alkyl-cobalamins are hexa-coordinate in their more stable "base-on"-constitution, the methylation by the SN2-mode takes place in a two step mechanism. In aqueous solution and at room temperature methylcobalamin (MeCbl) is more stable in the "base-on"-form by about four kcal/mol than "base-off" Coα-aquo-Coβ-methyl-cobalamin (Kräutler 1987, 1998).

The reverse process, the nucleophile-induced dealkylations of methyl-Co(III)-corrins, is less studied. Intramolecular coordination of the nucleotide base stabilises MeCbl and slows the demethylation down: thiolates demethylate MeCbl to Co(I)corrinoids at approximately 1,000 times lower rate than methyl-corrinoids lacking a nucleotide base (Hogenkamp et al. 1985), reflecting the effect of the coordinated nucleotide (Kräutler 1987). Considerable axial base effects are thus expected for methyl-group transfer reactions involving enzyme-bound Co(I)- and methyl-Co(III)-corrins (Drennan et al. 1994; Kräutler 1998).

The two most relevant modes of formation and cleavage of the Co–C bond of the cobalt centre differ significantly in their structural requirements:

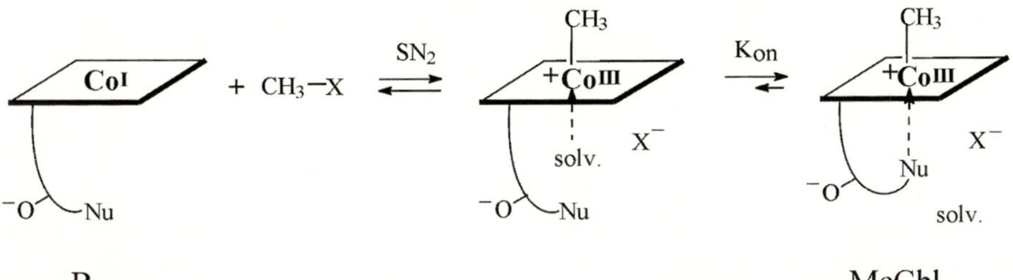

Fig. 17.7 The highly nucleophilic cob(I)alamin (B12s) reacts with methylating agents, to give methylcobalamin (MeCbl). For kinetic and thermodynamic reasons methylation of cob(I)alamin (B12s) occurs at the "upper" or β-face

- in the homolytic mode of cleavage, the complete cobalt-corrin part of organocobalamins (such as AdoCbl) hardly changes its structure (Fig. 17.6);
- in the heterolytic mode of cleavage and formation of the Co–C bond significant reorganization occurs at both faces of the corrin-bound cobalt centre. Cleavage of the Co–CH_3 bond is brought about by attack of a nucleophile at the readily accessible carbon of the cobalt-bound methyl group (Fig. 17.7).

Organocobalamins have long been know to be sensitive to visible light (Pailes and Hogenkamp 1968), which induces homolytic cleavage of the Co–C bond with a quantum yield of about 0.3 (Cole et al. 2002). Another mode of cleavage of the Co–C bond of organometallic B12-derivatives is the thermodynamically favourable radical-substitution at the cobalt-bound carbon centre, as observed in the demethylation of MeCbl by an organic radical (Kräutler et al. 1995; Mosimann and Kräutler 2000).

17.3 Occurrence of B12-Dependent Enzymes

B12-coenzymes are nowadays known to act as cofactors in four classes of distinct enzymes: B12-dependent methyl-transferases (Matthews 2001; Banerjee and Ragsdale 2003), B12-dependent isomerases and ribonucleotide reductases (both use AdoCbl) (Buckel and Golding 2006) and corrinoid dehalogenases (Siebert et al. 2002), some of which use a nor-vitamin B12-derivative (Kräutler et al. 2003). In addition, evidence for enzymatic catalysis of the biosynthetic incorporation of the organometallic groups of AdoCbl and of MeCbl (aside from such reactions in the methyl transferases and isomerases) is available (see below).

B12-dependent methyl transferases play an important role in amino acid metabolism in many organisms (including humans) as well as in one-carbon metabolism and CO_2 fixation in anaerobic microbes (Banerjee and Ragsdale 2003). The reactivity of the "supernucleophilic" Co(I)-corrins and of methyl-Co(III)-corrins make B12-derivatives ideal as cofactors in enzymatic methyl-group transfer reactions (Kräutler et al. 1998; Banerjee 1999; Kräutler and Ostermann 2003; Brown 2005). B12-dependent methionine synthase has been particularly well studied (see below and refs (Matthews 1999, 2001; Banerjee and Ragsdale 2003)) as have methyl transferases in methanogenesis (Sauer and Thauer 1999) and in aerobic acetogenesis (Ragsdale et al. 1998). In human metabolism, methionine synthase is crucial for the endogenous formation of methionine (Banerjee 1997).

About ten coenzyme B12-dependent enzymes are now known, which are distributed rather disproportionately in the living world (see refs (Kräutler et al. 1998; Banerjee 1999; Brown 2005)). Only one of these, the carbon skeleton mutase methylmalonyl-CoA mutase (hMMCM), is indispensable in human metabolism (Banerjee 1997). Three more microbial carbon skeleton mutases are known (besides MMCM (Banerjee and Chowdhury 1999)): glutamate mutase (Buckel and Golding 1996; Gruber and Kratky 2002), methylene glutarate mutase (Buckel and Golding

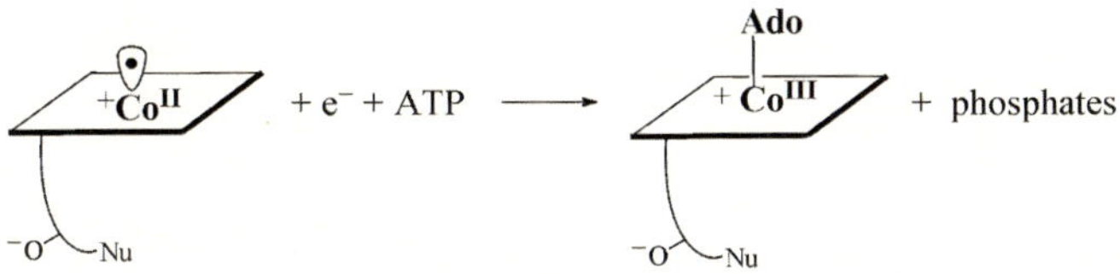

Fig. 17.8 Coenzyme B12 (AdoCbl, in its "base-off"-form) is biosynthesized (in humans) from "base-off" cob(II)alamin (B12r) by the mitochondrial enzyme adenosyltransferase

1996) and isobutyryl-CoA mutase (Zerbe-Burkhardt et al. 1999). In addition, five bacterial isomerases, diol and glycerol dehydratase, ethanolamine ammonia lyase and two amino mutases are B12-dependent, as is an anaerobic form of ribonucleotide reductase (Frey and Chang 1999; Stubbe 2000; Toraya 2003).

The labile and light sensitive coenzyme B12 (AdoCbl) is not typically taken up with the nutrition, but is enzymatically produced in the human mitochondrion by adenosyltransferase. A feature of this enzyme is its ability to bind cob(II)alamin (B12r) in a remarkable four coordinate "base-off"-constitution and, thus, to render the bound B12r more accessible to reduction by biological reducing agents (Stich et al. 2005). This biosynthetic enzyme catalyzes the reduction of B12r to cob(I)alamin (B12s) and adenosylation by ATP to give AdoCbl, which appears to be bound in a "base-off"-form also (Fig. 17.8) (Yamanishi et al. 2005). Direct delivery of AdoCbl in the "base-off"-constitution has been suggested from adenosyl-transferase to its target (apo-)enzyme, methylmalonyl-CoA mutase.

17.4 Methionine Synthase

Methionine synthase catalyzes the formation of methionine by methylation of homocysteine and demethylation of N5-methyl-tetrahydrofolate to tetrahydrofolate (Fig. 17.9). In B12-dependent methionine synthase, a methylcorrinoid, such as methyl-cobalamin (MeCbl) is the cofactor, whereas in cobalamin-independent methionine synthases, which have much slower turnover, homocysteine is merely activated by coordination to a zinc-centre (Matthews 2001).

17.4.1 B12-Dependent Methyl Group Transfer

In a catalytic cycle of B12-dependent methyl transferases (such as methionine synthase), the corrinoid cycles between enzyme-bound methyl-Co(III)- and Co(I)-forms (Matthews 2001). The transition between the hexa-coordinate methyl-Co(III)-form and the tetra-coordinate Co(I)-form must be accompanied by large structural changes (Figs. 17.5 and 17.7) and complementary changes in a protein environment.

This provides a means for controlling the organometallic reactivity of the bound cofactor at the same time (Kräutler 1998). The enzymatic methyl-group transfers

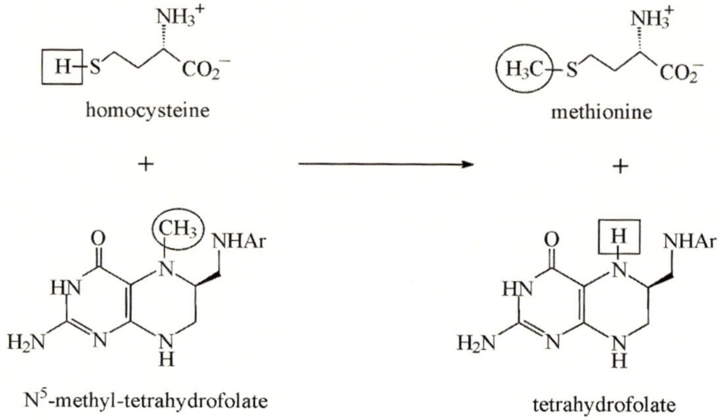

Fig. 17.9 Methionine synthase produces methionine and tetrahydrofolate from homocysteine and N5-methyl-tetrahydrofolate

(which formally involve nucleophile-induced heterolytic cleavage and formation of the organometallic Co–CH3 bond at the corrin bound cobalt centre) are thus expected to be subject to strict geometric control (mediated via the "regulatory" His-Asp-Ser-triad in methionine synthase (MetH) (Drennan et al. 1994; Matthews 2001), (Fig. 17.10 and below): The His-Asp-Ser-triad appears to have a dual role in MetH;

(i) it may participate in maintaining conformational control of the mutual placement of the corrinoid cofactors and the bound substrates (in their respective enzyme-modules) by a H^+-mediated conformational switch mechanism (Matthews 2001; Ludwig and Matthews 2002; Bandarian et al. 2003) and
(ii) it may exert a relevant thermodynamic effect on the strength of the Coβ–CH_3 bond by α-axial base-coordination (Drennan et al. 1994).

Solution studies had shown a significant thermodynamic trans-effect of the DMB-coordination in methyl-cobalamin (MeCbl) and stabilization of the methyl-Co(III)-corrin MeCbl by about 4 kcal/mol against nucleophilic abstraction of the methyl group (Kräutler 1987, 1998) (a similar effect on heterolytic methyl group transfer reactions was deduced for imidazole-coordination in methyl-imidazoyl-cobamide (Me-ImCba) (Fasching et al. 2000).

The methyl group may originate from N5-methyl-tetrahydropterins (such as N5-methyl-tetrahydromethanopterin or N5-methyl-tetrahydrofolate) and a variety of other methylated substrates (Stupperich et al. 1998; Matthews 2001). Typically, thiols are the methyl group acceptors in methionine synthesis (homocysteine) (Matthews 2001; Bandarian et al. 2003) and methanogenesis (coenzyme M) (Sauer and Thauer 1999). In the anaerobic biosynthesis of acetyl-coenzyme-A from one-carbon precursors the methyl group acceptor is suggested to be a nickel centre (Drennan et al. 2004).

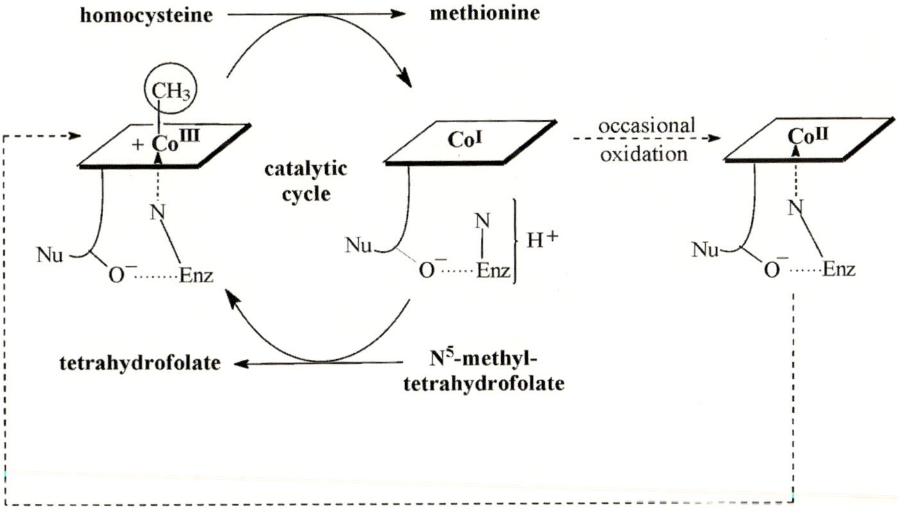

Fig. 17.10 Illustration of methionine formation catalyzed by MetH (Enz signifies the MetH-apoenzyme), where the bound corrinoid shuttles between MeCbl, in a "base-off/His-on" form, and cob(I)alamin (B12s) (Matthews 2001). In addition, reductive methylation with SAM and a reduced flavodoxin reactivates oxidized enzyme (with catalytically inactive cob(II)alamin)

17.4.2 Methionine Synthase from Escherichia coli

The methyl group transfer catalysed by methionine synthase from *E. coli* (MetH) (Matthews 2001; Ludwig and Matthews 2002) is indicated to proceed as a sequence of two nucleophilic displacement steps. This excludes free methyl cations or radicals as intermediates. The methyl group transfer relies on the inherent reactivities of enzyme-bound Co(I)corrins and methyl-Co(III)corrins (Matthews 2001). The great structural changes that must accompany the transitions from (tetra-coordinate) Co(I)corrins to (hexa-coordinate) methyl-Co(III)corrins (Kräutler 1998) also provide a means for control from the protein environment (on the methyl transfer, see above) (Ludwig and Matthews 2002).

MetH represents the most thoroughly studied B12-dependent methyl transferase (Matthews 1999, 2001; Bandarian et al. 2003). It is a modular enzyme containing four separate (and independently functioning) domains for binding (beginning at the N-terminus) homocysteine, N5-methyl-tetrahydrofolate, the B12-cofactor or S-adenosyl-methionine (SAM). The B12-binding domain interacts specifically with each of the other three domains (depending upon its three oxidation states): The Co(I) form with the N5-methyl-tetrahydrofolate domain, the Co(II) form with the SAM domain, and the CH3–Co(III) form with the homocysteine domain.

MetH catalyses the methylation of the bound and reduced cob(I)alamin cofactor by (N5-protonated) N5-methyltetrahydrofolate to give enzyme-bound MeCbl in a "base-off/His-on" form (see later) (Matthews 2001; Ludwig and Matthews 2002; Bandarian et al. 2003). The bound MeCbl is demethylated by homocysteine

(Fig. 17.10), whose sulfur is activated and deprotonated due to the coordination to a zinc ion (held by three cysteine residues) of the homocysteine domain. The two methyl-transfer reactions occur in a rapid sequential mechanism (kcat (MetH) = 27 s^{-1}) (Jarrett et al. 1996; Ludwig and Matthews 2002). The bound cob(I)alamin (B12s) may adventitiously be oxidized to inactive cob(II)alamin (B12r) and requires reactivation by reductive methylation with SAM and a reduced flavodoxin (Ludwig and Matthews 2002).

X-ray crystal analysis of the 28 kDa B12-binding domain of MetH provided the first insight into the three-dimensional structure of a B12-binding protein (Drennan et al. 1994, 1998; Ludwig and Matthews 2002). The cobalt-coordinating DMB-nucleotide tail of the protein-bound cofactor MeCbl was displaced by a histidine and extended into the core of the "Rossmann fold", two most astounding revelations of this work (Drennan et al. 1994). Consequently, in MetH the corrinoid cofactor is bound "base-off" and in a "base-off/His-on"-mode, due to histidine ligation to the metal centre. The crucial cobalt-ligating histidine residue is part of a Gly–X–X–His–X–Asp sequence, which was noticed to be a common sequence in some B12-binding proteins (Marsh and Holloway 1992). The B12-binding domain of MetH, therefore, provides both an anchoring site for the nucleotide tail and cobalt-ligation via the residues of the ("regulatory") His-Asp-Ser triad, holding the corrinoid cofactor with its "catalytic" β-side exposed at an inter-domain interface.

Further crystallographic studies provided more evidence for how the selection between the two ways of methylating the bound corrinoid (the "productive" methylation by N5-methyl-tetrahydrofolate or the "regenerative" methylation by enzyme bound SAM (Ludwig and Matthews 2002)) are achieved. This is controlled by a dynamic domain alternation (Bandarian et al. 2002). As shown by the crystal structure of the N-terminal (two domain) substrate-binding modules of MetH, the two substrates, homocysteine and N5-methyl-tetrahydrofolate, are bound and activated in orientations that position them for reaction with the bound corrinoid. However, the two active sites are separated by $\approx$ 50 Å and the B12-binding domain must shuttle back and forth between these distant active sites to complete the catalytic cycle (Evans et al. 2004).

The crystal structure and the finding of the "base-off/His-on" binding of the B12-cofactor in MetH explained ESR-spectroscopic evidence for histidine binding to the cobalt centre of p-cresolyl-cobamide (a "base-off" cobamide, with a non-coordinating cresolyl-"nucleotide") in the acetogen Sporomusa ovata (Stupperich et al. 1990).

17.4.3 Human Methionine Synthase

Human methionine synthase (hMS, E.C. 2.1.1.13) is a cytoplasmic enzyme with 4 domains, which shares 55% identity in deduced amino acid sequence with MetH (from *E. coli*). hMS shows extensive structural similarities with MetH and has all residues implicated in relevant B12-binding to MetH. hMS also catalyzes the methylation of homocysteine (with demethylation of N5-methyl-tetrahydrofolate) in an apparently related mechanism to that of MetH (Banerjee 1997; Banerjee 1998). In both methionine synthases occasional oxidation (to an inactive cob(II)alamin carrying form) occurs (in the micro-aerophilic conditions of mammalian cells, hMS is oxidized about once every 2,000 turnovers). In contrast to MetH, which is reactivated via reduced ferredoxin (Matthews 2001), hMS is reactivated by a specific reductase, called hMS-reductase (hMSR) (Yamada et al. 2006). hMSR is a dual flavoprotein homologous to flavodoxin and appears to function as a chaperone in the assembly of holo-hMS from apo-hMS and B12a in the presence of NADPH (it also achieves the reduction of aquo-cobalamin (B12a) to B12r) (Yamada et al. 2006) and to reactivate oxidized hMS.

17.5 Methylmalonyl-CoA Mutase

The reversible isomerisation of methylmalonyl-CoA to succinyl-CoA is an enzyme catalyzed radical process (Fig. 17.11). Related isomerisations (malonates to succinates) have been achieved in model radical processes in solution (Halpern 1982; Buckel and Golding 2006).

17.5.1 Coenzyme B12-Dependent Enzymes

Coenzyme B12-dependent enzymes, such as methylmalonyl-CoA mutase (Fig. 17.11), thus performs "difficult" transformations. With the exception of the enzymatic ribonucleotide reduction (Stubbe et al. 1998), the results of coenzyme B12-catalysed enzymatic reactions correspond to isomerisations with vicinal exchange of a hydrogen atom and of a larger moiety (Buckel and Golding 1996). Homolytic cleavage of the Co–C bond of the protein-bound AdoCbl to a 5′-deoxy-5′-adenosyl radical and cob(II)alamin (B12r) (Fig. 17.6) was shown to be the entry to H-abstraction reactions induced by the 5′-deoxy-5′-adenosyl radical (Rétey 1999). Therefore, homolysis of the Co–C bond of AdoCbl is its biologically most significant reactivity (Halpern 1985; Finke 1998). The coenzyme B12-catalysed enzyme reactions occur with maximal rates of approximately 100 s^{-1} (Licht et al. 1999; Marsh and Drennan 2001). With MMCM, rapid formation of

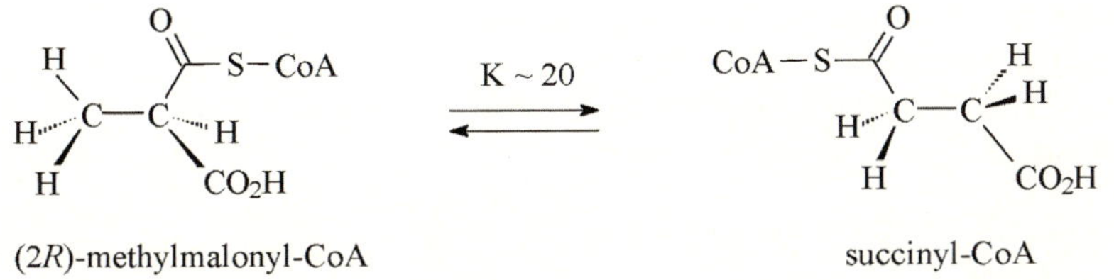

Fig. 17.11 (2R)-methylmalonyl-CoA is reversibly converted into succinyl-CoA (see below)

Co(II)corrins was observed when substrate was added to a solution of holoenzyme (or of apoenzymes and AdoCbl) (Rétey 1998).

Homolysis of the Co–C bond of protein-bound AdoCbl needs to be accelerated by a factor of $>10^{12}$ to relate the observed rates of catalysis by the coenzyme B12-dependent enzymes to the rate of homolysis of AdoCbl in aqueous solution (Halpern 1985; Finke 1998). This dramatic labilisation of the bound cofactor towards homolysis of the Co–C bond is an intriguing feature of coenzyme B12-dependent enzymes (Halpern 1985; Finke 1998). The mechanism of the (enzyme- and substrate-) induced labilisation of AdoCbl towards cleavage of its Co–C bond is a much discussed problem. A "conformational distortion" of the cobalt-corrin part of AdoCbl was suggested as its basis (Halpern 1985). However, the crystal structures of several coenzyme B12-dependent enzymes (Mancia et al. 1996; Reitzer et al. 1999; Shibata et al. 1999) and of cob(II)alamin (B12r) (Kräutler et al. 1989) have not supported this mode of action. Labilisation of the Co–C bond appears to mostly result from protein- and substrate-induced strain on the organometallic group, separation of the largely non-strained homolysis fragments and strong binding of the separated pair, 5′-deoxy-5′-adenosyl radical and B12r (in "base-off/His-on" or "base-on" form) (Kräutler et al. 1989; Toraya 2003; Gschösser et al. 2005). Fixed placement of the corrin moiety at the interface of the B12-binding and substrate-binding/activating domains is given a high significance and movements of the corrin moiety are not required. The nature of the axial trans ligand (histidine or DMB) has a lesser effect on the homolytic Co–C bond dissociation (Fasching et al. 2000). The "regulatory triad" appears not to be involved in proton-transfer steps and may conserve its structure largely during enzymatic turnover (Ludwig and Evans 1999).

The 5′-deoxy-5′-adenosyl radical is the established reactive partner in the actual coenzyme B12-dependent enzymatic reactions, so that AdoCbl functions as a "pre-catalyst" (or catalyst precursor) (Kräutler 1998). The further enzyme reactions all involve bound organic radicals, which are formed (directly or indirectly) by a H-atom abstraction by the 5′-deoxy-5′-adenosyl radical (that originates from AdoCbl). The rearrangement steps of B12-dependent enzymatic rearrangements are assumed to be accomplished by tightly protein-bound radicals that are controlled in their reaction space by the protein (Toraya 2003; Buckel and Golding 2006). The role of the Co(II)-corrin fragment of the coenzyme (as a "spectator" or a "conductor") is a matter of ongoing discussions (Buckel et al. 2006). The major functions of the enzyme concern not only the catalysis of its proper reactions but also the reversible generation of the radical intermediates and the effective protection from non-specific radical chemistry, called "negative catalysis" (Rétey 1990).

Four of the known coenzyme B12-dependent enzymatic reactions concern carbon skeleton rearrangements, in which two vicinal groups (a hydrogen atom and an organic substituent) exchange their positions in a (pseudo-) intra-molecular fashion (Marsh and Drennan 2001; Buckel and Golding 2006). The B12-cofactor is bound "base-off/His-on" at an interface between two modules, the B12-binding and substrate activating domains (or subunits) as revealed by the analysis of the crystal structures of methylmalonyl-CoA mutase (MMCM) from *Propionibacterium shermanii* (Hodgkin et al. 1956) and of glutamate mutase from Clostridium cochlearium

(Reitzer et al. 1999; Gruber and Kratky 2002). The B12-binding motif (Gly–X–X–His–X–Asp) (Marsh and Holloway 1992) is common to the carbon skeleton mutases (Marsh and Drennan 2001). The B12-binding domain of MMCM exhibits considerable sequence homology, even with the B12-binding domain of MetH (Ludwig and Evans 1999). Such homology can be rationalized by the related way of "base-off/His-on" binding the cobalamin cofactor and does not extend to the substrate binding domains (subunits) of the carbon skeleton mutases (Marsh 2000).

17.5.2 Methylmalonyl-CoA Mutase from Propionibacterium shermanii

Methylmalonyl-CoA mutase has been most thoroughly studied as the *P. shermanii* mutase (MMCM), which interconverts (2R)-methylmalonyl-CoA and succinyl-CoA (Banerjee and Chowdhury 1999) (Fig. 17.12). The main function of MMCM in *P. shermanii* concerns the formation of (2R)-methylmalonyl-CoA from succinyl-CoA. Binding of the substrate to holo-MMCM triggers a very fast homolysis of the Co–C bond of the bound AdoCbl. The overall turnover for the radical rearrangement occurs with a rate of about 60 s^{-1} at 25°C. The 5′-deoxy-5′-adenosyl (5′-Ado) radical induces the rearrangement reaction by abstracting an H-atom from the methyl group of enzyme-bound methylmalonyl-CoA. A large deuterium isotope effect of the abstracted H-atoms on the rate of homolysis of the Co–C bond was observed (Chowdhury and Banerjee 2000a). Labilisation of the Co–C bond towards homolysis by about 16 kcal/mol is due largely to a decrease of the activation enthalpy (Chowdhury and Banerjee 2000b). Recent studies (Brooks et al. 2004) indicated the stabilisation of the separated homolysis fragments, 5′-Ado radical and cob(II)alamin (B12r) to be a dominant contribution to the activation (for homolysis) of the Co–C bond of AdoCbl, consistent with an earlier suggestion, based on the crystal structure of B12r (Kräutler et al. 1989).

H-atom abstraction gives the 2-methylmalon-2′-yl-CoA radical which rearranges rapidly to the succin-3-yl-CoA radical (Golding et al. 1998; Banerjee and Chowdhury 1999). Both, fragmentation/ recombination and intramolecular addition/elimination, via a cyclopropyloxyl radical, pathways have been considered for this rearrangement (Buckel and Golding 2006). However, computational studies indicate that the energetic barrier for the addition/elimination pathway is lower than the dissociative pathway (Smith et al. 1999). Experimental evidence has also been used to support this theory, in which (i) the succin-3-yl-CoA radical arises from an intramolecular radical rearrangement and (ii) occurs without noticeable participation of the bound cob(II)alamin (B12r) (Golding et al. 1998). The succin-3-yl-CoA radical (resulting from the rearrangement) then re-abstracts an H-atom from 5′-deoxyadenosine to give succinyl-CoA and the 5′-deoxy-5′-adenosyl radical (which recombines with B12r to give enzyme-bound AdoCbl).

X-ray analysis of the 150 kDa heterodimeric MMCM from *P. shermanii* was the first accurate structure of a coenzyme B12-dependent enzyme. It showed the B12-cofactor to be bound "base-off/His-on". The cobalt centre was coordinated to

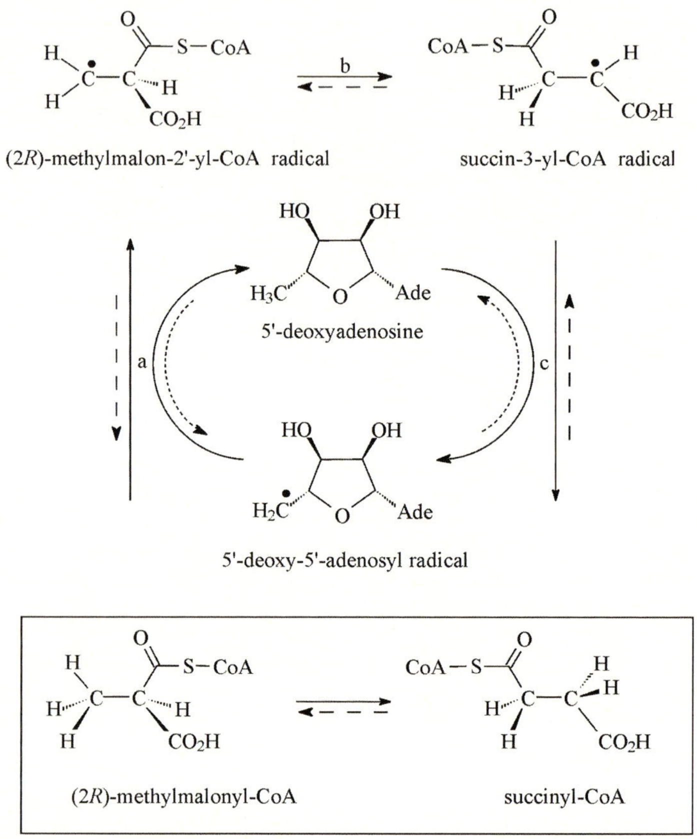

Fig. 17.12 Methylmalonyl-CoA mutase (MMCM) interconverts (2R)-methylmalonyl-CoA and succinyl-CoA. Proposed reaction mechanism, involving H-atom abstraction (**step a**), radical rearrangement (**step b**) and back transfer of H-atom (**step c**). The 5′-deoxy-5′-adenosyl radical (and cob(II)alamin (B12r)) originate(s) from homolysis of protein bound AdoCbl (Fig. 17.6) (Banerjee and Chowdhury 1999; Buckel and Golding 2006)

the histidine of the "regulatory triad" His-Asp-Lys. As in MetH the nucleotide tail of the corrinoid was tightly inserted into the protein and the corrinoid was bound at an interface between two domains (Evans and Mancia 1998; Ludwig and Evans 1999).

The question of how enzymes recognize and bind their corrinoid cofactor in a "base-off"-form is intriguing (in aqueous solution, AdoCbl and MeCbl prefer to be "base-on"). The solution structure of the cofactor-free B12-binding subunit of glutamate mutase from Cl. tetanomorphum (Tollinger et al. 1998, 2001), derived by NMR-spectroscopy, showed the B12-binding subunit to be largely pre-organized for B12-binding. However, the apo-protein was seen to include a flexible loop and a "nascent" helix, which were both suggested to structure upon binding of the B12-cofactor. A model for the events in binding of the "base-off/His-on" corrinoid by the

B12-binding subunit was derived, in which the "base-off"-form of the B12-cofactor was trapped by its nucleotide tail and the bound nucleotide moiety, in turn, stabilized the protein (Tollinger et al. 2001).

In crystallographic work, substrate free MMCM was investigated, as well as with bound pseudo-substrate: both structures indicated the organometallic group to be strained or detached from the cobalt centre (Mancia and Evans 1998; Mancia et al. 1999). Substrate binding appears to squeeze the adenosyl group off from the cobalt-corrin and to assist Co–C bond homolysis (Ludwig and Evans 1999).

17.5.3 Human Methylmalonyl-CoA Mutase

Human methylmalonyl-CoA mutase (hMMCM, EC 5.4.99.2) is a mitochondrial enzyme and catalyzes the isomerisation of (2R)-methylmalonyl-CoA to succinyl-CoA (as final part of the catabolic pathway of uneven numbered and branched chain fatty acids) (Banerjee and Chowdhury 1999). Impairment of hMMCM leads to methyl-malonic aciduria, a metabolic error that leads to developmental retardation and mortality in infants. hMMCM binds two AdoCbl per (homodimeric) enzyme (predicted mass: about 159 kDa) and is specific for the 2R-epimer of methylmalonyl-CoA. The (monomer-)chain of hMMCM and the B12-binding α-chain of MMCM are homologous; evidence for a related mechanism in both mutases has been provided (Buckel and Golding 2006). Apo-hMMCM has been suggested to be loaded with its cofactor, AdoCbl, in a direct and "targeted" interaction with adenosyl-transferase. In the presence of the latter enzyme, hMMCM exhibits a striking resistance towards deactivation (Yamanishi et al. 2005).

Acknowledgements This work was supported by the Austrian National Science Foundation (project No P13595) and by the European Commission (project No HPRN-CT-2002-00195).

References

Bandarian V, Ludwig ML, Matthews RG (2003) Factors modulating conformational equilibria in large modular proteins: a case study with cobalamin-dependent methionine synthase. Proc Natl Acad Sci USA 100:8156–8163

Bandarian V, Pattridge KA, Lennon BW, Huddler DP, Matthews RG, Ludwig ML (2002) Domain alternation switches B12-dependent methionine synthase to the activation conformation. Nat Struct Biol 9:53–56

Banerjee R (1997) The Yin-Yang of cobalamin biochemistry. Chem Biol 4:175–186

Banerjee R (1998) Spectroscopic and molecular genetic characterization of the two mammalian B12-dependent enzymes. In: Kräutler B, Arigoni D, Golding BT (eds) Vitamin B12 and B12-proteins, Wiley-VCH, Weinheim, pp. 189–197

Banerjee R (ed) (1999) Chemistry and biochemistry of B12. Wiley, New York

Banerjee R, Chowdhury S (1999) Methylmalonyl-CoA mutase. In: Banerjee R (ed) Chemistry and biochemistry of B12. Wiley, New York, pp. 707–730

Banerjee R, Ragsdale SW (2003) The many faces of vitamin B12: catalysis by cobalamin-dependent enzymes. Annu Rev Biochem 72:209–247

Battersby AR (1998) B12-biosynthesis in an aerobic organism: how the pathway was elicudated. In: Kräutler B, Arigoni D, Golding BT (eds) Vitamin B12 and B12-proteins. Wiley-VCH, Weinheim, pp. 47–61

Brooks AJ, Vlasie M, Banerjee R, Brunold TC (2004) Spectroscopic and computational studies on the adenosylcobalamin-dependent methylmalonyl-CoA mutase: evaluation of enzymatic contributions to Co–C bond activation in the Co_3^+ ground state. J Am Chem Soc 126:8167–8180

Brown KL (1999) NMR Spectroscopy of B12. In: Banerjee R (ed) Chemistry and biochemistry of B12. Wiley, New York, pp. 197–237

Brown KL (2005) Chemistry and enzymology of vitamin B12. Chem Rev 105:2075–2149

Brown KL, Zou X, Banka RR, Perry CB, Marques HM (2004) Solution structure and thermolysis of Coβ-5′-deoxyadenosylimidazolylcobamide, a coenzyme B12 analogue with an imidazole axial nucleoside. Inorg Chem 43:8130–8142

Buckel W, Golding BT (1996) Glutamate and 2-methyleneglutarate mutase: from microbial curiosities to paradigms for coenzyme B12-dependent enzymes. Chem Soc Rev 25:329

Buckel W, Golding BT (2006) Radical enzymes in anaerobes. Ann Rev Microbiol 60:27–49

Buckel W, Kratky C, Golding BT (2006) Stabilization of methylene radicals by Cob(II)alamin in coenzyme B12 dependent mutases. Chem Eur J 12:352–362

Chowdhury S, Banerjee R (2000a) Evidence for quantum mechanical tunneling in the coupled cobalt-carbon bond homolysis-substrate radical generation reaction catalyzed by methylmalonyl-CoA mutase. J Am Chem Soc 122:5417–5418

Chowdhury S, Banerjee R (2000b) Thermodynamic and kinetic characterization of Co–C bond homolysis catalyzed by coenzyme B12-dependent methylmalonyl-CoA mutase. Biochemistry 39:7998–8006

Cole AG, Yoder LM, Shiang JJ, Anderson NA, Walker LA, Holl MMB, Sension RJ (2002) Time-resolved spectroscopic studies of B12 coenzymes: a comparison of the primary photolysis mechanism in methyl-, ethyl-, n-propyl-, and 5′-deoxyadenosylcobalamin. J Am Chem Soc 124:434–41

Dolphin D (ed) (1982) B12, vols I and II. Wiley, New York

Drennan CL, Dixon MM, Hoover DM, Jarrett JT, Goulding CW, Matthews RG, Ludwig ML (1998) Cobalamin-dependent methionine synthase from *Escherichia coli*: structure and reactivity. In: Kräutler B, Arigoni D, Golding BT (eds) Vitamin B12 and B12-proteins. Weinheim, Wiley-VCH, pp. 133–155

Drennan CL, Doukov TI, Ragsdale SW (2004) The metalloclusters of carbon monoxide dehydrogenase/acetyl-CoA synthase: a story in pictures. J Biol Inorg Chem 9:511–515

Drennan CL, Huang S, Drummond JT, Matthews RG, Ludwig ML (1994) How a protein binds B12 – a 3.0 Å X-ray structure of B12-binding domains of methionine synthase. Science 266:1669–1674

Drennan CL, Matthews RG, Ludwig ML (1994) Cobalamin-dependent methionine synthase – the structure of a methylcobalamin-binding fragment and implications for other B12-dependent enzymes. Curr Opin Struct Biol 4:919–929

Ellenbogen L, Cooper BA (1984) Vitamin B12. In: Machlin LJ (ed) Handbook of vitamins, nutritional and clinical aspects. Food Science and Technology. Marcel Dekker, New York, pp. 491–536

Endicott JF, Netzel TL (1979) Early events and transient chemistry in the photohomolysis of alkylcobalamins. J Am Chem Soc 101:4000–4002

Eschenmoser A (1988) Vitamin-B12 – Experiments concerning the origin of its molecular structure. Angew Chem Int Ed 27:5–39

Evans JC, Huddler DP, Hilgers MT, Romanchuk G, Matthews RG, Ludwig ML (2004) Structures of the N-terminal modules imply large domain motions during catalysis by methionine synthase. Proc Natl Acad Sci USA 101:3729–3736

Evans PR, Mancia F (1998) Insights on the reaction mechanism of methylmalonyl-CoA mutase from the crystal structure. In: Kräutler B, Golding BT, Arigoni D (eds) Vitamin B12 and B12 proteins. Wiley-VCH, Weinheim, pp. 217–226

Fasching M, Schmidt W, Kräutler B, Stupperich E, Schmidt A, Kratky C (2000) Coα-(1H-imidazolyl)-Coβ-methylcob(III)amide: model for protein-bound corrinoid cofactors. Helv Chim Acta 83:2295–2316

Finke RG (1998) Coenzyme B12-based chemical precedent for Co–C bond homolysis and other key elementary step. In: Kräutler B, Arigoni D, Golding BT (eds) Vitamin B12 and B12-proteins. Wiley-VCH, Weinheim, pp. 383–402

Frey PA, Chang CH (1999) Aminomutases. In: Banerjee R (ed) Chemistry and biochemistry of B12. Wiley, New York, pp. 835–858

Friedrich, W. (1988) Vitamins. Walter de Gruyter, Berlin

Golding BT, Anderson RJ, Ashwell S, Edwards CH, Garnett I, Kroll F, Buckel W (1998) A mechanistic overview of B12 dependent processes. In: Kräutler B, Arigoni D, Golding BT (eds) Vitamin B12 and B12 proteins. Wiley-VCH, Weinheim, pp. 201–216

Gruber K, Kratky C (2002) Coenzyme B12 dependent glutamate mutase. Curr Opin Chem Biol 6:598–603

Gschösser S, Hannak RB, Konrat R, Gruber K, Mikl C, Kratky C, Kräutler B (2005) Homocoenzyme B12 and bishomocoenzyme B12, covalent structural mimics for homolyzed, enzyme-bound coenzyme B12. Chem Eur J 11:81–93

Halpern J (1982) Chemistry and significance of vitamin B12 model systems. In: Dolphin D (ed) B12. Wiley, New York, pp. 501–541

Halpern J (1985) Mechanisms of coenzyme B12-dependent rearrangements. Science 227:869–875

Hay BP, Finke RG (1986) Thermolysis of the Co–C bond of adenosylcobalamin. 2. products, kinetics, and Co–C bond-dissociation energy in aqueous-solution. J Am Chem Soc 108:4820–4829

Hodgkin DC, Kamper J, Mackay M, Pickworth J, Trueblood KN, White JG (1956) Structure of Vitamin-B12. Nature 178:64–66

Hodgkin DC, Pickworth J, Robertson JH, Trueblood KN, Prosen RJ, White JG (1955) Crystal structure of the hexacarboxylic acid derived from B12 and the molecular structure of the vitamin. Nature 176:325–328

Hoffmann B, Oberhuber. M., Stupperich E, Bothe H, Buckel W, Konrat R, Kräutler B (2000) Native corrinoids from *Clostridium cochlearium* are adeninylcobamides: spectroscopic analysis and identification of pseudovitamin B12 and factor A. J Bacteriol 182:4773–4782

Hogenkamp HPC, Bratt GT, Sun S (1985) Methyl transfer from methylcobalamin to thiols – a reinvestigation. Biochemistry 24:6428–6432

Jarrett JT, Amaratunga M, Drennan CL, Scholten JD, Sands RH, Ludwig ML, Matthews RG (1996) Mutations in the B12-binding region of methionine synthase: how the protein controls methylcobalamin reactivity. Biochemistry 35:2464–2475

Konrat R, Tollinger M, Kräutler B (1998) New NMR structural and dynamical probes of organometallic B12 derivatives. In: Kräutler B, Arigoni D, Golding BT (eds) Vitamin B12 and B12-proteins. Wiley-VCH, Weinheim, pp. 349–368

Kratky C, Kräutler B (1999) Molecular structure of B12 cofactors and other B12 derivatives. In: Banerjee R (ed) Chemistry and biochemistry of B12. Wiley, New York, pp. 6–41

Kräutler B (1987) Thermodynamic trans-effects of the nucleotide base in the B12 coenzymes. Helv Chim Acta 70:1268–1278

Kräutler B (1998) B12 coenzymes, the central theme. In: Kräutler B, Arigoni D, Golding BT (eds) Vitamin B12 and B12 proteins. Wiley-VCH, Weinheim, pp. 3–43

Kräutler B (1999) Electrochemistry and organometallic electrochemical synthesis. In: Banerjee R (ed) Chemistry and biochemistry of B12. Wiley, New York, pp. 315–339

Kräutler B, Arigoni D, Golding BT (eds) (1998) Vitamin B12 and B12-proteins. Wiley-VCH, Weinheim

Kräutler B, Caderas C (1984) Complementary diastereoselective cobalt methylations of the Vitamin-B12 derivative cobester. Helv Chim Acta 67:1891–1896

Kräutler B, Dérer T, Liu PL, Mühlecker W, Puchberger M, Kratky C, Gruber K (1995) Oligomethylene-bridged vitamin-B12 dimers. Angew Chem Int Ed 34:84–86

Kräutler B, Fieber W, Ostermann S, Fasching M, Ongania KH, Gruber K, Kratky C, Mikl C, Siebert A, Diekert G (2003) The cofactor of tetrachloroethene reductive dehalogenase of *Dehalospirillum multivorans* is norpseudo-B12, a new type of a natural corrinoid. Helv Chim Acta 86:3698–36716

Kräutler B, Keller W, Kratky C (1989) Coenzyme B12-chemistry: the crystal and molecular structure of Cob(II)alamin. J Am Chem Soc 111:8936–8938

Kräutler B, Konrat R, Stupperich E, Färber G, Gruber K, Kratky C (1994) Direct evidence for the conformational deformation of the corrin ring by the nucleotide base in vitamin-B12 – synthesis and solution spectroscopic and crystal-structure analysis of Coβ-cyanoimidazolylcobamide. Inorg Chem 33:4128–4139

Kräutler B, Ostermann S (2003) Structure, reactions and functions of B12 and B12-proteins. In: Kadish KM, Smith KM, Guilard R (eds) The porphyrin handbook. Elsevier Science, Oxford, pp. 229–276

Lenhert PG, Hodgkin DC (1961) Structure of 5,6-dimethylbenzimidazolylcobamide coenzyme. Nature 192:937

Lexa D, Savéant JM (1983) The electrochemistry of vitamin-B12. Acc Chem Res 16:235–243

Licht SS, Booker S, Stubbe J (1999) Studies on the catalysis of carbon-cobalt bond homolysis by ribonucleoside triphosphate reductase: evidence for concerted carbon-cobalt bond homolysis and thiyl radical formation. Biochemistry 38:1221–1233

Ludwig ML, Evans PR (1999) X-ray crystallography of B12 enzymes: methylmalonyl-CoA mutase and methionine synthase. In: Banerjee R (ed) Chemistry and biochemistry of B12. Wiley, New York, pp. 595–632

Ludwig ML, Matthews RG (2002) B12-dependent methionine synthase: a structure that adapts to catalyze multiple methyl transfer reactions. Struct Mech: from Ashes to Enzymes 827:186–201

Mancia F, Evans PR (1998) Conformational changes on substrate binding to methylmalonyl CoA mutase and new insights into the free radical mechanism. Structure 6:711–720

Mancia F, Keep NH, Nakagawa A, Leadlay PF, McSweeney S, Rasmussen B, Bösecke B, Diat O, Evans PR (1996) How coenzyme B12 radicals are generated: the crystal structure of methylmalonyl-coenzyme A mutase at 2 Å resolution. Structure 4:339–350

Mancia F, Smith GA, Evans PR (1999) Crystal structure of substrate complexes of methylmalonyl-CoA mutase. Biochemistry 38:7999–8005

Marsh ENG (2000) Coenzyme-B12-dependent glutamate mutase. Bioorg Chem 28:176–189

Marsh ENG, Drennan CL (2001) Adenosylcobalamin-dependent isomerases: new insights into structure and mechanism. Curr Opin Chem Biol 5:499–505

Marsh ENG, Holloway DE (1992) Cloning and sequencing of glutamate mutase component-S from *Clostridium tetanomorphum* – homologies with other cobalamin-dependent enzymes. Febs Lett 310:167–170

Matthews RG (1999) Cobalamin-dependent methionine synthase. In: Banerjee R (ed) Chemistry and biochemistry of B12. Wiley, New York, pp. 681–706

Matthews RG (2001) Cobalamin-dependent methyltransferases. Acc Chem Res 34:681–689

Mosimann H, Kräutler B (2000) Methylcorrinoids methylate radicals – Their second biological mode of action? Angew Chem Int Ed 39:393–200

Pailes WH, Hogenkamp HPC (1968) Photolability of Co-alkylcobinamides. Biochemistry 7:4160–4166

Pratt JM (1972) Inorganic chemistry of vitamin B12. Academic, New York

Ragsdale SW, Kumar M, Zhao S, Menon S, Seravalli J, Doukov T (1998) Discovery of a biological organometallic reaction sequence involving vitamin B12. In: Kräutler B, Arigoni D, Golding BT (eds) Vitamin B12 and B12-proteins. Wiley-VCH, Weinheim, pp. 167–178

Randaccio L, Geremia S, Nardin G, Würges J (2006) X-ray structural chemistry of cobalamins. Coord Chem Rev 250:1332–1250

Reitzer R, Gruber K, Jogl G, Wagner, UG, Bothe H, Buckel W, Kratky C (1999) Glutamate mutase from *Clostridium cochlearium*: the structure of a coenzyme B12-dependent enzyme provides new mechanistic insights. Structure 7:891–902

Rétey J (1990) Enzymatic-reaction selectivity by negative catalysis or how do enzymes deal with highly reactive intermediates. Angew Chem Int Ed 29:355–361

Rétey J (1998) Coenzyme B12-dependent enzymes and their models. In: Kräutler B, Arigoni D, Golding BT (eds) Vitamin B12 and B12-proteins. Wiley-VCH, Weinheim, pp. 273–288

Rétey J (1999) Stereospecifity of the coenzyme B12-catalyzed rearrangements and the role of negative catalysis. In: Banerjee R (ed) Chemistry and biochemistry of B12. Wiley, New York, pp. 271–288

Rickes EL, Brink NG, Koniuszy FR, Wood TR, Folkers K (1948) Crystalline vitamin B12. Science 107:396–397

Rosenblatt DS, Fenton WA (1999) Inborn errors of cobalamin metabolism. In: Banerjee R (ed) Chemistry and biochemistry of B12. Wiley, New York, pp. 367–384

Sauer K, Thauer RK (1999) The role of corrinoids in methanogenesis. In: Banerjee R (ed) Chemistry and biochemistry of B12. Wiley, New York, pp. 655–679

Schrauzer GN, Deutsch E (1969) Reactions of cobalt(I) supernucleophiles. The alkylation of vitamin B12s cobaloximes(I), and related compounds. J Am Chem Soc 91:3341–3350

Scott AI, Roessner CA, Santander PJ (2003) Genetic and mechanistic exploration of the two pathways of vitamin B12 biosynthesis. In: Kadish KM, Smith KM Guilard R (eds) Porphyrin handbook. Elsevier Science, Amsterdam, pp. 211–228

Shibata N, Masuda J, Tobimatsu T, Toraya T, Suto K, Morimoto Y, Yasuoka N (1999) A new mode of B12 binding and the direct participation of a potassium ion in enzyme catalysis: x-ray structure of diol dehydratase. Structure 7:997–1008

Siebert A, Neumann A, Schubert T, Diekert G (2002) A non-dechlorinating strain of Dehalospirillum multivorans: evidence for a key role of the corrinoid cofactor in the synthesis of an active tetrachloroethene dehalogenase. Arch Microbiol 178:443–449

Smith DM, Golding BT, Radom L (1999) Facilitation of enzyme-catalyzed reactions by partial proton transfer: application to coenzyme-B12-dependent methylmalonyl-CoA mutase. J Am Chem Soc 121:1383–1384

Smith EL, Parker LFJ (1948) Purification of anti-pernicious anaemia factor. Biochem J 43:R8–R9

Stich TA, Yamanishi M, Banerjee R, Brunold TC (2005) Spectroscopic evidence for the formation of a four-coordinate Co_2^+ cobalamin species upon binding to the human ATP: *Cobalamin adenosyltransferase*. J Am Chem Soc 127:7660–7661

Stubbe J (2000) Ribonucleotide reductases: the link between an RNA and a DNA world? Curr Opin Struct Biol 10:731–736

Stubbe J, Licht S, Gerfen G, Silva D, Booker S (1998) Adenosylcobalamin-dependent ribonucleotide reductases: still amazing but no longer confusing. In: Kräutler B, Arigoni D, Golding BT (eds) Vitamin B12 and B12-proteins. Wiley-VCH, Weinheim, pp. 321–331

Stupperich E, Eisinger HJ, Albracht SPJ (1990) Evidence for a super-reduced cobamide as the major corrinoid fraction in vivo and a histidine residue as a cobalt ligand of the p-cresolyl cobamide in the acetogenic bacterium *Sporomusa ovata*. Eur J Biochem 193:105–109

Stupperich E, Konle R, Lehle M (1998) Corrinoid-dependent methyl transfer reactions in *Sporomusa ovata*. In: Kräutler B, Arigoni, D, Golding BT (eds) Vitamin B12 and B12 proteins. Wiley VCH, Weinheim, pp. 179–187

Summers MF, Marzilli LG, Bax A (1986) Complete 1H and 13C assignments of coenzyme-B12 through the use of new two-dimensional NMR experiments. J Am Chem Soc 108:4285–4294

Tollinger M, Eichmüller C, Konrat R, Huhta MS, Marsh ENG, Kräutler B (2001) The B12-binding subunit of glutamate mutase from *Clostridium tetanomorphum* traps the nucleotide moiety of coenzyme B12. J Mol Biol 309:777–791

Tollinger M, Konrat R, Hilbert BH, Marsh ENG, Kräutler B (1998) How a protein prepares for B12 binding: structure and dynamics of the B12-binding subunit of glutamate mutase from *Clostridium tetanomorphum*. Structure 6:1021–1033

Toraya T (2003) Radical catalysis in coenzyme B12-dependent isomerization (eliminating) reactions. Chem Rev 103:2095–2127

Yamada K, Gravel, RA, Toraya T, Matthews RG (2006) Human methionine synthase reductase is a molecular chaperone for human methionine synthase. Proc Natl Acad Sci USA 103:9476–9481
Yamanishi M, Vlasie M, Banerjee R (2005) Adenosyltransferase: an enzyme and an escort for coenzyme B12? Trends Biochem Sci 30:304–308
Zerbe-Burkhardt K, Ratnatilleke A, Vrijbloed JW, Robinson JA (1999) Isobutyryl-CoA mutase. In: Banerjee R (ed) Chemistry and biochemistry of B12. Wiley, New York, pp. 859–870

Chapter 18
Physiological and Molecular Aspects of Cobalamin Transport

Sergey N. Fedosov

Abstract Minute doses of a complex cofactor cobalamin (Cbl, vitamin B12) are essential for metabolism. The nutritional chain for humans includes: (1) production of Cbl by bacteria in the intestinal tract of herbivores; (2) accumulation of the absorbed Cbl in animal tissues; (3) consumption of food of animal origin. Most biological sources contain both Cbl and its analogues, i.e. Cbl-resembling compounds physiologically inactive in animal cells. Selective assimilation of the true vitamin requires an interplay between three transporting proteins – haptocorrin (HC), intrinsic factor (IF), transcobalamin (TC) – and several receptors. HC is present in many biological fluids, including gastric juice, where it assists in disposal of analogues. Gastric IF selectively binds dietary Cbl and enters the intestinal cells via receptor-mediated endocytosis. Absorbed Cbl is transmitted to TC and delivered to the tissues with blood flow. The complex transport system guarantees a very efficient uptake of the vitamin, but failure at any link causes Cbl-deficiency. Early detection of a negative B12 balance is highly desirable to prevent irreversible neurological damages, anaemia and death in aggravated cases. The review focuses on the molecular mechanisms of cobalamin transport with emphasis on interaction of corrinoids with the specific proteins and protein-receptor recognition. The last section briefly describes practical aspects of recent basic research concerning early detection of B12-related disorders, medical application of Cbl-conjugates, and purification of corrinoids from biological samples.

Keywords Adsorbent · Affinity binding · Analogues · B12 · Cobalamin · Conjugates · Haptocorrin · Intrinsic factor · Receptor · Transcobalamin

18.1 Introduction

Cobalamin (Cbl, vitamin B12) is the most complex cofactor existing in the nature (Pratt 1972; Kräutler and Ostermann 2003), Fig. 18.1. Its molecule consists of two invariable moieties: (i) the corrin ring with cobalt ion; and (ii) the nucleotide loop

S.N. Fedosov (✉)
Protein Chemistry Laboratory, Department of Molecular Biology, University of Aarhus, Aarhus, Denmark
e-mail: snf@mb.au.dk

O. Stanger (ed.), *Water Soluble Vitamins*, Subcellular Biochemistry 56,
DOI 10.1007/978-94-007-2199-9_18,

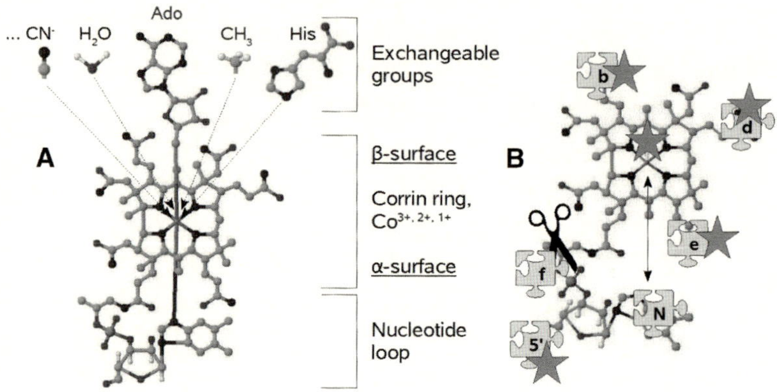

Fig. 18.1 Schematic structures of (**a**) cobalamins; and (**b**) cobalamin analogues containing substitutes (*puzzle symbols*). *Stars* indicate potential sites of chemical conjugation

coordinated to cobalt at the lower α-surface. The third component, the active group coordinated to cobalt at the upper β-surface, is exchangeable. A few typical β-groups are shown in Fig. 18.1, but only two of them, 5′-deoxy-adenosyl (Ado) and methyl (Me), form a physiologically active cofactor. The three major cobalamins in biological objects are AdoCbl (≈60%), MeCbl (≈20%) and H2OCbl (≈20%), (Mervyn 1971; Linnell et al. 1974; Quadros et al. 1979).

The most abundant natural cobalamin is AdoCbl. In animal cells it is involved in catabolism of fatty acids with a branched structure or an odd number of carbons (Banerjee and Ragsdale 2003). The process takes place inside mitochondria, where an AdoCbl-dependent enzyme converts a metabolite methylmalonic acid (MMA) to succinate. Inadequate Cbl concentration results in high level of MMA in blood plasma which is considered as a characteristic indication of Cbl-deficiency (Carmel et al. 2003).

The second cofactor, MeCbl, acts in cytoplasm and mediates transfer of Me-group from the inactive form of folate (methyl tetrahydrofolate, MeTHF) to homocysteine (Hcy), (Carmel et al. 2003; Toohey 2006). This process reactivates THF and produces methionine, precursor of a methylating agent S-adenosyl methionine (SAM). Thus, MeCbl controls most methylation reactions including THF-dependent synthesis of DNA and SAM-dependent methylation of lipids and creatine precursor guanidinoacetate. Low activity of MeCbl-dependent enzyme causes insufficient production of erythrocytes and components of nerve membranes. This metabolic disorder is exposed by elevated Hcy in blood plasma (Carmel et al. 2003).

The two cofactors are occasionally converted to H2OCbl because of abnormal termination of their catalytic cycles (Kräutler and Ostermann 2003; Banerjee and Ragsdale 2003). This makes H2OCbl the third ubiquitous cobalamin available from the natural sources. Cobalamins with the upper group differing from Ado and Me (e.g. water, cyanide, histidine, glutathione) cannot perform catalysis but are easily transformed to the cofactors inside the animal cell. Although all the above forms serve as vitamins, the notation "B12" corresponds, strictly speaking, to CNCbl, the main industrial form of the vitamin (Martens et al. 2002).

Beside true cobalamins, there is a large group of Cbl-like molecules of natural or synthetic origin called analogues (Pratt 1972; Kräutler and Ostermann 2003). The differences between these molecules and vitamin B12 are related to other structural elements than the upper group (Fig. 18.1b). The most typical natural analogues have either substituted or absent nucleotide (Fig. 18.1b, positions N, f). Structural changes in the nucleotide often reduce its affinity for cobalt ion (Kräutler and Ostermann 2003). The analogues are inactive inside the animal cells, but many of them can catalyse Cbl-dependent reactions in bacteria (Pratt 1972; Kräutler and Ostermann 2003). Both cobalamins and analogues belong to a common group called "corrinoids" (Cor).

18.2 Nutritional and Physiological Facets of Cobalamin Intake

18.2.1 Dietary Cbl and Its Gastrointestinal Absorption

The process of Cbl assimilation is rather complex. Firstly, this cofactor is produced exclusively by bacteria (Pratt 1972; Kräutler and Ostermann 2003; Martens et al. 2002), and secondly, the amount of Cbl available from natural sources is extremely low (Mervyn 1971; Fedosov et al. 1996; Ortigues-Marty et al. 2005). Yet, vertebrates manage to provide themselves with sufficient amount of Cbl from either internal or external sources. Thus, herbivores (Fig. 18.2a) obtain the vitamin from bacteria inhabiting their digestive tract (Pratt 1972; Ortigues-Marty et al. 2005). These microorganisms convert indigestible cellulose to monosaccharides and synthesize corrinoids, sharing the excessive products with their host. Cbl is accumulated in animal tissues, where it can be stored for years or utilized. It is known that low amount of cobalt in the soil causes malproduction of bacterial Cbl with accompanying undergrowth of microorganisms in the fore-stomach of a grassing cow (Pratt 1972; Ortigues-Marty et al. 2005). The consequence is a low velocity of grass fermentation and starvation of the animals.

Humans are ultimately dependent on animal sources of vitamin B12: meat, milk, eggs, etc. (Mervyn 1971; Fedosov et al. 1996; Ortigues-Marty et al. 2005), Fig. 18.2b. The highest concentration of Cbl ever reported (5,000 nmol/kg) was found in a Japanese salted and fermented salmon kidney "Mefun" (Adachi et al. 2005). The normal daily dose of vitamin B12 for an adult human corresponds to 1–2 nmol of Cbl (1 nmol = 1.58 μg of AdoCbl or 1.35 μg of CNCbl). This is an extremely low amount when compared to other vitamins, but limited availability of Cbl is compensated by an efficient and complex capturing mechanism. Three transporting proteins – intrinsic factor (IF), transcobalamin (TC) and haptocorrin (HC) – as well as a number of receptors, are the key players in this process (Allen 1975; Nexø 1998; Moestrup 2006). Gastrointestinal uptake of B12 includes several

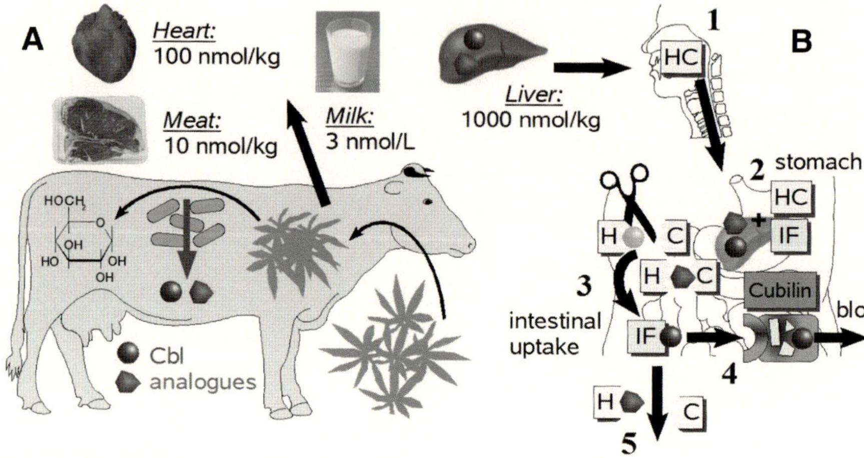

Fig. 18.2 Assimilation of cobalamin by (**a**) ruminant animals; and (**b**) humans

steps as depicted in Fig. 18.2b. First step might occur in the mouth, where the corrin-binding glycoprotein HC (called previously R-binder) is secreted to saliva at concentration of ≈50 nM (Nexø et al. 1988), Fig. 18.2b, step 1. Specificity of this carrier is low, and it binds both vitamin B12 and Cbl-analogues. Nevertheless, it hardly does so in the mouth, since the ligands are not yet released from the food components. HC of salivary origin proceeds to the stomach, where some amount of gastric HC is also produced. Total concentration of HC in gastric juice can be estimated as ≈10 nM (Allen et al. 1978). Simultaneous effect of low pH and proteolytic activity of pepsin causes liberation of 60–70% of corrinoids encapsulated in food (Del Corral and Carmel 1990; Kittang and Schjonsby 1987). They interact primarily with HC (Fig. 18.2b, step 2) despite high excess of another Cbl-binding glycoprotein, gastric IF ≈ 50 nM (Del Corral and Carmel 1990; Allen et al. 1978b; Kittang and Schjonsby 1987). Interaction of IF with Cbl is pH sensitive; therefore this carrier binds the vitamin only in the intestine after the digest is neutralized (Allen et al. 1978a, b; Gueant et al. 1986). Pancreatic enzymes accomplish liberation of the dietary Cbl; in addition, they cleave HC in two fragments (Allen et al. 1978a; Gueant et al. 1986). Nicked HC partially loses its affinity for Cbl, and the major part of the vitamin (≈80%) transfers to IF (Allen et al. 1978a, Gueant et al. 1986), Fig. 18.2b, step 3. The latter carrier has a preference for the true vitamin and serves as the first filter sorting out Cbl-analogues (Allen et al. 1978b; Gueant et al. 1984). The produced IF·Cbl complex is recognized by the receptor cubilin (Moestrup 2006) and internalized in the distal ileum, Fig. 18.2b, step 4. The unsaturated IF does not bind to cubilin and is gradually digested (Gueant et al. 1984; Gordon et al. 1991). The specific intestinal assimilation of Cbl is exceptionally efficient and guarantees uptake of the minute vitamin doses, passive diffusion accounts for only ≈1% of absorption (Mervyn 1971).

In contrast to a widespread citation, HC is not completely proteolysed and HC-fragments with the remaining Cbl-binding activity were found even in faeces (Gueant et al. 1984). They seem to retain the bound Cbl-analogues, since faecal excretions are enriched by corrinoids differing from Cbl (Gueant et al. 1984), Fig. 18.2b, step 5.

18.2.2 Transport of Cbl in Blood

Two carriers of Cbl, protein TC $\approx$ 0.8 nM and glycoprotein HC $\approx$ 0.4 nM, are present in blood due to internal secretion (Allen 1975; Nexø 1998; Moestrup 2006). Thus, practically all tissues synthesize TC, though the patterns vary among different species. The highest production is typically associated with kidney, lymphatic nodes and liver (Li et al. 1994; Fedosov et al. 1999). TC circulates mostly as the unsaturated apo-form (Allen 1975; Nexø 1998), which is fairly fast-metabolising with $t\frac{1}{2}$ $\approx$ 1 h (Allen 1975). The Cbl-saturated form holoTC $\approx$ 0.06 nM accounts for $\approx$10% of the TC-pool under steady-state (Hvas et al. 2006; Miller et al. 2006, Nexø et al. 2002; Ulleland et al. 2002). Clearance of TC·Cbl from blood plasma is biphasic, when monitoring radioactive Cbl associated with TC (Hom and Olesen 1969). The half-life times were estimated as $\approx$1 (Allen 1975; Hom and Olesen 1969) and $\approx$15 h (Carkeet et al. 2006; Hom and Olesen 1969).

The second plasma transporter, HC, exists in two variants, which are probably secreted by liver (Boisson et al. 1999) and granulocytes (Allen 1975). The essential difference between the two forms is composition of the attached carbohydrates (Allen 1975). Thus, high content of sialic acids in HC of the assumed liver origin (called also form I) slows down the protein metabolism to $t\frac{1}{2} \approx$ 10 days (Allen 1975; Hom and Olesen 1969). On the contrary, HC from granulocytes (form III) has low amount of sialic acids and is cleared very rapidly by liver ($t\frac{1}{2} \approx$ 5 min) without discrimination between HC·Cbl and HC (Allen 1975). Blood plasma under steady state conditions contains almost exclusively the slow metabolising form saturated by 80% with Cbl, which corresponds to holoHC $\approx$ 0.3 nM (Allen 1975; Mørkbak et al. 2006).

The scheme of Cbl transport cycle is depicted in Fig. 18.3. At the first step, intestinal IF·Cbl degrades in lysosomes of enterocytes, and the vitamin is released into blood (Fig. 18.3, step 1). Most of Cbl becomes attached to TC, and 7 h after ingestion of B12 the plasma level of TC·Cbl (holoTC) increases by 50% (Carkeet et al. 2006; Bor et al. 2004). Prevalence of TC·Cbl complex can be explained by either (i) high excess of the unsaturated TC in plasma (Allen 1975), or (ii) binding of Cbl to TC already inside the intestinal cells (Quadros et al. 1999). The main targets for the newly produced plasma holoTC are liver (Allen 1975) and kidney (Alsenz et al. 2000).

Hepatocytes bind TC·Cbl (Fig. 18.3, step 2) and to some extent TC via a ubiquitous TC-receptor (Moestrup 2006; Quadros et al. 2005, Seetharam and Li 2000; Seligman and Allen 1978; Nexø and Hollenberg 1980), which identity is still debated (Quadros et al. 2005, Seetharam and Li 2000). Absorbed holoTC degrades

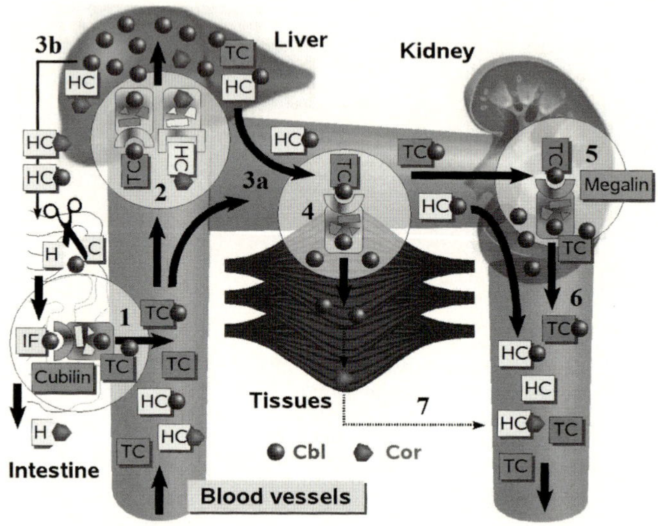

Fig. 18.3 Transportation of Cbl in blood

inside liver cells, and the liberated Cbl is partially stored, partially secreted back to blood bound to HC or TC of liver origin (Allen 1975; Hom and Olesen 1969; Bor et al. 2004), Fig. 18.3, step 3a. Primary absorption of TC·Cbl complex by the liver and its following resynthesis and recirculation may explain the biphasic clearance of TC·Cbl (Hom and Olesen 1969). Liver seems to verify the physiological usefulness of absorbed corrinoids, whereupon the TC-bound vitamin is distributed among other tissues and transferred to the enzymes, Fig. 18.3, step 4.

Cbl turnover in kidney stands apart from that in other organs. Thus, both TC and TC·Cbl are prone to glomerular filtration ($\approx$10% sieving) because of their relatively small size, $M \approx 46$ kDa, $r \approx 2.5$ nm (Moestrup 2006; Nexø 1998). The filtered TC·Cbl complex is reabsorbed from primary urine by the specific receptor megalin, Fig. 18.3 step 5, (Moestrup 2006; Birn 2006). The endocytosed protein degrades, but the vitamin returns to blood bound to freshly synthesized TC, Fig. 18.3 step 6. Therefore, kidney both produces significant quantities of TC (Li et al. 1994) and temporarily stores Cbl, which concentration reaches 500 nmol/kg (Mervyn 1971).

The role of the other Cbl-transporting protein, plasma HC, is not clarified. Thus, HC seems to be absorbed only by liver (Allen 1975). It is not subjected to glomerular filtration due to heavy glycosylation, which increases its apparent radius to $\approx$ 4.5 nm (Nexø et al. 1988). HC appears to be a scavenger occasionally capturing products of Cbl degradation, Fig. 18.3, step 7. These Cbl-analogues account for 10–20% of all corrinoids present in the tissues and plasma (Gueant et al. 1984; Djalali et al. 1990; Kanazawa and Herbert 1983; Kondo et al. 1980). Fast and slow metabolising forms of HC are expected to deliver all bound corrinoids to liver interacting with a galactose-specific receptor expressed on the surface of hepatocytes (Allen 1975; Nexø 1998). Absorbed HC generally degrades, and the corrinoids are stored or secreted to blood as TC·Cbl (Allen 1975) or HC·Cbl (Hom and Olesen 1969), Fig. 18.3, step 3a. Approximately 10% of the endocytosed HC-corrinoid complexes are excreted to bile without degradation, 2–4 nmol/day (Allen 1975; Fedosov et al. 2005b), Fig. 18.3, step 3b. The latter process seems to be associated with disposal of Cbl-analogues.

18.3 Molecular Aspects of Cbl Binding and Receptor Recognition

18.3.1 Structure of the Cbl-Transporting Proteins

Structural investigations of Cbl-transporting proteins have made an essential progress in recent years. Successful DNA cloning of HC (Allen 1975), IF (Hewitt et al. 1991) and TC (Platica et al. 1991) supplied information about their primary structures and pointed to homology between these proteins, Fig. 18.4. The Cbl-binders have similar molecular masses of 45.5 kDa (TC), 43.4 kDa (IF), 45.6 kDa (HC) and demonstrate significant pairwise identity of 23–30%. Expression of TC (Quadros et al. 1993; Fedosov et al. 1999, 2000) and IF (Gordon et al. 1991; Fedosov et al. 2003) in different recombinant organisms provided material for various protein-consuming experiments including crystallisation studies (Garau et al. 2001; Wuerges et al. 2006). The three-dimensional structures of bovine and human TCs in complex with H2OCbl are now established at atomic resolution (Wuerges et al. 2006), Fig. 18.5.

TC consists of two domains connected by a short flexible link. The ligand H2OCbl is sandwiched between the two domains (Fig. 18.5). The larger N-terminal unit of 300 amino acid residues contains two circular layers of α-helices, where the odd and even segments belong to the external and internal bundles, respectively (Wuerges et al. 2006), Figs. 18.4 and 18.5. The N-domain is stitched by three disulfide bridges (Fedosov et al. 1999), which are conservative in all known sequences. Disruption of bridges C98(99)–C291(303) and C147(151)–C187(193) negatively affects Cbl binding, while C3(4)–C249(261) does not seem to play any essential role (Kalra et al. 2003). The C-terminal domain of 100 residues is built mostly of β-strands and has no disulfide bridges. The two domains have 17+17 bonds with Cbl, where double contacts are found for the residues N267(279) and Y137(141) (Wuerges et al. 2006). The ligand is practically buried inside TC with only 7% of Cbl surface open to solvent around 5′OH-group of ribosyl.

One particular bond connects the central cobalt ion of Cbl with H173(179) of human TC (or H175 of bovine TC). This His-residue attacks a weakly associated water molecule at the upper coordination position of the TC-bound H2OCbl. The substitution occurs as a well-defined conformational transition $\left(k+2 = 0.02\ \mathrm{s}^{-1}\right)$ visualized via changes in the Cbl absorbance spectrum (Fedosov et al. 2000). His-coordination may be advantageous because of a vulnerable nature of H_2OCbl,

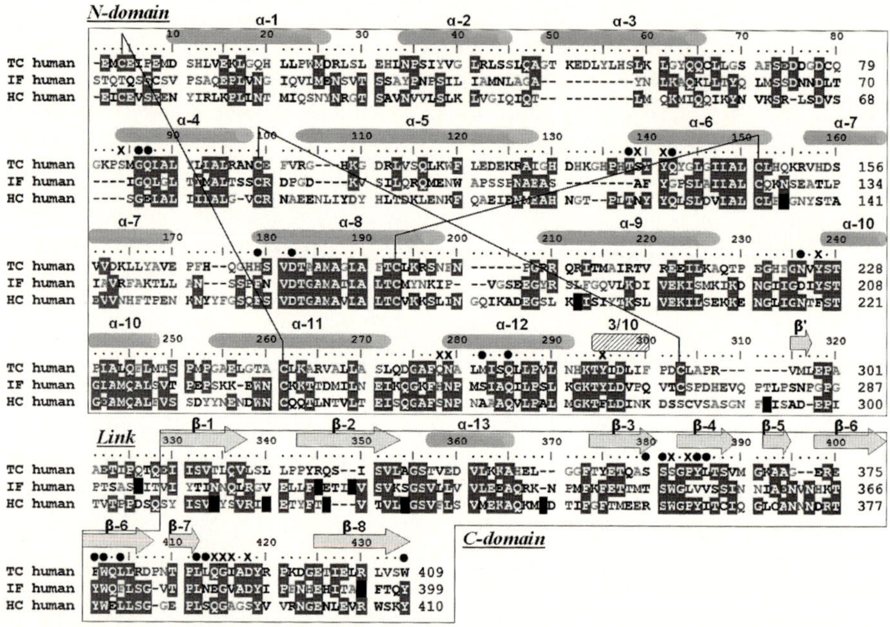

Fig. 18.4 Alignment of human TC, IF and HC (mature proteins), where N-/C-domains are *boxed*. The following notation is used: (α)-helices; (β)-strands; (3/10)-helix; (•) residues bound to Cbl directly and (x) via a solvent molecule; (■) Asn where carbohydrates can be attached. Links depict S–S-bonds. Numbering of residues in the main text (e.g. C148(151) of TC) refers to both individual protein sequences (C148 of TC) and the total alignment (C(151))

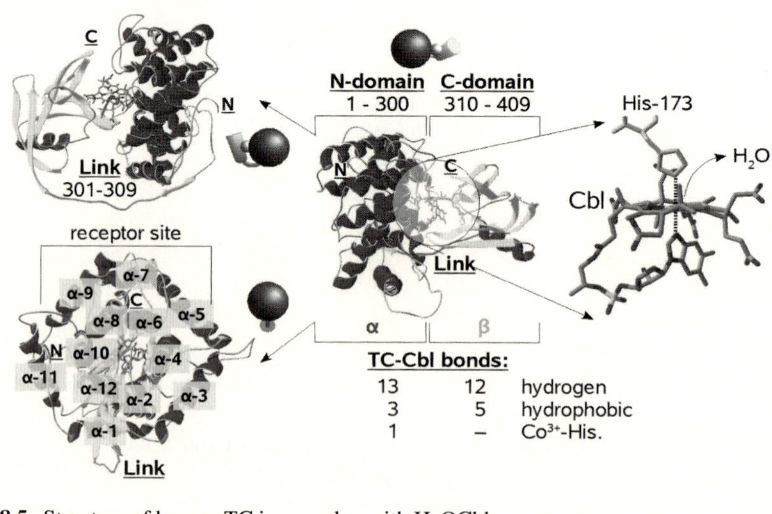

Fig. 18.5 Structure of human TC in complex with H_2OCbl

which readily reacts with external nucleofiles and reducing agents (Kräutler and Ostermann 2003; Pratt 1972). The His–Co^{3+} contact is absent in TC–CNCbl and TC–AdoCbl complexes, where the original upper group of Cbl is firmly attached to cobalt ion (Fedosov et al. 1999, 2002). Yet, the abnormal spectrum of AdoCbl bound to TC may point to essential elongation of the Ado–Cbl bond (Fedosov et al. 2002). Two other transporting proteins do not substitute water in H2OCbl (Fedosov et al. 2000, 2002), since H(179) is absent in IF and HC.

A two-domain organisation was demonstrated for TC (Fig. 18.5), and the same architecture can be conjectured for all Cbl-binding proteins. For instance, recombinant IF from plants is partially cleaved into two stable fragments – a 30 kDa peptide IF30 and a 20 kDa glycopeptide IF20 (Fedosov et al. 2004) – analogous to N- and C-terminal domains of TC. The split IF-fragments can be efficiently assembled by Cbl into a firm complex IF30·Cbl·IF20 (Fedosov et al. 2004), apparently with a "sandwich"-like structure, resembling TC·Cbl. This arcitecture was recently corroborated by the crystallographic analysis of the split IF in complex with Cbl (Mathews et al. 2007).

While TC is a true protein, IF and HC are glycoproteins supplemented with ≈10 kDa and ≈40 kDa of carbohydrates, respectively (Allen 1975; Nexø 1998). The potential N-glycosylation sites (Fig. 18.4, black bars) seem to be remote from the expected Cbl-protein contacts. All carbohydrates are placed on the C-terminal fragment of IF increasing its mass from 13 to 20 kDa, as established for the recombinant IF from plants (Fedosov et al. 2004). Carbohydrates of HC can be conjugated at three positions in the N-terminal domain and five positions in the C-terminal unit. The highest density of glycosylation sites is observed in the segment 310–350 of the total alignment.

Protruding sugar residues increase the apparent sizes of IF and HC, and they behave during gel-filtration as proteins of MwIF ≈ 60 kDa and MwHC ≈140 kDa (Allen 1975). Since oligosaccharides are not involved in the protein–Cbl and IF–receptor interactions (Gordon et al. 1991; Mørkbak et al. 2006), their role can be confined to (i) partial protection of HC and IF from proteolysis in the intestine (Gordon et al. 1991; Gueant et al. 1984); (ii) prevention of glomerular filtration of HC; and (iii) recognition of HC by a hepatic galactose-specific receptor (Allen 1975).

IF is the only Cbl-transporter prone to oligomerization. Thus, the full-length protein associates to dimers (IF·Cbl)2 after saturation with the ligand (Fedosov et al. 2004). Cleaved form IF30·Cbl·IF20 does not dimerize (Fedosov et al. 2004), suggesting involvement of the native inter-domain segment in the association of IF·Cbl monomers. Since the dimers have a dissociation constant of about 1 μM, they are hardly present under physiological concentrations (IF ≈ 50 nM), unless an external stabilising factor exists. If proven, the ability of IF to dimerize would have doubled the velocity of Cbl uptake from the intestine.

18.3.2 Cbl Binding and Ligand Specificity

It has long been known, that all Cbl-transporting proteins have a very high affinity for the true substrate B12. However, the reported dissociation constants KCbl = $10^{-9} - 10^{-15}$ M show exceptional dispersion (Allen 1975; Nexø 1998; Fedosov et al. 2002; Gordon et al. 1991; Brada et al. 2001). The sources of ambiguity were analysed in a number of publications (Fedosov et al. 2005a; Fedosov et al. 2006, 2007b), but the insufficient incubation time (10–20 min) during the equilibrium binding deserves to be mentioned specifically. The pseudo-saturation curve, produced after a non-equilibrium experiment, noticeably deviates from the correct shape, especially at the equivalent concentrations of the biding species. This leads to essential overestimation of Kd (Fedosov et al. 2005a). Examination of this effect led to the conclusion that the equilibrium methods cannot be used for analysis of Kd in a nearly irreversible binding reaction (Fedosov et al. 2005a). The alternative way is determination of the rate constants k+Cbl and k–Cbl in two separate reactions. The performed kinetic measurements pointed to extremely low dissociation constants (KCbl $\approx$ 5 fM) for all Cbl-transporting proteins (Fedosov et al. 2005a, 2006, 2007b). Note, that IF's affinity for Cbl heavily depended on the method of preparation of the apo-form (Fedosov et al. 2006).

Significant progress in elucidation of the mechanisms of Cbl – protein interaction was achieved during work with the fragments: IF30 (N-domain of IF), IF20 (C-domain of IF) (Fedosov et al. 2004, 2005a, 2006) and TC12 (C-domain of TC) (Fedosov et al. 2005b). The results are summarized in Fig. 18.6. It was found that the small C-terminal fragments IF20 or TC12 bound Cbl thousandfold stronger than the large N-terminal peptide IF30 (Fedosov et al. 2005a, b, 2006). These results agree with the atomic structures of TC·Cbl (Wuerges et al. 2006) and IF30·Cbl·IF20 (Mathews et al. 2007) as well as the epitope mapping of TC (Fedosov et al. 2005b), where high density of the ligand binding residues was found on a short C-terminal stretch, Fig. 18.4, residues 380–420 in the alignment. Preferential attachment of Cbl to the C-terminal fragment may be considered as the first binding step in the mixture IF30+IF20+Cbl. It is followed by assembly of the two protein units, as shown in the binding reaction IF20·Cbl + IF30 (Fedosov et al. 2004, 2005a, 2006), Fig. 18.6a. Stability of the produced complex IF30·Cbl·IF20 is impressive. Thus, the dissociation of IF30 from the assembled structure is characterized by Kd = 10–100 pM, depending on the refolding technique used for preparation of the unsaturated fragments (Fedosov et al. 2005a, 2006).

Interaction of Cbl with the full-length binding proteins seems to follow a two-step pattern. Thus, initial attachment of a ligand to the C-terminal domain is accompanied by assembly of N- and C-units subdivided into multiple transitions (Fedosov et al. 2005a, 2006, 2007b), Fig. 18.6b. Convergence of the two domains is facilitated by their attachment to each other via a flexible link.

The described above two-domain organization of Cbl-binding proteins may provide a structural basis for different ligand specificities observed for IF, HC and TC. Kinetic analysis of the interactions between a number of corrinoids and the three binding proteins revealed essential details of ligand recognition (Fedosov et al.

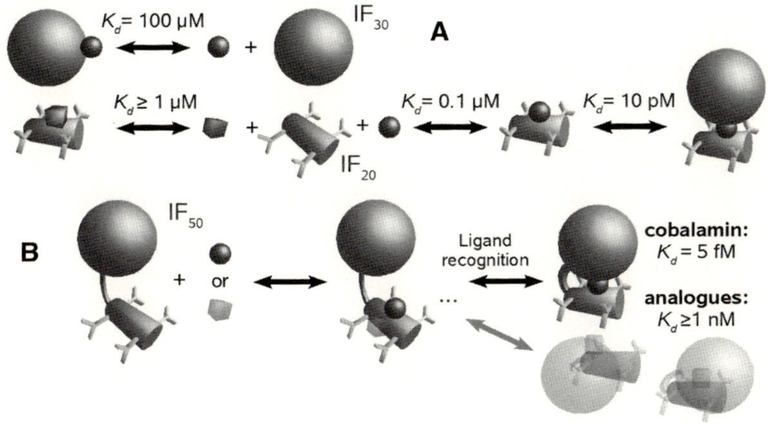

Fig. 18.6 Binding of Cbl (*small sphere*) or an analogue (*small cube*) to (**a**) IF-fragments; and (**b**) full-length IF

2006, 2007b). It was found that the rate constants of attachment k + Cor $\approx$ 70 $\mu M^{-1}s^{-1}$ were similar in all cases. Nevertheless, the following adjustments and fastening of the bound ligand obeyed different mechanisms for HC, TC and IF. A firm attachment to HC was observed for all corrinoids immediately after initial contact with the protein, though a slow conformational transition without any visible effect on affinity was found in some cases. On the other hand, several slow transitions were observed during fastening of Cbl-analogues bound to TC and IF. The number of conformational steps directly correlated with Cbl-specificity, i.e. the most complex kinetic patterns were observed for IF – Cor interactions (Fedosov et al. 2006, 2007b). The transitions were ascribed to the mutual re-arrangements of the protein domains, thus, the ligand selectivity was related to the internal flexibility of the protein molecule. It was concluded that recognition of the true substrate by, for example, IF is tightly associated with assembly of the two domains and following adjustments at the domain-domain interface (Fedosov et al. 2006, 2007b), Fig. 18.6b. This statement is corroborated by the fact that many Cbl-analogues are not able to induce association of the cleaved fragments IF30+IF20 despite successful binding to IF20 (Fedosov et al. 2007b). It seems that the sequence of bond formation (e.g. 1, 2, 3, 4...) is a crucial factor during ligand recognition. The correct order leads to firm binding of Cbl, while binding of the distorted corrinoids follows the ramified schemes giving multiple protein-ligand complexes of different dissociation stability (Fedosov et al. 2007b), Fig. 18.6b.

18.3.3 Receptor Recognition

It was already mentioned, that the efficient uptake of vitamin B12 by the organism is mediated by several receptors (Allen 1975; Moestrup 2006; Birn 2006; Nexø 1998;

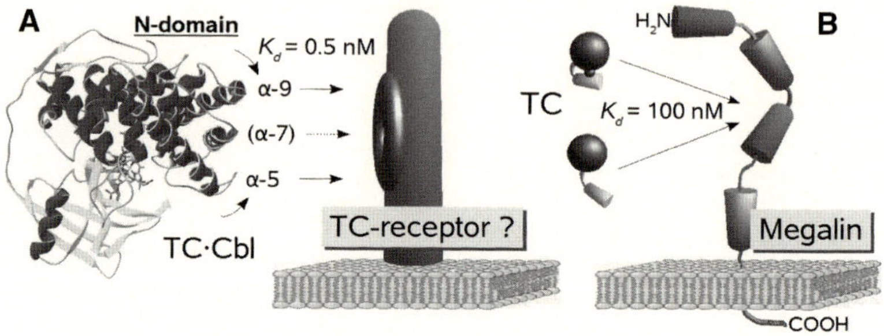

Fig. 18.7 Interaction of TC with (**a**) ubiquitous receptor; and (**b**) renal receptor megalin

Seetharam and Li 2000). The simplest mechanism describes absorption of HC from plasma by the liver cells. HC, with or without attached Cbl, uniformly interacts with a receptor, which binds in the presence of Ca^{2+} all glycoproteins containing the terminal galactose residues (Allen 1975).

TC, the main distributor of Cbl among all the tissues, binds in the presence of Ca^{2+} to a membrane-associated glycoprotein of still debated identity (Seligman and Allen 1978; Nexø and Hollenberg 1980; Seetharam and Li 2000; Quadros et al. 2005). The receptor-binding site on the surface of TC involves α-5, α-9 and possibly α-7 helices in the N-terminal domain (Figs. 18.5 and 18.7a) according to mapping with the monoclonal anti-TC antibodies (Fedosov et al. 2005b). The above external segments seem to converge after the neighbouring helices α-4, α-6 and α-8 in the inner bundle have attached Cbl. This causes formation of an assembled surface α-5, 7, 9 with better compatibility and a 2- to 30-fold higher affinity for the receptor (Seligman and Allen 1978; Nexø and Hollenberg 1980; Quadros et al., 2005; Moestrup 2006), Fig. 18.7a. Difference between conformations of apo- and holoTC was confirmed by existence of a monoclonal antibody characterized by low affinity for apoTC and high affinity for holoTC (Örning et al. 2006).

Another TC-related receptor is a well-described glycoprotein megalin present in a number of tissues (Moestrup 2006; Seetharam and Li 2000; Fedosov et al. 2002; Moestrup et al. 1996). This is a complex molecule of 600 kDa (Fig. 18.7b) which consists of four extracellular clusters enriched by cysteines and similar to those of low density lipoprotein (LDL) receptors. Megalin is associated with Cbl homoeostasis in kidney, where it binds filtered apo/holoTC and mediates return of the vitamin to the blood circulation (Fig. 18.3, step 5). Apart from TC, this receptor binds variety of other ligands (Birn 2006; Moestrup et al. 1996; Seetharam and Li 2000).

The most complex mechanism of recognition describes interaction between IF·Cbl and the intestinal receptor cubilin (Fig. 18.2, step 4). The latter is a 460 kDa glycoprotein constructed of 8 epidermal growth factor (EGF) repeats and 27 CUB domains with a characteristic β-folding (Birn et al. 1997; Birn 2006; Kristiansen et al. 1999; Moestrup 2006), Fig. 18.8. This receptor has no transmembrane segment of its own and is anchored to the cell surface via a protein amnionless (AMN)

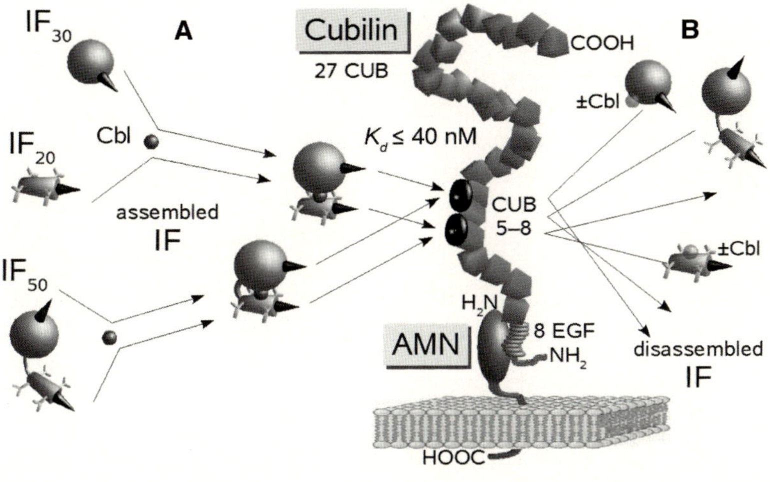

Fig. 18.8 Binding of IF to cubilin while IF-domains are (**a**) assembled; or (**b**) disassembled

(Fyfe et al. 2004). In the absence of AMN, cubilin is detached from the membranes, whereupon the cell loses the ability to absorb IF·Cbl (Fyfe et al. 2004). Apart from intestine, cubilin is also found in a number of tissues, where it mediates absorption of ligands other than IF·Cbl (Moestrup 2006). The domains CUB 5–8 of the receptor bind IF·Cbl with high affinity (Kristiansen et al. 1999), where estimations by different methods give Kd = 2 nM or 40 nM (Birn et al. 1997; Kristiansen et al. 1999). Apo-form of the full-length IF or its isolated fragments (with or without Cbl) do not interact with the receptor (Fedosov et al. 2005a), Fig. 18.8b.

Exclusive recognition of the complex IF·Cbl by cubilin is based on the essential difference in the structures of apo- and holo-forms of full-length IF. Receptor recognition site of IF seems to have a composite organisation, where the function of ligand is to induce association of the two domains, thereby assembling the scattered elements into a cubilin-compatible interface (Fedosov et al. 2005a).

All described receptors mediate endocytosis of the Cbl-transporting proteins, whereupon the receptors return to the cell surface, the Cbl-carriers are cleaved in lysosomes, and the vitamin is released to the cytoplasm or on the opposite side of the cell.

18.4 Practical Implementation of Recent Basic Research

18.4.1 Novel Methods for Estimation of Cbl-Status

Efficient uptake of dietary B12 is ensured by the complex transport mechanism. Yet, complexity has its drawbacks, and breakdown of the transportation chain at any link results in Cbl-malabsorption. Determination of the current Cbl-status is, nevertheless, not as easy as it might appear (Carmel et al. 2003). Thus, the normal level of total plasma B12 often coexists with the signs of deficiency, as well as a low total B12 does not necessarily cause metabolic or clinical manifestation of the disease (Carmel et al. 2003; Miller et al. 2006; Hvas and Nexø 2005; Herbert et al. 1990). The inconsistency between total B12 and clinical signs can be explained by inert character of the prevailing form of B12 in plasma – HC·Cbl. Other detection methods are based on the elevated levels of metabolites MMA and Hcy in plasma. They aim to reveal decreased catalytic activities of the Cbl-dependent enzymes due to low B12 balance. Yet, folate deficiency and renal dysfunction may provoke similar symptoms (Carmel et al. 2003; Hvas and Nexø 2005). Additionally, complexity of the measurements of MMA and Hcy hinders their widespread application. Eventually, low concentration of the Cbl-saturated TC (holoTC) was suggested as a sensitive marker for developing Cbl-deficiency (Herbert et al. 1990), though no accurate method of measurement existed at that time.

Recent advances in characterization of the anti-TC antibodies (Fedosov et al. 2005b) allowed a quantitative determination of holoTC. The available techniques (Fedosov et al. 2006; Nexø et al. 2002) are based on sequestration of the total TC-pool using an immobilized antibody (Fig. 18.9a, step 1) or a Cbl-matrix followed by an antibody-matrix (Fig. 18.9b, steps 1 and 2). Afterwards, Cbl is either released from TC and measured by a standard procedure (Fig. 18.9a, steps 2 and 3) or the adsorbed complex TC·Cbl is detected by a second antibody (Fig. 18.9b, step 3). The most recent method uses a holoTC-specific antibody as a capturing substance, whereupon the bound TC is measured by a labelled antibody (Örning et al. 2006), Fig. 18.9c. This setup seems to be very convenient, and we are looking forward to its practical implementation.

Measurements of holoTC in plasma were carried out and correlated with total plasma B12, MMA and Hcy, as well as with clinical manifestations of Cbl-deficiency, see recent references (Goringe et al. 2006; Hin et al. 2006; Hvas and Nexø 2005; Hvas et al. 2006; Miller et al. 2006). It seems, that none of the mentioned parameters taken alone is an absolutely reliable indicator of Cbl deficiency (Goringe et al. 2006; Hvas and Nexø 2005; Miller et al. 2006). Yet, there was a

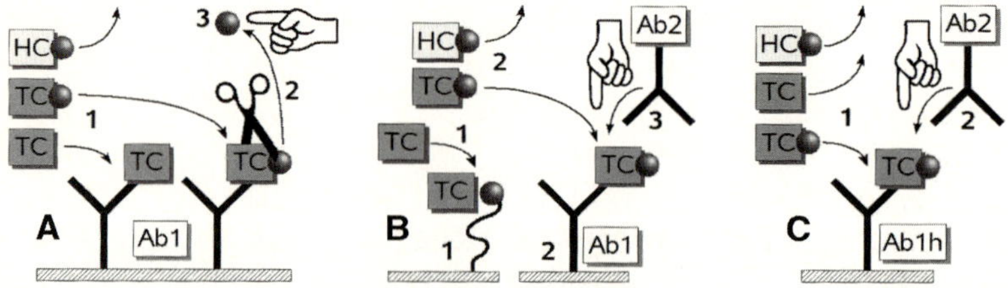

Fig. 18.9 Determination of holoTC (TC·Cbl) in blood plasma using (**a**) adsorption of total TC on anti-TC antibody Ab1 followed by liberation and measurement of TC-associated Cbl (Fedosov et al. 2006); (**b**) removal of apoTC by a Cbl-matrix followed by adsorption of the remaining holoTC on Ab1 and measurement of the sequestered TC by Ab2 (Nexø et al. 2002; (**c**) adsorption of holoTC on the holoTC-specific antibody Ab1h and measurement of the sequestered TC by Ab2 (Örning et al. 2006)

correlation between the elevated concentrations of both metabolites (MMA > 0.35 μM, Hcy > 13 μM) and low concentrations of Cbl in both Cbl-pools (B12 < 150 pM and holoTC < 35 pM) (Miller et al. 2006). In this regard, application of the combined coefficients $(B12\times holoTC)^{1/2} < 78$ pM and $(MMA\times Hcy)^{1/2} > 2.3$ μM as indicators of Cbl-deficiency seems to be convenient and mathematically justified (Fedosov 2006b).

Measurements of plasma holoTC also characterize the intestinal B12 uptake. Thus, a pronounced spike in plasma holoTC (+50%) was detected 1 day after ingestion of three sequential 9 μg (6.7 nmol) doses of B12 (Bor et al. 2004; Hvas et al. 2006). It seems to reflect the normal absorption of the vitamin. Examination of patients by this method might be a convenient alternative to the Schilling test.

18.4.2 Cbl-Conjugates and Their Medical Application

All animal cells require Cbl to maintain DNA synthesis. Hence, accumulation of Cbl is a characteristic feature of the fast growing tissues, and it may serve as a good marker of cancer cells (Russell-Jones 1998). Unfortunately, the radioactive vitamin [57Co]Cbl is not applicable for imaging in vivo because the half-life time of 57Co is too long (272 days). Cbl-conjugates with the attached imaging or therapeutic groups can be used instead of [57Co]Cbl to visualize and treat cancer cells (Russell-Jones 1998; Collins et al. 2000; McGreevy et al. 2003; Bagnato et al. 2004).

Chemical modification of Cbl molecule must not affect its affinity for IF or/and TC, since it would hinder targeted delivery of the synthesized conjugate to the tissues. A popular strategy of Cbl modification includes acidic hydrolysis of amides b, d or e (Fig. 18.1b), separation of monocarboxylic derivatives, and attachment of an active substance via amine-containing spacer (Russell-Jones 1998; Collins et al. 2000; Hogenkamp et al. 2000). Yet, the above method does not seem to be optimal in the frame of recent structural analysis of TC·Cbl, where each of the derivatized amides is involved in bonding with TC residues (Wuerges et al. 2006). At the same time, the 3D-structure shows that 5′OH-group of ribosyl (Fig. 18.1b, position 5′) is open to solvent (Wuerges et al. 2006) and can be used for conjugation of different probes without disturbing the Cbl-binding site of TC (Horton et al. 2003; Fedosov et al. 2006). This approach was examined after synthesis of a fluorescent 5′OH-conjugate CBC-244 (Fedosov et al. 2006). High affinities were established when binding this derivative to IF and TC, as well as IF·CBC complex to cubilin (Fedosov et al. 2006). Increase in the quantum yield of CBC-244 by 50–300%, induced by its binding to the Cbl-carriers, makes it a sensitive probe for investigations of protein-corrinoid interactions (Fedosov et al. 2006, 2007b).

Another way to attach a supplementary group to Cbl without loosing affinity is carbon–cobalt bonding at the upper position of corrin ring (Bagnato et al. 2004; Smeltzer et al. 2001), Fig. 18.1. The produced cobalamins are, however, light sensitive (Kräutler and Ostermann 2003; Pratt 1972; Smeltzer et al. 2001).

Synthesis of the Cbl-conjugated drugs, their targeted delivery and application in cancer treatment is another developing field. A toxic compound attached to

Cbl is expected to be physiologically neutral until it is liberated. Therefore, a Cbl-conjugate can be safely delivered by TC to cancer cells, where the active toxic component is released by influence of light (Smeltzer et al. 2001; Bagnato et al. 2004), low lysosomal pH (Bagnato et al. 2004) or mild radiation (Hogenkamp et al. 2000). In such case, the poisonous effect would be essentially confined to the target tissue with a relatively low negative impact on the rest of the organism.

A recently suggested strategy of the targeted delivery of therapeutic Cbl-conjugates to cancer cells is based on the binding to HC but not TC (Waibel et al. 2008). It was demonstrated that many tumours produce and accumulate high amounts of HC resulting in a significant uptake of the HC-specific conjugates by the malignant tissues in mouse.

18.5 Affinity Purification of Cbl

Purification of Cbl from biological sources or isolation of Cbl-derivatives after organic synthesis is a multi-step procedure. Thus, purification from a fermentation extract (Martens et al. 2002) typically requires adsorption on several unspecific materials, e.g. charcoal, Amberlite XAD/IRC(H^+), Dowex(Cl^-) (Adachi et al. 2005; Barchielli et al. 1960; Lamm et al. 1980; Martens et al. 2002; Vogelmann and Wagner 1974). Application of an affinity adsorbent would essentially simplify this process. Guidelines for the development of such material are, however, not so evident.

Analysis of the environment in the Cbl-binding site of TC provides several useful observations (Figs. 18.4 and 18.5). First of all, most of TC–Cbl contacts belong to the well-known hydrogen and hydrophobic bonds, which are also responsible for retention of Cbl inside the unspecific microporous resins. Much higher affinity of Cbl for TC is caused exclusively by spacial organization of TC-residues around the ligand and better compatibility between the interacting species. Clearly, the TC-like scaffolding of residues cannot be reproduced by organic synthesis. Yet, one bond between TC and H2OCbl has a completely different nature, and this is the coordination contact of His-173(179) with cobalt ion (Fig. 18.5). Coordination binding has never been exploited for Cbl purification, and this approach required a thorough examination.

It was found that H_2OCbl binds to heterocyclic tetrazole (TZ) derivatives with Kd $\approx 4\ \mu M$ (Fedosov et al. 2007a), i.e. 60-fold better than to His (Fedosov et al. 2003, 2007a). Treatment of the CNBr-activated Sepharose with a neutral solution of 50–100 mM azide easily initiates tetrazole groups in a solid material. After 2–10 h of incubation the matrix acquires the H_2OCbl-binding properties manifested at pH 3–8 (Fig. 18.10). Concentration of the active groups in the packed material corresponds to $\approx$10 mM (Fedosov et al. 2007a). Interaction of H_2OCbl with Sepharose-TZ occurs in two steps: (1) fast coordination of one TZ-group at the upper surface of Cbl ($K1 \leq 10\ \mu M$); and (2) slow substitution for the nucleotide base N at the low surface rendered by the second TZ-residue ($K1 \cdot K2 \leq 1\ \mu M$).

AdoCbl and MeCbl, prevailing in the biological extracts, both have the protected cobalt ion. Yet, illumination with visible light efficiently converts them to H2OCbl

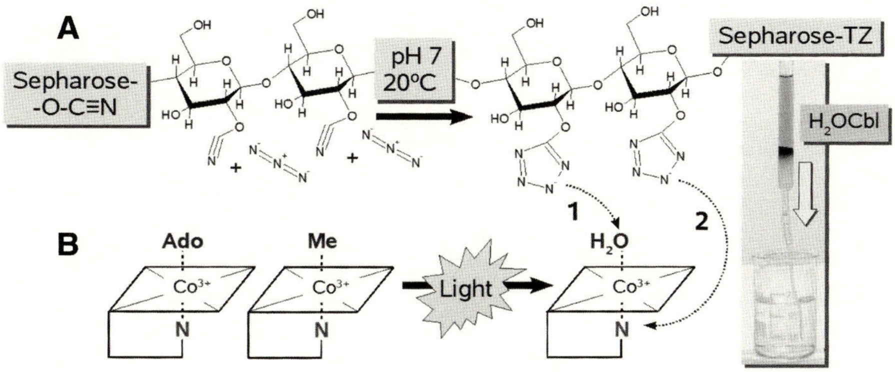

Fig. 18.10 Affinity adsorption of aquo-corrinoids. (**a**) Preparation of tetrazole-(TZ)-containing Sepharose; (**b**) conversion of Ado/MeCbl to H_2OCbl, and binding of the latter to Sepharose-TZ

when working in an oxygenated aqueous solution (Pratt 1972; Kräutler and Ostermann 2003), Fig. 18.10b. Some biological compounds (e.g. His-containing peptides, reduced glutathione) can substitute for water of H_2OCbl (Pratt 1972; Fedosov et al. 2007a) and hinder adsorption on TZ-material. In such case, the medium should be acidified (pH 3–4) to protonate the competing compounds. The developed affinity purification of Cbl included: (i) hot acidic extraction of biological cobalamins in the absence of cyanide; (ii) illumination of the solution followed by the binding to a TZ-column; (iii) washing and elution of Cbl with 20 mM KCN at alkaline pH (Fedosov et al. 2007a). Optional desalting on charcoal or XAD can be performed.

The tetrazole-containing matrix is also applicable for adsorption of other aquo-corrinoids (Co^{3+} and Co^{2+}), e.g. diaquo-cobinamide, cyano-aquo-cobinamide, aquo-forms of pseudo B12 and factor A (Fedosov et al. 2007a).

Acknowledgements This work was supported by the Lundbeck foundation and Cobento A/S.

References

Adachi S, Miyamoto E, Watanabe F, Enomoto T., Kuda T, Hayashi M, Nakano Y, (2005) Purification and characterization of a corrinoid compound from a Japanese salted and fermented salmon kidney "Mefun". J Liquid Chromatogr Related Technol 28:2561–2569

Allen RH (1975) Human vitamin B12 transport proteins. Prog Hematol 9:57–84

Allen RH, Seetharam B, Allen NC, Podell ER, Alpers DH (1978b) Correction of cobalamin malabsorption in pancreatic insufficiency with a cobalamin analogue that binds with high affinity to R protein but not to intrinsic factor. In vivo evidence that a failure to partially degrade R protein is responsible for cobalamin malabsorption in pancreatic insufficiency. J Clin Invest 61:1628–1634

Allen RH, Seetharam B, Podell E, Alpers DH (1978a) Effect of proteolytic enzymes on the binding of cobalamin to R protein and intrinsic factor. In vitro evidence that a failure to partially degrade R protein is responsible for cobalamin malabsorption in pancreatic insufficiency. J Clin Invest 61:47–54

Alsenz J, Russell-Jones GJ, Westwood S, Levet-Trafit B, de Smidt PC (2000) Oral absorption of peptides through the cobalamin (vitamin B12) pathway in the rat intestine. Pharm Res 17:825–832

Bagnato JD, Eilers AL, Horton RA, Grissom CB (2004) Synthesis and characterization of a cobalamin-colchicine conjugate as a novel tumor-targeted cytotoxin. J Org Chem 69:8987–8996

Banerjee R, Ragsdale SW (2003) The many faces of vitamin B12: catalysis by cobalamin-dependent enzymes. Annu Rev Biochem 72:209–247

Barchielli R, Boretti G, Di Marco A, Julita P, Migliacci A, Minghetti A, Spalla C (1960) Isolation and structure of a new factor of the vitamin B12 group: guanosine diphosphate factor B. Biochem J 74:382–387

Birn H (2006) The kidney in vitamin B12 and folate homeostasis: characterization of receptors for tubular uptake of vitamins and carrier proteins. Am J Physiol Renal Physiol 291:F22–F36

Birn H, Verroust PJ, Nexø E, Hager H, Jacobsen C, Christensen EI, Moestrup SK (1997) Characterization of an epithelial approximately 460-kDa protein that facilitates endocytosis of intrinsic factor-vitamin B12 and binds receptor-associated protein. J Biol Chem 272:26497–26504

Boisson F, Fremont S, Migeon C, Nodari F, Droesch S, Gerard P, Parache RM, Nicolas JP (1999) Human haptocorrin in hepatocellular carcinoma. Cancer Detect Prev 23:89–96

Bor MV, Nexø E, Hvas AM (2004) Holo-transcobalamin concentration and transcobalamin saturation reflect recent vitamin B12 absorption better than does serum vitamin B12. Clin Chem 50:1043–1049

Brada N, Gordon MM, Wen J, Alpers DH (2001) Transfer of cobalamin from intrinsic factor to transcobalamin II. J Nutr Biochem 12:200–206

Carkeet C, Dueker SR, Lango J, Buchholz BA, Miller JW, Green R, Hammock BD, Roth JR, Anderson PJ (2006) Human vitamin B12 absorption measurement by accelerator mass spectrometry using specifically labeled (14)C-cobalamin. Proc Natl Acad Sci USA 103:5694–5699

Carmel R, Green R, Rosenblatt S, Watkins D (2003) Update on cobalamin, folate, and homocysteine. Hematology, Am Soc Hematol Educ Program 62–81

Collins DA, Hogenkamp HP, O'Connor MK, Naylor S, Benson LM, Hardyman TJ, Thorson LM (2000) Biodistribution of radiolabeled adenosylcobalamin in patients diagnosed with various malignancies. Mayo Clin Proc 75:568–580

Del Corral A, Carmel R (1990) Transfer of cobalamin from the cobalamin-binding protein of egg yolk to R binder of human saliva and gastric juice. Gastroenterology 98:1460–1466

Djalali M, Gueant JL, Lambert D, el Kholty S, Saunier M, Nicolas JP (1990) High-performance liquid chromatographic separation and dual competitive binding assay of corrinoids in biological material. J Chromatogr 529:81–91

Fedosov SN, Berglund L, Nexø E, Petersen TE (1999) Sequence, S-S bridges, and spectra of bovine transcobalamin expressed in *Pichia pastoris*. J Biol Chem 274:26015–26020

Fedosov SN, Berglund L, Fedosova NU, Nexø E, Petersen TE (2002) Comparative analysis of cobalamin binding kinetics and ligand protection for intrinsic factor, transcobalamin and haptocorrin. J Biol Chem 274:26015–26020

Fedosov SN, Berglund L, Nexø E, Petersen TE (2007a) Tetrazole derivatives and matrices as novel cobalamin coordinating compounds. J Organometallic Chem 692:1234–1242

Fedosov SN, Fedosova NU, Berglund L, Moestrup SK, Nexø E, Petersen TE (2004) Assembly of the intrinsic factor domains and oligomerization of the protein in the presence of cobalamin. Biochemistry 43:15095–15102

Fedosov SN, Fedosova NU, Berglund L, Moestrup SK, Nexø E, Petersen TE (2005a) Composite organization of the cobalamin binding and cubilin recognition sites of intrinsic factor. Biochemistry 44:3604–3614

Fedosov, SN., Fedosova NU, Kräutler B, Moestrup SK, Nexø E, Petersen TE (2007b) Mechanisms of discrimination between cobalamins and their natural analogues during their binding to the specific B12-transporting proteins. Biochemistry 46:6446–6458

Fedosov SN, Fedosova NU, Nexø E, Petersen TE (2000) Conformational changes of transcobalamin induced by aquocobalamin binding. Mechanism of substitution of the cobalt-coordinated group in the bound ligand. J Biol Chem 275:11791–11798

Fedosov SN, Grissom CB, Fedosova NU, Moestrup SK, Nexø E, Petersen TE (2006) Application of a fluorescent Cbl-analogue for analysis of the binding kinetics. A study employing recombinant human transcobalamin and intrinsic factor. FEBS J 273:4742–4753

Fedosov SN, Laursen NB, Nexø E, Moestrup SK, Petersen TE, Jensen EØ, Berglund L (2003) Human intrinsic factor expressed in the plant Arabidopsis thaliana. Eur J Biochem 270:3362–3367

Fedosov SN, Örning L, Løvli T, Quadros EV, Thompson K, Berglund L, Petersen TE (2005b) Mapping the functional domains of human transcobalamin using monoclonal antibodies. FEBS J. 272:3887–3898

Fedosov SN, Petersen TE, Nexø E (1996) Transcobalamin from cow milk: isolation and physico-chemical properties. Biochim Biophys Acta 1292:113–119

Fyfe JC, Madsen M, Hojrup P, Christensen EI, Tanner SM, de la Chapelle A, He Q, Moestrup SK (2004) The functional cobalamin (vitamin B12)-intrinsic factor receptor is a novel complex of cubilin and amnionless. Blood 103:1573–1579

Garau G, Fedosov SN, Petersen TE, Geremia S, Randaccio L (2001) Crystallization and preliminary X-ray diffraction analysis of human transcobalamin, a vitamin B12-transporting protein. Acta Cryst D57:1890–1892

Gordon MM, Hu C, Chokshi H, Hewitt JE, Alpers DH (1991) Glycosylation is not required for ligand or receptor binding by expressed rat intrinsic factor. Am J Physiol 260: G736–G742

Goringe A, Ellis R, McDowell I, Vidal-Alaball J, Jenkins C, Butler C, Worwood M (2006) The limited value of methylmalonic acid, homocysteine and holotranscobalamin in the diagnosis of early B12 deficiency. Haematologica 91:231–234

Gueant JL, Djalali M, Aouadj R, Gaucher P, Monin B, Nicolas JP (1986) In vitro and in vivo evidences that the malabsorption of cobalamin is related to its binding on haptocorrin (R-binder) in chronic pancreatitis. Am J Clin Nutr 44:265–277

Gueant JL, Vidailhet M, Pasquet C, Djalali M, Nicolas JP (1984) Effect of pancreatic extracts on the faecal excretion and on the serum concentration of cobalamin and cobalamin analogues in cystic fibrosis. Clin Chim Acta 137:33–41

Herbert V, Fong W, Gulle V, Stopler T (1990) Low holotranscobalamin II is the earliest serum marker for subnormal vitamin B12 (cobalamin) absorption in patients with AIDS. Am J Hematol 34:132–139

Hewitt JE, Gordon MM, Taggart RT, Mohandas TK, Alpers DH (1991) Human gastric intrinsic factor: characterization of cDNA and genomic clones and localization to human chromosome 11. Genomics 10:432–440

Hin H, Clarke R, Sherliker P, Atoyebi W, Emmens K, Birks J, Schneede J, Ueland PM, Nexø E, Scott J, Molloy A, Donaghy M, Frost C, Evans JG (2006) Clinical relevance of low serum vitamin B12 concentrations in older people: the Banbury B12 study. Age Ageing 35:416–422

Hogenkamp HP, Collins DA, Live D, Benson LM, Naylor S (2000) Synthesis and characterization of nido-carborane-cobalamin conjugates. Nucl Med Biol 27:89–92

Hom BL, Olesen HA (1969) Plasma clearance of 57cobalt-labelled vitamin B12 bound in vitro and in vivo to transcobalamin I and II. Scand J Clin Lab Invest 23:201–211

Horton RA, Bagnato JD, Grissom CB (2003) Structural determination of 5′-OH α-ribofuranoside modified cobalamins via 13C and DEPT NMR. J Org Chem 68:7108–7111

Hvas AM, Mørkbak AL, Nexø E (2006) Plasma holotranscobalamin compared with plasma cobalamins for assessment of vitamin B(12) absorption; optimisation of a non-radioactive vitamin B(12) absorption test (CobaSorb). Clin Chim Acta 376:150–154

Hvas AM, Nexø E (2005) Holotranscobalamin – a first choice assay for diagnosing early vitamin B deficiency? J Intern Med 257:289–298

Kalra S, Li N, Seetharam S, Alpers DH, Seetharam B (2003) Function and stability of human transcobalamin II: role of intramolecular disulfide bonds C98-C291 and C147-C187. Am J Physiol Cell Physiol 285:C150–C160

Kanazawa S, Herbert V (1983) Noncobalamin vitamin B12 analogues in human red cells, liver, and brain. Am J Clin Nutr 37:774–777

Kittang E, Schjonsby H (1987) Effect of gastric anacidity on the release of cobalamins from food and their subsequent binding to R-protein. Scand J Gastroenterol 9:1031–1037

Kondo H, Kolhouse JF, Allen RH (1980) Presence of cobalamin analogues in animal tissues. Proc Natl Acad Sci USA 77:817–821

Kräutler B, Ostermann S (2003) Structure, reactions, and functions of B12 and B12-proteins. In: Kadish KM, Smith KM, Guilard R (eds) The porphyrin handbook, vol 11. Elsevier Science, New York, USA, pp. 229–276

Kristiansen M, Kozyraki R, Jacobsen C, Nexø E, Verroust PJ, Moestrup SK (1999) Molecular dissection of the intrinsic factor-vitamin B12 receptor, cubilin, discloses regions important for membrane association and ligand binding. J Biol Chem 274:20540–20544

Lamm L, Horig JA, Renz P, Heckmann G (1980) Biosynthesis of vitamin B12. Experiments with the anaerobe *Eubacterium limosum* and some labelled substrates. Eur J Biochem 109:115–118

Li N, Seetharam S, Rosenblatt DS, Seetharam B (1994) Expression of transcobalamin II mRNA in human tissues and cultured fibroblasts from normal and transcobalamin II-deficient patients. Biochem J 301:585–590

Linnell JC, Hoffbrand AV, Hussein HA, Wise IJ, Matthews DM (1974) Tissue distribution of coenzyme and other forms of vitamin B12 in control subjects and patients with pernicious anaemia. Clin Sci Mol Med 46:163–172

Martens J-H, Barg H, Warren MJ, Jahn D (2002) Microbial production of vitamin B12. Appl Microbiol Biotechnol 58:275–85

Mathews FS, Gordon MM, Chen Z, Rajashankar KR, Ealick SE, Alpers DH, Sukumar N (2007) Crystal structure of human intrinsic factor: cobalamin complex at 2.6-A resolution. Proc Natl Acad Sci USA 104:17311–17316

McGreevy JM, Cannon MJ, Grissom CB (2003) Minimally invasive lymphatic mapping using fluorescently labeled vitamin B12. J Surg Res 111:38–44

Mervyn L (1971) The metabolism of the cobalamins. In: Florkin M, Stotz EH (eds) Comprehensive biochemistry, vol 21. Elsevier, London, England, pp. 153–177

Miller JW, Garrod MG, Rockwood AL, Kushnir MM, Allen LH, Haan MN, Green R (2006) Measurement of total vitamin B12 and holotranscobalamin, singly and in combination, in screening for metabolic vitamin B12 deficiency. Clin Chem 52:278–285

Moestrup SK (2006) New insights into carrier binding and epithelial uptake of the erythropoietic nutrients cobalamin and folate. Curr Opin Hematol 13:119–123

Moestrup SK, Birn H, Fischer PB, Petersen CM, Verroust PJ, Sim RB, Christensen EI, Nexø E (1996) Megalin-mediated endocytosis of transcobalamin-vitamin-B12 complexes suggests a role of the receptor in vitamin-B12 homeostasis. Proc Natl Acad Sci USA 93:8612–8617

Mørkbak AL, Hvas AM, Lloyd-Wright Z, Sanders TA, Bleie O, Refsum H, Nygaard OK, Nexø E (2006) Effect of vitamin B12 treatment on haptocorrin. Clin Chem 52:1104–1111

Nexø E (1998) Cobalamin binding proteins. In: Kräutler B, Arigoni D, Golding BT (eds) Vitamin B12 and B12-proteins. Wiley-VCH, Weinheim, Germany, pp. 461–475

Nexø E, Christensen AL, Hvas AM, Petersen TE, Fedosov SN (2002) Quantification of holo-transcobalamin, a marker of vitamin B12 deficiency. Clin Chem 48:561–562

Nexø E, Hansen MR, Konradsen L (1988) Human salivary epidermal growth factor, haptocorrin and amylase before and after prolonged exercise. Scand J Clin Lab Invest 48:269–273

Nexø E, Hollenberg MD (1980) Characterization of the particulate and soluble acceptor for transcobalamin II from human placenta and rabbit liver. Biochim Biophys Acta 628:190–200

Örning L, Rian A, Campbell A, Brady J, Fedosov SN, Bramlage B, Thompson K, Quadros EV (2006) Characterization of a monoclonal antibody with specificity for holo-transcobalamin. Nutr Metab 3:3, PMID: 16393340

Ortigues-Marty I, Micol D, Prache S, Dozias D, Girard CL (2005) Nutritional value of meat: the influence of nutrition and physical activity on vitamin B12 concentrations in ruminant tissues. Reprod Nutr Dev 45:453–467

Platica O, Janeczko R, Quadros EV, Regec A, Romain R, Rothenberg SP (1991) The cDNA sequence and the deduced amino acid sequence of human transcobalamin II show homology with rat intrinsic factor and human transcobalamin I. J Biol Chem 266:7860–7863

Pratt JM (1972) Inorganic chemistry of vitamin B12. Academic, London, New York

Quadros EV, Matthews DM, Wise IJ, Linnell JC (1979) Tissue distribution of endogenous cobalamins and other corrins in the rat, cat and guinea pig. Biochim Biophys Acta 421:141–152

Quadros EV, Nakayama Y, Sequeira JM (2005) The binding properties of the human receptor for the cellular uptake of vitamin B12. Biochem Biophys Res Commun 327:1006–1010

Quadros EV, Regec AL, Khan KM, Quadros E, Rothenberg SP (1999) Transcobalamin II synthesized in the intestinal villi facilitates transfer of cobalamin to the portal blood. Am J Physiol 277:G161–G166

Quadros EV, Sai P, Rothenberg SP (1993) Functional human transcobalamin II isoproteins are secreted by insect cells using the baculovirus expression system. Blood 81:1239–1245

Russell-Jones GJ (1998) Use of vitamin B12 conjugates to deliver protein drugs by the oral route. Crit Rev Ther Drug Carrier Syst 15:557–586

Seetharam B, Li N (2000) Transcobalamin II and its cell surface receptor. Vitam Horm 59:337–366

Seligman PA, Allen RH (1978) Characterization of the receptor for transcobalamin II isolated from human placenta. J Biol Chem 253:1766–1772

Smeltzer CC, Cannon MJ, Pinson PR, Munger JD Jr, West FG, Grissom CB (2001) Synthesis and characterization of fluorescent cobalamin (CobalaFluor) derivatives for imaging. Org Lett 3:799–801

Toohey JI (2006) Vitamin B12 and methionine synthesis: a critical review. Is nature's most beautiful cofactor misunderstood? BioFactors 26:45–57

Ulleland M, Eilertsen I, Quadros EV, Rothenberg SP, Fedosov SN, Sundrehagen E, Örning L (2002) Direct assay for cobalamin bound to transcobalamin (holo-transcobalamin) in serum. Clin Chem 48:526–532

Vogelmann H, Wagner F (1974) Isolation and chromatographic separation of vitamin B12 and other corrinoids from biological sources. Biotechnol Bioeng Symp 4:969–975

Waibel R, Treichler H, Schaefer NG, van Staveren DR, Mundwiler S, Kunze S, Küenzi M, Alberto R, Nüesch J, Knuth A, Moch H, Schibli R, Schubiger PA (2008) New derivatives of vitamin B12 show preferential targeting of tumors. Cancer Res 68:2904–2911

Wuerges J, Garau G, Geremia S, Fedosov SN, Petersen TE, Randaccio L (2006) Structural basis for mammalian vitamin B12 transport by transcobalamin. PNAS 103:4386–4391

Index

Note: The letters 'f' and 't' following locators refer to figures and tables respectively.

O. Stanger (ed.), *Water Soluble Vitamins*, Subcellular Biochemistry 56,
DOI 10.1007/978-94-007-2199-9, © Springer Science+Business Media B.V. 2012

W

X